Nanopackaging

James E. Morris
Editor

Nanopackaging

Nanotechnologies and Electronics Packaging

 Springer

Editor
James E. Morris
Portland State University
Department of Electrical and Computer Engineering
1900 SW 4th Avenue
Portland, OR 97201
USA

ISBN 978-0-387-47325-3   e-ISBN 978-0-387-47326-0

Library of Congress Control Number: 2008923105

© 2008 Springer Science+Business Media, LLC
All rights reserved. This work may not be translated or copied in whole or in part without the written permission of the publisher (Springer Science+Business Media, LLC, 233 Spring Street, New York, NY 10013, USA), except for brief excerpts in connection with reviews or scholarly analysis. Use in connection with any form of information storage and retrieval, electronic adaptation, computer software, or by similar or dissimilar methodology now know or hereafter developed is forbidden.
The use in this publication of trade names, trademarks, service marks and similar terms, even if they are not identified as such, is not to be taken as an expression of opinion as to whether or not they are subject to proprietary rights.

Printed on acid-free paper

9 8 7 6 5 4 3 2 1

springer.com

Foreword

Semiconductors entered the nanotechnology era when they went below the 100 nm technology node a few years ago. Today the industry is shipping 65 nm technology wafers in high volume, 45 nm is in production, with 32 nm working at the development stage. While the predictions that Moore's Law has reached it practical limits have been heard for years, they have proven to be premature. And it is expected that the technology will continue to move forward unabated for some years before it comes close to the basic physical limits to CMOS scaling.

Consumers are becoming the dominant force for electronic products. The industry has learnt that the consumer market is driven by many factors other than CMOS scaling alone. Functional diversification, accomplished through integration of multiple circuit types, and different device types, such as MEMs, optoelectronics, chemical and biological sensors and others, provides electronic product designers with different functional capabilities meeting the needs, wants, and tastes of consumers. This functional diversification together with cost, weight, size, fashion and appearance, and time to market, are critical differentiators in the market place. These two technology directions are often described as "More Moore" and "More than Moore".

Packaging is the final manufacturing process transforming semiconductor devices into functional products for the end user. Packaging provides electrical connections for signal transmission, power input, and voltage control. It also provides for thermal dissipation and the physical protection required for reliability. Packaging governs the size, weight, and shape of the end product and is the enabler for functional diversification through package architecture and package design. In the new landscape of advancing device technology nodes, and a dynamic consumer market place, packaging can become either the enabling or limiting factor. This market force has resulted in an unprecedented acceleration of innovation. Design concepts, packaging architecture, material, manufacturing process, equipment, and system integration technologies, are all changing rapidly.

Materials are at the heart of packaging technology. Packaging material contributes significantly to the packaged device performance, reliability and workability as well as to the total cost of the package. With the driving forces from "More Moore" and "More than Moore," the challenges for packaging materials have broadened from traditional package requirements for future generation devices to include those for new package types, such as the system in package (SiP) families, wafer level packaging, integrated passive device (IPD), through silicon vias (TSV), die and wafer stacking, 3D packaging, and RF, MEMs, physical, chemical and biological sensors, and optoelectronics applications. It is believed that materials in use today cannot meet the requirements of future packaging requirements. This is particularly true for complex SiP structures where hot spots, high currents, mechanical stresses for very thin die and ever shrinking geometries would require electrical, thermal, and mechanical properties that are beyond those of existing materials and manufacturing processes.

Nanomaterials and nanotechnologies promise to offer significant solutions towards packaging technology challenges in coming years. Carbon nanotubes (CNTs), nanowires and nanoparticles, have shown unique electrical, thermal, and mechanical properties orders of magnitude superior to current packaging materials used today. They had fired up the imaginations of engineers and scientists alike. How to design the next generation packaging materials and develop materials processing and application methodologies utilizing the nanomaterials' unique physical properties is an important question for the electronic packaging community.

Do CNTs have a place in future generation low-dimensional thermal interface materials (TIM), smoothing out the hot spots and taking higher levels of thermal energy away from the die? How do we utilize the CNT electrical properties for future generation high density packages? What role will nanoparticles play in the new generation passives? How would macromolecules be designed into polymer materials to provide specific electrical, thermal and mechanical properties required for the package function? With advances on the science and technology of nanomaterials, one envisions that whole new classes of materials will be introduced into the packaging structure to enable high power, high density interconnects, and new package features such as embedded and integrated passives, stacked and thinned dies, wafer level process, TSVs, MEMS, sensors, and medical and bio-chip applications.

This book is a compendium of in depth reviews written by some of the leading practitioners in the field. They cover the broad aspects of the field from materials preparations, materials properties, surface modifications, engineering applications, mathematical simulations, and "More than Moore" technical issues. It is a timely and important contribution to the technical literature for practitioners and researchers in the electronic industry field.

The editor of this book is a member of the IEEE Nanotechnology Council. Many of the contributors are from the IEEE/CPMT Society membership. They are to be congratulated for bringing this very important topic forth in the timely manner for the benefit of the electronic packaging and materials community.

Santa Clara, CA   William T. Chen

Preface

Moore's Law has been remarkably effective over 40 years or so in predicting the march of CMOS technology, as device dimensions shrank to mils, to microns, to nanometers. With continued CMOS shrinkage projected to 20 nm, there is clearly continued life in the technology, despite past predictions of its demise which turn out to be, like Mark Twain's, greatly exaggerated. However, the day will clearly come when the physical device structure cannot be supported at near atomic dimensions, but despite concerted research, no obvious successor technology has yet emerged as a clear winner. One of the factors in identifying that technology must be consideration of packaging techniques and design for reliability. However, package design depends on the nature of the basic device technology, and the decision process goes in circles.

However, the rapid development of nanotechnologies in almost every branch of science and engineering is already yielding new approaches to packaging materials and techniques, and these should be well developed and compatible for the next generation of devices, whether they are single electron transistors, spintronics, carbon nanotube transistors, molecular electronic devices, or something not yet envisaged.

While the packaging of nanoelectronic devices has been slowed by uncertainty of which device technology will turn out to be commercially viable, nanotechnologies are being developed to address current packaging problems of microelectronic systems, with details showing up in many conference presentations, e.g., at the annual *IEEE Electronic Component and Technology Conference*. However, many experts in nanotechnologies are unaware of the possible applications in electronics packaging, and conversely many packaging engineers are unfamiliar with the potential of nanoscale materials and devices. This book is intended to bridge that gap, with Chap. 1 introducing the scope of the field with a literature survey.

Then three chapters deal with computer modeling in nanopackaging. Bailey et al. take a high-level approach to the modeling process in Chap. 2, backed up with multiple examples of nanoscale modeling in packaging, present and future, including nanoimprinting, solder paste printing, microwave heating, underfill, and anisotropic conductive film. Chapter 3 from Fan and Yuen and Chap. 4 from van der Sluis et al. both focus on the molecular modeling technique, especially for interfacial characterization, with applications to carbon nanotube (CNT) thermal performance, moisture diffusion and thermal cycling, and delamination failures. The intention in each case is to understand macroscale package properties by modeling at nanoscale dimensions, and emphasize the need to be able to transfer modeling results between software at different length scales.

The bulk of the book from here on splits naturally into nanoparticle and CNT applications.

Morris covers fundamental metal nanoparticle properties in Chap. 5, with introductions to melting point depression, the coulomb block, interface diffusion effects, optical absorption, sintering, etc. The references in this chapter intentionally include many from the earliest days on nanotechnology research, to make the point that much work was done before the current decade's surge of interest and funding. Nanoparticle fabrication is introduced in Chap. 6 by Hayashi et al., who concentrate on an ecologically friendly sonochemical technique. Other fabrication methods are touched on in other chapters, including Chaps. 7 and 14.

The next three chapters consider nanotechnologies for passive devices, which are moving into the substrate as embedded components. The development of nanoparticle based high-k dielectrics is covered by Lu and Wong in Chap. 7, with consideration of the effects of both metallic and ferroelectric nanoparticles on material performance. At higher metal loading levels, the cermet (ceramic–metal, or polymer–metal) material becomes resistive, and cermets have been used as resistors in various applications for decades. The basic principle of operation balances the nanoscale effects of activated tunneling and percolation, as explained by Wu and Morris in Chap. 8. Nanoparticle applications in passive components are rounded out by the Jha et al.'s Chap. 9 on inductors and antennas, which are essential to portable wireless systems. These are generally micron-sized devices with nanoscale features, e.g., size effects, surface roughness, and nano-granular materials (for which classical theory does not match the properties).

Nanoscale engineering of isotropic conductive adhesives (ICAs) in Chap. 10, by Lu et al., covers both nanoparticle additives (i.e., low temperature nanosintering, CNT additives, etc.,) and enhancements by surface treatments. Chapter 11 by Das and Egitto deals with printed wiring board (PWB) microvias, and especially nanoparticle loaded fillers. Completing this group of three chapters, Felba and Schäfer cover nanoparticle-based PWB interconnect developments in Chap. 12, including progress toward a printable solution, and sintering (or laser sintering) of nano-Ag.

Soldering is the core technology of circuit assembly, so it is not surprising that researchers would explore the possible benefits of nanoparticle or CNT additives. As it turns out, Co, Ni, or Pt nanoparticles have some dramatic effects in limiting intermetallic compound (IMC) growth and hence mechanical failure by brittle fracture. These effects and others are covered by Amagai in Chap. 13.

Lall et al. describe the use of ceramic nanoparticle additives to lower the coefficient of thermal expansion in underfill in Chap. 14, the final chapter on nanoparticles. To model this effect, they also consider the problems of random distributions, viscoelasticity, etc.

The cluster of CNT chapters is introduced by two from the same research group. Various CNT fabrication techniques are reviewed in Chap. 15, by Yadav et al., and then Chap. 16 follows up with a review of basic CNT properties, characterization, and applications from Kunduru et al., who provide a primer on some device research which parallels the work described in this book.

High thermal conductance CNT microchannel cooling is described by Liu and Wang in Chap. 17, where they also cover the thermal conductance of CNT bumps and a novel electro-spun thermal interface material incorporating metal nanoparticles.

High CNT conductance suggests CNT—polymer composites for light weight electromagnetic shielding, and Cheng et al. present their work on the effectiveness of this technique in Chap. 18.

Chapter 19 provides the CNT parallel to Chap. 13, with the account by Kumar et al. of the results of adding CNTs to both eutectic Sn–Pb and Pb-free solders, with the verdict that essentially every parameter of interest can be improved.

The subject moves from CNTs to nanowires in Chap. 20 by Fiedler et al. The chapter includes both applications and fundamental problems, with an extensive bibliographic review. Then Ma et al. introduce a novel stress-engineered cantilever technique to form free-standing interconnect wires (or springs) in Chap. 21. Micron-scale structures are described first, before demonstrating their reduction to the nanoscale.

There is very little in the current literature about the specific packaging problems of either extreme CMOS shrinkage (to 45 nm and below) or future disruptive nanoelectronics technologies. Chapter 22 by Mallik et al. is devoted to the shrinking CMOS issue, providing historical perspective and analysis of the nm-CMOS challenges, along with insights on the future.

Zhang rounds out the book in Chap. 23 with a broad top-down overview of future directions of the industry as microelectronics moves to nanoelectronics, with both "More Moore" and "More-than-Moore" applications beyond CMOS integration.

Most chapters include a focus on the authors' own research in each respective field, but all end with extensive reference listings. The intentions of the book are to present an overview of each topic area, usually with the deeper treatment of one particular aspect, and especially to provide the reader with a resource for future study of those of interest. Hopefully, the book will pique such interest.

Portland, OR   James E. Morris

Contents

1. Nanopackaging: Nanotechnologies and Electronics Packaging 1
   James E. Morris

2. Modelling Technologies and Applications 15
   C. Bailey, H. Lu, S. Stoyanov, T. Tilford, X. Xue,
   M. Alam, C. Yin, and M. Hughes

3. Application of Molecular Dynamics Simulation
   in Electronic Packaging 39
   Haibo Fan and Matthew M.F. Yuen

4. Advances in Delamination Modeling 61
   O. van der Sluis, C.A. Yuan, W.D. van Driel, and G.Q. Zhang

5. Nanoparticle Properties 93
   James E. Morris

6. Nanoparticle Fabrication 109
   Y. Hayashi, M. Inoue, H. Takizawa, and K. Suganuma

7. Nanoparticle-Based High-k Dielectric Composites:
   Opportunities and Challenges 121
   Jiongxin Lu and C.P. Wong

8. Nanostructured Resistor Materials 139
   Fan Wu and James E. Morris

9. Nanogranular Magnetic Core Inductors: Design, Fabrication,
   and Packaging 163
   Gopal C. Jha, Swapan K. Bhattacharya, and Rao R. Tummala

10	**Nanoconductive Adhesives**.. Daoqiang Daniel Lu, Yi Grace Li, and C.-P. Wong	189
11	**Nanoparticles in Microvias**.. Rabindra N. Das and Frank D. Egitto	209
12	**Materials and Technology for Conductive Microstructures**.............. Jan Felba and Helmut Schaefer	239
13	**A Study of Nanoparticles in SnAg-Based Lead-Free Solders**............ Masazumi Amagai	265
14	**Nano-Underfills for Fine-Pitch Electronics**... Pradeep Lall, Saiful Islam, Guoyun Tian, Jeff Suhling, and Darshan Shinde	287
15	**Carbon Nanotubes: Synthesis and Characterization**........................ Yamini Yadav, Vindhya Kunduru, and Shalini Prasad	325
16	**Characteristics of Carbon Nanotubes** **for Nanoelectronic Device Applications**... Vindhya Kunduru, Yamini Yadav, and Shalini Prasad	345
17	**Carbon Nanotubes for Thermal Management of Microsystems**....... Johan Liu and Teng Wang	377
18	**Electromagnetic Shielding of Transceiver Packaging** **Using Multiwall Carbon Nanotubes**.. Wood-Hi Cheng, Chia-Ming Chang, and Jin-Chen Chiu	395
19	**Properties of 63Sn-37Pb and Sn-3.8Ag-0.7Cu Solders** **Reinforced With Single-Wall Carbon Nanotubes**............................... K. Mohan Kumar, V. Kripesh, and Andrew A.O. Tay	415
20	**Nanowires in Electronics Packaging**.. Stefan Fielder, Michael Zwanzig, Ralf Schmidt, and Wolfgang Scheel	441
21	**Design and Development of Stress-Engineered** **Compliant Interconnect for Microelectronic Packaging**.................... Lunyu Ma, Suresh K. Sitaraman, Qi Zhu, Kevin Klein, and David Fork	465

| 22 | **Flip Chip Packaging for Nanoscale Silicon Logic Devices: Challenges and Opportunities** .... | 491 |

Debendra Mallik, Ravi Mahajan, and Vijay Wakharkar

| 23 | **Nanoelectronics Landscape: Application, Technology, and Economy** .... | 517 |

G.Q. Zhang

Index 537

Contributors

Mohammad Alam
School of Computing and Mathematical Sciences, University of Greenwich,
Old Royal Naval College, Greenwich, London SE10 9LS, UK
M.O.Alam@gre.ac.uk

Masazumi Amagai
Tsukuba Technology Center, Texas Instruments, 17 Miyukigaoka,
Tsukuba-shi, Ibaragi-ken 305-0841 Japan
amai@ti.com

Chris Bailey
School of Computing and Mathematical Sciences, University of Greenwich,
Old Royal Naval College, Greenwich, London SE10 9LS, UK
c.bailey@gre.ac.uk

Swapan K. Bhattacharya
School of Electrical and Computer Engineering, Georgia Institute of Technology,
Atlanta, GA 30332, USA
swapan@ece.gatech.edu

Chia-Ming Chang
Institute of Electro-Optical Engineering, National Sun Yat-sen University,
Kaohsiung 80424, Taiwan, ROC
d9135805@student.nsysu.edu.tw

Wood-Hi Cheng
Institute of Electro-Optical Engineering, National Sun Yat-sen University,
Kaohsiung 80424, Taiwan, ROC
whcheng@mail.nsysu.edu.tw

Jin-Chen Chiu
Institute of Electro-Optical Engineering, National Sun Yat-sen University,
Kaohsiung 80424, Taiwan, ROC
m933050043@student.nsysu.edu.tw

Rabindra N. Das
Endicott Interconnect Technologies, Inc., 1093 Clark Street, Endicott,
New York, NY 13760, USA
rabindra.das@eitny.com

Frank D. Egitto
Endicott Interconnect Technologies, Inc., 1093 Clark Street, Endicott, New York,
NY 13760, USA
egittofd@eitny.com

Haibo Fan
Department of Mechanical Engineering, Hong Kong University of Science
and Technology, Clearwater Bay, N.T., Hong Kong
mehaibo@ust.hk

Jan Felba
Faculty of Microsystem Electronics and Photonics, Wroclaw University
of Technology, ul. Janiszewskiego 11/17, 50-372 Wroclaw, Poland
Jan.Felba@pwr.wroc.pl

Stefan Fiedler
Dept. Module Integration and Board Interconnection Technologies,
Fraunhofer Institut Zuverlässigkeit und Mikrointegration (IZM),
13355 Berlin, Gustav-Meyer-Allee 25, Germany
stefan.fiedler@izm.fraunhofer.de

David K. Fork
Palo Alto Research Center, 3333 Coyote Hill Rd., Palo Alto, CA 94304, USA

Yamato Hayashi
Department of Applied Chemistry, Tohoku University, 6-6-07 Aoba Aramaki,
Aoba-ku, Sendai 980-8579, Japan
hayashi@aim.che.tohoku.ac.jp

Michael Hughes
School of Computing and Mathematical Sciences, University of Greenwich,
Old Royal Naval College, Greenwich, London SE10 9LS, UK
M.Hughes@gre.ac.uk

Masahiro Inoue
Nanoscience and Nanotechnology Center, The Institute of Scientific
and Industrial Research (ISIR), Osaka University, Mihogaoka 8-1, Ibaraki,
Osaka 567-0047, Japan
inoue@sanken.osaka-u.ac.jp

Saiful Islam
Intel Corporation, 5000 W. Chandler Blvd., Chandler, AZ 85226, USA
muhammad.s.islam@intel.com

Gopal C. Jha
Packaging Research Center, Georgia Institute of Technology,
Atlanta, GA 30332, USA
gopal.jha@gatech.edu

Kevin Klein
Computer-Aided Simulation of Packaging Reliability (CASPaR) Lab,
The George W. Woodruff School of Mechanical Engineering,
Georgia Institute of Technology, 813 Ferst Drive, Atlanta, GA 30332-0405, USA

Vaidyanathan Kripesh
Institute of Microelectronics, 11 Science Park Road, Science Park II,
Singapore, Singapore 117685
Kripesh@ime.a-star.edu.sg

Katta Mohan Kumar
Nano/Microsystems Integration Laboratory, Department of Mechanical
Engineering, National University of Singapore, 9 Engineering Drive 1,
Singapore, Singapore 117576
g0203709@nus.edu.sg

Vindhya Kunduru
Department of Electrical & Computer Engineering, Portland State University,
FAB Suite 160, 1900 SW 4th Avenue, Portland, OR 97207-0751, USA
vindhya.es@gmail.com

Pradeep Lall
Department of Mechanical Engineering, Auburn University, 270 Ross Hall,
Auburn, AL 36849, USA
lall@eng.auburn.edu

Grace Li
School of Materials Science and Engineering, Georgia Institute of Technology,
771 Ferst Dr. NW, Atlanta, GA 30332, USA
yi.li@gatech.edu

Johan Liu
Bionano Systems Laboratory, Department of Microtechnology and Nanoscience,
Chalmers University of Technology, Kemivägen 9 Room A517, Se 412 96
Gothenburg, Sweden
jliu@chalmers.se

Daniel D. Lu
Henkel Loctite (China) Co., Ltd, 90 Zhujiang Road, Yantai, ETDZ, Shandong,
China 264006
Daniel.d.lu@ieee.org

Hua Lu
School of Computing and Mathematical Sciences, University of Greenwich,
Old Royal Naval College, Greenwich, London SE10 9LS, UK
h.lu@gre.ac.uk

Jiongxin Lu
Georgia Institute of Technology, Atlanta, GA, USA

Lunyu Ma
Computer-Aided Simulation of Packaging Reliability (CASPaR) Lab,
The George W. Woodruff School of Mechanical Engineering,
Georgia Institute of Technology, 813 Ferst Drive, Atlanta,
GA 30332-0405, USA

Ravi Mahajan
Intel Corp, M/S CH5-157, 5000 W. Chandler Blvd., Chandler, AZ 85226, USA
ravi.v.mahajan@intel.com

Debendra Mallik
Intel Corp, M/S CH5-157, 5000 W. Chandler Blvd., Chandler, AZ 85226, USA
debendra.mallik@intel.com

James E. Morris
Department of Electrical & Computer Engineering, Portland State University,
P.O. Box 751, Portland, OR 97201, USA
j.e.morris@ieee.org

Shalini Prasad
Department of Electrical & Computer Engineering, Portland State University,
FAB Suite 160, 1900 SW 4th Avenue, Portland, OR 97207-0751, USA
prasads@cecs.pdx.edu

Helmut Schaefer
Fraunhofer Institut Fertigungstechnik Materialforschung (IFAM),
Wiener Strasse 12, 28359 Bremen, Germany
helmut.schaefer@ifam.fraunhofer.de

Wolfgang Scheel
Department of Module Integration and Board Interconnection Technologies,
Fraunhofer Institut Zuverlässigkeit und Mikrointegration (IZM), 13355 Berlin,
Gustav-Meyer-Allee 25, Germany
wolfgang.scheel@izm.fraunhofer.de

Ralf Schmidt
Department of Module Integration and Board Interconnection Technologies,
Fraunhofer Institut Zuverlässigkeit und Mikrointegration (IZM), 13355 Berlin,
Gustav-Meyer-Allee 25, Germany
ralf.schmidt@izm.fraunhofer.de

Darshan Shinde
Department of Mechanical Engineering, Auburn University, 270 Ross Hall,
Auburn, AL 36849, USA
shindda@auburn.edu

Suresh K. Sitaraman
Computer-Aided Simulation of Packaging Reliability (CASPaR) Lab,
The George W. Woodruff School of Mechanical Engineering,
Georgia Institute of Technology, 813 Ferst Drive, Atlanta, GA 30332-0405, USA
suresh.sitaraman@me.gatech.edu

Stoyan Stoyanov
School of Computing and Mathematical Sciences, University of Greenwich,
Old Royal Naval College, Greenwich, London SE10 9LS, UK
S.Stoyanov@gre.ac.uk

Katsuaki Suganuma
Nanoscience and Nanotechnology Center, The Institute of Scientific and
Industrial Research (ISIR), Osaka University, Mihogaoka 8-1, Ibaraki,
Osaka 567-0047, Japan
suganuma@sanken.osaka-u.ac.jp

Jeff Suhling
Department of Mechanical Engineering, Auburn University,
270 Ross Hall, Auburn, AL 36849, USA
suhling@eng.auburn.edu

Hirotsugu Takizawa
Department of Applied Chemistry Tohoku University, 6-6-07 Aoba Aramaki,
Aoba-ku, Sendai, 980-8579, Japan
takizawa@aim.che.tohoku.ac.jp

Andrew A.O. Tay
Nano/Microsystems Integration Laboratory, Department of Mechanical
Engineering, National University of Singapore, 9 Engineering Drive 1,
Singapore
mpetayao@nus.edu.sg

Guoyun Tian
Department of Mechanical Engineering, Auburn University,
270 Ross Hall, Auburn, AL 36849, USA
tianguo@auburn.edu

Tim Tilford
School of Computing and Mathematical Sciences, University of Greenwich,
Old Royal Naval College, Greenwich, London SE10 9LS, UK
T.Tilford@gre.ac.uk

Rao R. Tummala
Packaging Research Center, Georgia Institute of Technology, Atlanta,
GA 30332, USA
rtummala@ece.gatech.edu

O. (Olaf) van der Sluis
Department of Precision and Microsystems Engineering, Delft University
of Technology, Mekelweg 2 2628 CD Delft, The Netherlands
o.vandersluis@tudelft.nl
and
Philips Applied Technologies, High Tech Campus 7 5656 AE Eindhoven,
The Netherlands
olaf.van.der.sluis@philips.com

W.D. (Willem) van Driel
NXP Semiconductors, Gerstweg 2 6534 AE Nijmegen, The Netherlands
Department of Precision and Microsystems Engineering, Delft University
of Technology, Mekelweg 2 2628 CD Delft, The Netherlands
willem.van.driel@nxp.com

Vijay Wakharkar
Intel Corp, M/S- CH5-157, 5000 W. Chandler Blvd., Chandler, AZ 85226, USA
vijay.s.wakharkar@intel.com

Teng Wang
Bionano Systems Laboratory, Department of Microtechnology and Nanoscience,
Chalmers University of Technology, Kemivägen 9 Room A517, Se 412 96
Gothenburg, Sweden
teng.wang@chalmers.se

C.-P. Wong
School of Materials Science and Engineering, Georgia Institute of Technology,
771 Ferst Dr. NW, Atlanta, GA 30332, USA
cp.wong@mse.gatech.edu

Fan Wu
Zounds, Inc., 1910 S. Stapley Drive, Suite 202, Mesa, AZ 85204, USA
fan.wu@zoundshearing.com

Xiangdiong Xue
School of Computing and Mathematical Sciences, University of Greenwich,
Old Royal Naval College, Greenwich, London SE10 9LS, UK
x.xue@gre.ac.uk

Yamini Yadav
Department of Electrical & Computer Engineering, Portland State University,
FAB Suite 160, 1900 SW 4th Avenue, Portland, OR 97207-0751, USA
ysy@pdx.edu

Chunyan Yin
School of Computing and Mathematical Sciences, University of Greenwich,
Old Royal Naval College, Greenwich, London SE10 9LS, UK
c.yin@gre.ac.uk

C.A. (Cadmus) Yuan
Department of Precision and Microsystems Engineering, Delft University
of Technology, Mekelweg 2, 2628 CD Delft, The Netherlands
c.a.yuan@tudelft.nl
and
NXP Semiconductors, Gerstweg 2, 6534 AE Nijmegen, The Netherlands
Matthew
c.a.yuan@nxp.com

M.F. Yuen
Department of Mechanical Engineering, Hong Kong
University of Science and Technology, Clearwater Bay, N.T., Hong Kong
meymf@ust.hk

G.Q. (Kouchi) Zhang
Department of Precision and Microsystems Engineering, Delft University
of Technology, Mekelweg 2, 2628 CD Delft, The Netherlands
g.q.zhang@tudelft.nl
and
NXP Semiconductors, High Tech Campus 60, Room 203, 5656 AG Eindhoven,
The Netherlands
g.q.zhang@nxp.com

Qi Zhu
Computer-Aided Simulation of Packaging Reliability (CASPaR) Lab,
The George W. Woodruff School of Mechanical Engineering,
Georgia Institute of Technology, 813 Ferst Drive, Atlanta, GA 30332-0405, USA

Michael Zwanzig
Department of Module Integration and Board Interconnection Technologies,
Fraunhofer Institut Zuverlässigkeit und Mikrointegration (IZM), 13355 Berlin,
Gustav-Meyer-Allee 25, Germany
michael.zwanzig@izm.fraunhofer.de

Chapter 1
Nanopackaging: Nanotechnologies and Electronics Packaging

James E. Morris

1.1 Introduction

It often seems that the promise of nanotechnology's impact on everyone's quality of life is as overhyped as past promises of endless cheap energy from cold fusion and high-temperature superconductivity. But there are two major differences. While the term "nanotechnology" has caught the attention of industry, legislators, and research funding agencies, in most cases the technologies in question are rooted in steady research progress in the field in question, as fabrication and characterization techniques have steadily conquered ever smaller dimensions, with the parallel development of theory to explain and model the new phenomena exposed. Furthermore, nanotechnologies have already yielded everyday consumer benefits beyond stain-resistant clothing and transparent sunblock. So, it is hardly surprising to discover active research and development programs in nanotechnology applications to electronics packaging, with special nanotechnology sessions at electronics packaging research conferences and research journal papers demonstrating the range and progress of these applications.

The definition of nanotechnology is usually taken to be where the size of the functional element falls below 100 nm or 0.1 μm. Of course, according to this definition, and with 45-nm CMOS in production, the nanoelectronics era is already here. Furthermore, with metallic grain sizes typically below this limit, one might also argue that solder has always qualified as a nanotechnology, along with many thin film applications. So, the requirement that the specific function depends upon this nanoscale dimension is conventionally added to the definition. According to this *caveat*, MOSFET technology, for example, would not qualify by simple device shrink, but would at dimensions permitting ballistic charge transport.

Nanotechnology drivers are the varied ways in which materials properties change at low dimensions. Electron transport mechanisms at small dimensions

---

J.E. Morris
Department of Electrical & Computer Engineering, Portland State University, P.O. Box 751, Portland, OR, 97207-0751, USA

include ballistic transport, severe mean free path restrictions in very small nanoparticles, various forms of electron tunneling, electron hopping mechanisms, and more. Other physical property changes include:

- Melting point depression, i.e., the reduction of metal nanoparticle melting points at small sizes [1], although this is unlikely to be a factor in packaging applications with even 10% reductions typically requiring dimensions under 5 nm [2]
- Sintering by surface self-diffusion, which is thermally activated, with net diffusion away from convex surfaces of high curvature [3]
- The Coulomb blockade effect, which requires an external field or thermal source of electrostatic energy to charge an individual nanoparticle, and is the basis of single-electron transistor operation [4]
- Theoretical maximum mechanical strengths in single grain material structures [5]
- Unique optical scattering properties by nanoparticles that are one to two orders smaller than the wavelength of visible light [6]
- The enhanced chemical activities of nanoparticles, which make them effective as catalysts, and other effects of the high surface-to-volume ratio

New nanoscale characterization techniques will be applied wherever they can provide useful information, and the atomic force microscope (AFM), for example, is relatively commonly used to correlate adhesion to surface feature measurements. More recently, confocal microscopy has been applied to packaging research [7], but it is especially interesting to note the development of a new instrument, such as the atomic force acoustic microscope [8], which adapts the AFM to the well-known technique for package failure detection.

1.2 Computer Modeling

The use of composite materials is well established for many applications. But, while overall effective macroscopic properties are satisfactory for computer modeling of automotive body parts, for example, they are clearly inadequate for structures of dimensions similar to the particulate sizes in the composite. The modeling of such microelectronics (or nanoelectronics) packages must include two-phase models of the composite structure, and this general principle of inclusion of the nanoscale structural detail in expanded material models must be extended to all aspects of package modeling [9]. The extended computer models can be based on either the known properties of the constituent materials (and hopefully known at appropriate dimensions) or the measured nanoscale properties (e.g., by a nanoindenter [10, 11] or AFM [12]). Molecular Dynamics modeling software has been particularly useful in the prediction of macroscale effects from the understanding of nanoscale interactions [13].

1.3 Nanoparticles

1.3.1 Nanoparticles: Fabrication

Noble metal nanoparticles have been fabricated by an ultrasonic processing technique [14], and Ag/Cu with "polyol" [15]. Alternatively, a precursor may be used, e.g., $AgNO_3$ for Ag nanoparticles, and there are techniques to control the particle shapes, e.g., spherical, cubic, or wires [16].

1.3.2 Nanoparticles: High-k Dielectrics

The move toward embedded passive components at both on-chip and PWB levels has prompted a search for high dielectric constant materials for low area capacitors. High dielectric constants can be achieved by the inclusion of high dielectric constant particulates and minimal thickness. The latter requirement pushes one toward nano-scale particulates, with examples of the former covering ceramic [17–20], silicon [21], and metal [22–26]. The ceramic particles are generally barium titanate, e.g., applied to organic FETs with composite k around 35 [18]; in such materials, the particle surface energy must be reduced to avoid aggregation [19].

The target k is 50–200, and while $k \sim 150$ has been achieved, it is at the expense of high leakage (dielectric loss). Similar structures have been studied in the past as "cermets" (ceramic–metal composites) for high resistivity materials for on-chip resistors [27], which conduct by electron tunneling between particles. At low fields, the nanoparticles can act as Coulomb blocks to minimize DC leakage if they are sufficiently small [21], but still do not eliminate it at finite temperature [28]. It is the AC performance which is more important, however, and interparticle capacitance will bypass the block unless pseudoinductive effects develop at capacitor thicknesses which permit even short nanoparticle chains [29].

An alternative approach to leakage is to use aluminum particles, to take advantage of the native oxide coating [24], with $k \sim 160$ achieved [25]. Ag/Al mixtures have also been studied [26].

Note that thermally conductive materials have very similar structural requirements to the passive components, with metallic or SiC nanoparticles as fillers [30].

1.3.3 Nanoparticles: Electrically Conductive Adhesives

The addition of smaller μm diameter silver powder to 10-μm silver flakes in isotropic conductive adhesives reduces resistance by inserting bridging particles between the flakes. The simple addition of nanoparticles does not improve conductance, due to mean free path restrictions and added interface resistances, and the same principles limit the performance of alumina-loaded thermal composites [31]. The addition of silver nanoparticles does achieve dramatic reductions, however, by sintering wide area contacts between flakes [32], a principle also applicable to via fill [33]. Filler nanoparticle sintering can also improve anisotropic conductive adhesive performance [34], aided by contact conductance enhancement by the addition of self-assembly molecular surface treatments [32, 35, 36]. Sintering effects have also been shown to improve contacts in materials with sufficiently low filler content as to be regarded as nonconductive adhesives [37].

1.3.4 Nanoparticles: Interconnect

Surface electrical interconnect for board and package levels can be achieved by screen-printing nanoscale metal colloids in suspension [38], and there has been recent effort to achieve the same effect by ink-jet printing [39–41]. Electrical continuity is established by sintering, e.g., of 5–10-nm silver particles [42–46]. Sintered Ag nanoparticles can also be used for die attach [47]. As a variation, magnetic composite films (e.g., Co/SiO_2 in BCB and Ni/ferrite in epoxy) have been screen-printed for antennas [48]. Sn/Ni bumps have also been grown on Sn from a nm Ni slurry [49].

1.3.5 Nanoparticles: Silica Filler in Underfill

The key advantage of nanoscale silica particles in underfill formulations is that they resist settling [50]. They also scatter light less than the larger traditional fillers, permitting UV optical curing, and providing a dual photoresist function from a single material [51] and other advantages of optical transparency [52]. The higher viscosity of the nanofilled material can be reduced by silane surface treatments [53].

1.3.6 Nanoparticles: Solder

The addition of Pt, Ni, or Co nanoparticles to no-Pb Sn–Ag-based solders [54, 55] eliminates Kirkendall voids, reduces intermetallic compound (IMC) growth, and reduces IMC grain sizes, significantly improving drop test performance [56]. Similarly, Ni or Mo nanoparticles promote finer grain growth, increased creep resistance, and better contact wetting [57]. Nanoparticles in the grain boundaries also inhibit grain boundary sliding and thermomechanical fatigue, but a similar function can be provided by 1.5-nm $SiO_{1.5}$ polyhedral oligomeric silsesquioxane structures with surface-active Si-OH groups [58].

1.4 Carbon Nanotubes

1.4.1 Carbon Nanotubes: Solder

The addition of carbon nanotubes (CNTs) to solder can also have beneficial effects, e.g., 30–50% improvements in tensile strength [55, 59].

1.4.2 Carbon Nanotubes: Thermal

The high thermal conductivity of CNTs is being exploited for microelectronics chip cooling both directly in conductive cooling and indirectly in convective cooling systems [60, 61]. For conductive systems, the key is to establish CNT alignment [59], since the thermal conductivities of random arrays (of CNTs and carbon fibers alike) fall far short of expectation, showing no advantages over conventional materials, often also because of CNT fracture at the substrate [62]. In one of the most advanced techniques, vertical CNTs are first grown on both the aluminum heat sink and silicon chip surfaces, which are then positioned a few μm apart in a CVD furnace, enabling the CNTs from the two surfaces to grow further and connect with each other [59]. Composites incorporating CNTs have also been studied for thermal interface materials, e.g., CNT/carbon black mixtures in epoxy resin [63]. The use of a liquid crystal resin matrix can impose structural order on the CNT alignment to yield a sevenfold improvement in thermal conductivity [64]. Recently, electrospun polymer fibers filled with CNTs, or with SiC or metallic nanoparticles, have shown advances in both mechanical and thermal properties [65].

So far, convective CNT cooling has been limited to the use of μm-scale clusters of vertically grown nanotubes [66, 67]. These clusters define microchannels for coolant flow which look very much like the metal or silicon structures they aim to replace (Fig. 17.7) with similar thermal performances. The problem is that the flowing coolant is only in contact with the outermost CNTs of the clusters and the internal CNTs are not even in good contact with each other. The system has been modeled [60] and the solution is clearly to spread the CNTs apart by an optimal separation to permit coolant contact with each one [66]. The problem then is whether individual CNTs can withstand the coolant flow pressure without detaching from the substrate.

1.4.3 Carbon Nanotubes: Electrical

An important development has been the ability to open CNTs after growth [68], since the open ends permit better wetting by Sn/Pb (and presumably other metals) for improved electrical contact. Au and Ag incorporation into CNTs and fullerenes has also been studied for electrical contacts with minimal galvanic corrosion [69]. Metal- and carbon-loaded polymers have long been used for high-frequency conductors in electromagnetic shielding, and both carbon fibers [70] and multiwalled CNTs have been studied in polymer matrices for the purpose [71, 72], but CNT replacement of metal filler in isotropic conductive adhesives [73–75] does not even match the electrical conductivity of standard materials [75, 76]. However, 10–50-μm long Ag/Co nanowires of 200-nm diameter can be maintained in a parallel vertical orientation by a magnetic field while polymer resin flows around them [77], to form an anisotropic conductive film for z-axis contacts [78–80]. CNT interconnection schemes are also under intense study [81–84], with μm-scale CNT clusters successfully developed as flip-chip "nanobumps" [85].

1.4.4 Carbon Nanotubes: Fabrication

CNT growth can be accomplished for both electrical and thermal applications by chemical vapor deposition [86], with satisfactory solder wetting of the CNTs for electrical contacts.

1.5 Nanoscale Structures

The incorporation of nanodiamond particles into an electroless Ni film coating on an electrothermal actuator [87] can improve cantilever performance by changing the thermal and mechanical properties. In a truly impressive development, the microspring contacts originally developed at PARC-Xerox have been downsized to 10-nm wide cantilevers, still 10-μm long, for biological sensing [88] (Fig. 21.10). Nanoimprinting technology is also being used to fabricate optical interconnect waveguides in organic PCBs [89].

1.6 "Nanointerconnects"

The "nanointerconnect" terminology is applied to interconnect structures which are clearly μm-scaled [90–97]. The ITRS roadmap calls for 20–100-μm pitch interconnects for nanoelectronics systems of feature size under 100 nm [93], which has prompted studies of nanograin solders [90] or copper [92], nanocrystalline copper and nickel [93], and nanoscale via fillers [91], all for applications at around 30–35-μm pitch [90, 92]. Some nanointerconnect options are reviewed in [95]. Other technologies can be included in this group, too, e.g., metal-coated polymer posts on a similar scale [95], and embedded micro- or nanoelectrodes for biological flow sensing [97].

Control of the interfacial surface charge on the nanoelectrode in contact with the fluid can be used to control the flow [97].

1.7 Conclusion

The importance of nanoelectronics and "electro-nanotechnologies" in the future is sufficiently well recognized to have become the subject of industrial and government policy roadmaps [98]. Similarly, the academic world is responding with graduate level courses (although with few textbooks so far). As for electronics packaging, the field requires students to be "subject multilingual" [99].

One of the surprising observations to come out of this survey, in full agreement with prior comment [100], has been that there is almost no work reported on the development of packaging for future nanoelectronics technologies. The "nanointerconnect" work [90–95] is directed toward continued Moore's Law shrinkage of silicon, but only one paper specifically addresses the impact of the package on the device [101], specifically of organic flip-chip packaging of 110-nm CMOS. Candidate next-generation nanoelectronics technologies (e.g., single-electron transistors, quantum automata, molecular electronics, etc.) are generally hypersensitive to dimensional change, if based on quantum-mechanical electron tunneling, and this is just one example of how appropriate packaging will be essential to the success or failure of these technologies [102]. Packaging strategies must therefore be developed in parallel with the basic nanoelectronics device technologies to make informed decisions as to their commercial viabilities.

Another observation is that the work to date has been highly concentrated in a few laboratories, as reported in [103, 104], where the numbers of nanotechnology papers presented at the annual *IEEE Electronic Components & Technology Conference* are tracked.

New materials are emerging from small companies and university labs all the time, and with diverse applications beyond those discussed above [105].

References

1. H. Jiang, K. Moon, H. Dong, and F. Hua, Thermal Properties of Oxide Free Nano Non Noble Metal for Low Temperature Interconnect Technology, Proceedings of the 56th IEEE Electronic Components and Technology Conference, San Diego, CA, 2006, pp. 1969–1973
2. J.R. Sambles, An electron microscope study of evaporating gold particles: the Kelvin equation for liquid gold and the lowering of the melting point of solid gold particles, Proc. R. Soc. Lond. A 324 (1971) 339–351
3. M. Ohring, Materials Science of Thin Films: Deposition & Structure (2nd edition), Academic, San Diego, 2002, pp. 395–397
4. J.E. Morris, Single-electron transistors, in The Electrical Engineering Handbook (3rd edition): Electronics, Power Electronics, Optoelectronics, Microwaves, Electromagnetics, and Radar, R. C. Dorf (editor), CRC/Taylor & Francis, Boca Raton/London, 2006, pp. 3.53–3.64

5. R.A. Flinn and P.K. Trojan, Engineering Materials & Their Applications (2nd edition), Houghton-Mifflin, Boston, MA, 1981, pp. 75–77
6. T. Yamaguchi, M. Sakai, and N. Saito, Optical properties of well-defined granular metal systems, Phys. Rev. B 32(4) (1985) 2126–2130
7. M. Luniak, H. Hoeltge, R. Brodmann, and K.-J. Wolter, Optical Characterization of Electronic Packages with Confocal Microscopy, Proceedings of the 1st IEEE Electronics Systemintegration Technology Conference, Dresden, Germany, 2006, pp. 1318–1322
8. B. Koehler, B. Bendjus, and A. Striegler, Determination of Deformation Fields and Visualization of Buried Structures by Atomic Force Acoustic Microscopy, Proceedings of the 1st IEEE Electronics Systemintegration Technology Conference, Dresden, Germany, 2006, pp. 1330–1335
9. B. Michel, R. Dudek, and H. Walter, Reliability Testing of Polytronics Components in the Micro-Nano Region, Proceedings of the 5th International Conference on Polymers and Adhesives in Microelectronics and Photonics, Wroclaw, Poland, 2005, pp. 13–15
10. S. Koh, R. Rajoo, R. Tummala, A. Saxena, and K.T. Tsai, Material Characterization for Nano Wafer Level Packaging Application, Proceedings of the 55th IEEE Electronic Components and Technology Conference, Orlando, FL, 2005, pp. 1670–1676
11. S. Bansal, E. Toimil-Molares, A. Saxena, and R. Tummala, Nanoindentation of Single Crystal and Polycrystalline Copper Nanowires, Proceedings of the 55th IEEE Electronic Components and Technology Conference, Orlando, FL, 2005, pp. 71–76
12. C.K.Y. Wong, H. Gu, B. Xu, and M.M. Fyuen, A New Approach in Measuring Cu-EMC Adhesion Strength by AFM, Proceedings of the 54th IEEE Electronic Components and Technology Conference, Las Vegas, NV, 2004, pp. 491–495
13. E.D. Dermitzaki, J. Bauer, B. Wunderle, and B. Michel, Diffusion of Water in Amorphous Polymers at Different Temperatures Using Molecular Dynamics Simulation, Proceedings of the 1st IEEE Electronics Systemintegration Technology Conference, Dresden, Germany, 2006, pp. 762–772
14. Y. Hayashi, H. Takizawa, M. Inoue, K. Niihara, and K. Suganuma, Ecodesigns and applications for noble metal nanoparticles by ultrasound process, IEEE Trans. Electron. Packag. Manuf. 28(4) (2005) 338–343 (also Proc. Polytronic 2004)
15. H. Jiang, K. Moon, and C.P. Wong, Synthesis of Ag–Cu Alloy Nanoparticles for Lead-Free Interconnect Materials, Proceedings of the 10th IEEE/CPMT International Symposium on Advanced Packaging Materials, Irvine, CA, 2005
16. S. Pothukuchi, Yi Li, and C.P. Wong, Shape Controlled Synthesis of Nanoparticles and Their Incorporation into Polymers, Proceedings of the 54th IEEE Electronic Components and Technology Conference, Las Vegas, NV, 2004, pp. 1965–1967
17. J. Xu, J. Xu, S. Bhattacharya, K. Moon, J. Lu, B. Englert, and P. Pramanik, Large-Area Processable High k Nanocomposite-Based Embedded Capacitors, Proceedings of the 56th IEEE Electronic Components and Technology Conference, San Diego, CA, 2006, pp. 1520–1532
18. A. Rasul, J. Zhang, and D. Gamota, Printed Organic Electronics with a High K Nanocomposite Dielectric Gate Insulator, Proceedings of the 56th IEEE Electronic Components and Technology Conference, San Diego, CA, 2006, pp. 167–170
19. R. Das, M. Poliks, J. Lauffer, and V. Markovich, High Capacitance, Large Area, Thin Film, Nanocomposite Based Embedded Capacitors, Proceedings of the 56th IEEE Electronic Components and Technology Conference, San Diego, CA, 2006, pp. 1510–1515
20. J. Lu, K.-S. Moon, and C.-P. Wong, High-k Polymer Nanocomposites as Gate Dielectrics for Organic Electronics Applications, Proceedings of the 57th IEEE Electronic Components and Technology Conference, Reno, NV, 2007
21. R. Kubacki, Molecularly Engineered Variable Nanocomposites to Embed Precision Capacitors On-Chip, Proceedings of the 56th IEEE Electronic Components and Technology Conference, San Diego, CA, 2006, pp. 161–166
22. Yi Li, S. Pothukuchi, and C.P. Wong, Development of a Novel Polymer–Metal Nanocomposite Obtained Through the Route of In Situ Reduction and It's Dielectric Properties, Proceedings

of the 54th IEEE Electronic Components and Technology Conference, Las Vegas, NV, 2004, pp. 507–513
23. J. Lu, K. Moon, J. Xu, and C.P. Wong, Dielectric Loss Control of High-K Polymer Composites by Coulomb Blockade Effects of Metal Nanoparticles for Embedded Capacitor Applications, Proceedings of the 10th IEEE/CPMT International Symposium on Advanced Packaging Materials, Irvine, CA, 2005 pp. 237–242
24. J. Xu and C.P. Wong, High-K Nanocomposites with Core–Shell Structured Nanoparticles for Decoupling Applications, Proceedings of the 55th IEEE Electronic Components and Technology Conference, Orlando, FL, 2005, pp. 1234–1240
25. J. Xu and C.P. Wong, Effects of the Low Loss Polymers on the Dielectric Behavior of Novel Aluminum-Filled High-k Nano-Composites, Proceedings of the 54th IEEE Electronic Components and Technology Conference, Las Vegas, NV, 2004, pp. 496–506
26. J. Lu, K. Moon, and C.P. Wong, Development of Novel Silver Nanoparticles/Polymer Composites as High K Polymer Matrix by In-Situ Photochemical Method, Proceedings of the 56th IEEE Electronic Components and Technology Conference, San Diego, CA, 2006, pp. 1841–1846
27. F. Wu and J.E. Morris, Characterizations of $(SiO_xCr_{1-x})N_{1-y}$ thin film resistors for integrated passive applications, Proceedings of the 53rd Electronic Components and Technology Conference, New Orleans, 2003, pp. 161–166
28. J.E. Morris, Recent progress in discontinuous thin metal film devices, Vacuum 50(1–2) (1998) 107–113
29. J.E. Morris, F. Wu, C. Radehaus, M. Hietschold, A. Henning, K. Hofmann, and A. Kiesow, Single Electron Transistors: Modeling and Fabrication, Proceedings of the 7th International Conference on Solid-State and Integrated Circuit Technology (ICSICT), Beijing, October 2004, pp. 634–639
30. L. Ekstrand, H. Kristiansen, and J. Liu, Characterization of Thermally Conductive Epoxy Nano Composites, Proceedings of the 28th International Spring Seminar on Electronics Technology (ISSE'05), Vienna, 2005, pp. 19–23
31. L. Fan, B. Su, J. Qu, and C.P. Wong, Electrical and Thermal Conductivities of Polymer Composites Containing Nano-Sized Particles, Proceedings of the 54th IEEE Electronic Components and Technology Conference, Las Vegas, NV, 2004, pp. 148–154
32. H. Jiang, K. Moon, L. Zhu, J. Lu, and C.P. Wong, The Role of Self-Assembled Monolayer (SAM) on Ag Nanoparticles for Conductive Nanocomposite, Proceedings of the 10th IEEE/CPMT International Symposium on Advanced Packaging Materials, Irvine, CA, 2005
33. R. Das, J. Lauffer, and F. Egitto, Electrical Conductivity and Reliability of Nano- and Micro-Filled Conducting Adhesives for Z-Axis Interconnections, Proceedings of the 56th IEEE Electronic Components and Technology Conference, San Diego, CA, 2006, pp. 112–118
34. K. Moon, S. Pothukuchi, Yi Li, and C.P. Wong, Nano Metal Particles for Low Temperature Interconnect Technology, Proceedings of the 54th IEEE Electronic Components and Technology Conference, Las Vegas, NV, 2004, pp. 1983–1988
35. Yi Li, K. Moon, and C.P. Wong, Improvement of Electrical Performance of Anisotropically Conductive Adhesives, Proceedings of the 10th IEEE/CPMT International Symposium on Advanced Packaging Materials, Irvine, CA, 2005
36. Yi Li, K. Moon, and C.P. Wong, Electrical Property of Anisotropically Conductive Adhesive Joints Modified by Self-Assembled Monolayer (SAM), Proceedings of the 54th IEEE Electronic Components and Technology Conference, Las Vegas, NV, 2004, pp. 1968–1974
37. Yi Li and C.P. Wong, Novel Lead Free Nano Scale Non-Conductive Adhesive (NCA) Interconnect Materials for Ultra-Fine Pitch Electronic Packaging Applications, Proceedings of the 56th IEEE Electronic Components and Technology Conference, San Diego, CA, 2006, pp. 1239–1245
38. S. Joo and D.F. Baldwin, Demonstration for Rapid Prototyping of Micro-Systems Packaging by Data-Driven Chip-First Process Using Nano-Particles Metal Colloids, Proceedings of the 55th IEEE Electronic Components and Technology Conference, Orlando, FL, 2005, pp. 1859–1863

39. A. Moscicki, J. Felba, T. Sobierajski, J. Kudzia, A. Arp, and W. Meyer, Electrically Conductive Formulations Filled Nano Size Silver Filler for Ink-Jet Technology, Proceedings of the 5th International Conference on Polymers and Adhesives in Microelectronics and Photonics, Wroclaw, Poland, 2005, pp. 40–44
40. J. Kolbe, A. Arp, F. Calderone, E.M. Meyer, W. Meyer, H. Schaefer, and M. Stuve, Inkjettable Conductive Adhesive for Use in Microelectronics and Microsystems Technology, Proceedings of the 5th International Conference on Polymers and Adhesives in Microelectronics and Photonics, Wroclaw, Poland, 2005, pp. 160–163
41. J.G. Bai, K.D. Creehan, and H.A. Kuhn, Inkjet printable nanosilver suspensions for enhanced sintering quality in rapid manufacturing, Nanotechnology 18 (2007) 1–5
42. W. Peng, V. Hurskainen, K. Hashizume, S. Dunford, S. Quander, and R. Vatanparast, Flexible Circuit Creation with Nano Metal Particles, Proceedings of the 55th IEEE Electronic Components and Technology Conference, Orlando, FL, 2005, pp. 77–82
43. J.G. Bai, Z.Z. Zhang, J.N. Calata, and G.-Q. Lu, Low-temperature sintered nanoscale silver as a novel semiconductor device-metallized substrate interconnect material, IEEE Trans. Compon. Packag. Technol. 29(3) (2006) 589–593
44. M. Nakamoto, M. Yamamoto, Y. Kashiwagi, H. Kakiuchi, T. Tsujimoto, and Y. Yoshida, A Variety of Silver Nanoparticle Pastes for Fine Electronic Circuit Patter Formation, Proceedings of the 6th International Conference on Polymers and Adhesives in Microelectronics and Photonics, Tokyo, 2007
45. D. Wakuda, M. Hatamura, and K. Suganuma, Novel Room Temperature Wiring Process of Ag Nanoparticle Paste, Proceedings of the 6th International Conference on Polymers and Adhesives in Microelectronics and Photonics, Tokyo, 2007
46. A. Moscicki, J. Felba, P. Gwiazdzinski, and M. Puchalski, Conductivity Improvement of Microstructures Made by Nano-Size-Silver Filled Formulations, Proceedings of the 6th International Conference on Polymers and Adhesives in Microelectronics and Photonics, Tokyo, 2007
47. J.G. Bai, Z.Z. Zhang, J.N. Calata, and G.-Q. Lu, Characterization of Low-Temperature Sintered Nanoscale Silver Paste for Attaching Semiconductor Devices, Proceedings of the 7th IEEE CPMT Conference on High Density Microsystem Design and Packaging and Component Failure Analysis (HDP'05), Shanghai, 2005, pp. 272–276
48. P.M. Raj, P. Muthana, T.D. Xiao, L. Wan, D. Balaraman, I.R. Abothu, S. Bhattacharya, M. Swaminathan, and R. Tummala, Magnetic Nano-Composites for Organic Compatible Miniaturized Antennas and Inductors, Proceedings of the 10th IEEE/CPMT International Symposium on Advanced Packaging Materials, Irvine, CA, 2005
49. R. Doraiswami and R. Tummala, Nano-Composite Lead-Free Interconnect and Reliability, Proceedings of the 55th IEEE Electronic Components and Technology Conference, Orlando, FL, 2005, pp. 871–873
50. P. Lall, S. Islam, J. Suhling, and G. Tian, Nano-Underfills for High-Reliability Applications in Extreme Environments, Proceedings of the 55th IEEE Electronic Components and Technology Conference, Orlando, FL, 2005, pp. 212–222
51. Y. Sun, Z. Zhang, and C.P. Wong, Photo-Definable Nanocomposite for Wafer Level Packaging, Proceedings of the 55th IEEE Electronic Components and Technology Conference, Orlando, FL, 2005, pp. 179–184
52. Y. Sun and C.P. Wong, Study and Characterization on the Nanocomposite Underfill for Flip Chip Applications, Proceedings of the 54th IEEE Electronic Components and Technology Conference, Las Vegas, NV, 2004, pp. 477–483
53. Y. Sun, Z. Zhang, and C.P. Wong, Fundamental Research on Surface Modification of Nano-Size Silica for Underfill Applications, Proceedings of the 54th IEEE Electronic Components and Technology Conference, Las Vegas, NV, 2004, pp. 754–760
54. W. Guan, S.C. Verma, Y. Gao, C. Andersson, Q. Zhai, and J. Liu, Characterization of Nanoparticles of Lead Free Solder Alloys, Proceedings of the 1st IEEE Electronics Systemintegration Technology Conference, Dresden, Germany, 2006, pp. 7–12

55. K. Mohan Kumar, V. Kripesh, and A.A.O. Tay, Sn–Ag–Cu Lead-Free Composite Solders for Ultra-Fine-Pitch Wafer-Level Packaging, Proceedings of the 56th IEEE Electronic Components and Technology Conference, San Diego, CA, 2006, pp. 237–243
56. M. Amagai, A Study of Nano Particles in SnAg-Based Lead Free Solders for Intermetallic Compounds and Drop Test Performance, Proceedings of the 56th IEEE Electronic Components and Technology Conference, San Diego, CA, 2006, pp. 1170–1190
57. K.M. Kumar, V. Kripesh, and A.A.O. Tay, Sn-Ag-Cu Lead-Free Composite Solders for Ultra-Fine-Pitch Wafer-Level Packaging, Proceedings of the 56th IEEE Electronic Components and Technology Conference, San Diego, CA, 2006
58. A. Lee, K.N. Subramanian, and J.-G. Lee, Development of Nanocomposite Lead-Free Electronic Solders, Proceedings of the 10th IEEE/CPMT International Symposium on Advanced Packaging Materials, Irvine, CA, 2005
59. K. Zhang, M.M.F. Yuen, J.Y. Miao, N. Wang, and D.G.W. Xiao, Thermal Interface Material with Aligned CNT Growing Directly on the Heat Sink Surface and Its Application in HB-LED Packaging, Proceedings of the 56th IEEE Electronic Components and Technology Conference, San Diego, CA, 2006, pp. 177–182
60. T. Wang, M. Jonsson, E. Nystrom, Z. Mo, E.E.B. Campbell, and J. Liu, Development and Characterization of Microcoolers Using Carbon Nanotubes, Proceedings of the 1st IEEE Electronics Systemintegration Technology Conference, Dresden, Germany, 2006, pp. 881–885
61. J. Xu and T.S. Fisher, Enhanced Thermal Contact Conductance Using Carbon Nanotube Array Interfaces, IEEE Trans. Compon. Packag. Technol. 29(2) (2006) 261–267
62. H.A. Zhong, S. Rubinsztajn, A. Gowda, D. Esler, D. Gibson, D. Bucklet, J. Osaheni, and S. Tonapi, Utilization of Carbon Fibers in Thermal Management of Microelectronics, Proceedings of the 10th IEEE/CPMT International Symposium on Advanced Packaging Materials, Irvine, CA, 2005
63. K. Zhang, G.-W. Xiao, C.K.Y. Wong, H.-W. Gu, M.M.F. Yuen, P.C.H. Chan, and B. Xu, Study on Thermal Interface Material With Carbon Nanotubes and Carbon Black in High-Brightness LED Packaging with Flip-Chip Technology, Proceedings of the 55th IEEE Electronic Components and Technology Conference, Orlando, FL, 2005, pp. 60–65
64. T.-M. Lee, K.-C. Chiou, F.-P. Tseng, and C.-C. Huang, High Thermal Efficiency Carbon Nanotube–Resin Matrix for Thermal Interface Materials, Proceedings of the 55th IEEE Electronic Components and Technology Conference, Orlando, FL, 2005, pp. 55–59
65. J. Liu, M.O. Olorunyomi, X. Lu, W.X. Wang, T. Aronsson, and D. Shangguan, New Nano-Thermal Interface Material for Heat Removal in Electronics Packaging, Proceedings of the 1st IEEE Electronics Systemintegration Technology Conference, Dresden, Germany, 2006, pp. 1–6
66. Z. Mo, R. Morjan, J. Anderson, E.E.B. Campbell, and J. Liu, Integrated Nanotube Microcooler for Microelectronics Applications, Proceedings of the 55th IEEE Electronic Components and Technology Conference, Orlando, FL, 2005, pp. 51–54
67. L. Ekstrand, Z. Mo, Y. Zhang, and J. Liu, Modelling of Carbon Nanotubes as Heat Sink Fins in Microchannels for Microelectronics Cooling, Proceedings of the 5th International Conference on Polymers and Adhesives in Microelectronics and Photonics, Wroclaw, Poland, 2005, pp. 185–187
68. L. Zhu, Y. Xiu, D. Hess, and C.P. Wong, In-Situ Opening Aligned Carbon Nanotube Films/Arrays for Multichannel Ballistic Transport in Electrical Interconnect, Proceedings of the 56th IEEE Electronic Components and Technology Conference, San Diego, CA, 2006, pp. 171–176
69. R.T. Pike, R. Dellmo, J. Wade, S. Newland, G. Hyland, and C.M. Newton, Metallic Fullerene and MWCNT Composite Solutions for Microelectronics Subsystem Electrical Interconnection Enhancement, Proceedings of the 54th IEEE Electronic Components and Technology Conference, Las Vegas, NV, 2004, pp. 461–465
70. J. Ding, S. Rea, D. Linton, E. Orr, and J. MacConnell, Mixture Properties of Carbon Fibre Composite Materials for Electronics Shielding in Systems Packaging, Proceedings of the 1st

IEEE Electronics Systemintegration Technology Conference, Dresden, Germany, 2006, pp. 19–25
71. J.-C. Chiu, C.-M. Chang, W.-H. Cheng, and W.-S. Jou, High-Performance Electromagnetic Susceptibility for a 2.5Gb/s Plastic Transceiver Module Using Multi-Wall Carbon Nanotubes, Proceedings of the 56th IEEE Electronic Components and Technology Conference, San Diego, CA, 2006, pp. 183–186
72. C.-M. Chang, J.-C. Chiu, C.-Y. Yeh, W.-S. Jou, Y.-F. Lan, Y.-W. Fang, J.-J. Lin, and W.-H. Cheng, Electromagnetic Shielding Performance for a 2.5Gb/s Plastic Transceiver Module Using Dispersive Multiwall Carbon Nanotubes, Proceedings of the 57th IEEE Electronic Components and Technology Conference, Reno, NV, 2007
73. J. Li and J.K.. Lumpp, Electrical and Mechanical Characterization of Carbon Nanotube Filled Conductive Adhesive, Proceedings of the IEEEAC, 2006, P. 1519
74. L. Xuechun and L. Feng, The Improvement on the Properties of Silver-Containing Conductive Adhesives by the Addition of Carbon Nanotube, Proceedings of the 6th IEEE CPMT Conference on High Density Microsystem Design and Packaging and Component Failure Analysis (HDP'04), Shanghai, 2004, pp. 382–384
75. A.M. Bondar, A. Bara, D. Patroi, and P.M. Svasta, Carbon Mesophase/Carbon Nanotubes Nanocomposite – Functional Filler for Conductive Pastes, Proceedings of the 5th International Conference on Polymers and Adhesives in Microelectronics and Photonics, Wroclaw, Poland, 2005, pp. 215–218
76. A. Bara, A.M. Bondar, and P.M. Svasta, Polymer/CNTs Composites for Electronics Packaging, Proceedings of the 1st IEEE Electronics Systemintegration Technology Conference, Dresden, Germany, 2006, pp. 334–336
77. R.-J. Lin, Y.-Y. Hsu, Y.-C. Chen, S.-Y. Cheng, and R.-H. Uang, Fabrication of Nanowire Anisotropic Conductive Film for Ultra-Fine Pitch Flip Chip Interconnection, Proceedings of the 55th IEEE Electronic Components and Technology Conference, Orlando, FL, 2005, pp. 66–70
78. S. Fiedler, M. Zwanzig, R. Schmidt, E. Auerswald, M. Klein, W. Scheel, and H. Reichl, Evaluation of Metallic Nano-Lawn Structures for Application in Microelectronics Packaging, Proceedings of the 1st IEEE Electronics Systemintegration Technology Conference, Dresden, Germany, 2006, pp. 886–891
79. H.P. Wu, J.F. Liu, X.J. Wu, M.Y. Ge, Y.W. Wang, G.Q. Zhang, and J.Z. Jiang, High conductivity of isotropic conductive adhesives filled with silver nanowires, Int. J. Adhes. Adhes. 26 (2006) 617–621
80. H. Wu, X. Wu, J. Liu, G. Zhang, Y. Wang, Y. Zeng, and J. Jing, Development of a Novel Isotropic Conductive Adhesive Filled with Silver Nanowires, J. Compos. Mater. 40(21) (2006) 1961–1969
81. A. Naeemi, G. Huang, and J. Meindl, Performance Modeling for Carbon Nanotube Interconnects in On-Chip Power Distribution, Proceedings of the 57th IEEE Electronic Components and Technology Conference, Reno, NV, 2007
82. Y. Chai, J. Gong, K. Zhang, P.C.H. Chan, and M.M.F. Yuen, Low Temperature Transfer of Aligned Carbon Nanotube Films Using Liftoff Technique, Proceedings of the 57th IEEE Electronic Components and Technology Conference, Reno, NV, 2007
83. C.-J. Wu, C.-Y. Chou, C.-N. Han, and K.-N. Chiang, Simulation and Validation of CNT Mechanical Properties – The Future Interconnection Method, Proceedings of the 57th IEEE Electronic Components and Technology Conference, Reno, NV, 2007
84. A. Ruiz, E. Vega, R. Katiyar, and R. Valentin, Novel Enabling Wire Bonding Technology, Proceedings of the 57th IEEE Electronic Components and Technology Conference, Reno, NV, 2007
85. G.A. Riley, Nanobump Flip Chips, Advanced Packaging, April 2007, pp. 18–20
86. L. Zhu, Y. Sun, J. Xu, Z. Zhang, D.W. Hess, and C.P. Wong, Aligned Carbon Nanotubes for Electrical Interconnect an Thermal Management, Proceedings of the 55th IEEE Electronic Components and Technology Conference, Orlando, FL, 2005, pp. 44–50
87. L.-N. Tsai, G.-R. Shen, Y.-T. Cheng, and W. Hsu, Power and Reliability Improvement of an Electro-Thermal Microactuator Using Ni-Diamond Nanocomposite, Proceedings of the 54th IEEE Electronic Components and Technology Conference, Las Vegas, NV, 2004, pp. 472–476

88. K.M. Klein, J. Zheng, A. Gewirtz, D.S. Sarma, S. Rajalakshmi, and S.K. Sitaraman, Array of Nano-Cantilevers as a Bio-Assay for Cancer Diagnosis, Proceedings of the 55th IEEE Electronic Components and Technology Conference, Orlando, FL, 2005, pp. 583–587
89. B. Lee, R. Pamidigantham, and C.S. Premachandran, Development of Polymer Waveguide Using Nano-Imprint Method for Chip to Chip Optical Communication and Study the Suitability on Organic Substrates, Proceedings of the 56th IEEE Electronic Components and Technology Conference, San Diego, CA, 2006
90. P. Dixit and J. Miao, Fabrication of High Aspect Ratio 35 Micron Pitch Nano-Interconnects for Next Generation 3-D Wafer Level Packaging by Through-Wafer Copper Electroplating, Proceedings of the 56th IEEE Electronic Components and Technology Conference, San Diego, CA, 2006, pp. 388–393
91. S. Spiesshoefer, L. Schaper, S. Burkett, G. Vangara, Z. Rahman, and P. Arunasalam, Z-Axis Interconnects Using Fine Pitch, Nanoscale Through-Silicon Vias: Process Development, Proceedings of the 54th IEEE Electronic Components and Technology Conference, Las Vegas, NV, 2004, pp. 466–471
92. A.O. Aggarwal, P.M. Raj, V. Sundaram, D. Ravi, S. Koh, and R.R. Tummala, 50 Micron Pitch Wafer Level Packaging Testbed with Reworkable IC-Package Nano Interconnects, Proceedings of the 55th IEEE Electronic Components and Technology Conference, Orlando, FL, 2005, pp. 1139–1146
93. S. Bansal, A. Saxena, and R.R. Tummala, Nanocrystalline Copper and Nickel as Ultra High-Density Chip-to-Package Interconnections, Proceedings of the 54th IEEE Electronic Components and Technology Conference, Las Vegas, NV, 2004, pp. 1647–1651
94. A.O. Aggarwal, K. Naeli, P.M. Raj, F. Ayazi, S. Bhattacharya, and R.R. Tummala, MEMS Composite Structures for Tunable Capacitors and IC-Package Nano Interconnects, Proceedings of the 54th IEEE Electronic Components and Technology Conference, Las Vegas, NV, 2004, pp. 835–842
95. A.O. Aggarwal, P.M. Raj, I.R. Abothu, M.D. Sacks, A.A.O. Tay, and R.R. Tummala, New Paradigm in IC-Package Interconnections by Reworkable Nano-Interconnects, Proceedings of the 54th IEEE Electronic Components and Technology Conference, Las Vegas, NV, 2004, pp. 451–460
96. R. Doraiswami and M. Muthuswamy, Nano Bio Embedded Fluidic Substrates: System Level Integration Using Nano Electrodes for Food Safety, Proceedings of the 56th IEEE Electronic Components and Technology Conference, San Diego, CA, 2006, pp. 158–160
97. R. Doraiswami, Embedded Nano Nickel Interconnects and Electrodes for Next Generation 15 Micron Pitch Embedded Bio Fluidic Sensors in FR4 Substrates, Proceedings of the 56th IEEE Electronic Components and Technology Conference, San Diego, CA, 2006, pp. 1323–1325
98. G.Q.(Kouchi) Zhang, M. Graef, and F. van Roosmalen, The Rationale and Paradigm of "More than Moore", Proceedings of the 56th IEEE Electronic Components and Technology Conference, San Diego, CA, 2006, pp. 151–157
99. A.P. Malshe, Development of a Curriculum in Nano and MEMS Packaging and Manufacturing for Integrated Systems to Prepare Next Generation Workforce, Proceedings of the 54th IEEE Electronic Components and Technology Conference, Las Vegas, NV, 2004, pp. 1706–1711
100. T. Zerna and K.-J. Wolter, Developing a Course About Nano-Packaging, Proceedings of the 55th IEEE Electronic Components and Technology Conference, Orlando, FL, 2005, pp. 1925–1929
101. A. Govind and F. Gahghahi, Development of Organic Flip Chip Packaging Technology for Nanometer Silicon Incorporating Copper Metallization and Low-k Dielectric, Proceedings of the 54th IEEE Electronic Components and Technology Conference, Las Vegas, NV, 2004, pp. 347–351
102. J.E. Morris, Nanodot Systems Reliability Issues, Proceedings of the Smart Systems Integration Conference, Paris, 2007
103. J.E. Morris, Nanopackaging: Nanotechnologies in Electronics Packaging, Proceedings of the 8th IEEE CPMT Conference on High Density Microsystem Design and Packaging and Component Failure Analysis (HDP'06), Shanghai, 2006, pp. 109–115

104. J.E. Morris, Nanopackaging: Nanotechnologies in Electronics Packaging, Proceedings of the 1st IEEE Electronics Systemintegration Technology Conference, Dresden, Germany, 2006, pp. 873–880
105. E. Suhir, New Nano-Particle Material (NPM) for Micro- and Opto-Electronic Packaging Applications, Proceedings of the 10th IEEE/CPMT International Symposium on Advanced Packaging Materials, Irvine, CA, 2005

Chapter 2
Modelling Technologies and Applications

C. Bailey(✉), H. Lu, S. Stoyanov, T. Tilford, X. Xue, M. Alam, C. Yin, and M. Hughes

2.1 Introduction

Numerical modelling technology and software is now being used to underwrite the design of many microelectronic and microsystems components. The demands for greater capability of these analysis tools are increasing dramatically, as the user community is faced with the challenge of producing reliable products in ever shorter lead times.

This leads to the requirement for analysis tools to represent the interactions amongst the distinct phenomena and physics at multiple length and timescales. Multi-physics and Multi-scale technology is now becoming a reality with many code vendors. Figure 2.1 illustrates the interaction between physics-based modelling tools and optimisation in predicting the behaviour and reliability of microsystems devices from device fabrication, its packaging, test and qualification, and finally in-service performance.

This chapter discusses the current status of modelling tools that assess the impact of nano-technology on the fabrication/packaging and testing of microsystems. The chapter is broken down into three sections: Modelling Technologies, Modelling Application to Fabrication, and Modelling Application to Assembly/Packaging and Modelling Applied for Test and Metrology.

2.2 Modelling Technologies

All matter is made of atoms and molecules and its behaviour is ultimately governed by the law of quantum physics. However, in the macroscopic world, the fact that matter is a collection of discrete entities is often ignored because continuum theory and methods can be used to describe the material behaviour reasonably well at this length scale. The use of modelling tools across the length scales is classified in Fig. 2.2.

C. Bailey
School of Computing and Mathematical Sciences, University of Greenwich, Old Royal Naval College, Greenwich, London, SE10 9LS, UK

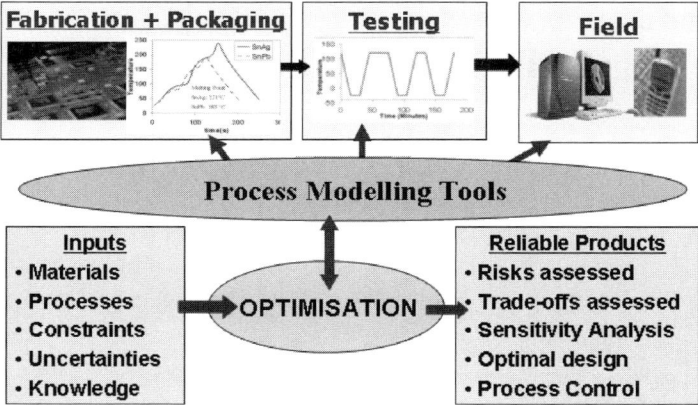

Fig. 2.1 Optimisation-driven numerical modelling for predicting reliable nano-packaging microsystems

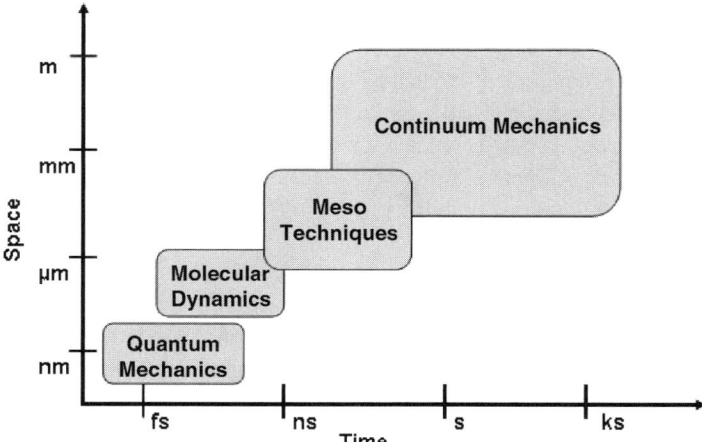

Fig. 2.2 Modelling across the length scales

2.2.1 Continuum Modelling

Continuum mechanics modelling tools can be classified as:

- *Computational fluid dynamics (CFD)*: solving phenomena such as fluid flow, heat transfer, combustion, solidification, etc.
- *Computational solid mechanics (CSM)*: solving deformation, dynamics, stress, heat transfer, and failures in solid structures
- *Computational electromagnetics (CEM)*: used to solve electromagnetics, electrostatics and magneto-statics

Table 2.1 Some continuum-based modelling tools

Software	Web address
ANSYS	http://www.ansys.com
COMSOL	http://www.comsol.com
ANSOFT	http://www.ansoft.com
FLOMERICS	http://www.flomerics.com
PHYSICA	http://www.physica.co.uk

Until recently, the majority of continuum mechanics codes focused on the prediction of distinct physics, but now there has been a strong push by software vendors to develop multi-physics or co-disciplinary tools that capture the complex interactions between the governing physics such as fluidics, thermal, mechanical and electrical.

The use of these continuum methods is justified for most electronic assemblies because the feature size in these assemblies is so large compared with the size of the atoms/molecules that there is an astronomically large number of atoms in any assembly. Table 2.1 details a number of commercial continuum mechanics codes as used by the microsystems packaging community.

2.2.2 Atomistic and Multi-Scale Modelling

To model a structure with a nano-scale dimension, modelling methods that take into account the structure and the interactions of the atoms and molecules have to be used. This kind of modelling is called atomistic modelling and the most frequently used atomistic modelling method is molecular dynamics (MD).

MD was first used by Alder and Wainwright [1] to simulate a system of hard spheres. The classic MD method uses simple potential functions to describe the interactions between atoms and molecules. The average effects of the electrons are assumed to be included in the potential. In the ab initio or the first-principle MD method, the interactions between the ions as well as the interactions between the electrons and the ions are taken into account and both the distribution of the electron and the movements of the ions are tracked in the modelling [2]. The embedded atom method (EAM) and its variants such as the modified embedded atom method (MEAM) enhance the classic MD method by including a separate potential term that can be attributed to the effects of the electron [3].

In a classic MD simulation, the most important input is the potential function. In general, the function depends on the location of many atoms, but in many situations the most important term is pairwise, i.e. the potential depends only on the distance between two atoms. The best-known example of this type of potential function is the Lenard–Jones potential [4]. But even with the use of this simple potential, the number of atoms that can be modelled using the MD method is still very small compared with the number of atoms in any small macroscopic object. Even in a large-scale MD simulation, the number of atoms is limited to a few million and the modelled time is in the order of pico-seconds to nano-seconds.

To bridge the length and time gaps between atomistic and continuum modelling, methods such as the particle-in-cell method can be used [5]. This method takes into account atomic interactions in a physical phenomenon that takes place in a macroscopic system. Atomistic–continuum mechanics (ACM) is another method that combines the atomistic nature of materials with continuum mechanics [6]. In this method, the lattice structure and the interactions between atoms are taken into account. The macroscopic mechanical properties can be derived but the dynamics of the atoms cannot be studied using this method.

2.2.3 Uncertainty and Optimisation Modelling

Simulation-based optimisation for virtual design prototyping of various electronic products and manufacturing processes has proven as an effective approach for process characterisation and product development at the early design stages [7].

Design for nano-fabrication and nano-packaging itself is an extremely complex engineering task. The complexity of the nano-structures often makes real prototyping and testing difficult or expensive. Therefore, it is essential to incorporate at the early design stage methods that exploit not only numerical simulation for the physical behaviour but also techniques that allow for quantification and optimisation of the risk and reliability of the systems. Deterministic and stochastic simulation models are now emerging as valuable tools in modern design and to help managing and mitigating the associated failure risks.

Computational optimisation techniques can aid the identification of the optimal design/process specification and the formulation of design rules for optimal performance/reliability of the fabricated nano-structures. However, in reality, such optimal package or process design, from deterministic point of view, may be far from a reliable and safe design solution. The reason for this is the presence of uncertainty which is inherent in various aspects of the nano-electronics. Natural variations can be found in the manufacturing and/or operational process parameters (e.g. operational temperature, humidity, etc.), the tolerances in the dimensions of the manufactured structures, the physical properties of the materials, etc.

Modelling and quantifying uncertainty can be undertaken using several different concepts and methods. The most popular approach is using probability theory. The key advantage is in the ability to quantify uncertainty of a design or process parameter using probability distribution functions (PDFs). This important concept relates to the definition of the so-called limit state function (failure surface) which quantifies the reliability metric. The limit state is a function of the uncertain (random) design, material or process parameters $x_1, x_2,\ldots,x_n$, each of these assigned with appropriate PDFs that characterise their uncertainty. The limit state function can be expressed as

$$g(x_1, x_2, \ldots, x_n) = 0 \tag{2.1}$$

The failure domain is defined by $g(x) < 0$ and the failure probability p_f is calculated by solving the following multi-dimensional integral:

$$p_f = \int_{g(x)<0} f_X(x_1, x_2, \ldots, x_n) \, dx_1 dx_2 \ldots dx_n, \tag{2.2}$$

where f is the joined PDF of the uncertainty input variables.

Commonly used techniques for uncertainty evaluation are the sampling methods such as the Monte Carlo simulation where samples of random variables are generated according to the parameter probability distribution and then the reliability function is directly evaluated and checked for failure [8]. The proportion of the sample points for which failure is indicated through the limit state function approximates the failure probability. The disadvantage of these methods is the huge number of sampling point evaluations. A single evaluation might be complex, expensive and time consuming if uses experiment or high-fidelity analysis. Reduced-order models are used to overcome this limitation. These models offer fast analysis of the process or design, and therefore a fast evaluation of the reliability function.

A different numerical approach to evaluate failure probability is based on the construction of approximations of the limit state function using first- or second-order Taylor series. These methods are known as first-order reliability methods (FORM) and second-order reliability methods [9].

There is an increased interest about non-probabilistic uncertainty modelling which can potentially overcome some of the limitations of the probabilistic approach and can handle in a better way 'subjective' uncertainty (e.g. lack of knowledge about the modelling process). Examples include the evidence theory [10], fuzzy sets and possibility theory [11] and interval-based approaches [12].

The uncertainty modelling is a key aspect in the procedure of reliability-based optimisation task formulation. Integrating reliability assessment into a design optimisation numerical framework results in a powerful and cost-effective design approach where probability constraints are handled and satisfied. A generic problem under uncertainty can be defined as

$$\min_x F(x)$$

$$\text{subject to: } P(g(x<0) \leq p, \tag{2.3}$$

where $F(x)$ denotes the objective function (aspect of the design or process we aim to improve) and p is the acceptable limit for the failure probability associated with the state limit function $g(x)$.

There are various numerical techniques that can be used to find out the solution of the above optimisation problem. The interested reader is referred to [13] for more details on the most common numerical optimisation techniques. Reliability-based design optimisation formulations offer convenient and automated virtual exploration of the design space defined by x to identify the best configuration of input parameters from the view point of the objective function and uncertainty (probabilistic) influenced requirements.

2.2.4 Future Challenges for Modelling Tools

There are a number of computational mechanics software tools now on the market. These technologies provide manufacturing engineers with the knowledge and design rules to help them deliver reliable products in time and at lower cost than could ever be achieved through physical prototyping alone. Although computational mechanics codes are now used in the design of manufacturing processes, there are still a number of challenges. These can be classified as:

1. *Multi-physics.* Many packaging processes are governed by close coupling between different physical processes. Computational mechanics tools are now addressing the need for multi-physics calculations, but more work is required to capture the physics accurately in these calculations.
2. *Multi-discipline.* Thermal, electrical, mechanical, environmental and other factors are important in the design and packaging of microsystems products. Computational mechanics tools that allow design engineers from different disciplines to trade-off their requirements early in the design process will dramatically reduce lead times.
3. *Multi-scale.* Nano-packaging processes are governed by phenomena taking place across the length scales (nano–micro–meso–macro). Techniques that provide seamless coupling between simulation tools across the length scales are required.
4. *Fast calculations.* Computational mechanics software that solves highly non-linear partial differential equations is computed intensive and slow. There is a need for reduced-order models (or compact models) for nano-fabrication and packaging processes. Although not as accurate as high-fidelity finite element or atomistic modelling techniques, they provide the design engineer with the ability to quickly eliminate many unattractive designs early in the design process.
5. *Life-cycle considerations.* Major life-cycle factors such as reliability, maintenance and end-of-life disposition receive limited visibility in computational mechanics analysis. Future models will include all life-cycle considerations, such as product greenness, reliability, recycling, disassembly and disposal.
6. *Variation risk mitigation.* Current product and process models used in computational mechanics usually ignore process variation, manufacturing tolerances and uncertainty. This will be very important for nano-packaging. Future models will include these types of parameters to help provide a prediction of manufacturing risk. This can then be used by the design engineer to enable them to implement a mitigation strategy.

2.3 Modelling Applied to Fabrication Processes

Fabrication of nano-structures that can be used in nano-packaging of electronic systems is considered in this section. Three techniques which illustrate a bottom-up (electro-deposition) and top-down (focused ion beam and print forming) approach to nano-fabrication are discussed.

2.3.1 Modelling of Focused Ion Beam Milling Process

Focused ion beam (FIB) is a milling process used to remove material from a defined area or to deposit material onto it at micro- and nano-scales. The principle of operation for FIB is bombardment of a target surface through high energy gallium Ga+ (or other) ions. As a result, small amounts of material sputter in the form of secondary ions, natural atoms and secondary electrons.

FIB process reduces dramatically the damage on the surface being subject to ion bombardment compared with other classical methods. In the FIB process, a critical variable to control is depth variation. This is essential to ensure suitable fabrication of 3D nano-features, miniaturised objects, masks and moulds for various microsystems.

A mathematical model that can be used to predict the etched shape or to calculate the dwell times required to achieve a predefined shape has been developed [14, 15]. This model assumes a square pixel matrix placed over the target surface. The sputtering model is then distributed over each element of the pixel matrix, so that a system of linear equations that relates the dwell times t_{ij} with the sputtering depth H_{ij} at any pixel (i, j) is constructed. A brief outline of the model is given below.

If (x_i, y_j) denotes the centre of the pixel (i, j), then the sputtering at this pixel in terms of depth due to material removal at that pixel can be expressed as

$$H_{ij} = \iint \frac{\Phi(x,y)}{\eta} f_{x,y}(x_i, y_j) Y\left(E_0, \alpha_{x_i, y_j}\right) t_{x,y} \, dx \, dy. \tag{2.4}$$

where H_{ij} is the sputtering depth at the point (x_i, y_j), $\Phi(x, y)$ is the ion flux at point (x, y) (cm^{-2} s^{-2}), η is the atomic density of the target material (atoms cm^{-3}), $Y(E_0, \alpha_{x_i, y_j})$ is the sputtered yield (atoms per incident ion at point (x_i, y_j)), $t_{x,y}$ is the dwell time of the ion beam at point (x_i, y_j) (s) and $f_{x,y}(x_i, y_j)$ is the ion beam density distribution function in two dimensions.

The sputtered yield in (2.4) is a function of the incident angle $\alpha_{xi,yj}$ of the ion beam at point (x_i, y_j) and the ion energy E_0 as well as the type of ion source and target material. Generally, the yield increases from perpendicular ion beam incidence to a maximum at angle 60°–85°, and then rapidly decreases due to the strong reflection at grazing incidence. A classical empirical formula for the angular dependence of the sputtered yield is given by (2.5) which was originally proposed by Yamamura [16]:

$$Y(E_0, \alpha) = Y(E_0, 0) \frac{\exp\left\{f\left[1 - \frac{1}{\cos\alpha}\right]\cos\alpha_{opt}\right\}}{(\cos\alpha)^f}, \tag{2.5}$$

where $Y(E_0, \alpha)$ is the sputtering yield at ion energy E_0 and nominal angle of incidence α. The quantities f and α_{opt} are parameters to fit the experimental data. In addition, α_{opt} is the nominal incidence angle at maximum sputtering yield.

The ion beam geometry in terms of density distribution $f_{x,y}(x_i, y_j)$ is also taken into account in the model (2.4). If Gaussian bivariate density function is assumed, then

$$f_{x,y}(x_i, y_j) = \left(\frac{1}{\sqrt{2\pi}\sigma}\right)^2 e^{-r^2/2\sigma^2} \qquad (2.6)$$

where $r^2 = (x_i - x)^2 + (y_j - y)^2$ defines the radial coordinate for an ion beam focused at (x, y).

Figure 2.3 shows an example of a FIB simulation predicting the dwell times and milling shape using the model defined above. The model is used to calculate the dwell times over each of the pixel cells, so that with given ion beam parameters a cavity with prior defined parabolic shape is sputtered. The parabolic predefined shape has a maximum depth of 2 μm. In this analysis, the ion beams are assumed to have normal distribution with standard deviation $\sigma = 0.075$ um, $\Phi(x, y) = 1 \times 10^{19}$ ions s^{-1} cm^{-2}, $\eta = 5 \times 10^{22}$ atoms cm^{-3}, pixel grid is 20 × 20 over target area 3 × 3 μm. The target surface is silicon and the sputtering yield is calculated using (2.5) and assuming 20-keV Ga ions.

The above model for FIB sputtering can be enhanced by considering re-deposition. The mathematical model given in (2.7) assumes that the amount of the sputtered atoms or ions from a source pixel cell (i, j) which is re-deposited onto another target pixel cell (k, l) is dependent on the relative locations between the two cells and their orientations [17]. The re-deposited volume of material Rij as a function of the sputtered volume S_{ij} can be calculated as

$$R_{ij} = \frac{F(\beta) - F(\gamma)}{F(180°)} S_{ij}, \text{ where } F(x) = \frac{\pi r^3}{3}[\cos^3(x) - 3\cos(x) + 2]. \qquad (2.7)$$

In (2.7), β and γ are the minimum and maximum angles that are measured from the centre of the source cell (i, j) to any possible locations within the target unit cell (k, l),

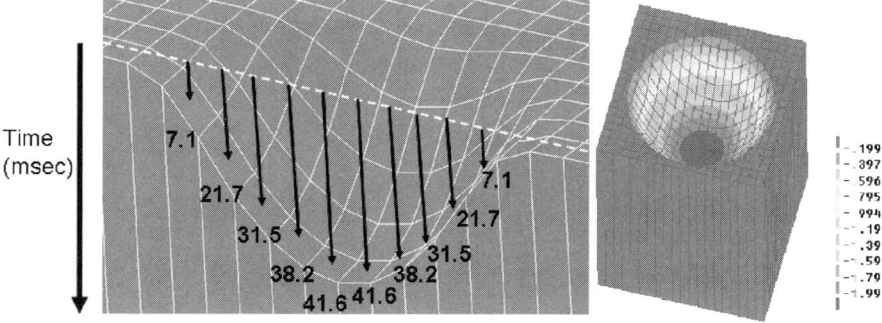

Fig. 2.3 Modelling of FIB milling of a parabolic feature – milling times along the cross-sectional pixels (*left*) and contour depth levels (μm)

respectively. For a cell (k, l), the re-repositioned volume can be found as the summation of contribution from all other source cells (i, j). This model assumes that the total displacement after re-deposition is normal to the surface of the unit cell.

Some recent efforts in the modelling of the FIB process have focused on simulating the non-linear dynamics of ripple formation as a result of the ion beam sputtering process [18]. Modelling and understanding the ripple formation phenomenon is gaining interest as a result of the potential to use it in various nano-technology applications.

2.3.2 Modelling of Nano-Imprint Lithography Process

Among the most attractive and promising nano-fabrication processes is the nano-imprint lithography (NIL). This method offers low cost and high yield nano-scale patterning using various materials at dimensions as small as 6 nm [19].

Thermal NIL is one of the most typical methods for NIL. The thermal imprint process is based on the utilisation of thermo-plastic polymers and comprises several steps as outlined schematically in Fig. 2.4. A polymer and a nano-fabricated master tool (mould) are pre-heated above the polymer glass transition temperature (T_g) and then the fine mould, patterned according to the required specification, is pressed into the polymer forming a negative relief of the master. While the mould is pressed and held down, the polymer is cooled down below the T_g and hardens thus retaining the profile of the mould pattern. Finally, the imprinting pressure is removed and the mould is released.

Typical issues associated with NIL are related to the mechanical stresses and the large deformations of the polymer films and the residual thickness after imprinting. Numerical simulations of the cross-sectional profiles as a function of process parameters – such as the imprinting pressure, polymer initial thickness and the nano-cavity size/aspect ratio – can provide valuable knowledge on the imprint process. There are two main modelling approaches that can be utilised to model the NIL process.

The first method involves modelling of the mechanical deformation process using hyperelastic large strain finite element analysis [20]. For this type analysis,

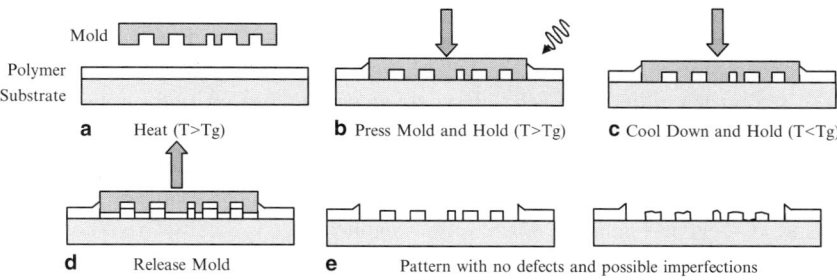

Fig. 2.4 Typical steps in thermal nano-imprint lithography

the polymer is modelled as a rubber elastic body above its T_g with large strain and assumed to be a non-compressive material. A suitable model to represent this behaviour is the Mooney–Rivlin model [21, 22]. According to this model, the stress is expressed as

$$\sigma_i = \lambda_i \frac{\partial W}{\partial \lambda_i}, \qquad (2.8)$$

where λ_i is the expansion strain rate (deviatoric strain) and W is a strain energy density function defined as

$$\begin{aligned} W &= C_{10}(I_1-3) + (C_{01}(I_2-3), \\ I_1 &= \lambda_1^2 + \lambda_2^2 + \lambda_3^2 \\ I_2 &= \lambda_1^2\lambda_2^2 + \lambda_2^2\lambda_3^2 + \lambda_3^2\lambda_1^2 \end{aligned} \qquad (2.9)$$

In (2.9), C_{10} and C_{01} are the Mooney material constants characterising the deviatoric deformation of the material.

The Mooney–Rivlin model for rubber elastic non-linear stress–strain behaviour is incorporated in commercially available software such as ANSYS and MARC. The other critical feature of this type analysis is the simulation of large deformations in materials under contact boundary conditions which is required to represent the interaction between the mould and the polymer. This analysis is based on some assumptions such as that no air bubbling, trapping or absorption into the polymer occurs. This does not impose any major setback because the imprint pressure is very high compared with the air ambient pressure; hence, no major impact on polymer deformation at the macro-level will take place.

The above modelling approach can be applied to study in detail the imprint process sequence. As an example, Hirai's group [23] uses this approach to undertake defect analysis in thermal NIL and to study the dynamics of the deformation process. As part of their work, the authors have found very good agreement between simulation results and the experiments. The numerical analyses have identified correctly a stress concentration site near the polymer corner due to the applied pressure below T_g, which subsequently led to the defect of polymer fracture during the mould release step. The simulation results for the cross-sectional profiles from the analysis of the resist deformation process [24] have been shown to agree quantitatively very well with the experimental results for various geometric and pressure conditions.

The second modelling approach is based on modelling the flow of the polymer using CFD analysis. The key features of this type of simulation include modelling the polymer as a non-Newtonian fluid with free (moving) boundary. In such two-phase large free boundary deformation flow analysis, phenomena such as the polymer capillary surface with surface tension boundary condition are explicitly considered. These continuum simulations can capture the underlying physics of the nano-imprint process from 10-nm to 1-mm scale and are capable of predicting accurately the polymer deformation mode and surface dynamics. A non-dimensionalised calculation procedure that follows this modelling strategy is presented in the work by Rowland and co-authors [25].

2.3.3 Modelling of Electroforming Processes

Attempts to numerically model the electro-deposition process are challenging as they must solve a system of coupled non-linear equations with the added complication that the governing equation set changes under different physical situations; for example as the deposition current varies from primary to secondary, tertiary or diffusion-limited regimes [26]. Additionally, the representation of electrode kinetics, the driving force for deposition, is of key importance and is complicated by its influence from the electrode surface over-potential and the concentration of reacting ions in the immediate vicinity of the depositing interface. Figure 2.5 illustrates the process taking place for trench or via filling.

The governing equations may therefore include all or a combination of the momentum, heat, concentration and electric potential equations with various degrees of inter-coupling by electro-migration, convection and importantly through the reaction rate boundary condition at the electrode surface. Standard continuum equations for momentum, electric field and ion concentration are solved except at the thin layers adjacent to the electrode boundaries, the electrical double layer which is of the order of <~100 nm in width. In these thin layers, the deposition current is accounted for by an electrode kinetic function, typically the Butler–Volmer equation [27]. Figure 2.6 illustrates the evolution of a deposition layer using a coupled simulation approach, a free surface tracking algorithm [28].

2.4 Modelling Applied to Assembly Processes

Modelling of typical assembly or packaging process is discussed in this section. The processes discussed are solder pasting and its reflow, and microwave heating to cure polymer materials as used in electronic packaging.

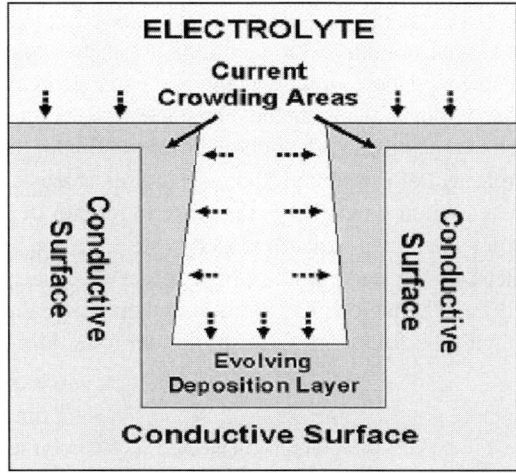

Fig. 2.5 Evolving deposition layer on a conductive surface

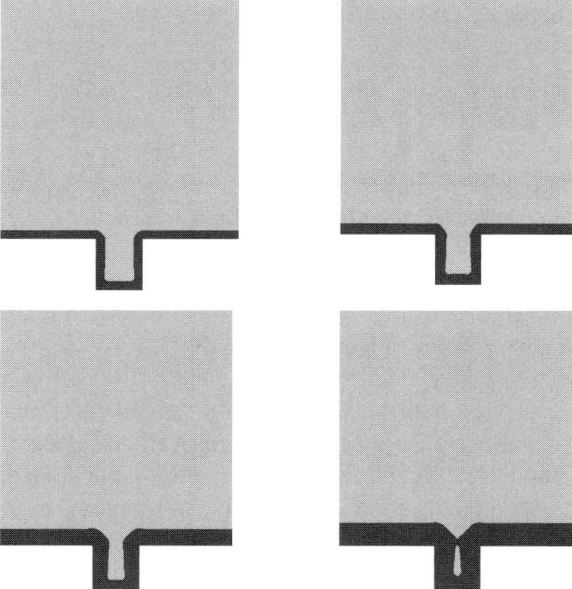

Fig. 2.6 Prediction deposition front filling a via

2.4.1 Solder Paste Printing

The stencil printing process is used to deposit solder paste at precise locations on the PCB pads to prepare for the placement and bonding of electronic components. In this process, a squeegee blade moves the solder paste over the surface of a stencil with a particular pattern of apertures. As a result of the high pressure in the solder paste, induced by the squeegee blade, the solder paste is forced to fill the stencil apertures.

The overall solder paste composition exhibits non-Newtonian rheological properties with shear-thinning behaviour, i.e. the viscosity decreases with increasing shear rate. This behaviour is what enables the paste to flow into the apertures with a low viscosity when the shear rate is high due to the action of the moving squeegee blade. After the removal of the stencil, the viscosity increases again in the absence of shearing, a phenomenon which helps the paste to remain in place.

CFD can predict the movement of solder paste across a stencil surface. For example, simulation of the paste motion of a solder material with nano-particles and characterised bulk behaviour can be undertaken using the classical Navier–Stokes equations with the following viscosity model for solder paste [29]:

$$\frac{\eta - \eta_\infty}{\eta_0 - \eta_\infty} = \frac{1}{1 + K\lambda^m}, \qquad (2.10)$$

where η is the apparent viscosity, η_0 and are the viscosity at zero and infinite shear rate, respectively, λ is the strain rate and K and m are experimentally obtained constants. Figure 2.7 shows the schematic of the printing process and associated CFD predictions for solder flow using a classical continuum approach.

Traditional CFD simulations based on the continuum simulation approach assume homogeneous fluid and may not provide realistic answers about the transport of the individual solder nano-particles. To understand the flow of the solder paste into the stencil apertures, coupled continuum-particle computational methods are required.

Among the most attractive discrete particle-based fluid dynamic computational techniques are Stokesian dynamics [30] and mesoscopic approaches such as lattice Boltzmann methods (LBM) [31] or dissipative particle dynamics (DPD) [32].

Stokesian dynamics is a method in which only forces between the particles of the solid phase are considered and the detailed flow of the suspending fluid is not simulated. The inter-particle forces are based on lubrication theory, where the drag on a particle is dependent on the position and velocity relative to that of its neighbouring particles or solid walls as well as the average local velocity of the suspending fluid [30]. Some major drawbacks of this method include the lack of detail for the flow of the suspending fluid and difficulties with mass conservation. Stokesian dynamics is also considered to be inefficient for suspension flows in comparison to mesoscopic methods.

Mesoscopic approaches are similar to molecular dynamics but replace the fluid molecules with much larger fictitious particles that can be considered to represent accumulations of the real underlying molecules. Computer memory requirements are therefore smaller and collision timescales are closer to that of the evolution of the macroscopic flow. The properties of these fictitious particles are set, so that they mimic the flow behaviour of the underlying real fluid at the macroscopic scale. These methods are therefore referred to as mesoscopic since they lie somewhere between microscopic atomistic and macroscopic continuum approaches. DPD methods are based on attractions and repulsions which are dependent on the relative positions and velocities of the particles relative to each other.

Hybrid models for coupled nano-scale dynamics with macroscopic continuum flow behaviour try to benefit from a multi-scale approach for simulating solder paste printing

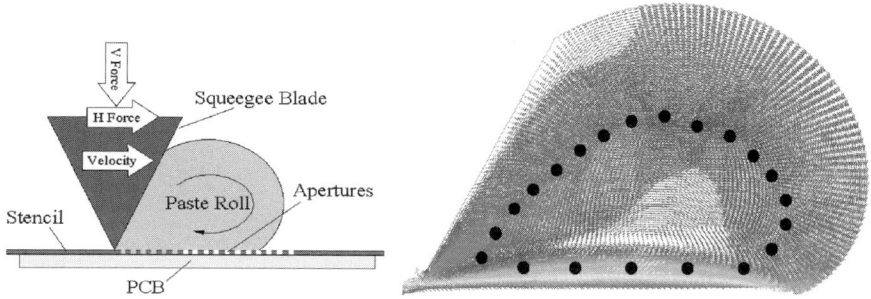

Fig. 2.7 Modelling predictions for the flow of solder material in the stencil printing

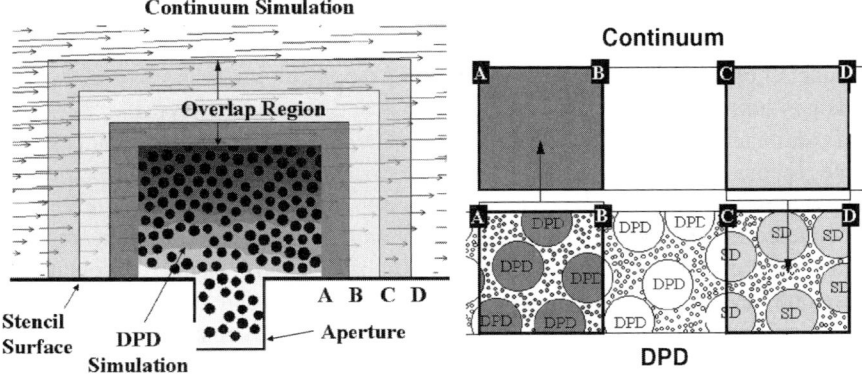

Fig. 2.8 Flow field results of coupled DPD-continuum 2D simulation

processes [33]. Figure 2.8 (left) shows simulation results for solder paste printing using a hybrid analysis approach. The velocity field of the DPD region is represented by filled contours. The continuum and particle regions overlap by a narrow band A–D (Fig. 2.8, right) with two defined sub-regions A–B and C–D. Strip A–B of the particle domain coincides with the boundary of the continuum region. Similarly, strip C–D of the continuum domain coincides with the boundary of the particle simulation. Mass and momentum flux densities are measured and coarse-grained in band A–B of the particle field and imposed at the boundary of the continuum field.

2.4.2 Molecular Dynamics Calculation of Solder Reflow Process

In the lead-free soldering technology area, Dong et al. [34] used the modified embedded atom method and the molecular dynamics method to study the collapsing and merging behaviour of tin and silver particles. One of the aims of this work was to find out if the mechanical mixture of tin and silver nano-particles can replace the Sn–Ag alloy for low reflow temperature applications. The modelled particles were all 4 nm in diameter and contain 1,257 and 1,895 atoms, respectively. The total simulated time was 3 ns. It turned out that the silver sphere kept its crystalline structure and no significant diffusion between tin and silver was observed within the simulation time.

2.4.3 Microwave Heating in Microelectronics and Nano-Packaging Applications

Microwave energy fundamentally accelerates the cure kinetics of polymer materials [35], providing a route to focus heat into the polymer materials, minimising the temperature increase and associated thermal stresses in the surrounding materials.

A number of systems using microwave energy to cure microelectronic components are in use today. The novel 'FAMOBS' (http://www.famobs.org) system proposed by Sinclair et al. [36] uses a open-ended oven mounted on a pick-and-place machine which is capable of heating/curing a single component at a time, thus reducing/eliminating the issues related to generation of unnecessary thermal stresses.

To accurately model the process of microwave polymer curing, a holistic approach must be taken. The process cannot be considered to be a sequence of discrete steps, but must be considered as a complex-coupled system combining electromagnetic and thermo-physical behaviour of the whole system because each of these processes fundamentally influences the other, as illustrated in Fig. 2.9.

A significant problem in the analysis of microwave heating is the disparity in timescales between the electromagnetic and thermo-physical problems. Microwave sources operate in the range 1–30 GHz. Therefore, substantial variation in the electric field distribution is apparent in sub-picosecond timescales. Significant variation in the thermo-physical properties is only apparent in timescales of seconds (one trillion times the duration). A method for linking pico-scale to macro-scale analyses is critical to solution of the problem.

A number of methods have been employed to determine a suitable steady state electric field distribution. These methods generally rely on the electric fields reaching a time-harmonic state in which the field magnitude varies rapidly but the distribution of the modal structure remains invariant. If a time-harmonic state is reached (or assumed to have been reached), the field magnitude can be assessed through using a time-averaged or root-mean-square value or through use of a (normally discrete) Fourier transform. The transformed electric field is used to determine the power absorbed by the dielectric load, and the differences between successive values of absorbed power at successive Fourier transfer analyses are used to determine if a converged time-harmonic solution has been obtained.

2.5 Modelling Applied to Reliability Predictions

Modelling applications for final reliability prediction of an electronic package are discussed below. The impact of how nano-technology may influence reliability and how this can be modelled is discussed for both underfills and anisotropic conductive films. The impact of very small joints and current crowding effects is also outlined.

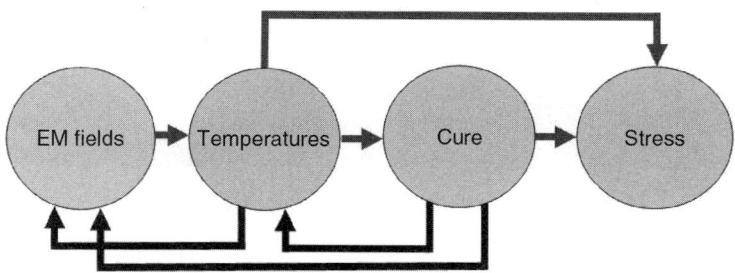

Fig. 2.9 Process coupling in microwave heating of polymers

2.5.1 The Effects of Underfills on Solder Joint Reliability

Underfills are widely used in the packaging industry to offset the damaging effects of CTE mismatch on interconnections such as solder joints between a chip and the substrate in a flip-chip assembly. The effectiveness of an underfill in reducing the impact of CTE mismatch mainly depends on its thermal–mechanical properties such as the Young's modulus and the CTE. Other important properties of an underfill include the heat conductivity, moisture absorption, the viscosity, etc. All these properties may be modified by adding filler particles such as silica into the polymer matrix. The underfill properties can be modified by changing the filler particle properties [37] or by changing the filler content.

The filler content is an important issue especially for no-flow underfills because, on the one hand, it is desirable to have high filler content to achieve low CTE, but, on the other hand, high filler content degrades the flow properties of the underfill making the process prone to defects formation in the underfill and at the solder-pad interface [38]. Figure 2.10 shows one defect developed during a no-flow underfill process.

Because of the low filler content, no-flow underfills have higher CTE than traditional capillary underfills. To investigate the effect of this on the lifetime of flip-chip solder joint, Lu et al. [38] modelled the lifetime of a flip chip's fatigue lifetime under cyclic thermal–mechanical loading for range of underfill properties. Figure 2.11 shows the 3D FEA model used in the modelling and Fig. 2.12 shows the predicted lifetimes. The results show that flip chips using no-flow underfills have significantly lower lifetimes than traditional underfills. To achieve the highest reliability, the CTE of the underfill needs to be brought down to about 20 ppm/°C.

The solution to this problem may be the use of nano-sized filler particles [39, 40] because this technology may increase the filler content without compromising the solder joint quality. Lall et al. [41] have developed a method based on representative volume element (RVE) and modified random spatial adsorption to predict the temperature-dependent underfill properties (see Chap. 14). Their results

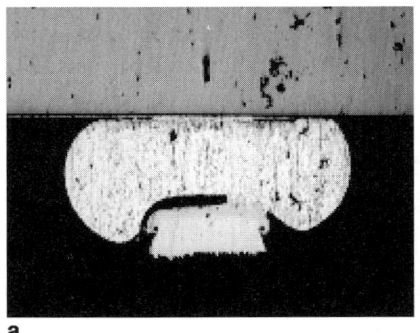

Fig. 2.10 Solder joint cross sections: (**a**) underfill trapping between solder bump and pad, (**b**) good bonding between solder bump and pad

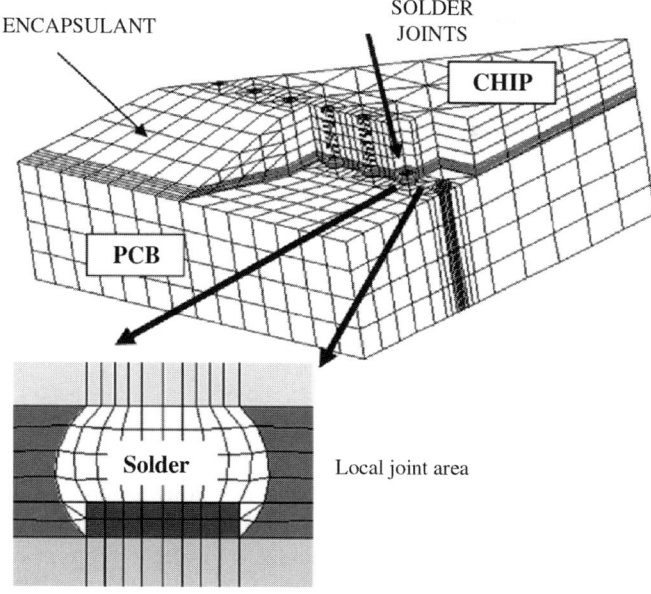

Fig. 2.11 Flip-chip computer model that has been used in the study of the effects of the no-flow underfill material

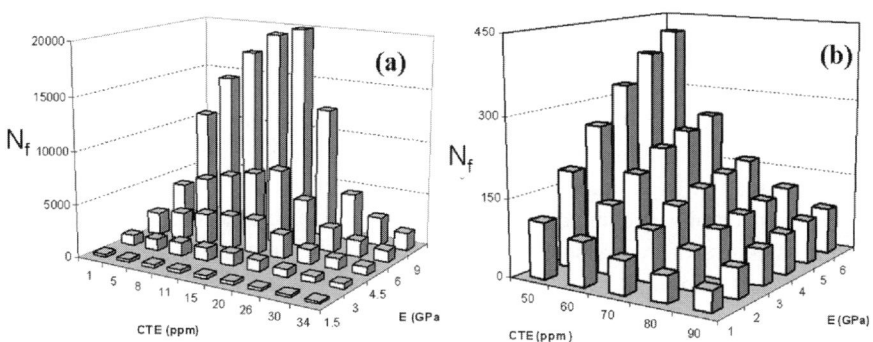

Fig. 2.12 Predicted lifetime of the flip-chip solder joint: (**a**) the traditional underfill, (**b**) no-flow underfill

show that when the volume fraction of the filler content is higher than 30%, the CTE is lower than 40 ppm/°C which is close to the capillary underfill's value.

2.5.2 Modelling of Anisotropic Conductive Films

Anisotropic conductive films (ACFs), with many distinct advantages such as extreme fine-pitch capability, being lead free and environmental friendly, are being widely used in the fine-pitch flip-chip technologies [42]. A typical ACF flip chip is

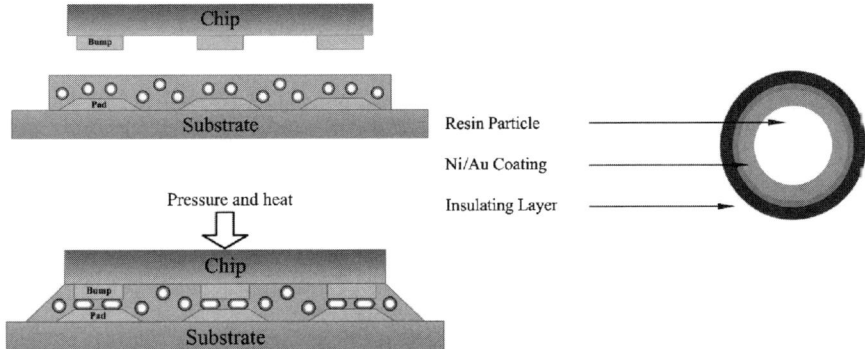

Fig. 2.13 ACF flip chip and structure of a conductive particle

shown in Fig. 2.13. The conductive particle is nickel–gold coated polymer ball with a diameter of 3.5 μm. To improve the electrical performance of ACF assembly, nano-scale conductive fillers are being considered for the next generation of high performance fine-pitch packaging applications [43].

Computer modelling analysis, in particular the finite element analysis, is being used as a powerful tool to predict the behaviour and responses of ACA particles during bonding process and reliability testing. However, previous modelling work [44, 45] has been mostly limited to the analysis of simplified two-dimensional models. Three-dimensional models have focused on the micro-domain and ignored the global effects at the package level, or they have modelled the whole package and used gross assumptions at the micro-interconnect level [46–48]. The recognised difficulty here is due to the vast range of length scales in an ACF flip-chip assembly, and the large number of conductive particles.

The diameter of the conductive particles in the ACF material is several micrometers and the thickness of the particle metallization is on the nano-scale, at about 50 nm. If the die is 11 mm in its length, the ratio of the two is approximately 1:200,000. In addition, there are thousands of conducting particles in a typical ACF material used to bond a flip-chip component to a substrate. This means that an 'exact' model which includes all the particles and interconnections would require millions if not billions of mesh elements to be used in a finite element model. This is simply not achievable with today's computer technology.

Therefore, a 3D macro–micro modelling technique is required to provide the ability to accurately model the behaviour of the conductive particles during the reliability test. Two models, one macro and one micro, with very different mesh densities were built (see Fig. 2.14). The macro-model is used to predict the overall behaviour of the whole assembly during reliability testing. The displacements obtained from this macro-model are then used as the boundary conditions for the micro-model, so that the detailed stress analysis in the region of interest could be carried out. This macro–micro modelling technique enables more detailed 3D modelling analysis of an ACF flip chip than previously.

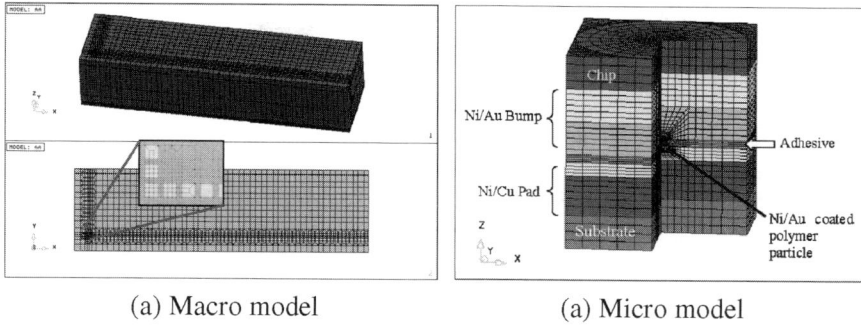

Fig. 2.14 Mesh details of the macro- and micro-models of an ACF assembly

Using this modelling technique, the moisture diffusion and induced stresses inside an ACF flip chip when subjected to autoclave test environment were predicted [49, 50]. Modelling results were consistent with the findings in the experimental work.

2.5.3 Electro-Migration and Thermo-Migration Related Damages in Nano-Packaging

Electro-migration in the on-chip interconnection/metallization of Al or Cu has been the subject of intense study over the last 40 years [51, 52]. Recently, because of the increasing trend of miniaturization, high current density-induced damages are becoming a growing concern for off-chip interconnection where low melting point solder joints are commonly used. Electro-migration is atom transfer due to a high current density.

Unlike Al and Cu metallization, Joule heating from the interconnect line can severely affect the damage characteristics of solder joints. Moreover, current crowding at the contact interface between the solder ball and the metal pad/under bump metallization (UBM) increases the local current density and local resistance of the solder alloy that further increases Joule heating and yields a localised hot spot [53]. In that case, atoms migrate from the hot spot to the remaining cooler region and this is known as thermo-migration.

Thermo-migration may assist electro-migration if the hot side coincides with the cathode side [53–55].

In summary, there are seven major phenomena associated with electro- and thermo-migration [56]. They are (1) electro-migration (mass transfer due to electron bombardment), (2) thermo-migration (mass transfer due to thermal gradient), (3) enhanced intermetallic compound (IMC) growth, (4) enhanced UBM dissolution, (5) enhanced current crowding, (6) high Joule heating and (7) solder melting.

Multi-physics models are required to predict the combined effect of electro-migration, thermo-migration, current crowding, Joule heating, thermal stress, and

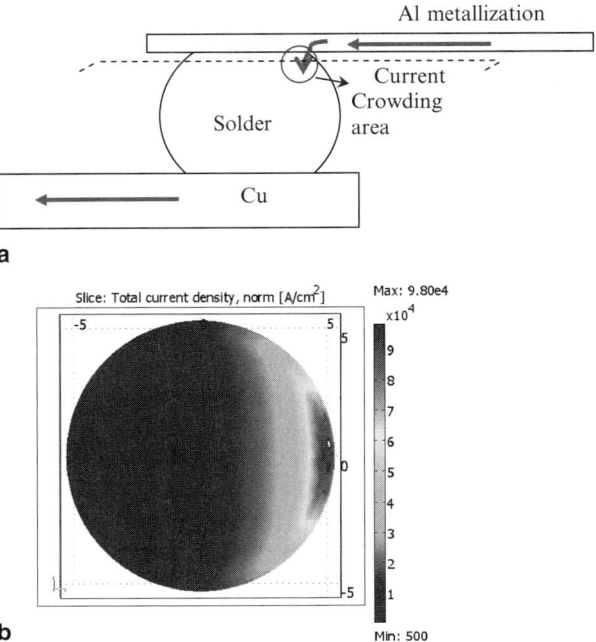

Fig. 2.15 Typical example of current crowding in the flip-chip solder joint: (**a**) flip-chip solder joint and current crowding region; (**b**) current density distribution in a slice through the line AB shown in (**a**), considering 1-A current passing through the joint. Current density at the current crowding area is ten times higher than that of the other area for the same slice

local melting phenomena. These need to capture IMC growth and UBM dissolution for solder interfaces under high current density.

Current crowding is always found to play a vital role in such a sensitive package level interconnect, in particular such as flip-chip solder joint. High growth/dissolution of IMC/UBM at the current crowding area will be encountered. This is because of typical flip-chip solder joint design where a spherical solder is connected with the thin film metallization (see Fig. 2.15). With the shrinking trends of interconnects in nano-packages, this current crowding effect increases exponentially.

2.5.4 Carbon Nanotubes for Thermal Management

Carbon nanotubes are potentially very important packaging materials because of their unique thermal and mechanical properties. They have been studied extensively using molecular dynamics ever since their discovery was popularised by Sumio Iijima in 1991 [57]. Recently, carbon nanotubes as electronic packaging materials have also attracted a lot of interest. For example, Fan et al. [58] studied the thermal conduction of single-walled carbon nanotubes (SWCNTs), and the interfacial thermal resistance between carbon nanotubes and the copper substrate was investigated. It was found that functional groups between the nanotubes and the copper substrate enhanced the interfacial heat transfer but degraded the heat conductivity in the carbon nanotube because of phonon scattering at the interface. Wu et al. [59] used the ACM to study the mechanical properties of CNTs. The predicted Young's modulus is consistent with ab initio, molecular dynamics and experimental results.

Nano-indentation is a powerful experimental technique that can be used to understand material responses and it is another interesting area where MD finds its use (see, e.g. a review paper by Szlufarska [60]). The technique is based on the measurement of forces acting on an indenter as it moves towards a contact surface (see Schuh's review paper [61]). In electronic packaging research, nano-indentation has been used frequently in materials characterisations. For example, in a recent work, Gao et al. [62] studied the mechanical properties of Sn–Ag lead-free solder and confirmed that the properties are dependent on solder joint size.

2.6 Conclusions

Modelling technology is now used extensively in industry and research institutions as a key tool to help produce reliable products and reduce cost and lead times from conceptual design to product introduction to the market.

The move towards nano-packaging poses a number of challenges for modelling tools. The multi-physics/scale challenge is being addressed by a number of software vendors although much needs to be done to validate these tools. The addition of uncertainty analysis within a physics-based modelling environment is very important as the scales of nano-packaging are not deterministic and risk-based approach is required to fully understand how input design variables will impact packaging assembly and reliability. Integrating modelling results within a whole life-cycle analysis approach are also very important and will be a major requirement for modelling tools in the future.

Acknowledgements The authors would like to acknowledge the financial support from the Engineering and Physical Sciences Research Council (EPSRC) that has supported some of the modelling development work at Greenwich as detailed in the illustrations above. We also acknowledge the 3D-Mintegration consortium (supported under EPSRC fund EP/C534212/) in the UK which is developing fabrication, assembly, packaging and testing procedures for 3D miniaturised products which is also supporting some of the work detailed above.

References

1. Alder BJ, Wainwright TE, Phase transition for a hard sphere system. J Chem Phys 27 (1957) 1208–1211
2. Car R, Parrinello M, Phys Rev Lett 55 (1985) 2471
3. Daw MS, Baskes MI, Phys Rev Lett 50 (1983) 1285
4. Allen MP, Tildesley DJ, Computer Simulation of Liquids (Clarendon, Oxford, 1987)

5. Hawlow FH, The particle in cell computing method for fluid dynamics. In Methods in Computational Physics, vol. 3, Fundamental Methods in Hydrodynamics, edited by B. Alder, S. Fernbach, M. Rotenberg (Academic, New York, 1964), pp 319–343
6. Lin CT, Chiang KN, Investigation of nano-scale single crystal silicon using the atomistic–continuum mechanics with Stillinger–Weber potential function. In IEEE Conference on Emerging Technologies – Nanoelectronics, January 10–13, 2006, pp 5–9
7. Zhang GQ, Maessen P, Bisschop J, Janssen J, Kuper F, Ernst L, Virtual thermo-mechanical prototyping of microelectronics – the challenges for mechanics professionals. In Proceedings of EuroSIME, 2001, pp 21–24
8. Melchers RE, Structural Reliability Analysis and Prediction (Wiley, Chichester, 1999)
9. Haldar A, Mahadevan S, Probability, Reliability and Statistical Methods in Engineering Design (Wiley, New York, 2000)
10. Shafer G, A Mathematical Theory of Evidence (Princeton University Press, Princeton, NJ, 1976)
11. Zadeh LA, Fuzzy sets. Inform Control 81 (1965) 338–353
12. Thacker B, Huyse L, Probabilistic assessment on the basis of interval data. In AIAA/ASME/ASCE/AHS Structures, Structural Dynamics, and Material Conference, AIAA-2003-1753, Norfolk, Virginia, 2003
13. Vanderplaats GN, Numerical Optimization Techniques for Engineering Design: with Applications (VR&D, Colorado Springs, 1999)
14. Vasile M, Niu Z, Nassar R, Zhang W, Liu S, Focussed ion beam milling: depth control for three-dimensional microfabrication. J Vac Sci Technol B 15(6) (1997) 2350–2354
15. Nassar R, Vasile M, Zhang W, Mathematical modelling of focused ion beam microfabrication. J Vac Sci Technol B 16(1) (1998) 109–115
16. Yamamura Y, Itikawa Y, Itoh N, Report IPPJ-AM-26. Institute of Plasma Physics, Nagoya University, 1983
17. Tseng A, Leeladharan B, Li B, Insua I, Fabrication and modelling of microchannel milling using focused ion beam. Int J Nanosci 2(4–5) (2003) 375–379
18. Munoz-Garcia J, Castro M, Cuerno R, Nonlinear ripple dynamics on amorphous surfaces patterned by ion beam sputtering. Phys Rev Lett 96 (2006) 086101
19. Chou SY, Krauss PR, Zhang W, Guo L, Zhuang L, Sub-10 nm imprint lithography and applications. J Vac Sci Technol B 15(6) (1997) 2897–2904
20. Hirai Y, Konishi T, Yoshikawa T, Yoshida S, Simulation and experimental study of polymer deformation in nanoimprint lithography. J Vac Sci Technol B 22(6) (2004) 3288–3293
21. Mooney M, A theory of large elastic deformation. J Appl Phys 11 (1940) 582–592
22. Rivlin RS, Large elastic deformation of isotropic materials. Phil Trans R Soc Lond A 241 (1948) 379–397
23. Hirai Y, Yoshida S, Takagi N, Defect analysis in thermal nanoimprint lithography. J Vac Sci Technol B 21(6) (2003) 2765–2770
24. Hirai Y, Fujiwara M, Okuno T, Tanaka Y, Study of the resist deformation in nanoimprint lithography. J Vac Sci Technol B 19(6) (2001) 2811–2815
25. Rowland H, King W, Sun A, Schunk P, Impact of polymer film thickness and cavity size on polymer flow during embossing: towards process design rules for nanoimprint lithography. SANDIA Report SAND2006-4864, 2006
26. Ritter G, McHugh P, Wilson G, Ritzdorf T, Three dimensional numerical modelling of copper electroplating for advanced ULSI metallisation. Solid State Electron 44 (2000) 797–807
27. Griffiths SK et al., Modeling electrodeposition for LIGA microdevice fabrication, SAND98-8231, 1998, Distribution Category UC-411
28. Osher S, Sethian JA, Fronts propagating with curvature-dependent speed: algorithms based on Hamilton–Jacobi formulations. J Comput Phys 79 (1988) 12–49
29. Nguty TA, Ekere NN, The rheological properties of solder and solder pastes and the effect on stencil printing. Rheol Acta 39 (1999) 607–612
30. Brady JF, Bossis G, Stokesian dynamics. Annu Rev Fluid Mech 20 (1988) 111–157

31. McNamara G, Zanetti G, Use of the Boltzmann equation to simulate lattice-gas automata. Phys Rev Lett 61(20) (1988) 2332–2335
32. Hoogerbrugge PJ, Koelman JM, Simulating microscopic hydrodynamic phenomena with dissipative particle dynamics. Europhys Lett 19(3) (1992) 155–160
33. Flekkoy EG, Wagner G, Feder J, Hybrid model for combined particle and continuum dynamics. Europhys Lett 52(3) (2000) 271–276
34. Dong H, Fan L, Moon K, Wong CP, Molecular dynamics simulation of lead free solder for low temperature reflow applications. In Proceedings of the 55th Electronic Components and Technology Conference, 2005, pp 983–987
35. Wang T, Fu Y, Becker M, Zhou M, Liu J, Microwave heating of metal-filled electrically conductive adhesive curing. In Proceedings of the IEEE Electronic Components and Technology Conference, 2001, pp 593–597
36. Sinclair KI, Desmulliez MPY, Sangster AJ, A novel RF-curing technology for microelectronics and optoelectronics packaging. In Proceedings of the IEEE Electronics Systemintegration Technology Conference, 2006, vol 2, pp 1149–1157
37. Sun Y, Zhang Z, Wong CP, (2005)Study on mono-dispersed nano-size silica by surface modification for underfill applications. J Colloid Interf Sci 292 436–444
38. Lu H, Hung KC, Stoyanov S, Bailey C, Chan YC, No-flow underfill flip chip assembly – an experimental and modelling analysis. Microelectron Reliab 42 (2002) 1205–1212
39. Shi SH, Wong CP, Recent advances in the development of no-flow underfill encapsulants – a practical approach towards the actual manufacturing application. IEEE Trans Electron Packag Manuf 22 (1999) 331–339
40. Liu J, Kraszewshi R, Lin X, Wong L, Goh SH, Allen J, New developments in single pass reflow encapsulant for flip chip application. In Proceedings of the International Symposium on Advanced Packaging Materials, Atlanta, GA, 2001, pp 74–79
41. Lall P, Islam S, Suhling J, Tian GY, Nano-underfills for high-reliability applications in extreme environments. In Proceedings of the 55th Electronic Components and Technology Conference, 2005, pp 212–222
42. Liu J, Conductive Adhesive for Electronics Packaging (Electrochemical Publications, Port Erin, UK, 1999), pp 234–248
43. Li Y, Yim MJ, Moon KS, Wong CP, Nano-scale conductive films with low temperature sintering for high performance fine pitch interconnect. In Proceedings of the 57th Electronic Components and Technology Conference (ECTC), Nevada, 2007, pp 1350–1355
44. Mercodo LL, White J, Sarihan V, Lee TYT, Failure mechanism study of anisotropic conductive film (ACF) packages. IEEE Trans Compon Packag Technol 26(3) (2003) 509–516
45. Wei Z, Waf LS, Loo NY, Koon EM, Huang M, Studies on moisture-induced failures in ACF interconnection. In Proceedings of the 7th Electronics Packaging Technology Conference (EPTC), Singapore, 2002, pp 133–138
46. Kim JW, Jung SB, Effects of bonding pressure on the thermo-mechanical reliability of ACF interconnection. J Microelectron Eng 83(11–12) (2006) 2335–2340
47. Rizvi MJ, Chan YC, Bailey C, Lu H, Study of anisotropic conductive adhesive joint behaviour under 3-point bending. J Microelectron Reliab 45(3–4) (2005) 589–596
48. Wu CML, Liu J, Yeung NH, The effects of bump height on the reliability of ACF in flip-chip. J Solder Surf Mount Technol 13(1) (2001) 25–30
49. Yin CY, Lu H, Bailey C, Chan YC, Moisture effects on the reliability of anisotropic conductive films. In Proceedings of the 6th International Conference on Thermal, Mechanical and Multiphysics Simulation and Experiments in Micro-Electronics and Micro-Systems, Berlin, 2005, pp 162–167
50. Yin CY, Lu H, Bailey C, Chan YC, Macro–micro modeling analysis for an ACF flip chip. J Solder Surf Mount Technol 18(2) (2006) 27–32
51. Tu KN, Recent advances on electromigration in very large scale integration of interconnects. J Appl Phys 94(9) (2003) 5451–5473
52. Lloyd JR, Electromigration and mechanical stress. Microelectron Eng 49(1–2) (1999) 51–64

53. Ye H, Basaran C, Hopkins D, Thermomigration in Pb–Sn solder joints under joule heating during electric current stressing. Appl Phys Lett 82(7) (2003) 1045–1047
54. Alam MO, Wu BY, Chan YC, Tu KN, High electric current density-induced interfacial reactions in micro ball grid array solder joints. Acta Mater 54(3) (2006) 613–621
55. Dan Y, Alam MO, Wu BY, Chan YC, Tu KN, Thermomigration and electromigration in solder joint. In Proceedings of the 8th Electronics Packaging Technology Conference (EPTC 2006), 2006, Singapore, pp 565–569
56. Alam MO, Bailey C, Wu BY, Yang D, Chan YC, High current density induced damage mechanisms in electronic solder joints – a state-of-art-review. In Proceedings of the High Density Packaging Conference, Shanghai, China, 2007, pp 93–99
57. Iijima S, Helical microtubules of graphite carbon. Nature 354 (1991) 56–58
58. Fan H, Zhang K, Yuen MMF, Thermal performance of carbon nanotube-based composites investigated by molecular dynamics simulation. In Proceedings of the 57th Electronic Components and Technology Conference, 2007, pp 269–272
59. Wu CJ, Chou CY, Han CN, Chiang KN, Simulation and validation of CNT mechanical properties – the future interconnection material. In Proceedings of the 57th Electronic Components and Technology Conference, 2007, pp 447–452
60. Szlufarska I, Atomistic simulations of nanoindentation. Mater Today 9(5) (2006) 42–50
61. Schuh CA, Nanoindentation studies of materials. Mater Today 9(5) (2006) 32–40
62. Gao F, Nishikawa H, Takemoto T, Nano-scale mechanical responses of Sn–Ag based lead-free solders. In Proceedings of the 57th Electronic Components and Technology Conference, 2007, pp 205–210

Chapter 3
Application of Molecular Dynamics Simulation in Electronic Packaging

Haibo Fan(✉) and Matthew M.F. Yuen

3.1 Introduction

With the increasing need for high input/output (I/O) counts and miniaturization, novel electronic packages are continuously being developed. Chip scale packages are beginning to replace older leadframe technology because of low cost, size, and performance advantages. Wafer-level packaging technology is becoming popular due to low cost and higher electronic performance. At same time, more and more functional materials at the nanoscale are used in electronic packaging for the improvement of the adhesion and thermal conductivity, such as carbon nanotube (CNT), thermal interface material (TIM), and self-assembly monolayer (SAM). To obtain good performance of these materials and guide the experimental research, it is important for us to find methods to understand material behavior at a fundamental level. Obviously, a traditional method like finite element analysis widely used in electronic packages is not suitable for modeling the behavior of these materials at a nanoscale level. Molecular dynamics (MD) simulation is now one of the fastest growing research areas and can reproduce material behavior at atomic level. Therefore, investigation of material behavior using MD simulation in electronic packaging is both necessary and attractive.

3.2 Molecular Dynamics Simulation

Molecular modeling is the science of representing molecular structures numerically and simulating their behavior with the equations of quantum and classical physics and it is one of the fastest growing fields in science. The MD simulation was first introduced by Alder and Wainwright [1, 2] to study the interactions of hard spheres

---

H. Fan
Department of Mechanical Engineering, Hong Kong University of Science and Technology, Clearwater Bay, N.T., Hong Kong

in the late 1950s. The method is now a well-established and important tool that endeavors to simulate the material measurement on an atomistic scale to understand the basic origins of material performance [3–8] in a wide variety of topics including modulus, adhesion, thermal conductivity, solubility, diffusion, and reactivity.

The basic theories of MD simulation can be found in many handbooks [9–11]. A brief introduction on the MD simulation is given here. In the classical MD method, the equations of motion for atoms are described by Newton's equations as follows:

$$F_i = m_i \frac{d^2 r_i}{dt^2}, \quad F_i = -\nabla_i \Phi, \tag{3.1}$$

where F_i, m_i, and r_i are, respectively, the force, mass, and position vector of molecule i and Φ is the potential energy function of the system.

Normally, atoms in MD simulations are modeled as point masses interacting through potentials. The accuracy of the MD simulation is directly related to these potentials, which are usually characterized experimentally. The potential energy of the system provides the forces on each atom, which can be used to determine the acceleration, velocity, and positions of each atom. Two kinds of methods, quantum mechanics (QM) and molecular mechanics (MM), are used in MD simulations. The QM method solves the Schrodinger equation from first principles, which accounts for the positions of nuclei and electrons and provides structural, electronic, and dynamic properties of the system in high accuracy. It can also describe the process of a chemical reaction involving bond forming, bond breaking, charge transfer, etc. However, due to long calculation times and costly calculations in QM, the QM method is only suitable for simulating small systems consisting of up to several hundreds of atoms. Electronic effects are averaged out in the MM method, so it cannot describe the evolution of electrons in a chemical reaction. However, it can be used to describe the performance of large and complex organic, inorganic, and solid state systems due to its inexpensive and fast calculation. Compared with QM, MM is more applicable for the investigation of material performance in the multi-interface systems commonly found in electronic packages.

The total potential in MM is represented by the superposition of valence and nonbond interactions. The valence terms consist of bond stretch, bond angle bending, and dihedral angle torsion terms, while nonbond interactions consist of Van der Waals and electrostatic terms. Most of the force fields have the characteristic that the energy is a function only of the atomic positions with some constants. For example, a simple generic force field [12] is given here:

$$V = V_{bond} + V_{angle} + V_{torsion} + V_{elec} + V_{vdw}. \tag{3.2}$$

The valence terms are given as follows:

$$E_{bond} = k_b (r_{ij} - r_0)^2, \quad E_{angle} = k_a (\theta_{ijk} - \theta_0)^2, \quad E_{torsion} = k_t (1 + \cos(3\varphi_{ijkl})), \tag{3.3}$$

where E_{bond} and E_{angle} are the energy due to bond stretching or compression and angle bending and $E_{torsion}$ is the energy due to torsion alternations.

Nonbond terms include the electrostatic and Van der Waals forces:

$$E_{elec} = \frac{q_i q_j}{4\pi\varepsilon_0 r_{ij}}, \quad E_{vdw} = \alpha \left(\frac{R_0}{r_{ij}}\right)^{12} - \beta \left(\frac{R_0}{r_{ij}}\right)^6, \quad (3.4)$$

where E_{elec} and E_{vdw}, respectively, are energy terms due to electrostatic and Van der Waals interactions.

We just showed a simple potential with several parameters. Commonly used force fields in MD simulation include AMBER, CHARMM, CVFF/PCFF, and COMPASS, which have been widely used for description of organic, inorganic, and biomaterial system. Parameters in these potentials are derived from experimental data or ab initio and semiempirical quantum mechanical theory. Considering the structural constraints, the forces on the atoms can be calculated from the potentials. Accelerations of atoms can be derived based on Newton's second law and the corresponding velocities and new positions of the atoms will be obtained. Based on statistical mechanics, macroscopic properties of a thermal equilibrium system can be obtained from motion of atoms and molecules of the system.

The Monte Carlo (MC) method is another simulation method used in the evaluation of potential energy. It only considers configuration space without the momentum part of the phase space. The method is suitable to model low or medium density systems but not for high density systems. Compared with MD simulation, the MC method cannot be used to simulate dynamic processes of the system.

MD is a very useful tool to provide a wealth of detailed information on the structure and dynamics of systems in electronic packaging. It can be used to predict material performance of new material in synthesis, such as Young's modulus, thermal expansion coefficients, glass transition temperature, interfacial adhesion, thermal conductivity, etc. Issues involved in packaging design, such as crack generation, moisture-induced failure, and interfacial adhesion, can also be studied by MD simulation. Although the MD simulation method has been widely used in the fields of physics, chemistry, and biology for material properties, and now in the field of engineering, few research efforts have been dedicated to the investigation on the performance of materials in electronic packaging until recently [13–33]. Iwamoto [13–22] has applied MD simulation to the investigation of property trend analysis, surface energy modulus, and the adhesive formulation effect in the microelectronics packaging industry. We have focused our research on the investigation of reliability, moisture diffusion, epoxy resin properties, and thermal conductivity in electronic packaging [23–33].

3.3 MD Simulation of the Thermal Cycling Test in Electronic Packaging

Thermal cycling test is one of the key qualification tests to be conducted for the reliability testing of electronic packages. The high interfacial stresses resulted from the mismatch in coefficient of thermal expansion between the different layers will threaten the package under environmental thermal loading conditions. Thermal cycling test is one of the standard accelerated test methods to investigate reliability of electronic packages. To predict material behavior during the thermal cycling, finite element analysis has been widely applied to model electronic packages subjected to a thermal cycling load [34, 35]. The failure mechanism occurring in electronic packages during reliability tests can be investigated by the finite element analysis; however, some phenomena in the tests are still unknown. Moreover, as interfacial material properties being quite different from the bulk materials, it is necessary to investigate the material behavior at an atomic level.

Molecular modeling techniques were successfully applied to investigate the fundamentals of the interfacial adhesion between a cured epoxy and other substrates and revealed the molecular mechanisms of the adhesion formulation and failure [6–8]. MD simulations of the adhesive formulation effect in the microelectronics packaging industry were conducted by Iwamoto [13, 14], and she also presented some significant and interesting research results in electronic packaging applications [15, 16]. However, little attention was paid to investigations of the material behavior under thermal cycling conditions, especially the behavior of the EMC–Cu system widely used in electronic packaging.

Iwamoto [15] developed a MD simulation procedure to predict material performance. Both stress cycling and process analyses were studied to give the probable behavior trend of materials for the understanding of the failure mechanisms. The same method was used for the investigation of the adhesion between the epoxy molding compound (EMC) and cuprous oxide substrate during the thermal cycling test using MD simulation by Fan et al. [23]. The detail was presented as follows.

The Material Studio software for Accelrys was used to conduct the thermal cycling test MD simulations. The condensed-phase optimized molecular potential for atomistic simulation studies (COMPASS) force field was used in the MD simulations. COMPASS force field enables accurate prediction of material properties for a broad range of materials under different conditions. The COMPASS force field can accurately be applied on the systems of polymers, metals, and their interfaces.

Typically, the core chemical structure of the EMC, governing the adhesion properties of the EMC material, was formed by the reaction of epoxy resin with curing agents. EMC modeled in this study consists of epoxy resin and curing agent. The model did not include filler and pigments, which need large-scale models that are beyond the current MD simulation scope. In this study, diglycidyl ether of bisphenol-A (DGEBA) epoxy resin and methylene diamine dianilene (MDA) curing agent were used as the basic components of the EMC formulation. Figure 3.1 shows the process of the curing reaction of DGEBA epoxy resin and MDA curing agent.

It has been observed that oxidation occurred on the metal surface at the EMC/Cu interface during the molding process, which has a strong influence on the interfacial adhesion. Cho et al. [36] found that the chemical content of the copper oxide on a copper surface changed from metal copper to cuprous oxide. Chung et al. [37] conducted a thermal cycling test and found that there was only cuprous oxide for all the sheared samples and the content of cuprous oxide varied with thermal cycles.

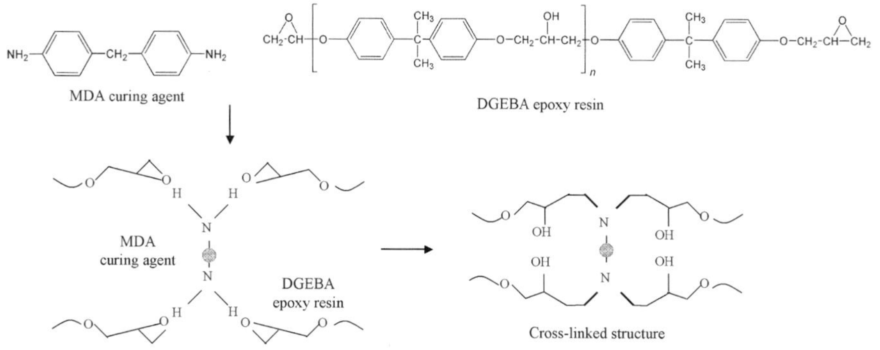

3 Application of Molecular Dynamics Simulation in Electronic Packaging

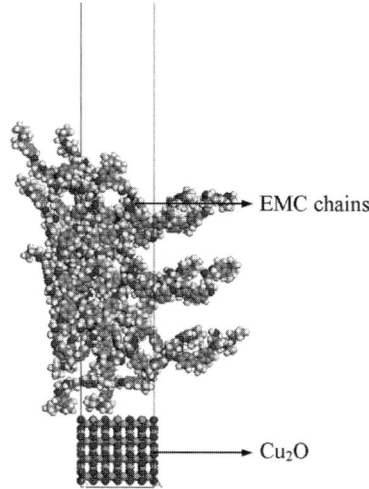

Fig. 3.1 Process of the curing agent and epoxy resin reaction

Fig. 3.2 Molecular model of the EMC–cuprous oxide system

Therefore, only the cuprous oxide was considered in the MD simulations. To investigate the effect of the content of the cuprous oxide on the copper substrate on the interfacial adhesion between the EMC and copper substrate, two MD models were built for the thermal cycling test by Fan et al. [23]. One is the MD model with the amount of the cuprous oxide atoms varying with the thermal cycles as that in the experiment conducted by Chung et al. [37]. The other is the MD model with a constant content of the cuprous oxide throughout the thermal cycling test. Figure 3.2 showed the MD model built with a rectangular simulation cell 17.9 × 17.9 Å² in the x and y directions, periodic in the plane perpendicular to the EMC/cuprous oxide interface. The bimaterial system composed of a fragment of EMC and cuprous oxide atoms was used in the molecular dynamics simulations. There is a large vacuum at the top of the epoxy chains in the simulation cell to avoid interaction across the mirror image in the z direction. In the model, the height of the cell is 100 Å and the cuprous oxide thickness is 20 Å. All the atoms of cuprous oxide were fixed, while all the EMC chains could move freely.

According to experimental conditions presented by Chung et al. [37], the MD simulations were conducted at 175°C with a pressure of 4.3 MPa, using the ensemble of the constant number of particles, pressure, and temperature (NPT). The whole structure was then cooled to room temperature using the ensemble of the constant number of particles, volume, and temperature (NVT).

The thermal cycling test was conducted by Chung et al. [37] with the standard JEDEC temperature profile (JESD22-A104-B Condition M). In this study, the coefficient of thermal expansion (CTE) of the EMC and Cu_2O is 45 and 8 ppm, respectively. The EMC and cuprous oxide CTE mismatch is used to help define the resulting deformation in the respective heating and cooling step within a thermal cycle. The thermal cycling process in MD simulations was accomplished using a different strain target during each simulation step. Different strains were applied to the MD model, making the EMC material pushed toward and pulled away from the cuprous oxide substrate in the heating and the cooling step, respectively. The whole system was relaxed at the cooling temperature and the heating temperature. The above cooling and heating procedure was repeated using different strains to reproduce the whole thermal cycling test.

The molecular structures of the system after different thermal cycles were shown in Fig. 3.3. It was found that a large void formed at the interface during the thermal cycling test. This resulted from the interfacial stresses due to CTE mismatch of the EMC and cuprous oxide during the thermal cycling test. Therefore, after 1,800 thermal cycles, a reduction in the adhesion strength between the EMC and cuprous oxide substrate was observed.

The interaction between the EMC and the cuprous oxide substrate is governed by the electrostatic and Van der Waals forces. The interfacial bonding energy γ is

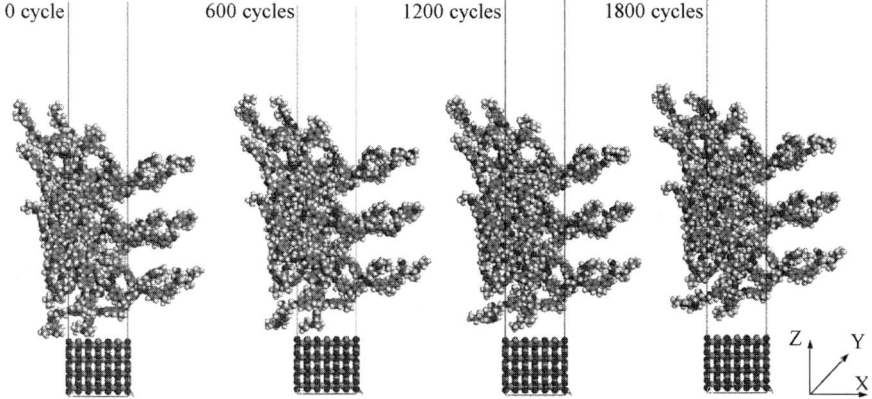

Fig. 3.3 MD simulation snapshots of the system after different cycles

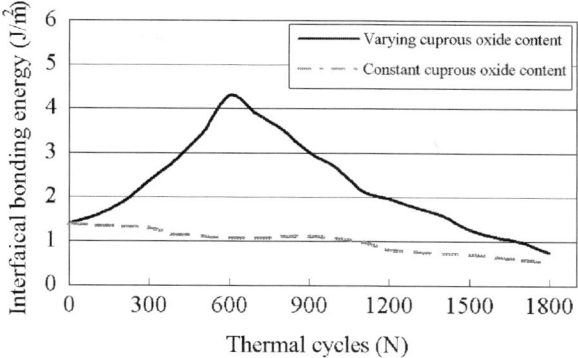

Fig. 3.4 Interfacial bonding energy as a function of thermal cycles

evaluated from the energy difference between the total energy of the whole system and the sum of the energy of the individual subsystems as follow [23]:

$$\gamma = \Delta E / 2A, \quad \Delta E = E_{total} - \left(E_{EMC} + E_{Cu_2O}\right), \tag{3.5}$$

where A is the contact area between the EMC and cuprous oxide substrate, E_{total} is the total energy of the whole system, E_{EMC} is the energy of the EMC without the cuprous oxide substrate, and E_{Cu_2O} is the energy of the cuprous oxide substrate without the EMC.

The interfacial bonding energies between the EMC and cuprous oxide substrate for the both MD models were calculated and plotted against the thermal cycles, as shown in Fig. 3.4. For the first MD model, the interfacial bonding energy increased first and then decreased for the remainder of the thermal cycles. For the second MD model, the interfacial bonding energy generally decreased with the increase of the thermal cycles. The difference between two kinds of simulations indicated that the interfacial adhesion between the EMC and cuprous oxide substrate was dominated by the change of the cuprous oxide content on the copper substrate. The higher the content of cuprous oxide on the copper substrate, the higher is the adhesion strength between EMC and copper substrate.

The above MD simulation results could be verified by the experimental results by Chung et al. [37]. A similar trend showed that the interfacial adhesion was governed by the change of the content of oxide on the copper surface. Su and Shemenski [38] also showed that the cuprous oxide could be used to achieve better adhesion between the rubber and copper wire. Kendall [39] argued that the adhesion between surfaces was governed the bonding surface properties rather than the bulk material properties of the two bonded materials. All of these were consistent with the observation in the MD simulation of the thermal cycling test. At this stage, the MD simulation results could not be evaluated quantitatively against the experimental measurements, but qualitative trends predicted by MD simulations are consistent with the experimental observations.

3.4 MD Study of Moisture Diffusion in Electronic Packaging

Moisture-induced reliability was one of the concerns extensively studied in package design. Moisture diffusion in plastic-encapsulated packages adversely affected the EMC/Cu interfacial adhesion and significantly reduces the reliability of the package. To obtain high reliable packages, understanding of the mechanism of moisture transport to the epoxy/copper interface at an atomic level is becoming necessary.

Normally, there are two moisture transport paths to the EMC/Cu interface, diffusion in the bulk epoxy resin as well as via seepage along the interface. Distinguishing the two mechanisms of moisture transport to the interface is rather important for the understanding of moisture effect on the interfacial adhesion. Attention has been paid to both the moisture diffusion in bulk materials [40–42] and moisture wicking along an interface [43–45]. However, it is very difficult to experimentally determine moisture seepage along the interface. Moreover, it cannot be predicted by the traditional FEA simulation method. MD simulation is the proper method to provide information about the dynamics of water molecules in polymer materials and at their interfaces. Although some researchers had successfully applied MD simulations in the investigation of the diffusion of water molecules in some kinds of polymers [46–50], little attention was paid to the moisture diffusion in a crosslinked epoxy resin and its interface in electronic packaging. Fan et al. [24] conducted MD simulations to investigate the respective moisture diffusion into the EMC and the EMC/Cu interface. The MD results showed that the seepage along the EMC/Cu interface was more prevalent than moisture diffusion in the bulk EMC, which obviously was dominant mechanism causing moisture-induced interfacial delamination in plastic packages.

We built two kinds of models to investigate the moisture diffusion in the bulk epoxy as well as seepage along the EMC/Cu interface. The EMC modeled in this study consisted of epoxy resin and curing agent, which is the same structure as shown in Fig. 3.1. The bulk MD models composed of a fragment of fully cured epoxy resin network and different amounts of water molecules were built using the amorphous module in Accelrys. All the bulk models were built with a rectangular simulation cell $1.97 \times 1.97 \times 1.97$ nm^3 in all directions. The interface models with different amount of water molecules at the interface were built with a rectangular simulation cell 1.81×1.81 nm^2 in the x and y directions, and periodic in the plane perpendicular to the EMC/Cu interface. The mass ratio of water molecules to the EMC in both kinds of MD models varied from 1.1 to 2.2%. Figure 3.5 shows the morphological configurations for the two kinds of MD moisture model.

At a temperature of 85°C, all the simulations were then performed with a presumed moisture concentration value using the NPT ensembles. MD simulations were conducted to equilibrate the whole systems for about 80 ps under a pressure of 0.1 MPa at 85°C. The velocity Verlet algorithm [51] was used for integration with the nonbonded interactions including Van der Waals and electrostatic forces in all MD simulations. The Ewald summation with a cutoff distance of 0.95 nm was used for the nonbonded interactions.

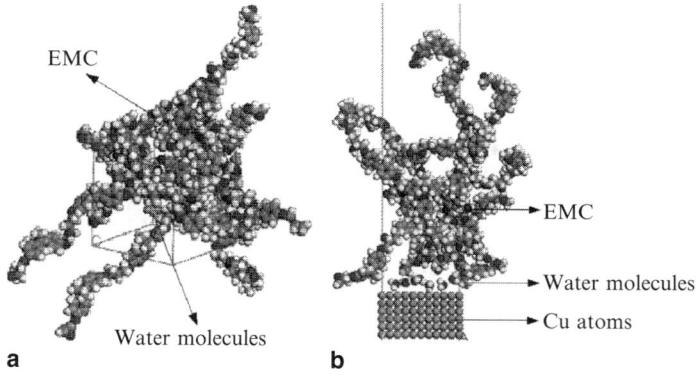

Fig. 3.5 Morphological configuration of (**a**) the bulk EMC and (**b**) the EMC/Cu system

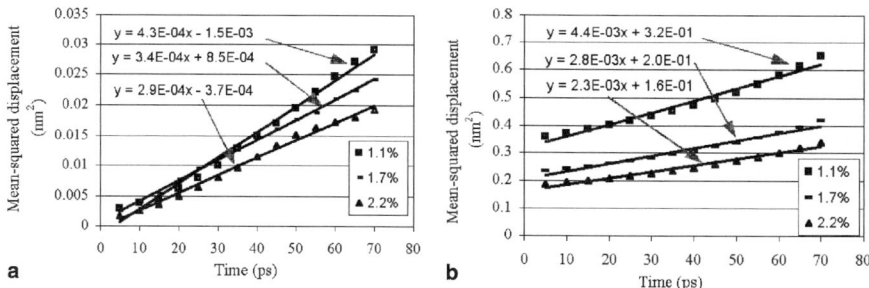

Fig. 3.6 Mean-squared displacement of water molecules as the function of the simulation time (**a**) in the bulk EMC and (**b**) at the EMC/Cu interface and the fitted lines for different mass ratios of water molecules to the EMC

The moisture diffusion constants were calculated based on the diffusion trajectories $r(t)$ of water molecules from an MD simulation. From the Einstein relation, the diffusion coefficients for the water molecules were calculated from the mean-squared displacements, $s(t) = \langle |r(t) - r(0)|^2 \rangle$, of the water molecules averaged over time as follows [24]:

$$D = \frac{1}{6N} \lim_{t \to \infty} \frac{d}{dt} \sum_{i=1}^{N} \langle [r_i(t) - r_i(0)]^2 \rangle, \quad (3.6)$$

where D is the moisture diffusion coefficient, $ri(t)$ is the coordinate of the center of the mass of the ith water molecule, and N is the number of water molecules in the system.

The mean-squared displacements of water molecules against time for both the bulk EMC and EMC/Cu interface models, respectively, are shown in Fig. 3.6. All the data in each graph were fitted using the straight line with the slope of a. Equation (3.6) can be simplified as follows [24]:

$$D = a/6. \tag{3.7}$$

The slope of the fitted line for different mass ratios of water molecules to the EMC was used to calculate the moisture coefficients for different cases by (3.7), and the calculated values are listed in Table 3.1.

From the above MD simulations, we found that the predicted moisture diffusion coefficient decreased with the increase in the mass ratio of water molecules to the EMC. At a given temperature and relative humidity (RH), the moisture diffusion coefficient was dependent on the moisture concentration in the system with initial low moisture concentration, a high amount of moisture is absorbed in the system. However, increase in moisture concentration resulted in a lower moisture diffusion coefficient. Moreover, water clusters could be formed from increasing water molecules present in the EMC materials or at the EMC/Cu interface. The clustering would reduce the mobility of water molecules resulting in low moisture diffusion coefficient. This phenomenon was also proved experimentally and numerically by other researchers [47, 52].

At the same environmental condition, the moisture diffusion coefficient for the bulk EMC material was smaller (almost one order of magnitude) than that for the EMC/Cu interface. The pore sizes at the EMC/Cu interface may be larger than those within the EMC material, which caters for higher water molecule mobility at the EMC/Cu interface. Comyn et al. [43] and Zanni-Deffarges and Shanahan [44] found that the amount of water diffusion at the interface was higher in magnitude than that in the bulk epoxy. They concluded that capillary diffusion along the interface exacerbated water ingress and surface tension effects near the polymer–metal interface increased the effective driving force for water penetration. These experimental results confirmed with the MD simulation results, stipulating that moisture could more readily penetrate into the package along the EMC/Cu interface than in the bulk EMC. On the other hand, cracks existing at the EMC/Cu interface after encapsulation were another potential factor for enhancing moisture diffusion along the EMC/Cu interface. Therefore, we concluded that the seepage along the interface was the dominant mechanism for moisture diffusion into the EMC/Cu interface in plastic packages. The conventional understanding of moisture diffusion into the interface via the bulk EMC material is only a secondary moisture penetration path

Table 3.1 Moisture diffusion coefficients in both the bulk EMC material and the EMC/Cu interface predicted by MD simulation under different conditions

	Mass ratio of water molecules to the EMC		
	1.1%	1.7%	2.2%
Moisture diffusion coefficient in the bulk EMC material (mm^2 s^{-1})	7.12×10^{-5}	5.58×10^{-5}	4.87×10^{-5}
Moisture diffusion coefficient at the EMC/Cu interface (mm^2 s^{-1})	7.29×10^{-4}	4.71×10^{-4}	3.79×10^{-4}

to the interface [24]. From MD simulation results, it is concluded that the seepage of moisture along the EMC/Cu interface is an important factor in the design against moisture-related failures in plastic packages. Although the MD simulation can only provide a qualitative prediction of the moisture diffusion coefficient, the approach would be a useful tool to investigate the performance of EMCs under different moisture conditions.

3.5 Material Properties of Epoxy Resin Compound Predicted by MD Simulation

One of the popular research topics in electronic packaging is the investigation of material properties, especially the crosslinked epoxy resin compounds widely used in electronic packaging. Prediction and better understanding of its properties would be necessary and significant to the design and development of the crosslinked epoxy resin compounds in electronic packaging. MD simulation has been used to predict material properties of different materials [53–57]. Most of these studies have been focused on thermoplastic materials, such as poly(methyl methacrylate), polyethylene, and polyarylethersulfones. However, due to the complexity of the crosslinking reactions, little attention was paid to the material properties of the thermoset materials, especially the crosslinked epoxy resin compounds widely used in electronic packaging. MD simulations were conducted to estimate the material properties of the crosslinked epoxy resin compound by Fan et al. [25]. The detail was presented as follows.

MD simulations were conducted using the polymer consistent force field (PCFF). The velocity Verlet algorithm was used for integration in all MD simulations. The nonbond interactions include Van der Waals and electronic static forces, and the Ewald summation with a cutoff distance of 9.5 Å was used for the dispersion interactions. The fully cured epoxy network was composed of EPON 862 resin and TETA-Triethylenetetramine curing agent. The chemical structures of the monomers (EPON 862 and TETA) were given in Fig. 3.7. During the curing reaction, the hydrogen atoms in the amine groups of curing agent molecules reacted with the epoxide groups of epoxy reins. The crosslinking activity expands in all directions and forms a network of macromolecules.

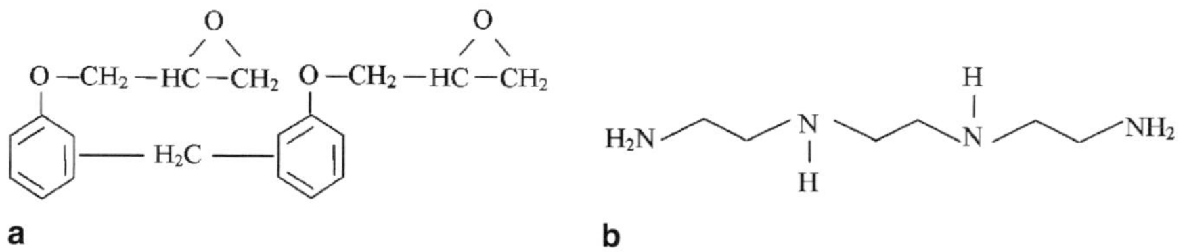

Fig. 3.7 Chemical structure of the monomers: (**a**) EPON 862 and (**b**) TETA-Triethylenetetramine

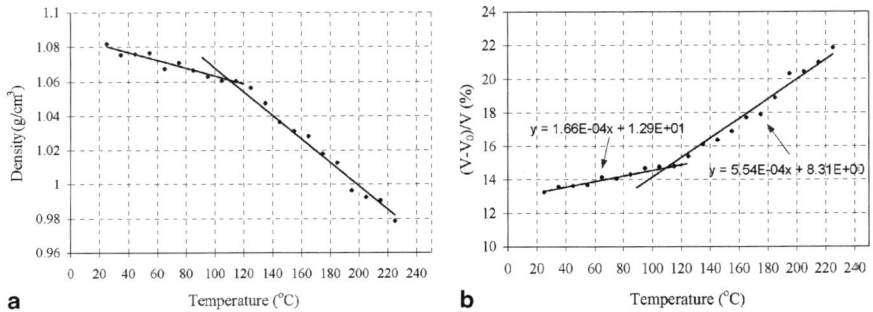

Fig. 3.8 (a) A plot of the density against temperature and (b) volume change against temperature and the fitted lines

Table 3.2 Material properties of the epoxy resin compound from both MD simulation and experimental data in [58]

Material properties	MD simulation	Ref. [58]
Young's modulus (GPa)	3.75	3.43
T_g (°C)	109	105
β_1 (below T_g) (ppm/°C)	55.3	61.0
β_2 (above T_g) (ppm/°C)	184.7	195.0

Based on the experimental condition in the literature [58], a fragment of fully cured network consisting of 12 molecules of EPON 862 epoxy resin and 4 molecules of TETA was built and an amorphous structure was constructed with a density equal to the experimental value of 1.23 g cm^{-3}. The simulation cell was periodic in all directions. MD simulations were then conducted starting at 225°C under a pressure of 0.1 MPa using the NPT. Temperature was cooled to room temperature at a rate of 10°C/200 ps by the nose-controlled method. Each subsequent simulation was started from the final configuration obtained at the preceding temperature. The MD simulation was conducted with an interval of 1 femtosecond (fs) in each MD simulation step.

Figure 3.8a shows the density of the cured epoxy at each temperature. The density increased steadily with decreasing temperature and the change of the slope of the density curve was observed. The change in the slope of the density curve defines the value of the glass transition temperature (T_g), at approximately 109°C. The predicted value of T_g from molecular dynamics simulation is very close to the experimental value from the literature [58] listed in Table 3.2.

The initial volume of the system from MD simulations was 36.20 nm^3 and the corresponding volume of the system at each temperature was calculated from MD simulations. The change in the volume of the cell structure against the temperature was given in Fig. 3.8b. The volumetric thermal expansion coefficients for both the glassy and rubbery state were obtained from the two slopes of the curve which discontinued at the glass transition temperature. The linear thermal expansion coefficient was related to the volumetric thermal expansion coefficient by

$$\beta = \frac{1}{3}\alpha. \tag{3.8}$$

The linear thermal expansion coefficients for both the glassy and rubbery state were calculated and are listed in Table 3.2.

The mechanical behavior of a system at the atomic scale is described using the equivalent-continuum constitutive relation. For the isotropic material, the stress–strain behavior can be described by only two independent coefficients, Lame coefficients, λ and μ, which are the same as those described at a macroscale level. Therefore, the mechanical properties of material from MD simulation, such as Young's modulus, can be given as follows:

$$E = \mu\left(\frac{3\lambda + 2\mu}{\lambda + \mu}\right). \tag{3.9}$$

Young's modulus of the epoxy resin compound was calculated from the MD simulations and is listed in Table 3.2. Compared with the experimental results from the literature [58], we found that the predicted values of both CTE and Young's modulus from the MD simulations were very close to those values with a small variation within 10%. The small difference between the MD simulation and experimental values can be attributed to the lower crosslink density of the epoxy resin compound in the experimental results from the literature [58]. During experiment, it is most likely that a certain amount of components (EPON 862 and TETA) are not fully crosslinked together, which allows the sample to produce larger strain than the idealized situation stipulated in the MD simulation modeling a fully crosslinked network. Therefore, the epoxy resin compound is rather soft and bears large deformation, which results in lower modulus and thermal expansion coefficients. Moreover, these material properties can also be degraded by voids or impurity inside the epoxy compound [25].

The predicted material properties from MD simulations were close to the experimental values from the literature [58], which confirms the accuracy of the cured epoxy resin model, the propriety of the force field and the cooling rate. The predicted material properties of an epoxy resin compound prior to laboratory design can guide the design and development of new epoxy resin compounds with higher quality in electronic packaging.

3.6 Thermal Performance of Carbon Nanotubes Investigated by MD Simulations

With the increasing need of electronic devices with smaller size, materials with higher performance are widely used in electronic packaging. CNTs have been widely applied in many fields due to their excellent mechanical and electrical properties. CNTs are now attracting more attention of researchers as promoter of TIMs, which are used to dissipate heat from die to heat sink in electronic packaging. CNTs are dispersed or aligned in polymer composites to enhance the thermal conductivity of the composite. However, it is still technically difficult to measure the thermal conductivity of an individual CNT. MD simulation provides a viable tool to investigate thermal transport in nanostructures. Basically, there are two kinds of frequently used MD simulation methods on heat transfer in CNT. One is the equilibrium molecular dynamics (EMD) method based on the Green–Kubo relation. The other is the nonequilibrium molecular dynamics (NEMD) method based on Fourier's law. Both methods have been applied to predict the thermal conductivity of single-walled CNT (SWCNT) or multiwalled CNT (MWCNT) [59–62]. The predicted thermal conductivity of SWCNT or MWCNT varied from several hundreds to thousands W K^{-1} m^{-1}. Large variation for the predicted values may result from the methods used for calculation, potential function, CNT length, etc. In spite of lots of these challenges in MD simulations on CNTs, it is still necessary for us to investigate the performance of CNTs at a fundamental level. Some useful information can be extracted from MD simulation for use as a guide for experimental design of CNT-based TIMs, which enhance heat dissipation and prevent temperature-related failure occurring in electronic packaging. Therefore, the thermal conductivity of SWCNT with different lengths was calculated, and the effects of voids on the material properties of CNTs were investigated in this study.

MD simulations were conducted using Material Studio (Accelrys, Inc.) software. COMPASS force field was used in the simulation. Most of previous studies of CNTs used Tersoff–Brenner potential whose parameters were taken from experiments on diamond and graphite. However, nonbonding energy, such as Van der Waals force, is not included in this potential. There will be some issues occurring during the MD simulation, such as mentioned by Bi et al. [60] that the Tersoff–Brenner potential excludes the Van der Waals force among the atoms resulting in bending of longer tubes, which cause failure of the thermal conductivity calculation. Moreover, the potential without nonbonding forces is not suitable to the MWCNT as well as the later study of CNT-based polymer composites. Comparing with Tersoff–Brenner potential, COMPASS contains not only valence bonding energy but also nonbonding energy, and can reproduce CNT behavior reliably. Due to the difficulty of converging the heat flux and the complexity of the autocorrelation function in EMD method, NEMD method was used in the following simulations.

A SWCNT model with a finite length was built with the periodic simulation cell periodical in x and y directions. The model was divided into N regions along the z axial direction of the CNT, including a hot and a cold region at the two ends, respectively, as shown in Fig. 3.9. Based on the NEMD algorithm proposed by Ikeshoji and Hafskjold [63], a constant energy was added to the hot region and the same amount of energy was removed from the cold region. The velocities of atoms in both hot and cold regions were scaled to accomplish the energy transfer from the hot region to the cold region, in which both the energy and momentum are conserved for the whole system. The heat flux was given as follows:

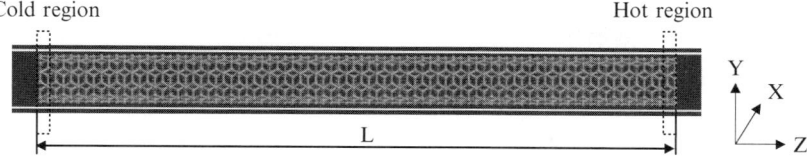

Fig. 3.9 MD model of the CNT with different lengths

$$J = \frac{\Delta E}{S \Delta t}, \qquad (3.10)$$

where J is the heat flux of the CNT, S is the cross-section area of the CNT, and Δt is the simulation time step.

The instantaneous local temperature in each region is given as follows [64]:

$$T_k = \frac{1}{3 n_k k_B} \sum_{i=1}^{n_k} m_i v_i^2, \qquad (3.11)$$

where nk is the number of atoms in region k, k_B is the Boltzmann's constant, and m_i and v_i are, respectively, the mass and velocity of atom i.

The temperature gradient along the CNT was then set up due to the heat flux imposed on the system. The thermal conductivity of the CNT can be calculated by the Fourier law:

$$\lambda = \frac{\langle J \rangle}{\langle \partial T / \partial Z \rangle}, \qquad (3.12)$$

where $\partial T / \partial Z$ is the temperature gradient along the CNT and the brackets denote a statistical time average.

To investigate the length effect on the thermal conductivity, MD models of the (5, 5) SWCNT with length varying from 16.6 to 37.1 nm were built. At the given temperature of 298 K, all the systems were initially equilibrated for 80 ps using NVT ensemble. An energy, $\Delta E = 3.0 \times 10^{-22}$ J, was then imposed on the hot region and removed from the cold region in the system and NEMD simulations were conducted based on the above method using NVT ensemble. NEMD simulation ran for a long time until a steady state temperature distribution was achieved along the SWCNT. All the simulations were performed with an interval of 1 fs in each MD simulation step. Velocities of atoms in each region were averaged over the last 1-ps period to obtain temperature distribution along the CNT. Figure 3.10a shows the temperature distribution along the SWCNT with the length of 16.6 nm.

Temperature is distributed linearly along the CNT for all the models. The temperature gradient $\partial T / \partial Z$ was obtained from the slope of the fitted straight line of the temperature profile. The cross-section area with a 3.4-Å thick annular ring was used to calculate heat flux. Based on the above equations, the thermal conductivities of the SWCNTs were calculated for different lengths and were plotted against

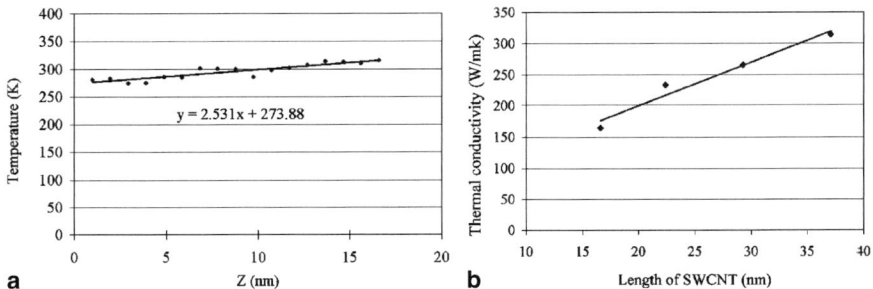

Fig. 3.10 (a) Temperature profile along a (5, 5) SWCNT and (b) thermal conductivity of SWCNT against CNT length for 298 K

SWCNT length, as shown in Fig. 3.10b. The value of the thermal conductivity increased with the increasing SWCNT length, which means that thermal conductivity of SWCNT is length dependent. There would be larger phonon scattering for the shorter SWCNT. Larger phonon scattering results in lowering the value of the thermal conductivity when the length of the SWCNT is shorter than the phonon mean free path (MFP). It is believed that the thermal conductivity would converge when the length of the SWCNT was larger than the MFP. The predicted values of the thermal conductivity were close to those values given by Maruyama [62]. Some other MD simulation results showed that thermal conductivity was temperature dependent [60, 61] and indicated that higher temperature was a dominating cause in enhancing phonon scattering [60].

The existence of defects in a CNT can affect both the stiffness and thermal conductivity of the CNT. There will be a strong impact on the performance of CNT as TIM in electronic packaging. Therefore, investigation of the effect of defects on the performance of CNT at a fundamental level is very important for the package design. For the SWCNT with the length of 16.6 nm, MD models with different kinds of defects in the SWCNT wall were built [33]. The configurations of these defects were presented in Fig. 3.11. The SWCNTs with different kinds of defects exhibit some degrees of discontinuity in the temperature profile at the defect location. Amongst them, the largest jump in the temperature profile results from the oxidation defect, while the smallest jump results from the Stone–Wales defect. Based on the above equations, the thermal conductivities of the SWCNTs were calculated for the SWCNT with different kinds of defects and were plotted, as shown in Fig. 3.12. It was found that all the values of the thermal conductivity for the SWCNT with defects were smaller than that of the defect-free SWCNT. The degradation of the thermal conductivity resulted from the increase of phonon scattering and the large temperature gradient caused by defects. MD simulation results showed consistency with the discussion by Che et al. [59] and Mingo and Broido [65] that defects scattered the long wavelength phonon more efficiently, and significantly reduced the thermal conductivity.

Although defects were also generated to achieve some designed functionalities [66–68], the thermal conductivity of CNTs can be drastically reduced by defects.

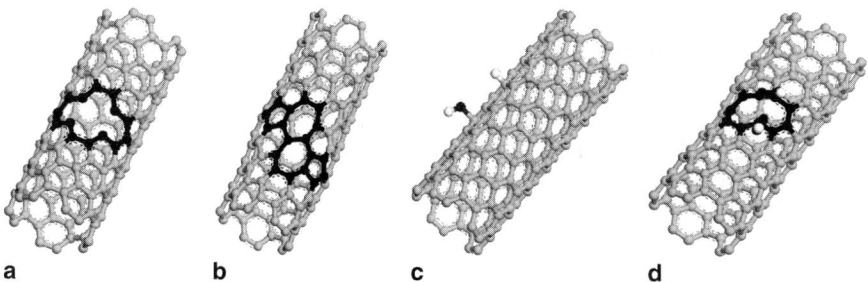

Fig. 3.11 Typical defects in a SWCNT: (**a**) vacancy defect; (**b**) Stone–Wales defects, i.e., seven- and five-membered rings instead of the normal six-membered ring in the CNT; (**c**) sp^3-hybridized defects which refer to the change from sp^2 to sp3 of a C–C bond and functionalized with –OH and –H group; and (**d**) C framework damaged by oxidation which leaves a hole with –COOH groups

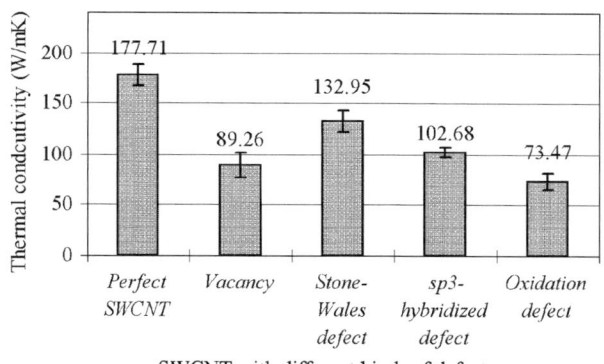

Fig. 3.12 Predicted thermal conductivity of the SWCNT with different kinds of defects at room temperature from MD simulation

Padgett and Brenner [69] found that the thermal conductivity of the SWCNT was drastically reduced by functionalized carbon atoms. They argued that increasing the heat transfer in a CNT-based polymer composite by chemical crosslinking may be counterproductive. Shenogin et al. [70] also found that chemical bonding between the polymer matrix and the CNT reduced not only the tube-matrix resistance but also the intrinsic tube conductivity. These results were all consistent with the results from MD simulations in this study, confirming that defects in CNT could heavily reduce the intrinsic thermal conductivity of SWCNT. There would be a significant effect on the thermal performance of CNT-based assemblies, especially CNT array as TIM in electronic packaging. The defect-induced ineffective thermal dissipation could threaten the reliability of the electronic packages. Therefore, special experimental treatments are needed to purify CNT and prevent defects occurring in the CNT wall, which is more important to enhance the thermal performance of CNT-based TIM in electronic packaging.

MD simulation results showed that thermal conductivity of SWCNT is dependent on the CNT length due to long phonon MFP. It was anticipated that the thermal conductivity would not converge until the CNT length is sufficiently greater than its MFP. It was also found that the thermal conductivity was heavily affected by defects in the CNT wall. The performance of CNT as TIM in electronic packages will be much lower due to the defects in CNT. What we have done is the first step in understanding the CNT performance, which is useful in experimental design of CNT-based TIMs used in electronic packaging. Further MD simulations will be focused on material performance of CNT-based epoxy compound and CNT assembly with metals or silicon in electronic packaging. MD simulation can provide information on performance of these CNT-based materials for the optimization of the structure before experimental trials.

3.7 Summary

With proper formulation of the MD model and appropriate use of boundary conditions, potential functions, and simulation procedure, MD simulation can provide good understanding of the material performance and interaction under different conditions. In spite of the large number of interfaces involved and scale issues, MD simulation has demonstrated provision of additional insight into the local interaction at material interfaces in microelectronic devices. The above sections have also illustrated that MD simulations can produce estimates of material properties in good agreement with the experimental values, such as Young's modulus, CTE, and T_g. It can also investigate phenomena which cannot be handled by experimental methods, such as moisture diffusion along the interface, the effect of the molecular structure causing structural weakness, defects, and defect generation.

Therefore, the fundamental knowledge obtained from MD simulation can contribute toward the development of a methodology that can provide detailed molecular interactions at local material interfaces in electronic packages. The knowledge is of vital importance to the development of next generation of IC packages with ever-decreasing feature size and more complex interconnect design, as in the case of system on chip (SOC) or system in package (SIP) designs. While the MD simulations are capable of generating in depth insight into the local molecular interactions, a consistent approach to relate the MD results to the results in an equivalent-continuum model is needed to provide a full understanding of the material performance across the scale range.

References

1. Alder BJ, Wainwright TE (1957) Phase Transition for a Hard Sphere System. J Chem Phys 27:1208–1211
2. Alder BJ, Wainwright TE (1959) Studies in Molecular Dynamics. I. General Method. J Chem Phys 31:459–466

3. Fukuda M, Kuwajima S (1997) Molecular-Dynamics Simulation of Moisture Diffusion in Polyethylene Beyond 10 ns Duration. J Chem Phys 107:2149–2159
4. Tsige M, Taylor PL (2002) Simulation Study of the Glass Transition Temperature in Poly(methyl methacrylate). Phys Rev E 65:021805(1–8)
5. Wang J, Li TL, Bateman SD, Erck R, Morris KR (2003) Modeling of Adhesion in Tablet Compression. I. Atomic Force Microscopy and Molecular Simulation. J Pharm Sci 92:798–814
6. Gou J, Minaie B, Wang B, Liang ZY, Zhang C (2004) Computational and Experimental Study of Interfacial Bonding of Single-Walled Nanotube Reinforced Composites. Comput Mater Sci 31:225–236
7. Yarovsky I, Evans E (2002) Computer Simulation of Structure and Properties of Crosslinked Polymers: Application to Epoxy Resin. Polymer 43:963–969
8. Yarovsky I (1997) Atomic Simulation of Interface in Materials: Theory and Applications. Aust J Phys 503:407–424
9. Haile JM (1992) Molecular Dynamics Simulation: Elementary Methods. New York: Wiley
10. Rapaport DC (2004) The Art of Molecular Dynamics Simulation (2nd edition). Cambridge: Cambridge University Press
11. Burghaus U (2006) A Practical Guide to Kinetic Monte Carlo Simulations and Classical Molecular Dynamics Simulations: An Example Book. New York: Nova Science Publishers
12. Mayo SL, Olafson BD, Goddard III WA (1990) DREIDING: A Generic Force Field for Molecular Simulations. J Phys Chem 94:8897–8909
13. Iwamoto N, Pedigo J (1998) Property Trend Analysis and Simulations of Adhesive Formulation Effects in the Microelectronics Packaging Industry Using Molecular Modeling. In Proceedings of the 48th IEEE Electronic Components and Technology Conference, USA, pp 1241–1246
14. Iwamoto N (2000) Applying Polymer Process Studies Using Molecular Modeling. In Proceedings of the 4th International Adhesive Joining and Coating Technology in Electronics Manufacturing Conference, pp 182–187
15. Iwamoto N (2000) Advancing Materials Using Interfacial Process and Reliability Simulations on the Molecular Level. Advanced Packaging Materials: Processes, Properties and Interfaces, pp 14–17
16. Iwamoto N (2000) Advancing Polymer Process Understanding in Package and Broad Applications Through Molecular Modeling. In Proceedings of the 50th IEEE Electronic Components and Technology Conference, USA, pp 1354–1359
17. Iwamoto N, Moro L, Bedwell B, Apen P (2002) Understanding Modulus Trends in Ultra Low k Dielectric Materials Through the Use of Molecular Modeling. In Proceedings of the 52nd Electronic Components and Technology Conference, San Diego, CA, pp 1318–1322
18. Iwamoto N, Lee E, Truong N (2004) New Metal Layers for Integrated Circuit Manufacture. Thin Solid Films 469–470:431–437
19. Iwamoto N (1999) Simulating Stress Reliability Using Molecular Modeling Methodologies. In Proceedings of the 32nd International Symposium on Microelectronics, Chicago, IL, pp 415–420
20. Iwamoto N (2000) Advancing Polymer Process Understanding in Package and Board Applications Through Molecular Modeling. In Proceedings of the 50th Electronic Components and Technology Conference, Las Vegas, NV, pp 1354–1359
21. Iwamoto N (2004) Molecular Modeling of IC Barrier Concerns. In Proceedings of the EuroSime 2004, Brussels, Belgium, pp 573–586
22. Iwamoto N, Bonne U (2006) Molecular Modeling of Analyte Adsorption on MEMS GC Stationary Phases. In Proceedings of the EuroSime 2006, Como, Italy, pp 749–757
23. Fan HB, Chan EKL, Wong CKY, Yuen MMF (2007) Molecular Dynamic Simulation of Thermal Cycling Test in Electronic Packaging. ASME J Electron Packag 129:35–40
24. Fan HB, Chan EKL, Wong CKY, Yuen MMF (2006) Investigation of Moisture Diffusion in Electronic Packages by Molecular Dynamics Simulation. J Adhes Sci Technol 20:1937–1947

25. Fan HB, Yuen MMF (2007) Material Properties of the Cross-Linked Epoxy Resin Compound Predicted by Molecular Dynamics Simulation. Polymer 48:2174–2178
26. Fan HB, Wong CKY, Yuen MMF (2006) Prediction of Material Properties of Epoxy Using Molecular Dynamic Simulation. In Proceedings of the EuroSimE 2006, Como, Italy
27. Fan HB, Wong CKY, Yuen MMF (2005) Reliability Prediction in Electronic Package Using Molecular Simulation. In Proceedings of the 55th Electronic Components and Technology Conference, Florida, USA, pp 1314–1317
28. Fan HB, Chan EKL, Wong CKY, Yuen MMF (2005) Thermal Cycling Simulation in Electronic Packages Using Molecular Dynamic Method. In Proceedings of the EuroSimE, Germany, pp 36–40
29. Wong CKY, Fan HB, Yuen MMF (2005) Investigation of Adhesion Properties of Cu–EMC Interface by Molecular Dynamic Simulation. In Proceedings of the EuroSimE, Berlin, pp 31–35
30. Fan HB, Chan EKL, Wong CKY, Yuen MMF (2006) Moisture Diffusion Study in Electronic Packaging Using Molecular Dynamic Simulation. In Proceedings of the 56th Electronic Components and Technology Conference, San Diego, USA, pp 1425–1428
31. Chan EKL, Fan HB, Yuen MMF (2006) Effect of Interfacial Adhesion of Copper/Epoxy Under Different Moisture Level. In Proceedings of the EuroSimE, Italy, pp 1–5
32. Zhang K, Fan HB, Yuen MMF (2006) Molecular Dynamics Study of the Thermal Conductivity of CNT-Array-Thermal Interface Material. In Proceedings of the EMAP, Hong Kong, pp 113–116
33. Fan HB, Zhang K, Yuen MMF (2006) Effect of Defects on Thermal Performance of Carbon Nanotube Investigated by Molecular Dynamics Simulation. In Proceedings of the EMAP, Hong Kong, pp 451–454
34. Fan XJ, Wang HB, Lim TB (2001) Investigation of the Underfill Delamination and Cracking in Flip-Chip Modules Under Temperature Cyclic Loading. IEEE Trans Compon Packag Technol 24:84–91
35. Basaran C, Chandaroy R (1997) Finite Element Simulation of the Temperature Cycling Tests. IEEE Trans Compon Hybr Manuf Technol A 20:530–536
36. Cho K, Cho EC (2000) Effect on the Microstructure of Copper Oxide on the Adhesion Behavior of Epoxy/Copper Leadframe Joints. J Adhes Sci Technol 14:1333–1353
37. Chung PW, Yuen MMF, Chan PCH, Ho NKC, Lam DCC (2002) Effect of Copper Oxide on the Adhesion Behavior of Epoxy Molding Compound–Copper Interface. In Proceedings of the 52nd IEEE Electronic Components and Technology Conference, pp 1665–1670
38. Su YY, Shemenski RM (2000) The Role of Oxide Structure on Copper Wire to the Rubber Adhesion. Appl Surf Sci 161:355–364
39. Kendall K (2001) Molecular Adhesion and Its Applications: The Sticky Universe. Kluwer/Plenum, New York
40. Van Landingham MR, Edujee RF, Gillespie JW (1999) Moisture Diffusion in Epoxy System. J Appl Polym Sci 71:787–798
41. Soles CL, Chang FT, Gidley DW, Yee AF (2000) Contributions of the Nanovoid Structure to the Kinetics of Moisture Transport in Epoxy Resins. J Appl Polym Sci Part B: Polym Phys 38:776–791
42. Soles CL, Yee AF (2000) A Discussion of the Molecular Mechanisms of Moisture Transport in Epoxy Resins. J Appl Polym Sci Part B: Polym Phys 38:792–802
43. Comyn J, Groves C, Saville R (1994) Durability in High Humidity of Glass to Lead Alloy Joints Bonded with and Epoxide Adhesive. Int J Adhes Adhes 14:15–20
44. Zanni-Deffarges M, Shanahan M (1995) Diffusion of Water into an Epoxy Adhesive: Comparison Between Bulk Behavior and Adhesive Joints. Int J Adhes Adhes 15:137–142
45. Bowditch MR (1996) The Durability of Adhesive Joints in the Presence of Water. Int J Adhes Adhes 16:73–79
46. Tanmai Y, Tanaka H, Nakanishi K (1994) Molecular Simulation of Permeation of Small Penetrants Through Membranes. 1. Diffusion Coefficients. Macromolecules 27:4498–4506

47. Fukuda M, Kuwajima S (1997) Molecular-Dynamics Simulation of Moisture Diffusion in Polyethylene Beyond 10 ns Duration. J Chem Phys 107:2149–2159
48. Braesicke K, Steiner T, Saenger W, Knapp EW (2000) Diffusion of Water Molecules in Crystalline -Cyclodextrin Hydrates. J Mol Graphics Model 18:143–152
49. Müller-Plathe F (1998) Diffusion of Water in Swollen Poly (Vinyl Alcohol) Membranes Studied by Molecular Dynamic Simulation. J Membrane Sci 141:147–154
50. Hofmann D, Fritz L, Ulbrich J, Schepers C, Bohning M (2000) Detailed-Atomistic Molecular Modeling of Small Molecule Diffusion and Solution Processes in Polymeric Membrane Materials. Macromol Theory Simul 9:293–327
51. Swope WC, Andersen HC, Berens PH, Wilson KR (1982) A Computer Simulation Method for the Calculation of Equilibrium Constant for the Formation of Physical Cluster of Molecules. J Chem Phys 76:637–649
52. Williams JL, Hopfenberg HB, Stannett V (1969) Water Transport and Clustering in Poly-(Vinylchloride), Poly-(Oxymethylene) and Other Polymers. J Macromol Sci Phys B 3:711–725
53. Fan CF, Hsu SL (1992) Application of the Molecular Simulation Technique to Characterize the Structure and Properties of an Aromatic Polysulfone System. 2. Mechanical and Thermal Properties. Macromoleculars 25:266–270
54. Hamerton I, Heald CR, Howlin BJ (1996) Molecular Modelling of a Polyarylethersulfone Under Bulk Conditions. Model Simul Mater Sci Eng 4:151–159
55. Xie J, Gironcoli SD, Baroni S, Scheffler M (1999) First-Principles Calculation of the Thermal Properties of Silver. Phys Rev B 59:965–969
56. Yang L, Srolovitz DJ, Yee AF (1999) Molecular Dynamic Study of Isobaric and Isochoric Glass Transitions in a Model Amorphous Polymer. J Chem Phys 110:7058–7069
57. Yoshioka S, Aso Y, Kojima S (2003) Prediction of Glass Transition Temperature of Freeze-Dried Formulations by Molecular Dynamic Simulation. Pharmaceut Res 20:873–878
58. Literature from Resolution Performance Products, http://www.resins.com/resins/am/pdf/SC0772.pdf (2005)
59. Che J, Cagin T, Goddard III WA (2000) Thermal Conductivity of Carbon Nanotube. Nanotechnology 11:65–69
60. Bi K, Chen Y, Yang J, Wang Y, Chen M (2006) Molecular Dynamics Simulation of Thermal Conductivity of Single-Wall Carbon Nanotube. Phys Lett A 350:150–153
61. Osman MA, Srivastava D (2001) Temperature Dependence of the Thermal Conductivity of Single-Wall Carbon Nanotube. Nanotechnology 12:21–24
62. Maruyama S (2003) A Molecular Dynamics Simulation of Heat Conduction of a Finite Length Single-Walled Carbon Nanotube. Microscale Thermophys Eng 7:41–50
63. Ikeshoji T, Hafskjold B (1994) Non-Equilibrium Molecular Dynamics Calculation of Heat Conduction in Liquid and Through Liquid–Gas Interface. Mol Phys 81:251–261
64. Müller-Plathe F (1997) A Simple Nonequilibrium Molecular Dynamics Method for Calculating the Thermal Conductivity. J Chem Phys 106:6082–6085
65. Mingo N, Broido DA (2005) Length Dependence of Carbon Nanotube Thermal Conductivity and the "Problem of Long Waves". Nano Lett 5:1221–1225
66. Hu Y, Jang I, Sinnott SB (2003) Modification of Carbon Nanotube-Polystyrene Composites Through Polyatomic-Ion Beam Deposition. Compos Sci Technol 63:1663–1669
67. Miko C, Milas M, Seo JW, Couteau E, Barisic N, Gaal R, Forro L (2003) Effect of Electron Irradiation on the Electrical Properties of Fibers of Aligned Single-Walled Carbon Nanotubes. Appl Phys Lett 83:4622–4624
68. Bockrath M, Liang W, Bozovic D, Hafner JH, Lieber CM, Tinkham M, Park H (2001) Resonant Electro Scattering by Defects in Single-Walled Carbon Nanotube. Science 291:283–285
69. Padgett CW, Brenner DW (2004) Influence of Chemisorption on Thermal Conductivity of Single-Walled Carbon Nanotubes. Nano Lett 4:1051–1053
70. Shenogin S, Bodapati A, Xue L, Ozisik R, Keblinski P (2004) Effect of Chemical Functionalization on Thermal Transport of Carbon Nanotube Composites. Appl Phys Lett 85:2229–2231

Chapter 4
Advances in Delamination Modeling

O. van der Sluis(✉), C.A. Yuan, W.D. van Driel, and G.Q. Zhang

4.1 Introduction

Today's microelectronic packages are typically composed of various materials, like silicon, metals, oxides, glues, and compounds (polymers). In Fig. 4.1, cross sections of a leadframe- and substrate-based package are depicted. Due to the dissimilar nature of these materials and the inherent presence of a large number of interfaces in each component, various failure modes, such as interface delamination, chip cracking, and/or solder fatigue, will occur during processing (qualification), testing, or usage. The occurring thermomechanically related failures in these components account for more than 65% of the total reliability issues [1].

Future generations of these electronic components will be subject to increasing miniaturization down to the nanoscale, increasing levels of system and function integration, introduction of new materials, cost reduction, and shorter time to market. In this context, virtual prototyping and virtual qualification provide a framework to generate optimized designs, resulting in less trial-and-error-based design cycles, thus reducing development costs. Reliable and efficient numerical models and advanced simulation-based optimization methods are therefore required [2, 3].

This chapter discusses the first topic: *toward reliable and efficient numerical models*. Continuum mechanical failure models will be discussed first. Typically, these models are implemented in a finite element (FE) framework from which the resulting thermomechanical response of a product with appropriate boundary and processing conditions can be calculated. However, due to the decreasing dimensions of the materials, the characterization of their "bulk" and interface properties becomes more critical [4]. In this respect, atomistic-based modeling methods could provide a means to predict and understand these properties, and ultimately, to develop materials, interfaces, and components with tailored properties. The basics and challenges of molecular dynamics (MD) will also be discussed in this chapter. Although MD will provide

---

O.van der Sluis
Department of Precision and Microsystems Engineering, Delft University of Technology,
Mekelweg 2 2628 CD, Delft, The Netherlands

Philips Applied Technologies, High Tech Campus 7, 5656 AE Eindhoven, The Netherlands

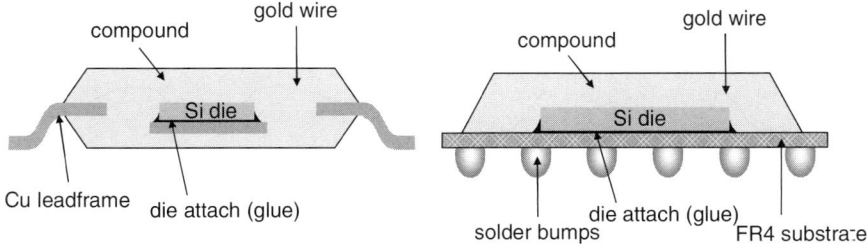

Fig. 4.1 Cross section of two typical package families: leadframe- (*left*) and substrate-based (*right*)

valuable information regarding these properties, the size of the models is still restricted to a very small region of the components under consideration. Hence, the combination of both modeling methods will provide a very powerful multiscale framework for the virtual prototyping of micro/nanoelectronic components.

4.2 Delamination-Related Failures in Micro/Nanoelectronics

As already mentioned above, various failure modes can occur in micro/nanoelectronic devices. A division can be made between failures at device level (the die) and failures at package level. Examples of failures at device level are electromigration, electrostatic discharge (ESD) damage, pattern shift (or metal shift), back-end structure delamination, and Kirkendahl voiding. At the package level, examples are stitch break, die lifting, body cracking, solder joint fatigue, and delamination at various interfaces. Figure 4.2 shows several examples of actually observed failures that originate from delaminating interfaces. As mentioned in the introduction, the major trends in the electronics industry are miniaturization and function integration. The smallest feature sizes on today's dies are already falling to 90 nm and lower, and from the packaging point of view a similar trend is visible. Die thickness will decrease below 50 μm, wire diameter below 15 μm, interconnect pitch below 20 μm, copper film thickness below 10 μm, and via diameter in substrates below 20 μm. If the industry is going to rapidly decrease these sizes at high-volume production during the next 15 years, then the need for delamination prediction needs our attention today. Not only those material combinations having sufficient interface toughness should be selected, but also design choices should be made resulting in components that are able to withstand increased forces due to the occurrence of delamination.

4.3 Continuum-Based Interface Delamination Modeling

In this section, an overview and typical applications of continuum-based failure models are given.

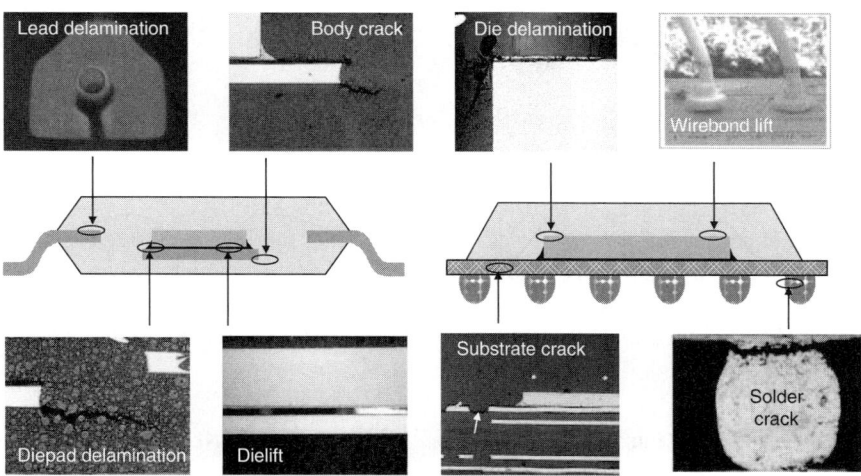

Fig. 4.2 Examples of failures for the leadframe-based (*left*) and substrate-based (*right*) families

4.3.1 Interface Fracture Mechanics

Fracture mechanics theory provides a way to evaluate if an already present (or assumed) crack of given geometry and location in a sample is critical. To this end, the energy release rate (ERR), also called crack driving force, as formulated in Griffith's energy balance, is calculated (e.g., [5]):

$$\frac{dW}{da} - \frac{dU}{da} = \frac{d\Gamma}{da}. \tag{4.1}$$

Here, W is the external work, U is the elastic energy, and Γ is the energy required for crack growth, while a is the crack length. Dividing the left-hand side by the specimen thickness B yields the energy release rate G (J m^{-2}). The right-hand side of (4.1) is also called the crack resistance force (after division by B). The criterion for crack growth is $G \geq G_c$, where G_c is the fracture toughness (singular) stress field at the crack tip. In interface fracture mechanics, appropriate when dealing with delamination issues, this factor is a complex variable: $K = K_1 + iK_2$ [6]. As a result, the crack-tip stresses are (e.g., [7])

$$\sigma_{22} = \frac{\Re\{Kr^{i\varepsilon}\}}{\sqrt{2\pi r}}, \quad \sigma_{12} = \frac{\Im\{Kr^{i\varepsilon}\}}{\sqrt{2\pi r}}. \tag{4.2}$$

Here, σ_{22} is the normal stress, σ_{12} is the shear stress, r is the distance from the crack tip, and ε is the bimaterial constant, defined as

$$\varepsilon = \frac{1}{2\pi}\ln\left(\frac{1-\beta}{1+\beta}\right), \tag{4.3}$$

in which β is the second Dundur's parameter. The ERR is determined from the stress intensity factor according to

$$G = \frac{1-\beta^2}{E_*}(K_1^2 + K_2^2), \qquad (4.4)$$

where $E_*^{-1} = 0.5(\overline{E}_1^{-1} + \overline{E}_2^{-1})$.

From experiments, it is now well established that the interface toughness is a function of the direction of the loading at the crack tip $G_c(\psi)$, which is characterized by the mode angle ψ (e.g., [7, 8]). The mode angle is defined as the ratio of shear to normal tractions, transmitted across the interface at a reference distance l ahead of the crack tip:

$$\Psi = \tan^{-1}\left[\frac{\sigma_{12}(\ell,0)}{\sigma_{22}(\ell,0)}\right] = \tan^{-1}\left[\frac{\Im(K\ell^{i\varepsilon})}{\Re(K\ell^{i\varepsilon})}\right] = \arg(K\ell^{i\varepsilon}). \qquad (4.5)$$

Clearly, the calculation of the mode angle at an interface requires a choice for the reference length value, and should therefore always be supplied when reporting interface toughness values. However, since from (4.5), it can be shown that $\Psi_2 = \Psi_1 + \varepsilon \ln(\ell_2/\ell_1)$ (in which ψ_i is associated with ℓ_i), toughness values can always be transformed.

Within a FE framework, for linear elastic materials, the ERR can be calculated by the so-called J-integral [5]. For accurate calculation of the J-integral value, the singularity, denoted by r^λ where λ is the order of the singularity, should be captured properly. For homogeneous materials, for which λ equals 0.5, Barsoum [9] has shown that the singularity can be described exactly when using so-called quarter point elements. However, as pointed out by Abdel-Wahab and de Roeck [10], this element cannot be used for λ values other than 0.5, unless an extremely fine mesh is used. This is confirmed by He et al. [11], who show that for interface cracks, convergence upon mesh refinement is obtained.

Several alternative methods exist to calculate G, like the virtual crack extension method [12] and the virtual crack closure method [13, 14]. For the calculation of the stress intensity factors, necessary for the determination of the mode angle, several methods have been developed: the interaction integral method proposed by Shih and Asaro [15] and the crack surface displacement method by using the crack-tip opening displacements [16]. Due to the scope of this chapter, these will not be discussed here.

As input for the finite element models, the interface toughness function $G(\psi)$ should be measured. In standard tests like double cantilever beam (DCB), three-point bending (TPB), and four-point bending (FPB), the mode mixity can be altered by changing the dimensions of the sample. Clearly, this requires different samples for different mode angles. For the purpose of characterizing interfaces in the full range of mode mixity using a single sample, a mixed-mode bending setup has been developed by Thijsse et al. [8]. The apparatus proposed by Reeder and Crews [17] has been modified to measure small forces accurately, due to the small dimensions of typical semiconductor samples. The modified setup is depicted in Fig. 4.3.

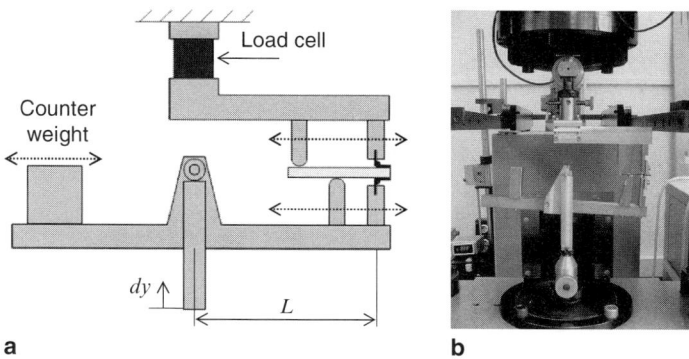

Fig. 4.3 Mixed-mode bending setup: (**a**) schematic and (**b**) picture of the actual setup

By changing the position of the force acting on the loading beam L, the mode mixity can be controlled and consequently, $G(\psi)$ can be determined. For more details, the reader is referred to [8]. More recently, an improved version of the mixed-mode bending setup has been suggested by Xiao et al. [18].

4.3.2 Cohesive Zone Elements

In contrast with fracture mechanics theory, cohesive zone modeling, based on the pioneering ideas of Dugdale [19] and Barenblatt [20], does not require an initial crack to be present. Furthermore, this technique can be used to describe both crack initiation and propagation. Finally, it is versatile in the sense that it can be applied to brittle and ductile failure behavior. The idea for developing cohesive zone models is that infinite stresses at the crack tip do not exist. Instead, the crack is divided into a stress-free region and a stressed region around the crack tip, loaded by so-called cohesive stresses. Fracture is regarded as a gradual process in which failure occurs across an extended crack tip, or cohesive zone, and is resisted by cohesive tractions [21]. In a finite element context, cohesive zone elements are placed between continuum elements. Hence, this method is extremely appealing when considering interface delamination issues [21, 22].

The cohesive tractions are calculated from a so-called traction–separation law (TSL), providing the relation between the separation vector $\vec{\delta}$ at the interface and the traction vector $\vec{t}$ (see Fig. 4.4).

Here, we follow van Hal et al. [23], in which a Smith–Ferrante model is employed. Furthermore, irreversibility is taken into account based on the framework provided by Ortiz and Pandolfi [21]. The TSL is given by

$$t = t_{max} \frac{\delta}{\delta_c} \exp\left(1 - \frac{\delta}{\delta_c}\right). \tag{4.6}$$

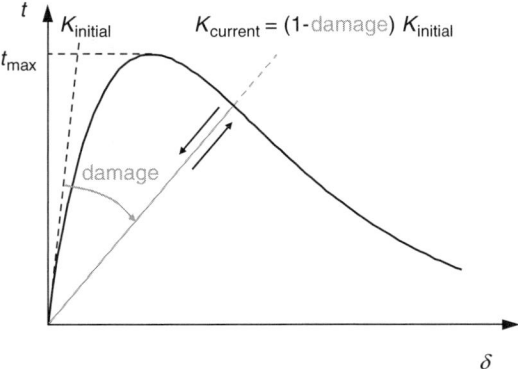

Fig. 4.4 A typical (exponential) traction–separation law. The current stiffness defines the elastic loading/unloading behavior

Here, t is the equivalent traction and δ is the effective separation, defined as

$$\delta = \sqrt{\langle \delta_n \rangle^2 + \beta^2 \delta_s^2}. \qquad (4.7)$$

The parameter δ_c is the effective displacement at which the maximum traction t_{max} is reached. The constitutive parameter β defines the ratio between the shear and normal critical tractions. As a result of the MacAuley brackets in (4.7), defined as $\langle x \rangle = 0.5(x+|x|)$ only positive normal separations influence the effective separation. It can be shown that the work of separation can be expressed as [21]

$$G_c = \int_0^\infty t \, d\delta. \qquad (4.8)$$

This cohesive zone element has been implemented as user element in the commercial finite element package MSC.Marc. For more detail, the reader is referred to [23, 24].

In several papers, it is emphasized that, to get meaningful results with cohesive zone elements, the size of the process zone should be accurately captured by the finite element discretization [4, 25]. In [25], it is argued to use ten elements within the process zone. However, when dealing with brittle interface behavior, as is the case in typical semiconductor applications, this severely limits the applicability of this method due to the mesh requirements, both in 2D and 3D applications. In addition, possible limit points (snap-through and snap-back) could be present caused by the combination of the brittle nature of the interfaces and the localized nature of the delamination. However, it has been shown by van Hal et al. [24] that, when using a conventional arc-length control method like the cylindrical one [26], sharp limit point as a result of the small localization zones (i.e., the interfaces) cannot be passed. However, when applying a *local* arc-length control method, limit points can be passed efficiently and accurately [24].

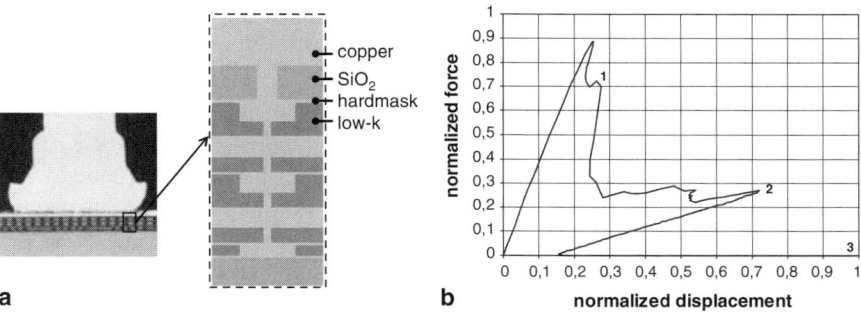

Fig. 4.5 Two-dimensional IC back-end model including cohesive zones: (**a**) cross section of a back-end structure and corresponding (enlarged) model and (**b**) resulting force–displacement curves showing several limit points

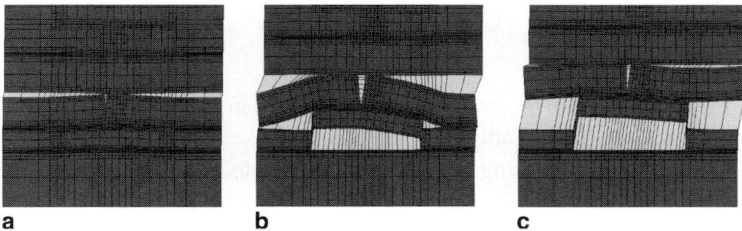

Fig. 4.6 Zoomed region of the back-end model showing several deformation stages of the delamination behavior of the back-end structure. The stages (**a**)–(**c**) correspond to (1)–(3) in Fig. 4.5

As an example, a two-dimensional model of an IC back-end structure has been generated in which all interfaces have been taken into account by cohesive zone elements (of course, with appropriate interface toughness values), see Fig. 4.5a. Figure 4.5b shows the resulting force–displacement curve illustrating the severe snap-back behavior. Several typical deformation stages (1, 2, 3) are depicted in Fig. 4.6 as deformed stages.

Due to the promising results of the developed local arc-length control method, currently, this methodology is being extended to three dimensions. In this way, the method can be applied to realistic geometries.

4.3.3 Area Release Energy Criterion

In Sect. 4.3.2, methods based on fracture mechanics and cohesive zones have been discussed. The disadvantage of fracture mechanics-based approaches lies in the fact that a crack with given size and location has to be defined a priori whereas the application of cohesive zone models to real 3D structures is not yet possible due to the fact that the current implementation of the robust local arc-length control is in two dimensions only. Alternatively, for the purpose of analyzing, ranking, and optimizing complex three-dimensional structures (e.g., a back-end structure), an energy-based method has been developed, called the area release energy (ARE) [27, 28]. This ARE criterion has the following advantages compared with the already discussed approaches (1) damage sensitivity analysis of complex three-dimensional structures is possible: it allows an instant overview of the critical areas within these structures; (2) it is not required to assume a pre-existing defect at a specific location with a specific size, as is the case in traditional fracture mechanics approaches.

For explanation purposes, a two-dimensional schematic is depicted in Fig. 4.7. For each node i on the interface, a number of nodes n will be released that lie within the area bounded by 2ℓ In a 3D setting, this area equals $\pi\ell^2$, corresponding to a circular area (i.e., a penny-shaped crack). The ARE value for each node i is calculated as,

$$G_i^{ARE} = \frac{1}{2A_i} \sum_{j=1}^{n} \vec{F}_j^T [\vec{u}]_j, \qquad (4.9)$$

in which $\vec{F}_j$ is the force vector acting on node j (before release) and $[\vec{u}]_j$ is the resulting separation vector (after release). The released surface is denoted by A_j. Clearly, the reason for releasing a predefined area instead of simply releasing individual nodes separately is to prevent mesh dependency of the energy release values in nonuniform grids. As a result, convergence of the energy values is obtained upon mesh refinement, as shown in [28].

From (4.9), it can be concluded that the ARE value does not correspond to the ERR value, the latter being defined as the energy released from a crack length a to $a + da$. For this reason, the ARE method has been extended to calculate ERR values in an accurate way while retaining the original advantages of the ARE method. Preliminary results are reported in [29].

The authors would like to point out that currently, another numerical method has emerged which, in essence, facilitates modeling of any kind of discontinuities (e.g., cracks or material interfaces) without the need for special remeshing steps. This method is based on the partition-of-unity principle and is called X-FEM [30]. More specifically, for crack propagation analysis, both the crack discontinuity and the singular crack-tip field are implicitly embedded in the element formulation.

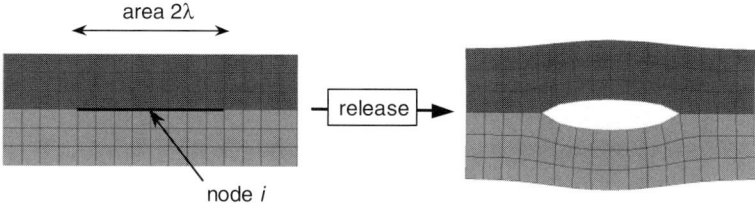

Fig. 4.7 Two-dimensional representation of the ARE method [27, 28] (with permission)

4.3.4 J-Integral Application: Delamination-Related Problems in Exposed PadPackages

Exposed pad packages were introduced in the late 1980s–1990s because of their excellent thermal and electrical performance. An exposed pad package is a package composed of a die attached to an exposed pad and in a later stage encapsulated with an epoxy molding compound. Figure 4.8 shows two examples for a gull wing-exposed pad package and a quad flat no-lead (QFN) package. Exposed pads increase the maximum power dissipation of packages due to their increased thermal performance. In most applications, the exposed pad is used as an electrical ground. To do so, so-called down-bonded wires are attached from the die to the exposed die pad.

Despite these advantages, a lot of delamination-related reliability problems are observed during qualification and testing of the exposed pad family, such as die lift (see examples in Fig. 4.2). These reliability problems are driven by the mismatch between the different material properties, hygroswelling, vapor pressure-induced expansion, and degradation of the interfacial strength due to moisture absorption which can be taken into account by numerical models that are able to predict the moisture diffusion, deformation, stress, and interfacial loading history as functions of processes, temperature and moisture loading [31, 32]. From experimental point of view, we used the four-point bending test to determine the interface fracture toughness between molding compound and leadframe. We found that at 20°C and dry conditions, the adhesion strength between compound and leadframe is approximately 6 J m^{-2} under a mode mixity of 38°C [33].

For predicting delamination growth, the *J*-integral method is applied to calculate the ERR at the interface as a function of hygrothermomechanical loading at different interfaces within the package (see Fig. 4.9).

Figure 4.10 shows the fully parametric 2D FE model with a typical crack-tip mesh. Through a mesh sensitivity analysis, the eventually used mesh size is determined.

Figure 4.11 shows the calculated *J*-integral values at the different locations as a function of the loading conditions. In this figure, 0% is the starting point at the side of the pad, 20% is the corner point, 40% is the point of the die attach fillet, 60% is exactly below the die, and 100% is at the symmetry line.

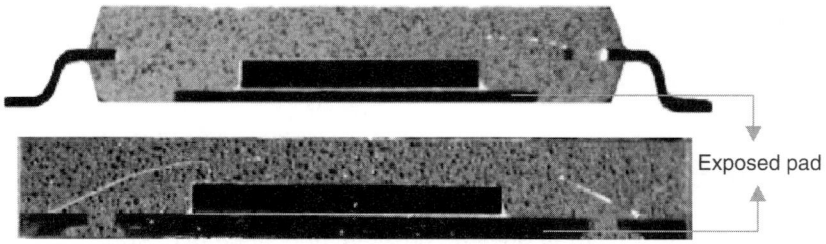

Fig. 4.8 Examples for exposed pad packages: gull wing leads (*top*) and QFN version (*bottom*)

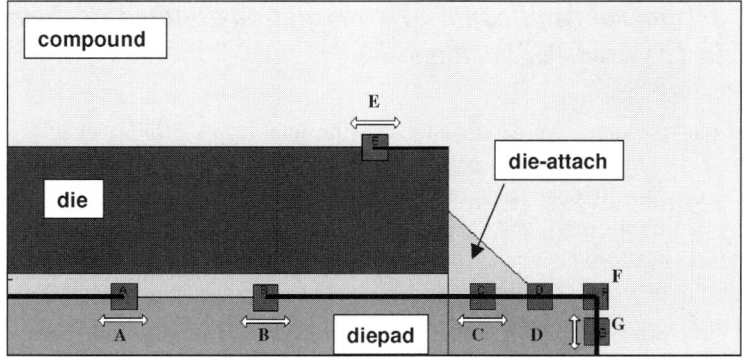

Fig. 4.9 *J*-integral values are calculated at locations A–G

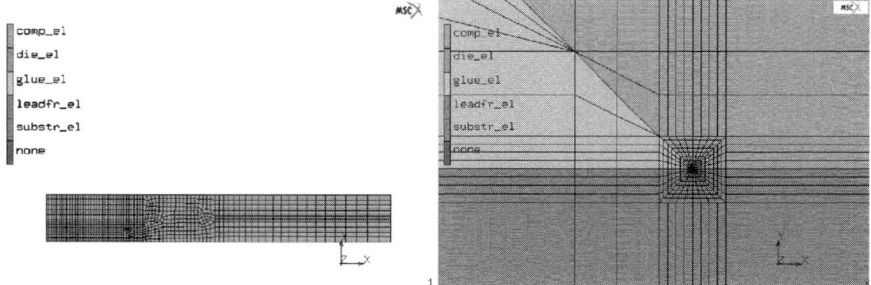

Fig. 4.10 FE mesh for the low profile quad flat package (HLQFP) (*left*) and a typical crack-tip mesh at location D (*right*)

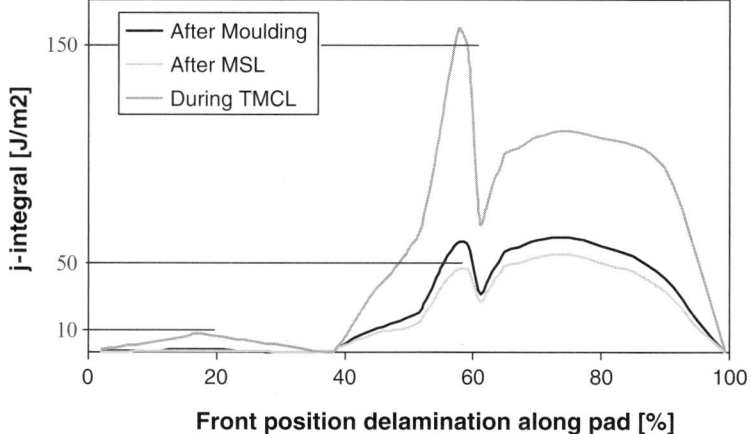

Fig. 4.11 ERR values at the different process conditions as a function of the location

The following can be concluded from this figure. At the side of the pad (locations G and F), the ERR values are below 5 J m^{-2} during processing, indicating that this interface will not fail from a thermomechanical point of view. During thermal cycling (TMCL), the ERR values increase to 10 J m^{-2} and are getting closer to the toughness values. During cool down from the molding temperature, the ERR values at the locations B and C (i.e., interface die attach with leadframe) increased dramatically and seem to exceed the measured values. Especially at location C, directly below the die corner, the ERR values rise until 50 J m^{-2} after molding and 150 J m^{-2} during TMCL testing and are beyond the measured toughness values. During moisture sensitivity level (MSL) testing, the ERR values drop due to the expansion of the compound as a result of moisture uptake. Upon swelling, the compound closes the interface and ERR values decrease. The effect of the moisture is purely degrading the interface toughness with 20–40%. The results indicate that delamination will occur at the die attach border and has the tendency to progress until point B, but not until point A.

The results indicate that when die pad delamination is present, cracks are likely to grow beneath the die and die lift will occur. The interaction between die lift and other failure modes, such as lifted ball bonds, is not found to be very significant. To prevent delamination, it is vital to secure processing conditions of the exposed pad family by proper curing of molding compound and/or die attach materials to obtain sufficient interface toughness.

4.3.5 Cohesive Zone Application: Buckling-Driven Delamination in Flexible Electronics

A frequently encountered failure mode in flexible electronics as illustrated in Fig. 4.12a is buckling-driven delamination between a thin layer (e.g., ITO or SiNx) and another functional layer or substrate (e.g., BCB, PI), as reported in [34] (see Fig. 4.12b). From an initial region of bad adhesion between film and substrate,

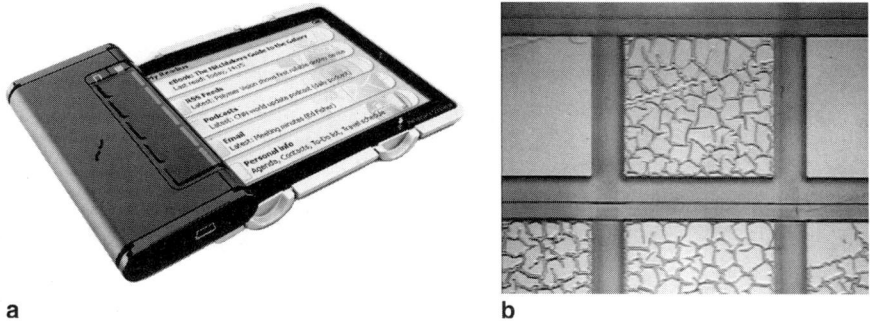

Fig. 4.12 (a) Example of a flexible display from Polymer Vision (with permission from http://www.polymervision.com) and (b) observed buckling-driven delamination within pixels [34] (with permission)

buckling of the film occurs due to the compressive stresses. Consequently, the delaminated region propagates due to the loading from the buckle. This failure mode is caused by the coupling of buckling and interfacial delamination [7]. In the past two decades, a huge amount of research has been devoted to the failure behavior of thin solid films; see for instance [7, 35–37].

Typical dimensions of the flexible samples are: the thickness of the thin layer is in the order of 100–800 nm, whereas the substrate has a thickness of 100–500 μm. To realize fully functional flexible displays (in fact, this is the goal of the European Project Flexidis [38]), these occurring failure modes should be understood and prevented. For this purpose, interface characterization methods and numerical tools are developed.

Due to the fact that the used materials are thin, standard available experimental methods (e.g., peel testing) cannot be applied. For this reason, a two-point bending setup has been developed by Bouten [34] and Abdallah et al. [35], which is illustrated in Fig. 4.13a. As a result, the buckling morphology can be studied as a function

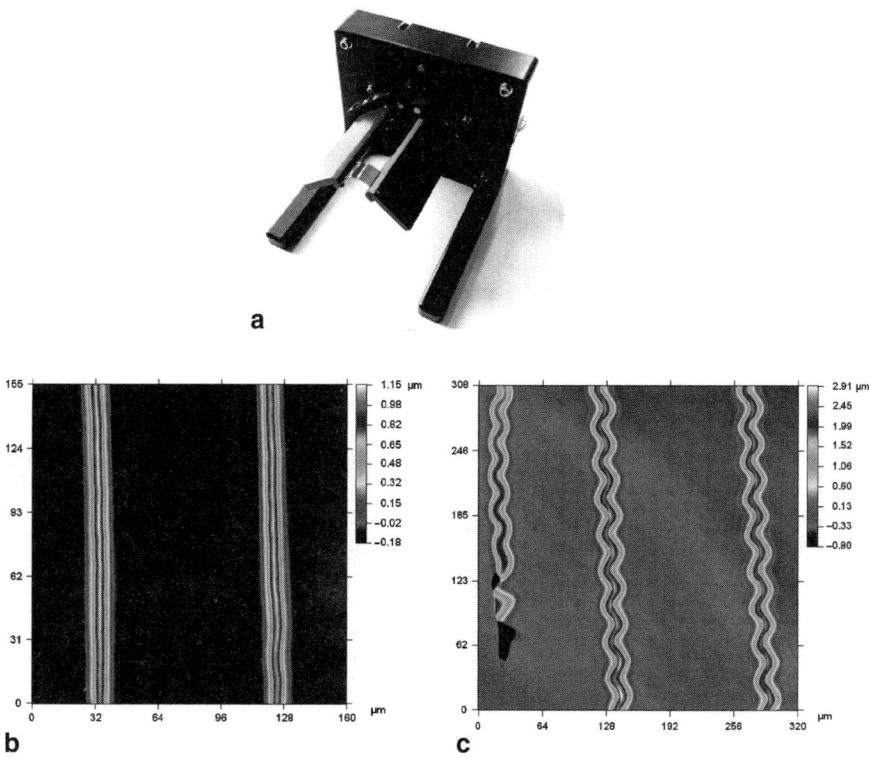

Fig. 4.13 (**a**) The mechanical two-point bending device of Bouten and Abdallah; confocal microscope images of (**b**) straight and (**c**) telephone cord buckle morphology. The bending strain is increased from (**b**) to (**c**). Pictures are courtesy of A. Abdallah and C. van Rekum

of prescribed strain [35], as indicated in Fig. 4.13b, c. By increasing the curvature, and thus the compressive strain in the sample, the straight buckles transform, under certain circumstances, into the telephone cord morphology.

Using approximate analytical solutions, based on the models of Hutchinson and Suo [7], an estimate can be made for the value of the interface strength. However, due to the assumptions made in these analytical models (e.g., the numbers of layers and elastic material behavior), their applicability is restricted. Therefore, numerical models have been developed that do not suffer from these restrictions [39]. An example of a two-dimensional two-layer system is depicted in Fig. 4.14. In fact, this model is a local model of the compressed area in the multilayered system. Initiation and propagation of interface failure is modeled using the above-mentioned cohesive zone elements. To trigger buckling, an initial geometric imperfection has been inserted in the model. To validate the numerical model, a benchmark problem has been simulated and is reported in [39]. It was found that the numerical model captures the buckling-driven delamination accurately.

This model will be used to estimate the interface work of separation and residual strains in the layer by comparing calculated buckle geometries with measured experimental values of the buckle height (w) and buckle width ($2b$). As was already remarked, the used experimental setup is a two-point bending testing device, as illustrated in Fig. 4.13a. Currently, the measuring of the buckle geometry is performed after straightening of the specimen as it is not yet possible to accurately measure the buckle geometry during and after loading. Therefore, the simulations should take into account the history dependency of the materials that define the loading/unloading behavior of the specimen: the plastic behavior of the substrate is

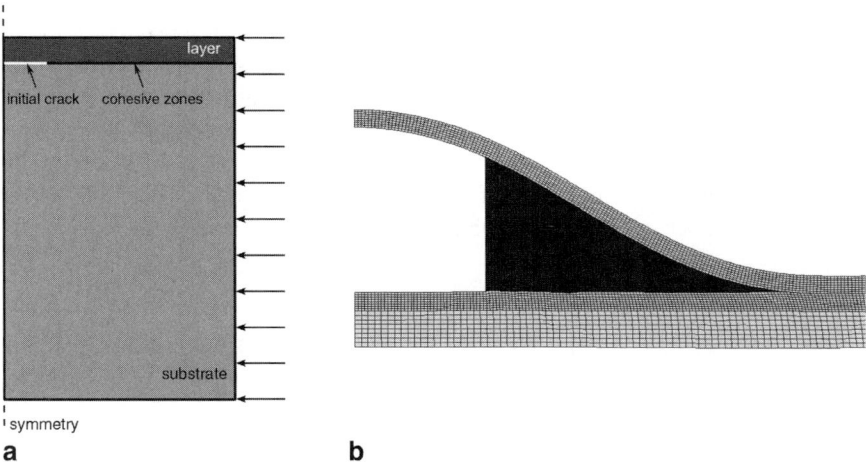

Fig. 4.14 (**a**) Two-dimensional buckling delamination model and (**b**) zoom of deformed geometry including the delamination region. The contours indicate the damage value in the cohesive zone elements: a value of 1.00 corresponds to a crack

Fig. 4.15 Buckle geometry at the loaded state and the straightened state

taken from [36] whereas the irreversible behavior of the cohesive zones is modeled according to [21]. Furthermore, the layer is assumed to deform elastically.

To illustrate the effect of loading and subsequent straightening, the buckle geometry at loaded and straightened state is depicted in Fig. 4.15 for t_{max} = 100 MPa. It can be observed that the decrease in buckle height as a result of the straightening is rather significant. Furthermore, the value of t_{max} influences the results, caused by the fact that the plastic zone at the delamination front is determined by the substrate, the ITO layer, and the cohesive zones. It is found that higher values of t_{max} result in higher plastic strain values in the substrate due to the higher stress levels transmitted through the cohesive zones and a smaller buckle width. For t_{max} = 100 MPa, the maximum plastic strain is 7.5% whereas for t_{max} = 200 MPa, this maximum value is 32%.

The experimentally measured buckle width $2b$ = 3.3 μm and height w = 0.14 μm. Comparing these values with the numerically determined results, the work of separation and total compressive strain values are G = 37 J m^{-2} and = 1.82% (for t_{max} = 100 MPa). These results are obtained with an initial imperfection length $2b_0$ = 2 μm. Thus, the developed experimental–numerical methodology enables estimation of the interface work of separation of thin film structures, taking into account the history-dependent material behavior and the loading–straightening conditions.

Future work is focused on:

1. The explanation of the occurring transition from straight to telephone cord buckling, as reported by Abdallah et al. [35] (see Fig. 4.13). For this purpose, in analogy with Jensen and Sheinman [40], numerical models will be developed which should take into account the dependency of the interface toughness on the mode angle and the compressive stress levels, which is, besides the increasing compressive stress, hypothesized as being the main driver for this transition (e.g., [35, 40]).
2. Following Hutchinson et al. [41] and Moon et al. [42], the influence of the geometry (i.e., shape and size) of initial imperfections on the buckling-driven delamination behavior in our flexible multilayer systems.

4.3.6 ARE Application: Reliability Modeling of Cu/Low-k Back-End Structures

For the development of current and future Cu/low-k CMOS technologies, the introduction of low-k dielectrics is a major issue due to their low stiffness and weak interfacial adhesion [43]. To evaluate the structural integrity of Cu/low-k bond pads, the wire pull qualification test, depicted in Fig. 4.16, has been used extensively [27].

4 Advances in Delamination Modeling

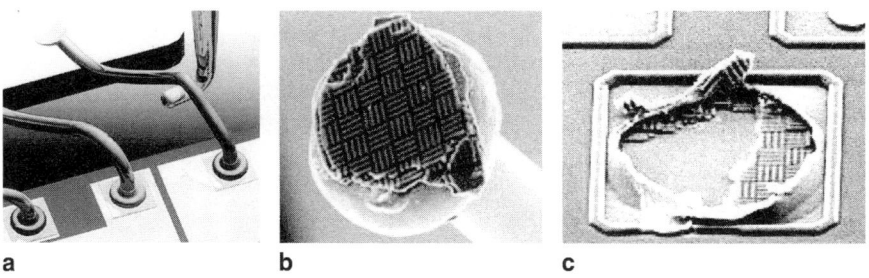

Fig. 4.16 (**a**) Wire pull test; (**b**) and (**c**) actual bond pad delamination

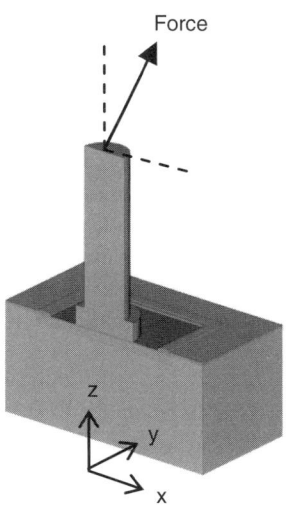

Fig. 4.17 Symmetric wire pull model

To investigate the effect of different back-end designs underneath the bond pad, a numerical multiscale framework including the ARE criterion has been developed [27, 28]. The wire pull model, depicted in Fig. 4.17, consists of a bond pad with an aluminum layer, passivation layers, the back-end structure, the ball bond, and a gold wire of 25-μm diameter. The wire pull qualification test is modeled by applying a force under an angle of 20° with respect to the z-axis in the (x–z) plane. The applied force equals 3 g, which is a typical qualification loading of such a ball bond. Several bond pad structures are considered as visualized in Fig. 4.18. The layout of the unit cells is according to the CMOS090 design rule manual.

The most critical structure as well as the critical interface will be identified by the maximum occurring ARE values. For the structures, the following (normalized) maximum values are obtained: structure A: 0.86; structure B: 0.80; and structure C: 1.00. This indicates that structure C is the least favorable one. This has been confirmed by experimental results. For this structure, the maximum normalized interface ARE values are given in Fig. 4.19a, whereas the maximum normalized peel

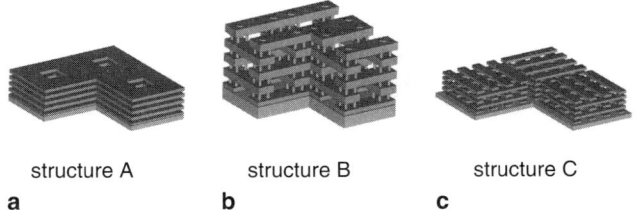

Fig. 4.18 Local back-end structures (hard mask, dielectrics, and top metal layer are not shown; one quarter of the structures is removed from the picture)

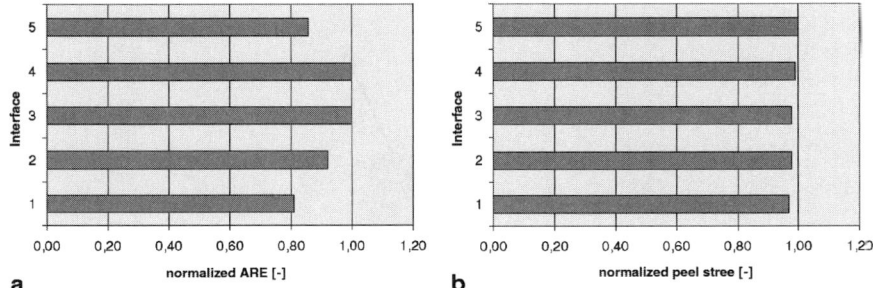

Fig. 4.19 Interface values for structure C: (**a**) normalized maximum ARE values and (**b**) normalized maximum peel stresses

stresses are given in Fig. 4.19b. According to ARE, fourth interface is the critical interface, whereas the peel stresses, still generally accepted as delamination criterion, show hardly any difference for all interfaces. Experimental FIB/SEM observations show that interface 4 is indeed the most critical: delamination is initiated at this interface. From these results, we can conclude that the ARE criterion indeed identifies the most critical back-end structure (C) as well as the most sensitive interface (fourth), in contrast with stress-based values.

It is important to note that these failure sensitivity analyses are not feasible with J-integral methods and cohesive zone models. However, in [29], it is proposed to perform local cohesive zone simulations at the most critical interface locations, identified by the ARE approach, from which the delamination propagation can indeed be analyzed.

4.4 Nanoscale-Based Modeling Techniques

As the geometry of electronic components continuously shrinks from mm to nm, continuum mechanics reaches the boundary of the atomic scale. From a physics point of view, interfacial strength is defined by physical binding, chemical bonds, and mechanical interlocking. MD, which is based on statistical mechanics, is a promising method with the capability of modeling the time evolution of interacting atoms. However, large CPU times limit the applicability of MD simulations to approximately 100 nm. Hence, continuum models are still required. To bridge the gap between the continuum models and MD simulations, so-called handshaking methods are required.

4.4.1 Fundamentals of Molecular Dynamics

From quantum mechanics point of view, matters have dual natures: particle and wave. However, while the geometry of the system is large enough, the wave nature of individual components becomes unapparent and the system becomes determined. When the wave nature of the particle will be ignored or considered implicitly by the potential functions, MD exhibits high efficiency in the simulation of the molecules. MD, widely used in organic chemistry, is a framework for many-particle problems [44]. This method assumes the atom(s) as rigid particles of which the movement is described by coordinate variables. The interactions between the particles are described by the potential functions (or force fields). A typical mechanical MD model is illustrated in Fig. 4.20.

MD is based on Newton's second law of motion,

$$\vec{F}_i = m_i \vec{a}_i \qquad (4.10)$$

for each particle i in a system constituted by N particles. In (4.10), m_i is the mass of particle i, $\vec{a}_i = d^2\vec{r}_i / dt^2$ is its acceleration, and $\vec{F}_i$ is the force acting on the particle.

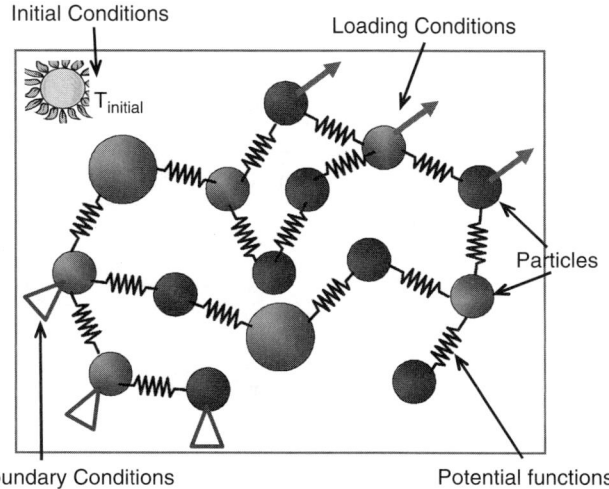

Fig. 4.20 Mechanical molecular dynamics model

Therefore, MD is a deterministic technique: given an initial set of positions and velocities, the subsequent time evolution can be determined. The interaction force between particles, which is required in (4.10), can be defined by potential functions or force fields:

$$\vec{F}_i = -\frac{\partial}{\partial \vec{r}_i} U(\vec{r}_1, \ldots, \vec{r}_N), \qquad (4.11)$$

where U is the potential function and $\vec{r}_k$, $k = 1, \ldots, N$, is the atomic coordinate. Potentials can be categorized broadly as (1) pair potentials, (2) empirical many-body potentials, and (3) quantum mechanical potentials. Two-body potentials – such as Lennard-Jones: $U(r) = 4\varepsilon[(\sigma/r)^{12} - (\sigma/r)^6]$ (ε is the position with minimized energy and σ is the distance where energy equals zero [45]) and Morse potentials: $U(r) = D\{\exp[-\alpha(r - r_0)] - 1\}^2$ (r_0 is the equilibrium distance, α is the elastic modulus, and D is the cohesion energy [46]) – are used for large-scale simulations where computational efficiency is a significant factor. A typical two-body potential is illustrated in Fig. 4.21. When the distance of the two-body system is larger than the equilibrium distance (denoted by r_0), the attractive energy, governed by Coulomb's law, is significant. The attractive energy will fade out when the distance approaches infinity. On the contrary, when the distance of the two-body system decreases, the repulsive energy which is governed by the Pauli exclusion principle will increase. Moreover, the energy approaches infinity if the distance approaches zero.

For systems where multibody interactions are important, the Stillinger–Weber [47], Tersoff [48], and Brenner potentials are often used. Such potentials are empirical in that they are parameterized by fitting either to a set of experimental measurements or to quantum mechanical calculations. However, large local departures from the coordination or bonding used for the parameterization can take such potentials outside

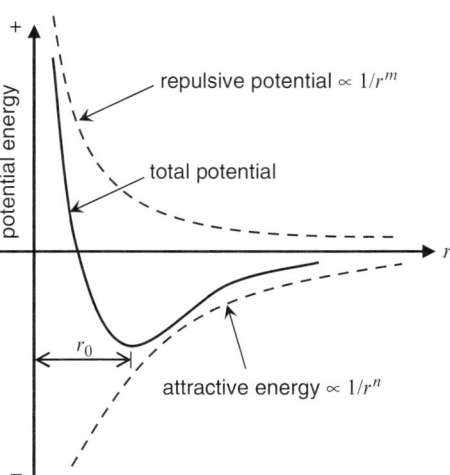

Fig. 4.21 Two-body potential function

their domain of validity and lead to unreliable results. This has fostered efforts for deriving interatomic potentials directly from quantum mechanical principles.

Equation (4.10) represents the system at specific time t. To understand the interaction of the particle and the mechanical response, a method regarding the time integration should be used. The most common time integration algorithms are based on finite difference methods. Two popular integration methods for MD are the Verlet and predictor–corrector algorithms [49, 50]. The integration time step must be small enough to capture the dynamics of the vibration modes of the system, with frequencies in the order of 10^{13} s^{-1}. Each particle which is described by MD has three degrees of freedom, which can be either fixed or free. For a system with billions of particles, periodic boundary conditions (PBC) [44] can be used to reduce the computational effort.

The initial conditions of the MD simulation include the definition of initial coordinates and velocities of particles in the system. The coordinates can be obtained from stoichiochemistry, measured by experiments or ab initio computational methods. The initial velocity can be defined by the temperature, which is directly related to the kinetic energy by the equipartition formula $K = 1.5 N k_B T$, where N is the number of particles and k_B is a Boltzmann's constant. The summation of the kinetic energy of each particle in the system obeys the equipartition formula. Moreover, the centroid velocity equals the given system velocity or zero if the system is at rest.

In principle, MD is based on the theory of classical mechanics while considering the interaction between atoms. MD can be applied when the wave nature can be ignored or represented implicitly. A simple test of the validity of the classical approximation is based on the de Broglie thermal wavelength, which is given by

$$\Lambda = \sqrt{\frac{2\pi h^2}{M k_B T}}, \qquad (4.12)$$

where M is the atomic mass and h is a Plank's constant. If $\Lambda \quad a$, where a is the nearest neighbor separation, the assumption of the classical approach is satisfied. However, for very light systems, like H_2, He, Li, or a system with sufficiently low temperature, the criterion is not satisfied and hence, quantum effects will become important. Hence, MD is not recommended to simulate these systems.

4.4.2 Nanoscale Interface Adhesion

Consider a system with an interface and a set of external loading applied onto that system. Assume that interfacial failure is observed by experiments. Therefore, according to the conservation of the total energy of the system, the work which is done by the external loading will equal the summation of material deformation energy, heat, and surface energy (intrinsic adhesion) of the interface [4]. The intrinsic adhesion of the interface consists of three potential contributors: chemical interaction, physical interaction, and mechanical interlocking.

Chemical interactions. The chemical interactions between the interfaces often refer to the covalent bond, ionic bond, or the metallic bond. These bonds are relatively strong and termed primary bonds. Ionic bonds in salts form as a result of electron transfers and the covalent bonds form as sharing electrons. In a metal, the valency electron of the atoms would form an electron sea (also called electron gel) which is around the atoms, and the atoms "sink" into the sea. However, in electronic or packaging laminated structures, the most common chemical interaction at the interface is the covalent bond besides the alloy system. The interaction scale of the chemical interaction is approximately 0.2–1.0 nm. As an example, consider the interface between SiO_2 and SiOC(H) film (low-k material): SiO_2 is deposited onto the SiOC(H) film by PECVD at O_2 atmosphere and 200–400°C with a precursor of TEOS (tetra-ethyl-ortho-silicate, R–Si–R, where R equals–O–CH_2–CH_3). Due to the fact that the processing temperature is sufficiently high and plasma is applied, the methyl group (CH_3) at the surface of the SiOC(H) will vaporize and is substituted by oxygen. As a result, the silicon–oxygen bond is formed at the interface. Hence, the interfacial strengths of SiOC(H)–SiCN and SiO_2–SiOC(H) are stronger than the one of Ta/TaN–SiOC(H) [51].

Physical interactions. The physical interaction often refers to the weak bonds between the interfaces, like the Coulomb force or van der Waals force. Although the magnitude of the physical interaction is weaker than the one of the chemical bond, it can be formed at most interfaces whereas chemical interaction requires certain chemical conditions to be formed. The scale of the physical interaction is approximately 5–10 nm. Moreover, considering a polymer interface system, two chained polymers can entangle together when the polymer chains have enough energy to move through the interface which contributes to the interfacial strength.

Mechanical interlocking. At the macroscopic scale, mechanical interlocking can be applied (cf. surface treatments applied to metals) to increase the surface roughness and to increase the adhesion strength. The pattern and distribution of the surface roughness is controlled by the processes and materials [52]. Strictly speaking, mechanical interlocking does not belong to the intrinsic material adhesion mechanisms.

To further explain the potential of the MD method, three application cases are described.

4.4.3 Application: Predictions of Silica and Amorphous Nanostructure Material Properties

It is known that the mechanical characteristics of nanoscaled structures differ from the ones of bulk-scaled structures [53]. These mechanical characteristics are difficult to measure directly due to the limitations of nanoscale experimental techniques. Alternatively, MD methods are often used to describe the physical response of nanoscaled materials and will be applied here to demonstrate the prediction of the mechanical stiffness.

The nanoscaled specimens are simulated by MD with an additional energy minimization procedure. Due to the assumed small deformation assumption in elasticity theory [54], the longitudinal elongation should be less than 1.0%. Moreover, based on St. Venant's principle, a model with high aspect ratio is required to prevent boundary effects, as illustrated in Fig. 4.22. The loading and boundary conditions are applied in longitudinal direction.

The reaction forces $\vec{F}_i$ (i represents the ith substep) at the fixed end (Fig. 4.22b) can be extracted by the total energy gradient:

$$\vec{\nabla} U^i_{\text{fixed end}} = F^i_x \vec{i} + F^i_y \vec{j} + F^i_z \vec{k} = \vec{F}_i, \tag{4.13}$$

where $\vec{i}$, $\vec{j}$, and $\vec{k}$ represent three orthogonal vectors.

According to linear elasticity theory [54], the mechanical deformation of the one-dimensional bar can be represented as $\Delta d = FL / EA$, where F, E, L, and A represent the external mechanical force, Young's modulus, length, and cross-sectional area of the specimen, respectively. Accordingly, we can obtain Young's modulus by

$$E = \frac{\Delta F^{ij}}{\Delta d^{ij}} \frac{L_0}{A_{\text{avg}}}, \quad A_{\text{avg}} = V_0 / L_0, \tag{4.14}$$

where $\Delta F^{ij} = F^j - F^i, j > i$, and $\Delta d^{ij} = d^j - d^i \ j > i$, V_0 and L_0 are the initial volume and initial length, respectively.

This method has been applied to calculate the properties of a silicon molecule. The model comprises 2,688 Si atoms having a volume of 58.3 nm³. The Young's modulus is determined as 130 GPa. The density is calculated as the ratio of the atomic mass and molecular volume, resulting in 2.5 g cm⁻³.

For an amorphous and/or porous material, the above-mentioned method can also be used when the chemical structure of the molecule is generated. There are several methods to predict the chemical structure. One can simulate the whole fabrication process of the amorphous material. Another method is to generate an approximate structure based on measured chemical information; this method composes the following steps:

- Assume that the material consists of several basic building blocks.
- Obtain the ratio of the building blocks by experimental methods (e.g., nuclear magnetic resonance, NMR).

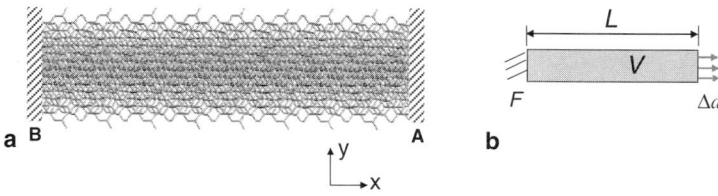

Fig. 4.22 Bar model: (**a**) the model of silicon and (**b**) illustration of the loading/boundary condition

- Locate the building blocks into a predefined framework with the distribution ratio obtained by the experiment.
- Apply the geometrical optimization method to the molecular topology from which an approximate amorphous molecular model is obtained.

One example is the amorphous SiOC(H) material. As feature sizes for advanced IC structures continue to decrease, the semiconductor industry is focusing on technologies to minimize the intrinsic time delay for signal propagation, quantified by the resistance–capacitance (RC) delay. The increasing demands for the electronic performance of the IC wiring have recently driven the replacement from aluminum to copper traces, and the alternative materials for SiO_2 film with lower dielectric constant [4, 55]. SiOC(H), also called black diamond (BD), is favored by the industry because the fabrication process of BD is highly compatible to existing IC processes. From the chemical structure of BD (Fig. 4.23a), four basic building blocks of BD

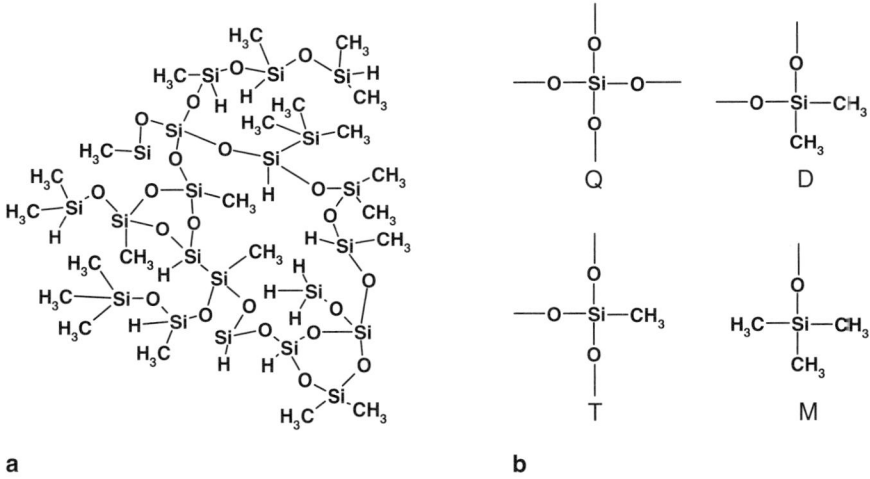

Fig. 4.23 SiOC(H): (**a**) chemical structure and (**b**) basic building blocks Q, D, T, and M

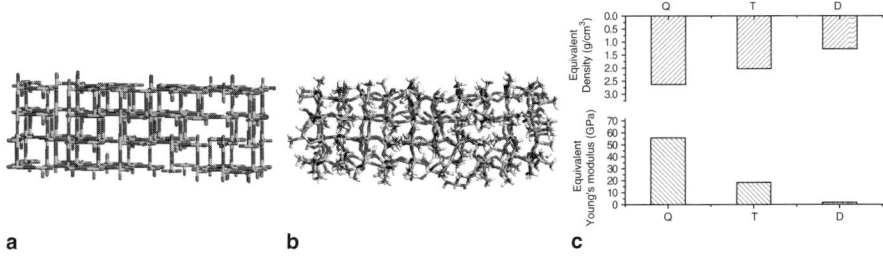

Fig. 4.24 SiOC(H) model: (**a**) topology of BD, (**b**) ideal topology, and (**c**) the calculated sensitivity of the building blocks Q, T, D, and void

can be defined, as illustrated in Fig. 4.23b. Because BD is modified from SiO_2, this framework is assumed. The size of the voids is similar to a building block which is confirmed by experiments. Based on NMR results [55], the Q, T, D, M, and void are distributed into the silicon dioxide framework as the ratio of 20, 50, 18, 2, and 10%, respectively.

Consequently, the molecular topology of BD is obtained (Fig. 4.24a). Hence, an approximate stereochemical structure is acquired after the geometrical optimization process (Fig. 4.24b). The MD computation results in the sensitivity of each building block on the overall Young's modulus of BD (see Fig. 4.24c).

4.4.4 Application: MD Simulation of an Organic Oligomer on a SilicaSubstrate

Two different interfaces are considered: an inorganic molecule on silica and an organic molecule on silica. Figure 4.25 shows a MD simulation of the adhesion problem [56, 57]. The simulation reveals that the adhesion energy of inorganic molecules is higher than in organic molecules. The simulation reveals that the peak adhesion energy of inorganic molecules is higher than in organic molecules but the decay of the inorganic adhesion energy is faster.

Figure 4.26 shows a simulation on the interface between amorphous silica and BD. The simulation results indicate that the delamination does not always result in

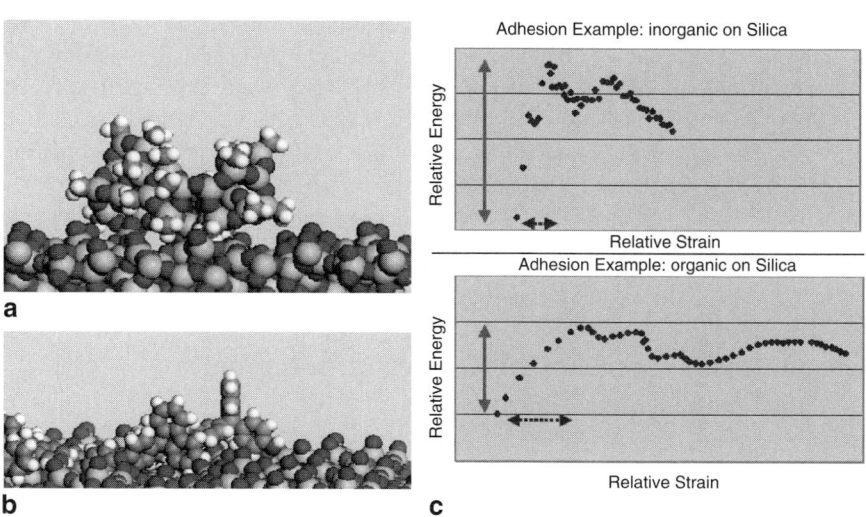

Fig. 4.25 Adhesion of inorganic and organic polymer on silica (**a**), (**b**) shows the organic and inorganic molecule on the silicon, and (**c**) shows the adhesion energy of organic and inorganic molecule [57] (with permission)

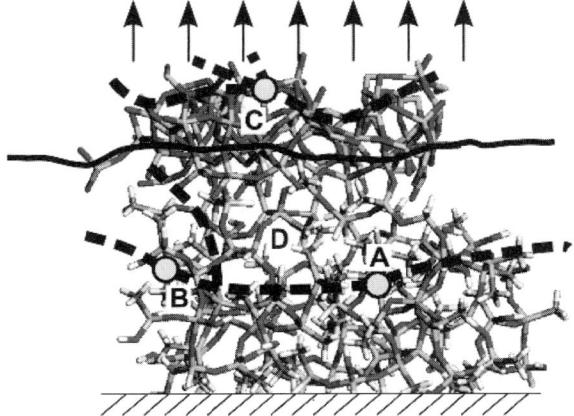

Fig. 4.26 Simulated crack propagation (*dashed lines*) through the interface (*solid line*). Points A, B, and C are the initial crack locations and point D indicates a pore within the low-k material

broken covalent bonds at the interface. Instead, the failure manifests itself as multiple crack propagation paths.

4.4.5 Application: Interfacial Strength of Polymer–Metal

For metal and epoxy interfacial systems, the oxidation of the metal might also play an important role. The oxidized aluminum substrate Al_2O_3 on the surface may transfer to boehmite (AlOOH) or bayerite ($Al(OH)_3$) when exposed to oxygen-containing environments, such as O_2, H_2O, CO_2, etc. [58]. The hydroxyl-rich layer of the Al substrate will chemically react with epoxy to form chemical bonding. Figure 4.27 demonstrates a molecular simulation on the effect of hydration. The simulation results reveal that the bonding energy densities with and without the chemical bond (–OH) are 1 and 3.5 J m^{-2} [58].

4.5 Outlook: Handshaking Between Continuum and Nanoscaled Methods

Continuum and nanoscale techniques perform very well at their particular application scale, dictated by the successive elimination of the original degrees of freedom. While wave nature is not apparent but atomistic resolution is still necessary, molecular dynamics offers many computational advantages over a full density functional calculation. Continuum equations provide a reduced description in terms of continuous

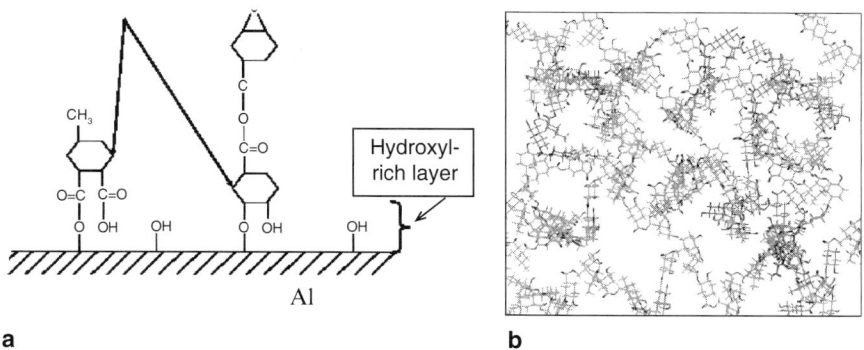

Fig. 4.27 Simulation of the oxidation of the metal layer: (**a**) illustration of the model and (**b**) molecular model of sites reactable with OH [52, 58] (with permission)

fields for the coarse-grained evolution of the system. Delamination in micro/nanoelectronics is a multiscale problem in nature. Regarding an interface at the molecular level, three main contributors to its strength are (1) chemical interaction, (2) physical interaction, and (3) mechanical interlocking. The chemical interactions refer to the strong bonds, such as covalent bond and metallic bond, formed at the interface. The physical interactions are the weak bonds, such as van der Waals force and hydrogen bonds. The mechanical interlocking occurs due to the fact that the materials lock each other due to their roughness. The chemical and physical interactions regard to the atomic structure and binding characteristics with the interaction distance approximately 1–10 nm. These two phenomena can be simulated by MD. However, the geometrical scale of roughness is approximately 10–100 nm, and consequently, MD is not an efficient method for this roughness problem due to the computational time limitation. Therefore, multiscale modeling techniques which combine MD and FEM are required.

The fundamental tenet of multiscale modeling is that information at each scale is systematically incorporated in a manner that transcends the single-scale description. There are two basic strategies to achieve this goal: the global–local method and the hybrid method. In the global–local approach [59], information from a calculation over particular length and time scales is used as input into a more coarse-grained method. This approach presumes that the phenomenon of interest can be separated into processes that operate at distinct length and time scales, as stated in St. Venant's principle. In hybrid multiscale modeling [60, 61], these disparate scales are combined within a single hybrid scheme, typically involving atomistic and continuum calculations. The main theoretical challenge is to merge the two descriptions in a manner that avoids any spurious effects due to this heterogeneity. This approach is well suited to the simulation of fracture, where the complex feedback between the atomic-scale interactions and macroscopic stresses pre-empts a clear-cut separation of scales.

To handshake between FEM and MD, three main challenges can be identified:

Simulations at finite temperature. Many of the methods which are described are confined to zero temperature, wherein the basic quantity is a Hamiltonian for the system expressed in terms of the appropriate degrees of freedom. In principle, Hamiltonian methods can be extended to equilibrium at finite temperatures by using the free energy, but inherently nonequilibrium situations are fundamentally different and a general approach is not clear. The temperature (energy level) change of the atomic system due to application of the external loading (external energy) might influence the potential functions, especially the specific temperatures of which could change the physical state of the molecules.

Time scales accessible by means of molecular dynamics. The bottleneck of the macroscopic time simulations remains because of the small time step in the classical and quantum molecular dynamics. While this may be sidestepped in certain applications, and acceleration strategies are available for particular situations, a general acceleration methodology would have revolutionary implications that would stretch across many disciplines. Theoretically speaking, two candidate techniques could resolve this issue with lack of the particular accuracy. One is to apply a faster loading speed, which is often 100–10,000 times faster than the real loading, onto the system to reduce the total simulation. However, the characteristics of the stress/strain wave propagation would be different from the slower one. Second one

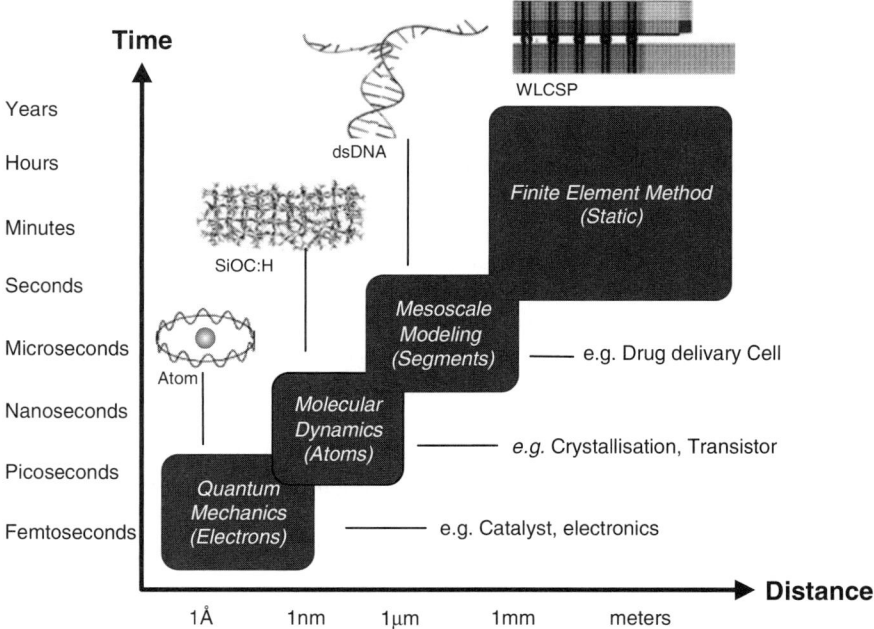

Fig. 4.28 Handshaking between continuum- and nanoscale techniques

is to use a heavier atomic group to replace the independent atoms. The clustered atomic group often contains 10–1,000 atoms, and the atomic groups are assumed as rigid body or homogeneous deformable bodies. However, the mechanical information inside the clustered atomic groups could not be achieved.

Wave propagation across atomistic/continuum interfaces. In methods with a specific interface between atomistically resolved and (finite element) continuum regions, high frequency modes emanating from the molecular region must be accommodated by the continuum region. For example, in the hybrid method of the molecular dynamics and the finite element method, the critical region would be modeled as molecular dynamics and the rest are the finite element methods. To efficiently reduce the computation time, the mesh density of the finite element area would be much coarser than the one in the molecular dynamics area. However, the finite elements are unable to resolve the small wavelengths which are generated by the atomistic region. Therefore, the high frequency response from the atomistic region would bounce back to the atomistic region, and this could lead to spurious results or the critical region.

To illustrate the handshaking approach, an overview of several modeling methods and their respective scales of applicability is given in Fig. 4.28.

4.6 Conclusion

Due to the rapid development of micro/nanoelectronic technology (e.g., mems), structures are approaching submicron and nanoscales. To design a reliable structure with low cost, one should understand the crack initiation and propagation as a function of chemical composition, fabrication process, and/or temperature and moisture conditions. However, due to the experimental techniques currently available, the interface properties, such as mode angle, energy release rate (G) or stress intensity factor (K), cannot be obtained efficiently. Hence, a numerical model, which is validated by the experimental results and has the capability to predict the mode angle, G or K, is required.

To understand/predict the initiation and propagation of delamination, handshaking between continuum-scale FEM simulations and MD methods is a promising and challenging way. The influence scales of the three main contributors (chemical interaction, physical interaction, and mechanical interlocking) are approximately 0.5, 10, and 100 nm, respectively. Continuum-scale failure models, like fracture mechanics, cohesive zones, and the area release method, are applicable from approximately 500 μm onward. However, for reasons of computational efficiency, handshaking can only be obtained by using multiscale simulation schemes.

There is an enormous challenge to predict the initiation and propagation of an interfacial system at the nanoscale range, due to:

- *Defects of material and interface.* The fabrication process of material and interface (e.g., deposition, curing, and plating) will cause the defects (e.g., point

defect, dislocation or grain boundary). Theoretically speaking, these distribution patterns of deflects should be carefully measured to represent the mechanical response of the material in reality. These patterns can be material and process dependent, and it is sometimes difficult to measure, especially at the interface.
- *Lack of sufficient force field.* The potential functions (force fields) currently available are obtained for the material and they can perform well in the prediction of the material. However, there are few force fields which are designed for the interfacial mechanism.
- *Lack of robust multiscale modeling technique.* Using MD to simulate the roughness scale (~100 nm) results in large computation time. On the contrary, using continuum-based method, it is hard to describe the chemical/physical interaction at the interface.
- *Lack of the exact chemical configuration of the interface.* Regardless of the mechanical interlocking, the magnitude of chemical bond strength is much higher than the physical bond. Therefore, the ratio of chemical bonds at the interface can influence the total interfacial strength and can depend on the surface treatment, process, material, etc. This ratio is difficult to characterize experimentally.

Acknowledgments Thanks to our colleagues at Delft University of Technology for the inspiring atmosphere: Leo Ernst, Fred van Keulen, Kaspar Jansen, Lingen Wang, and An Xiao; at Philips Applied Technologies: Richard van Silfhout, Peter Timmermans, Roy Engelen, and Rene Kregting; at NXP Semiconductors: John Janssen and Marcel van Gils. OvdS appreciates the financial support of the EC under contract IST-2004-4354 (Flexidis) on the work of buckling-driven delamination and the cooperation on this topic with Amir Abdallah (Eindhoven University of Technology) and Piet Bouten (Philips Research). Furthermore, thanks to Bas van Hal, Ron Peerlings, Marc Geers, and Marco van den Bosch from Eindhoven University of Technology for the cooperation on the development of the cohesive zone algorithms. CAY thanks Dr. F. Iacopi (IMEC, Belgium) for sharing her experimental results; Dr. N. Iwamoto (Honeywell, USA), Dr. C. Menke (Accelrys, Germany), and Dr. J. Wescott (Accelrys, England) for discussing numerical simulation techniques. CAY also thanks Prof. B.J. Thijsse for valuable discussion on the MD simulation techniques and Marcel Kouters (Eindhoven University of Technology) for the interface toughness measurements.

References

1. Zhang GQ, van Driel WD, Fan XJ (2006) Mechanics of Microelectronics. Springer, Berlin Heidelberg New York, ISBN 1-4020-4934-X
2. van Driel WD, Zhang GQ, Janssen JHJ, Ernst LJ, Su F, Chian KS, Yi S (2003) Prediction and verification of process induced warpage of electronic packages. Microelectronics Reliability 43:765–774
3. Zhang GQ (2003) The challenges of virtual prototyping and qualification for future microelectronics. Microelectronics Reliability 43:1777–1785
4. Ernst LJ, van Driel WD, van der Sluis O, Corigliano O, Tay AAO, Iwamoto N, Fan H, Yuen MMF (2006) Fracture and delamination in microelectronics. In: Proceedings of the Asian-Pacific conference for fracture and Strength (APCFS'06), Singapore
5. Kanninen MF, Popelar CH (1985) Advanced Fracture Mechanics. Oxford University Press, New York
6. Rice JR (1988) Elastic fracture mechanics concepts for interfacial cracks. Journal of Applied Mechanics 55:98–103

7. Hutchinson JW, Suo Z (1992) Mixed mode cracking in layered materials. Advances in Applied Mechanics 29:63–191
8. Thijsse J, van der Sluis O, van Dommelen JAW, van Driel WD, Geers MGD (2008) Characterization of semiconductor interfaces using a modified mixed mode bending apparatus. Microelectronics Reliability 48:401–407
9. Barsoum RS (1976) On the use of isoparametric finite elements in linear fracture mechanics. International Journal for Numerical Methods in Engineering 10:25–37
10. Abdel-Wahab MM, de Roeck G (1995) A 2-D five-noded finite element to model power singularity. International Journal of Fracture 74:89–97
11. He MY, Evans AG, Hutchinson JW (1994) Crack deflection at an interface between dissimilar elastic materials: role of residual stresses. International Journal of Solids and Structures 31:3443–3455
12. Hellen TK (1975) On the method of virtual crack extension. International Journal for Numerical Methods in Engineering 9:187–207
13. Rybicki EF, Kanninen MF (1977) A finite element calculation of stress intensity factors by a modified crack closure integral. Engineering Fracture Mechanics 9:931–938
14. Krueger R (2002) The virtual crack closure technique: history, approach and applications. NASA Report CR-2002-211628
15. Shih CF, Asaro RJ (1988) Elastic–plastic analysis of cracks on bimaterial interfaces. Part I. Small scale yielding. Journal of Applied Mechanics 55:299–316
16. Xuan ZC, Khoo BC, Li ZR (2006) Computing bounds to mixed-mode stress intensity factors in elasticity. Archive of Applied Mechanics 75:193–209
17. Reeder JR, Crews JH (1990) Mixed-mode bending method for delamination testing. AIAA Journal 28:1270–1276
18. Xiao A, Wang LG, van Driel WD, van der Sluis O, Yang DG, Ernst LJ, Zhang GQ (2007) Thin film interface fracture properties at scales relevant to microelectronics. Proceedings of EuroSimE 2007, London, pp. 350–355
19. Dugdale D (1960) Yielding of steel sheets containing slits. Journal of the Mechanics and Physics of Solids 8:100–104
20. Barenblatt G (1962) The mathematical theory of equilibrium cracks in brittle fracture. Advances in Applied Mechanics 7:55–129
21. Ortiz M, Pandolfi A (1999) Finite-deformation irreversible cohesive elements for three-dimensional crack-propagation analysis. International Journal for Numerical Methods in Engineering 44:1267–1282
22. Alfano G, Crisfield MA (2001) Finite element interface models for the delamination analysis of laminated composites: mechanical and computational issues. International Journal for Numerical Methods in Engineering 50:1701–1736
23. van Hal BAE, Peerlings RHJ, Geers MGD, van der Sluis O (2006) Cohesive zone modeling for structural integrity analysis of copper/low-k IC interconnects. Microelectronics Reliability 47:1251–1261
24. van Hal BAE, Peerlings RHJ, Geers MGD (2007) Local arc-length control method for cohesive zone modelling. Computer Methods in Applied Mechanics and Engineering (submitted)
25. Tomar T, Zhai J, Zhou M (2004) Bounds for element size in a variable stiffness cohesive finite element. International Journal for Numerical Methods in Engineering 61:1894–1920
26. Crisfield M (1981) A fast incremental/iterative solution procedure that handles 'snap-throughs'. Computers & Structures 13:55–62
27. van Gils MAJ, van der Sluis O, Zhang GQ, Janssen JHJ, Voncken RMJ (2006) Analysis of Cu/low-k bond pad delamination by using a novel failure index. Microelectronics Reliability 47:179–186
28. van der Sluis O, Engelen RAB, van Silfhout RBR, van Driel WD, van Gils MAJ (2007) Efficient damage sensitivity analysis of advanced Cu low-k bond pad structures by means of the Area Release Energy criterion. Microelectronics Reliability 47:1975–1982
29. van der Sluis O, van Silfhout RBR, Engelen RAB, van Driel WD, Zhang GQ (2007) Multi-scale energy-based failure modeling of bond pad structures. Proceedings of EuroSimE 2007, London, pp. 678–683

30. Belytschko T, Black T (1999) Elastic crack growth in finite elements with minimal remeshing. International Journal for Numerical Methods in Engineering 45:601–602
31. van Driel WD, van Gils MAJ, van Silfhout RBR, Zhang GQ (2005) Prediction of delamination related IC & packaging reliability problems. Microelectronics Reliability 45:1633–1638
32. van Gils MAJ et al. (2004) Characterization and modeling of moistures driven interface failures. Microelectronics Reliability 44:1317–1322
33. van Driel WD, Habets PJJHA, van Gils MAJ, Zhang GQ (2005) Characterization of interface strength as function of temperature and moisture conditions. Proceedings of ICEPT Conference, pp. 687–692
34. Bouten PCP, van Gils MAJ (2004) Buckling failure of compressive loaded layers in flexible devices. In: Basu SN, Krzanowski JE, Patscheider J, Gogotsi Y (eds) Surface Engineering 2004 – Fundamentals and Applications. Materials Research Society Symposium Proceedings 843, Warrendale, PA, 2005
35. Abdallah AA, Kozodaev D, Bouten PCP, den Toonder JMJ, Schubert US, de With G (2006) Buckle morphology of compressed inorganic thin layers on a polymer substrate. Thin Solid Films 503:167–176
36. Jansson NE, Leterrier Y, Månson J-AE (2006) Modeling of multiple cracking and decohesion of a thin film on a polymer substrate. Engineering Fracture Mechanics 73:2614–2626
37. Suo Z (2003) Reliability of interconnect structures. In: Gerberich W, Yang W (eds) Interfacial and Nanoscale Failure. Elsevier, London, ISBN 0-08-044151-3
38. FLEXIDIS European Project, www.flexidis-project.org
39. van der Sluis O, Abdallah AA, Bouten PCP, Timmermans PHM, den Toonder JMJ, de With G (2007) Effect of elastic mismatch on buckle delamination of thin Indium tin oxide layers on a compliant substrate (in preparation)
40. Jensen HM, Sheinman I (2002) Numerical analysis of buckling-driven delamination. International Journal of Solids and Structures 39:3373–3386
41. Hutchinson JW, He MY, Evans AG (2000) The influence of imperfections on the nucleation and propagation of buckling driven delaminations. Journal of the Mechanics and Physics of Solids 48:709–734
42. Moon MW, Chung J-W, Lee K-R, Oh KH, Wang R, Evans AG (2002) An experimental study of the influence of imperfections on the buckling of compressed thin films. Acta Materialia 50:1219–1227
43. Hartfield CD, Ogawa ET, Park Y-J, Chiu T-C (2004) Interface reliability assessments for copper/low-k products. IEEE Transactions on Device and Materials Reliability 4:129–141
44. Ercolessi F (1997) A Molecular Dynamics Primer. , Spring College in Computational Physics, ICTP, Trieste, Italy
45. Jones JE (1924) On the determination of molecular fields. II. From the equation of state of a gas. Proceedings of the Royal Society of London Series A 106:463–477
46. Morse PM (1929) Diatomic molecules according to the wave mechanics. II. Vibrational levels. Physical Review 34:57–64
47. Stillinger FH, Weber TA (1985) Computer simulation of local order in condensed phases of silicon. Physical Review B 31:5262–5271
48. Tersoff J (1986) New empirical model for the structural properties of silicon. Physical Review Letters 56:632–635
49. Ciccotti G, Hoover WG (1986) Molecular-Dynamics Simulation of Statistical–Mechanical Systems. North-Holland, The Netherlands
50. Haile JM (1992) Molecular Dynamics Simulation. Wiley, New York
51. Kouters MHM (2006) Characterisation of interfacial strength of low-k dielectric materials used in ICs. , MT 06.37 (Master Thesis) Eindhoven University of Technology, The Netherlands
52. Lee HY, Qu J (2003) Microstructure, adhesion strength and failure path at a polymer/roughened metal interface. Journal of Adhesion Science and Technology 17:195–215
53. Sun L, Murthy JY (2006) Domain size effects in molecular dynamics simulation of phonon transport in silicon. Applied Physics Letters 89:171919
54. Love AEH (1927) A Treatise on the Mathematical Theory of Elasticity, . 4th ed. Dover, New York

55. Iacopi F et al. (2006) Short-ranged structural rearrangement and enhancement of mechanical properties of organosilicate glasses induced by ultraviolet radiation. Journal of Applied Physics 99:053511
56. Iwamoto N (1994) A property trend study of polybenzimidazole using molecular modeling. Polymer Engineering and Science 34:434–437
57. Iwamoto N (2003) Material response prediction and understanding through the use of molecular modeling. Proceedings of EuroSimE 2003
58. Qu J (2003) Thermomechanical reliability of microelectronic packaging. In: Gerberich W, Yang W (eds) Comprehensive Structure Integrity – Fracture of Materials from Nano to Macro. Elsevier Science, London, pp. 219–240
59. Chiang KN, Yuan CA, Han CN, Chou CY, Cui Y (2006) Mechanical characteristic of ssDNA/dsDNA molecule under external loading. Applied Physics Letters 88:023902
60. Srivastava D, Atluri SN (2002) Computational nanotechnology: a current perspective. Computer Modeling in Engineering & Science 3:531–538
61. Xiao SP, Belytschko T (2004) A bridging domain method for coupling continua with molecular dynamics. Computer Methods in Applied Mechanics and Engineering 193:1645–1669

Chapter 5
Nanoparticle Properties

James E. Morris

5.1 Introduction

As the radius r of a spherical particle shrinks, the surface/volume ratio $3/r$ and the proportion of its constituent atoms at the surface both increase. The stable interatomic bonding arrangements which exist within large crystals are not satisfied for surface atoms, which therefore become more mobile and more reactive, and nanoparticle properties become dominated by surface properties.

The nanoparticles discussed below will all be metallic, but most of the phenomena described will also apply to nonmetallic materials. They will appear in three contexts:

1. In an aqueous environment [1]
2. In a "cermet" (i.e., in a two-phase ceramic–metal [2] or polymer–metal [3] mixture)
3. On an insulating substrate surface, as a discontinuous (island) metal thin film (DMTF) array of nanoparticles [4]

In the first two cases, the nanoparticles are usually modeled as spherical, but shape becomes an issue for the DMTF.

Classical nucleation theory covers the initial formation and growth of the nanoparticles, but the predicted critical nucleus sizes (beyond which the nuclei can grow as stable units) in the sub-nm range are clearly inconsistent with the classical model's use of bulk thermodynamic properties. For nuclei containing only a few atoms, as is typically the case in all these systems, the "atomistic" nucleation theory is necessary [5].

J.E. Morris
Department of Electrical & Computer Engineering, Portland State University, P.O. Box 751, Portland, OR, 97207-0751, USA

5.2 Structure

Nanoparticles can exist as perfect crystals, since impurities and lattice defects alike can migrate to the surface in relatively short times. Debye–Scherrer broadening of electron diffraction rings provides a means of determining nanocrystallite sizes, with their radii giving lattice spacings. The weakening of the internal lattice structure at finite sizes leads to contraction Δd of lattice spacings d:

$$\frac{\Delta d}{d} = \beta\left(\frac{2\alpha}{r}\right),$$

due to surface tension α, where β is the (anisotropic) linear compressibility [5]. However, surface energy σ also varies with r as

$$\sigma(r) = \sigma(\infty)\left(1 + \frac{3}{8}\frac{r_m}{r}\right),$$

where r_m is the screening radius (<5 nm), so the more complete form is

$$\frac{\Delta d}{d} = \beta\left(\frac{2\alpha(\infty)}{r}\right)\left(1 + A\frac{r_m}{r}\right),$$

where A is a constant [6]. (A counter effect of lattice parameter increase with decreasing size has also been reported [7].)

In the absence of any other information, nanoparticles in cermets and aqueous suspensions are usually assumed to be spherical, at concentrations sufficiently below the percolation threshold that particle contacts and coalescence can be ignored. (The percolation threshold is the minimum concentration for metallic electrical conduction through the medium.) However, in liquid environments anyway, the size [1] and shape (i.e., spheres, cubes, or rods [8]) of the nanoparticles can be controlled by varying the precipitation conditions.

In the absence of other considerations, the minimum energy configuration for a nanoparticle with bulk and surface energies would be a sphere, but the equilibrium minimum energy shape of a charged particle is actually an ellipsoid of rotation [9]. Looking through a DMTF, the nanoparticle islands appear to be slightly prolate in shape [10]. However, it is well established that the island size varies during DMTF deposition as r^2 proportional to time (or deposited mass) not r^3 as expected for quasispherical growth [11]. So, the dominant particle shape is oblate, with possible causes being electrostatic, as mentioned above, or substrate adatom capture with insufficient thermal energy to reach a spherical equilibrium.

Another factor in the nanoparticle's shape may be its degree of crystallization, and the relatively weak binding of the surface atoms in particular permits rapid motion and continual abrupt crystallographic reconstructions of 1–10nm nanoparticles on a time scale of seconds [12, 13], with the relative stabilities of different crystallographic forms determined by the bulk and surface energies [14, 15].

The relatively rare electron micrographs of nanoparticle islands from the side [16] show that they are neither spherical nor hemispherical, as often assumed in simplistic models. For most metals on insulators, the contact angle θ is greater than 90° and can be reasonably calculated from bulk surface energies σ_{sv}, σ_{sc}, and σ_{cv}, where s, v, and c refer to substrate, vapor, and condensate, by

$$\sigma_{sv} = \sigma_{sc} + \sigma_{cv} \cos\theta$$

giving $\theta \approx 136°$ for Au on glass, in good agreement with observation [4].

5.3 Electrical Properties

If the surface atoms are characterized by incompletely satisfied "dangling" chemical bonds, and the surface can be considered to be disordered, then this disorder extends into the crystal interior as dimensions decrease. If metallic conductivity is associated with the band structure of a regular crystal, the question arises as to whether metallic properties can be maintained at nanoscale dimensions. It has been shown for Pd that metallic properties persist at room temperature down to a cluster size of about 12 atoms (≤1-nm diameter), with the metal–insulator transition occurring at smaller sizes as the temperature goes up (Fig. 5.1) [17]. Kreibig [18] applied a different criterion, the experimental observations of dielectric absorption in glass containing Ag or Au nanoparticles, to conclude that the "cluster–solid state" transition occurs at ~500 atoms/particle (i.e., ~2.5-nm diameter).

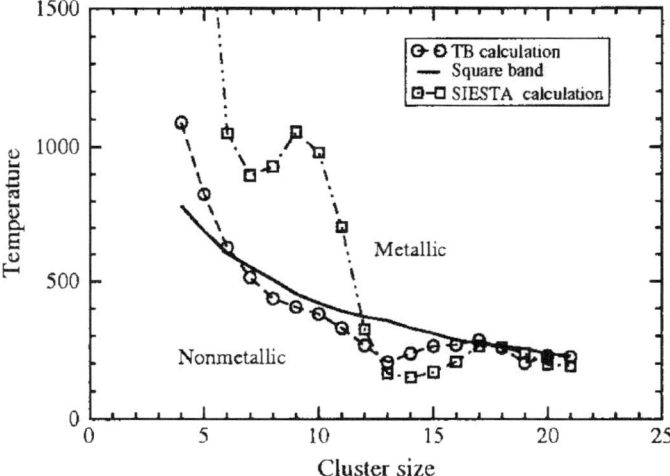

Fig. 5.1 Metallic–insulator transition as a function of number of atoms and temperature. Calculations were based on tight-binding (TB), pseudopotential (SIESTA), and simpler (square band) models [17] (with permission)

The nanoparticle Fermi energy varies with size, leading to increasing work function φ as dimensions shrink

$$\varphi(r) = \varphi(\infty) + \frac{B}{r}$$

where B is a constant [19], e.g., a shift from 4.50 to 4.53 eV for a 10-nm radius W nanoparticle [20]. At very small sizes, the work function changes fluctuate due to the changes in finite small numbers of surface atoms [21].

5.4 Catalysts

Catalysts are used to speed up chemical reactions, typically by one of two mechanisms [22]:

1. The provision of a new reaction path of lower activation energy
2. The provision of a surface to which the chemical reactants can adhere and react more readily than, for example, in the gas phase

The rapid advances of nanotechnologies have spawned many new innovations in nanoparticle catalysts, including their application in biomedical applications [23], as "seeds" for the vapor–liquid–solid (VLS) growth process in chemical vapor deposition (CVD) of both carbon nanotubes (CNTs) [24] and nanowires [25, 26], and the use of carbon nanoparticles and CNTs as support structures for nanoparticle catalysts [27], ensuring maximum active surface exposure.

5.5 Melting Point Depression

At small sizes, nanoparticles may melt at temperatures significantly below the bulk melting point (MP) [28–32], due to increasing surface energy at small sizes [28]. This phenomenon has been studied for decades, along with the parallel rapid evaporation of such nanoparticles due to their increased vapor pressure at high surface curvatures [33]. An example plot is shown in Fig. 5.2 [29], for three different metals, illustrating that a unified theory may be possible and that significant reductions require nanoparticle dimensions of ~5 nm or less.

Different electron microscopy techniques have been used to determine the MP. Sambles [29], for example, monitored the evaporation of small particles at controlled temperatures and noted the size when the evaporation rate changed, interpreting this as the melting point. Others have noted the transition from sharp to diffuse electron diffraction rings [34], or the loss of diffraction ring intensity [35]. Allen et al. [34] used dark-field images, interpreting the disappearance of the image as the melting point. One might also look for the disappearance of crystal facets to indicate

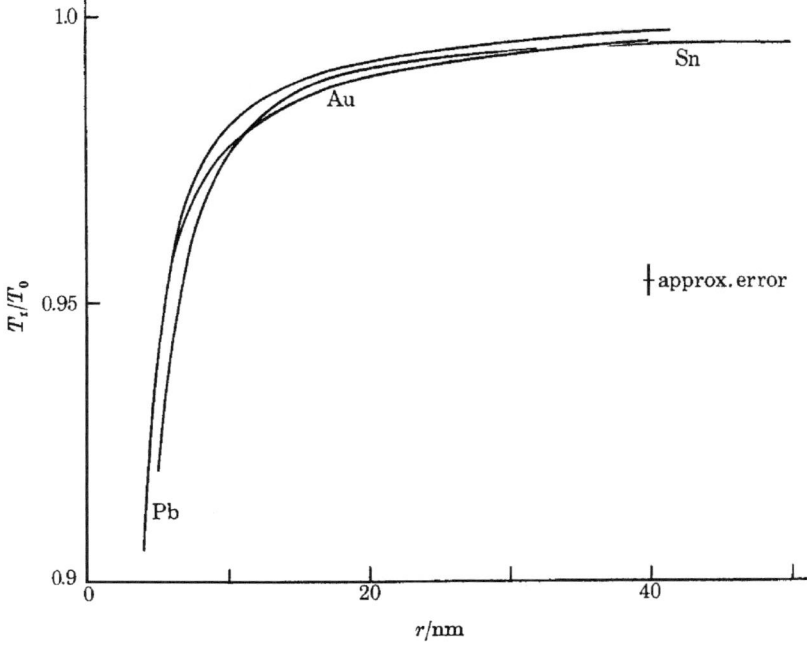

Fig. 5.2 Experimental melting point depression for Au, Sn, and Pb, normalized to the bulk melting points [29] (with permission)

melting, but the method fails at small sizes when solid particles may not exhibit faceting [34]. Comparisons of dark-field and bright-field images reveal a two-step melting process, with the abrupt appearance at the "solidus" temperature of a surface liquid sheath, surrounding a solid crystalline core which shrinks as the temperature continues to increase, until it disappears abruptly at the higher "liquidus" temperature [36].

To a first order, the MP T_M for particle of radius r may be related to the bulk value T_∞ by an empirical, experimental equation [34, 37]:

$$\frac{T_M}{T_\infty} = 1 - \frac{r_0}{r},$$

where r_0 is interpreted as the limiting radius for which the particle remains liquid at 0 K [38]. The two dominant approaches to the theory are the thermodynamic [34, 35, 39–41] and the "surface layer" model [37, 42], compared in [38]. In general, the thermodynamic models vary in the formulae for r_0 (although one [40] suggests the form $T_M/T_\infty = 1 - [r_0/r]^2$). The existence of a liquid-like shell on the nanoparticle has been demonstrated [42], and a good match of theory to experiment can be obtained by adjustment of the unknown layer thickness t_0 to an equation of the form [42]:

$$\frac{T_M}{T_\infty} = 1 - \left[A + \frac{B}{r} + \frac{C}{r-t_0} \right],$$

where A, B, and C are thermodynamic constants. The two-step melting process leads to complexities in alloy systems, due to compositional phase changes [36].

The high MPs of no-Pb solders lead to higher thermomechanical stresses than for conventional eutectic Sn–Pb solder, and melting point depression may be one mechanism to reduce the process temperatures and thermomechanical failure rates. The MP of Sn–Ag alloy, for example, has been shown to be reduced from 222 to 193°C for 5-nm radius particles [43].

5.6 Sintering

The three mechanisms of coalescence of adjacent nanoparticles are shown in Fig. 5.3 [44]. "Ostwald ripening" results from the fact that the equilibrium rate of atomic "escape" for nanoparticles is inversely proportional to the radius of curvature. So for two adjacent particles of unequal size as shown in Fig. 5.3a, the rate of atomic

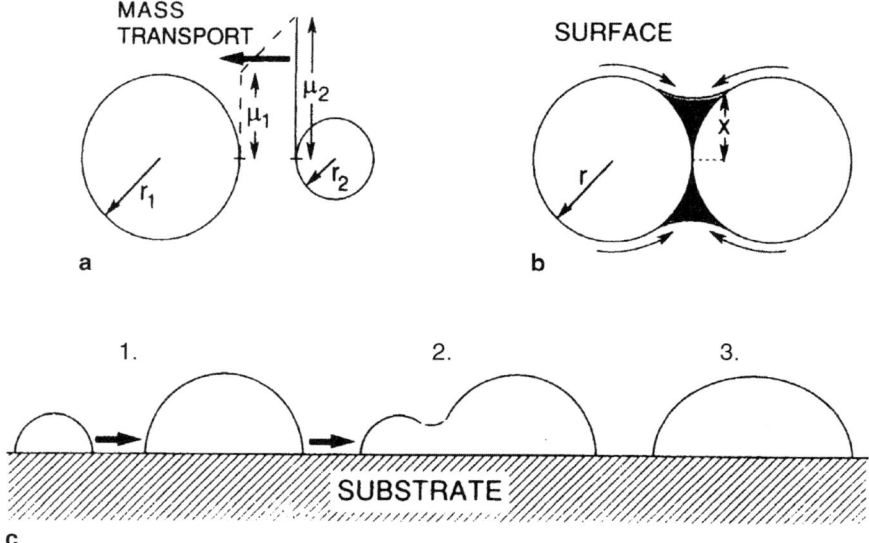

Fig. 5.3 Coalescence of nanoparticle "islands," due to (**a**) Ostwald ripening, (**b**) sintering, and (**c**) cluster migration [44] (with permission)

escape from particle 2 exceeds that from particle 1, resulting in net growth of particle 1 at the expense of the smaller particle 2. Clusters may also migrate freely about a surface (or in other environments) with thermal energy, and may collide and coalesce with other similar clusters (Fig. 5.3c). It has been suggested that the driving force behind such coalescences may be electrostatic, due to random polarizations of the particles.

In this section, we are primarily concerned with sintering, as illustrated in Fig. 5.3b. The process is dependent upon the local radius of curvature, as

$$\frac{X^n}{r^m} = A(T)E,$$

where $A(T)$ is a temperature-dependent constant, t is the time, and $n = 7$ and $m = 3$ for surface diffusion. The sintering process has been modeled by molecular dynamics [45].

Sintering is an essential step in effective use of nanoparticles to enhance conductivities of isotropic conductive adhesives (ICAs), and in the application of ink-jettable conductive connectors in flexible electronics [46, 47].

5.7 Mechanical Properties

The effects of nanoscale particulates on the mechanical properties of thin films have been well studied [48, 49] and generally improve as film thicknesses and particulate sizes decrease. At an elementary level, one can consider the improvement to be due to the relative lack of grain boundaries and defects in nanocrystals [50]. For metallic thin films, for example, yield strength is proportional to $r^{-1/2}$ (the Hall–Pecht relation) [48] and granular cermets display discontinuities in mechanical properties at the percolation threshold, with increased hardness figures observed for discrete particulate structures [49]. The different behaviors above and below percolation are due to the ability of dislocations to move along continuous metallic percolation paths above the threshold, whereas they are confined to the nanoparticles below [49].

In new applications, however, and for the effective modeling of nanocomposites at the nanoscale, the mechanical properties of the nanoparticles themselves must be known. The physical problems in making such measurements on individual nm-scale particles will be obvious, but progress is being made. A theoretical basis for the increase of Young's modulus, compressibility, etc., of nanodimension materials has been established [51] and shows that the dramatic increases in such properties begin (for Cu as an example) at 20 nm, accelerating below about 5 nm.

Direct measurement of the hardness of individual Si nanoparticles of radii 20–50 nm gives values around five times the bulk value for the smaller sizes, increasing with successive measurements as dislocations accumulate within the particles [52].

5.8 Coulomb Block

The energy of a charged conducting spherical particle of radius r is

$$\Delta E = \frac{q^2}{4\pi\varepsilon}\frac{1}{r},$$

where q is the charge and Δ is the effective dielectric constant of the surrounding medium, ΔE being the work done in removing charge $-q$ from the initially neutral sphere to infinity. If the charge only needs to be removed to distance s from the sphere, e.g., to a contact or adjacent particle "island," the work done is reduced (Fig. 5.4) to [4, 53]

$$\Delta E = \frac{q^2}{4\pi\varepsilon}\left[\frac{1}{r} - \frac{1}{r+s}\right].$$

In an assembly of $N\infty$ nanoparticles, Maxwell-Boltzmann (MB) statistics predict that n particles will be charged, where [4, 53]

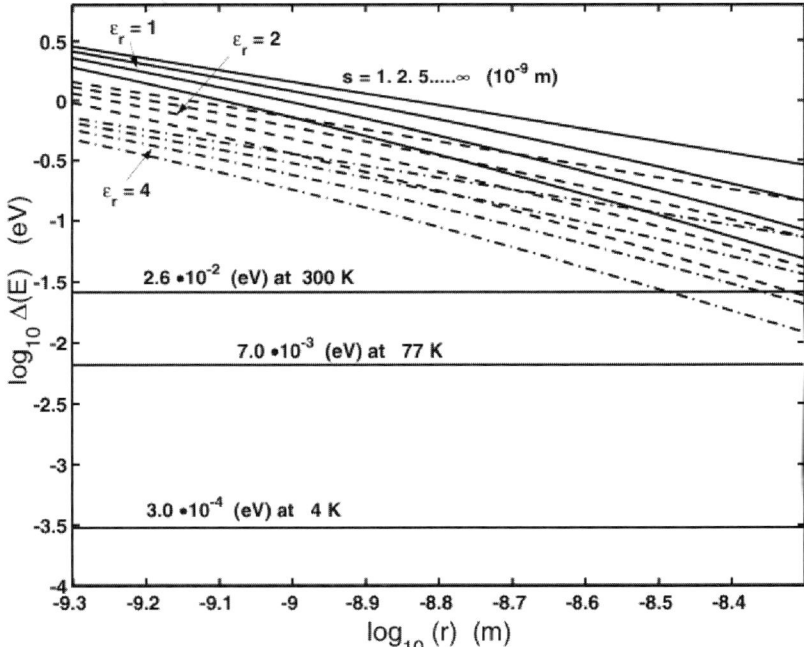

Fig. 5.4 Electrostatic charging energy ΔE as a function of island radius r. The three clusters of lines correspond to $\varepsilon_r = 1$ (*top*), 2, and 4. Within each cluster, the four lines correspond to gap width, $s = 1$ (*bottom*), 2, 5,…∞ nm. Thermal energy kT is shown for comparison at $T = 4$, 77, and 300 K [54]

$$n = N_\infty \exp-\frac{\Delta E}{kT}.$$

Similarly, a single nanoparticle would be randomly thermally charged for a proportion $\exp(-\Delta E/kT)$ of the time [54, 55].

The thermal charging energy is reduced by the application of an electric field F [56], which can supply part of the energy required (Fig. 5.5) [54]. At sufficiently high fields, the electrostatic "barrier" may disappear entirely, and this is the condition typically quoted for conduction to occur through a "coulomb blockade," consisting of a conducting nanoisland between source and drain terminals, at separations sufficiently small for electron tunneling to occur. At 0 K, the abrupt threshold voltage is at $V = \Delta E/q$, but at finite temperatures the I–V characteristics are rounded, due to thermal charging effects, until all nonlinearity vanishes when $T \sim \Delta E/k$. (These effects may be seen in Figs. 8.15 and 8.16.)

In practice, as mentioned above, small metallic islands on an insulating substrate are oblate ellipsoids of eccentricity e, and the charging energies must be modified (writing $R = 2r + s$ and $p = s/R$) to [57]:

- For

$$F < F_{min} = (q^2/4\pi\varepsilon R)4p(1+p)^{-1}[(1+p)^2 - e^2(1-p)^2]^{-1/2},$$
$$\Delta E = q^2/C = (q^2/4\pi\varepsilon R)(2/e)[\sin^{-1} e - \sin^{-1}(e(1-p)/(1+p))]/(1-p) - qRF$$

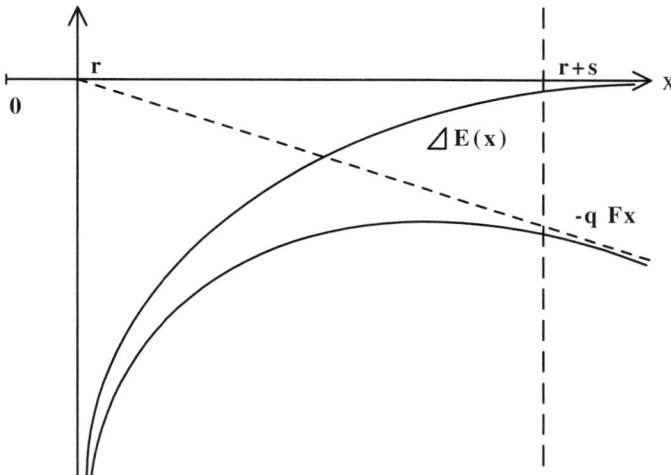

Fig. 5.5 Electrostatic charging energy for island of radius r. The composite function $\Delta E(x) - qFx$ develops a maximum at high fields, where $\Delta E(x) = q^2/4\pi\varepsilon x$ [54]

- For

$$F_{min} < F < F_{max} = (q^2/4\pi\varepsilon R)4p(1-p)^{-2}(1-e^2)^{-1/2},$$

$$\Delta E = (q^2/4\pi\varepsilon R)(2/e)[\sin^{-1}e - \sin^{-1}(e(1-p)R/((1-p)R+2x))]/(1-p) - qFx/p,$$

where

$$x = \frac{1}{2}\text{Re}(1-p)\{[([\{2qp/\pi\varepsilon F(1-p)^2 R^2 e^2\}^2 + 1]^{1/2} + 1)/2]^{1/2} - e^{-1}\}.$$

- $\Delta E = 0$ for $F > F_{max}$.

Conduction across an assembly of nanoparticles takes place by electron tunneling [4], which leads to thermally activated conduction of the form

$$\sigma = \sigma_0 \exp-[\Delta E/kT]^{1/n},$$

where $1 \le n \le 3$ [4]. The MB distribution of charged islands is swamped by injected charge, which sets up a space charge distribution and a nonlinear field distribution across the film [58–60]. The resistance of such films is sensitive to environmental gases, which modulate the metallic work function, and hence the tunneling barrier [61, 62]. The Pd/H_2 combination is unique, in that the dissolution of H_2 in the Pd lattice changes the tunneling barrier width as well as height [63–65]. Applied stress will change the tunneling gap width, too, leading to a very high gage factor, which is linear in s until the stress also affects ΔE [66]. Polarization or ion drift in the DMTF substrate [67] (or cermet insulator [3]) can lead to long-term drift in the characteristics and residual currents or hysteresis effects. Islands evolve slowly toward their thermal equilibrium shapes by surface self-diffusion, also causing long-term drift in the electrical properties [68]. And, as a material with a negative temperature coefficient, island films and cermets are subject to thermal runaway, which can lead to thermal switching [69]. Reproducible switching has also been observed in such films [3].

5.9 Diffusion

Metal atoms may diffuse into the insulating phase from metal-to-ceramic/polymer interfaces, creating a localized "cermet" phase. A proportion of the migrating atoms may nucleate into stable nanoparticles that still diffuse away from the interface, but at a much reduced rate [70–73].

5.10 Optical Properties

The existence of a structure-dependent optical absorption peak in discontinuous thin metal films and other nanoparticle aggregates is well documented, with many papers in the research literature [74, 75]. The colors of Ag aggregates in solution can change from yellow to blue with nanoparticle shape changes or solution pH variation [76], and discontinuous Au films can change from blue to green to magenta as the structure changes with thermal annealing [77]. These effects are unlikely to impinge upon electronics packaging unless possibly with optical interconnect systems, but are included here to make the point that nanotechnologies have been with us for a long time and that a full appreciation of their historical context goes back to Faraday [78] and Rayleigh [79], who correctly attributed the colors of gold glass, gold leaf, and aqueous suspensions to the nanoparticle dimensions in the 1800s. The formal theory was first developed by Maxwell Garnett (MG) [80, 81] and Mie [82] in terms of electron resonances in spherical metal nanoparticles in the 1900s, with contributions by Bruggeman [83] and David [84] in the 1930s, and by Schopper [85] in 1951, before effective medium approaches [86] gained ground in the 1970s.

Yamaguchi et al. [87] and Niklasson and Craighead [88] both showed that better agreement between experiment and theory was possible by treating the discontinuous thin film nanoparticles as oblate spheroids (with the minor axis perpendicular to the substrate) rather than as spheres. Norrman et al. [89, 90] similarly modified the MG theory for prolate spheroids (with the major axis parallel to the substrate), while Granqvist and Hunderi [91] considered both models before concluding more complex mixtures of shapes are necessary to match experiment.

Comparisons of experiment and theory have been made for varied structures: discontinuous thin metal films [92–96], cermets [97–99], and other colloidal forms [100–102]. Comparisons are made to both the MG [92–95, 97, 98, 102] and Mie [96, 98, 100–102] theories. Doremus compared the optical properties of Ag and Au nanoparticles in sol and cermet forms [103, 104] and discontinuous Au films [105] to the Mie theory, but finally concludes that the MG theory correctly predicts the absorption peak position, but that the width depends upon mean free path limitations in the nanoparticles [7, 105]. Other authors have concluded that deviations of experiment from theory were due to the inappropriate use of bulk parameters.

References

1. C. Tian, B. Mao, E. Wang, Z. Kang, Y. Song, C. Wang & L. Xu, One-step, size-controllable synthesis of stable Ag nanoparticles, Nanotechnology 18 (2007) 285607
2. J.E. Morris, Structure and electrical properties of Au–SiO thin film cermets, Thin Solid Films 11 (1972) 299–311
3. A. Kiesow, J.E. Morris, C. Radehaus & A. Heilmann, Switching behavior of plasma polymer films containing silver nanoparticles, J. Appl. Phys. 94 (2003) 6988–6990
4. J.E. Morris & T.J. Coutts, Electrical conduction inn discontinuous metal films; a discussion, Thin Solid Films 47 (1977) 3–65
5. C.A. Neugebauer, in *Handbook of Thin Film Technology*, L.I. Maissel & R. Glang (editors), McGraw-Hill, New York, 1970
6. Yu.F. Komnik, V.V. Pilipenko & L.A. Yatsuk, Changes in lattice spacings in bismuth and zinc island films, Thin Solid Films 52 (1978) 313–327

7. Y.K. Mishra, S. Mohapatra, D.K. Avasthi, D. Kabiraj, N.P. Lalla, J.C. Pivin, H. Sharma, R. Kar & N. Singh, Gold–silica nanocomposites for the detection of human ovarian cancer cells: a preliminary study, Nanotechnology 18 (2007) 345606
8. S. Pothukuchi, Y. Li & C.-P. Wong, Shape controlled synthesis of nanoparticles, and their incorporation into polymers, Proceedings of the 54th Electronic Components and Technology Conference, Las Vegas, 2004, pp. 1965–1967
9. J.E. Morris, Non-ohmic properties of discontinuous thin metal films, Thin Solid Films 11 (1972) 81–89
10. S. Norrman, T. Andersson & C.G. Granqvist, Optical properties of discontinuous gold films, Phys. Rev. B 18(2) (1978) 674–695
11. R.M. Hill, J. Appl. Phys. 37 (1966) 4590
12. D.J. Smith, A.K. Petford-Long, L.R. Wallenberg & J.-O. Bovin, Dynamic atomic-level rearrangements in small gold particles, Science 233 (1986) 872–875
13. C.P. Poole & F.J. Owens, *Introduction to Nanotechnology*, Wiley, New York, 2003, p. 86
14. Y. Fukano & C.M. Wayman, Shapes of nuclei of evaporated fcc metals, J. Appl. Phys. 40(4) (1969) 1656–1664
15. W. Romanowski, Equilibrium forms of very small metallic crystals, Surf. Sci. 18 (1969) 373–388
16. H. Terajima, S. Ozawa & S. Fujiwara, Nucleus shape of vacuum-deposited bismuth films, Thin Solid Films 18 (1973) S7–S9
17. F. Aguilera-Granja, A. Vega, J. Rogan & G. Garcia, Metallic behavior of Pd atomic clusters, Nanotechnology 18 (2007) 365706
18. U. Kreibig, The transition cluster–solid state in small gold particles, Solid State Commun. 28 (1978) 767–769
19. D.R. Snider & R.S. Sorbello, Variational calculation of the work function for small metal spheres, Solid State Commun. 47(10) (1983) 845–849
20. M.S. Sodha & P.K. Dubey, Dependence of Fermi energy on size, J. Phys. D: Appl. Phys. 3 (1970) 139–144
21. V.V. Kolesnikov, E.V. Polozhentsev, V.P. Sachenko & A.P. Kovtun, Size fluctuations of the work function in small metallic clusters, Sov. Phys. Solid State 19(5) (1977) 883–884
22. R. O'Connor, *Fundamentals of Chemistry* (2nd edition), Harper & Row, New York, 1977, pp. 398–399
23. V.L. Colvin & K.M. Kulinowski, Nanoparticles as catalysts for protein fibrillation, Proc. Natl Acad. Sci. USA 104(21) (2007) 8979–8980
24. Y. Huh, M.L.H. Green, Y.H. Kim, J.Y. Lee & C.J. Lee, Control of carbon nanotube growth using cobalt nanoparticles as catalyst, Appl. Surf. Sci. 249 (2005) 145–150
25. H.K. Edwards, E. Evans, S. McCaldin, P. Blood, D.H. Gregory, M. Poliakoff, E. Lester, G.S. Walker & P.D. Brown, Hydrothermally synthesized Fe2O3 nanoparticles for the CVD production of graphitic nanofibres, J. Phys.: Confer. Series 26 (2006) 195–198
26. F. Sammy, The growth of GaN nanowires using nano particles as catalyst, 2004 NNIN REU Research Accomplishments, pp. 112–113
27. H. Yoon, S. Ko & J. Jang, Nitrogen-doped magnetic carbon nanoparticles as catalyst supports for efficient recovery and recycling, Chem. Commun. (2007) 1468–1470
28. J.R. Sambles, L.M. Skinner & N.D. Lisgarten, An electron microscope study of evaporating small particles: the Kelvin equation for liquid lead and mean surface energy of solid silver, Proc. R. Soc. London A 3184 (1970) 507–522
29. J.R. Sambles, An electron microscope study of evaporating gold particles: the Kelvin equation for liquid gold and the lowering of the melting point of solid gold particles, Proc. R. Soc. London A 324 (1971) 339–351
30. S.J. Peppiatt & J.R. Sambles, The melting of small particles. I. Lead, Proc. R. Soc. London A 345 (1975) 387–399
31. S.J. Peppiatt, The melting of small particles. II. Bismuth, Proc. R. Soc. London A 345 (1975) 401–412
32. C.L. Reynolds, P.R. Couchman & F.E. Karasz, On the relation between surface energy, melting temperature, and interatomic separation for metals, Philos. Mag. 34(4) (1976) 659–661

33. B. Lewis, The enhanced vapour pressure of small clusters, Thin Solid Films 9 (1972) 305–308
34. G.L. Allen, R.A. Bayles, W.W. Gile & W.A. Jesser, Small particle melting of pure metals, Thin Solid Films 144 (1986) 297–308
35. P.-A. Buffat, Lowering of the melting temperature of small gold crystals between 150 Å and 25 Å diameter, Thin Solid Films 32 (1976) 283–286
36. W.A. Jesser, R.Z. Shneck & W.W. Gile, Solid–liquid equilibria in nanoparticles of Pb–Bi alloys, Phys. Rev. B 69 (2004) 144121
37. G. Nimtz, P. Marquardt, D. Stauffer & W. Weiss, Raoult's law and the melting point depression in mesoscopic systems, Science 242 (1988) 1671–1672
38. V.N. Bogomolov, A.I. Zadorozhnii, A.A. Kapanadze, E.L. Lutsenko & V.P. Petranovskii, Effect of size on the temperature of "melting" of 9 Å metallic particles, Sov. Phys. Solid State 18(10) (1976) 1777–1778
39. Z.M. Ao, W.T. Zheng & Q. Jiang, Size effects on the Kauzmann temperature and related thermodynamic parameters of Ag nanoparticles, Nanotechnology 18 (2007) 255706
40. H.H. Farrell & C.D. Van Siclen, Binding energy, vapor pressure, and melting point of semiconductor nanoparticles, J. Vac. Sci. Technol. B 25(4) (2007) 1441–1447
41. S.C. Hendy, A thermodynamic model for the melting of supported metal nanoparticles, Nanotechnology 18 (2007) 175703
42. S.L. Lai, J.Y. Guo, V. Petrova, G. Ramanath & L.H. Allen, Size-dependent melting properties of small tin particles: nanocalorimetric measurements, Phys. Rev. Lett. 77(1) (1996) 99–102
43. H. Jiang, K. Moon, F. Hua & C.-P. Wong, Thermal properties of tin/silver alloy nanoparticles for low temperature lead-free interconnect technology, Proceedings of the 57th Electronic Components and Technology Conference, Reno, 2007, pp. 54–58
44. M. Ohring, *Materials Science of Thin Films: Deposition & Structure* (2nd edition), Academic, New York, 2002, pp. 395–397
45. H. Zhu & R.S. Averback, Sintering processes of two nanoparticles: a study by molecular-dynamics simulations, Philos. Mag. Lett. 73(1) (1996) 27–33
46. C. Eberspacher, C. Fredric, K. Pauls & J. Serra, Thin-film CIS alloy PV materials fabricated using non-vacuum, particles-based techniques, Thin Solid Films 387 (2001) 18–22
47. S.H. Ko, H. Peng, C.P. Grigoropoulos, C.K. Luscombe, J.M.J. Frechet & D. Poulikakos, All-inkjet-printed flexible electronics fabrication on a polymer substrate by low-temperature high-resolution selective laser sintering of metal nanoparticles, Nanotechnology 18 (2007) 345202
48. D.A. Hardwick, Mechanical properties of thin films: a review, Thin Solid Films 154 (1987) 109–124
49. R.C. Cammarata, Mechanical properties of nanocomposite thin films, Thin Solid Films 240 (1994) 82–87
50. C.A. Wert & R.M. Thomson, *Physics of Solids* (2nd edition), McGraw-Hill, New York, 1970
51. R. Dingreville, J. Qu & M. Cherkaoui, Effective elastic modulus of nano-particles, Proceedings of the 9th International Symposium on Advanced Packaging Materials, Atlanta, 2004, pp. 187–192
52. C.R. Perrey, W.M. Mook, C.B. Carter & W.W. Gerberich, Characterization of mechanical deformation of nanoscale volumes, Mater. Res. Soc. Symp. Proc. 740 (2003) 13.13.1–13.13.6
53. C.A. Neugebauer & M.B. Webb, Electrical conduction mechanism in ultrathin, evaporated metal films, J. Appl. Phys. 33 (1962) 74–82
54. J.E. Morris, Single-electron transistors, in *The Electrical Engineering Handbook* (3rd edition): *Electronics, Power Electronics, Optoelectronics, Microwaves, Electromagnetics, and Radar*, R.C. Dorf, CRC/Taylor & Francis, Boca Raton/London, 2006, pp. 3.53–3.64
55. J.E. Morris, C. Radehaus, M. Hietschold, A. Kiesow & F. Wu, Single electron transistors & discontinuous thin films, in *The World of Electronic Packaging and System Integration*, B. Michel & R. Aschenbrenner (editors), dpp goldenbogen, 2004, pp. 84–93
56. J.E. Morris, Calculation of activation energy in discontinuous thin metal films, J. Appl. Phys. 39 (1968) 6107–6109

57. J.E. Morris, Non-ohmic properties of discontinuous thin metal films, Thin Solid Films 11 (1972) 81–89
58. F. Wu & J.E. Morris, Modeling conduction in asymmetrical discontinuous thin metal films, Thin Solid Films 317 (1998) 178–182
59. J.E. Morris, Recent progress in discontinuous thin metal film devices, Vacuum 50(1–2) (1998) 107–113
60. J.E. Morris, Electrical conduction in discontinuous thin metal films, in *Metal/Non-Metal Microsystems: Physics, Technology & Applications*, Vol. 2780, B. Licznerski & A. Dziedzic (editors), SPIE International Society for Optical Engineering, pp. 64–714
61. J.E. Morris, Resistance changes of discontinuous thin gold films in air, Thin Solid Films 5 (1970) 339–353
62. J.E. Morris & M. O'Krancy, Resistance increase of discontinuous gold films by substrate absorption of oxygen, Thin Solid Films 10 (1972) 319–320
63. J.E. Morris & F. Wu, The effects of hydrogen absorption on the resistance of discontinuous palladium films, Thin Solid Films 246 (1994) 17–23
64. J.E. Morris, A. Kiesow, M. Hong & F. Wu, The effect of hydrogen absorption on the electrical conduction of discontinuous palladium thin films, Int. J. Electron. 81(4) (1996) 441–447
65. J.E. Morris, A. Kiesow, M. Hong & F. Wu, The effect of hydrogen absorption on the electrical conduction of discontinuous palladium thin films, in *Metal/Non-Metal Microsystems: Physics, Technology & Applications*, Vol. 2780, B. Licznerski & A. Dziedzic (editors), SPIE International Society for Optical Engineering, pp. 245–248
66. J.E. Morris, The effect of strain on the electrical properties of discontinuous thin metal films, Thin Solid Films 11 (1972) 259–272
67. J.E. Morris, The influence of soda-lime substrate ion drift on the resistance of discontinuous thin gold films, J. Vac. Sci. Technol. 9 (1972) 1039–1040
68. J.E. Morris, The post-deposition resistance increase in discontinuous metal films, Thin Solid Films 28 (1975) L21–L23
69. J.E. Morris, Self-heating effects in discontinuous metal films, Thin Solid Films 35 (1975) 165–168
70. J.H. Das & J.E. Morris, Diffusion and self-gettering of ion-implanted copper in polyimide, J. Appl. Phys. 66(12) (1989) 5816–5820
71. J.E. Morris & J.H. Das, Diffusion and aggregation of copper in polymers, in *Electronics Packaging Forum*, Vol. 3, J.E. Morris, IEEE, New York, 1994, pp. 41–71
72. J.E. Morris & J. Das, Metal diffusion in polymers, IEEE Trans. CPMT-B: Adv. Packag. 17 (1994) 620–625
73. J.H. Das & J.E. Morris, Diffusion and gettering simulations of ion implanted copper in polyimide, in *Metallized Plastics 2*, K.L. Mittal, Plenum, New York, 1991, pp. 114–161
74. H.G. Craighead & G.A. Niklasson, Characterization and optical properties of arrays of small gold particles, Appl. Phys. Lett. 44(12) (1984) 1134–1136
75. N.L. Dmitruk, O.S. Kondratenko, S.A. Kovalenko & I.B. Mamontova, Size effects in optical properties of thin metal films, Proceedings of the 1st International Workshop on Semiconductor Nanocrystals, Budapest, 2005, pp. 227–230
76. Y. Chen, C. Wang, Z. Ma & Z. Su, Controllable colours and shapes of silver nanostructures based on pH: application to surface-enhanced Raman scattering, Nanotechnology 18 (2007) 325602
77. J.E. Morris, unpublished (1969)
78. M. Faraday, Philos. Trans. R. Soc. Lond. 147 (1857) 145
79. L. Rayleigh, Philos. Mag. 44 (1897) 28–52
80. J.C. Maxwell Garnett, Colours in metal glasses and in metallic films, Philos. Trans. R. Soc. Lond. A 203 (1904) 385–420
81. J.C. Maxwell Garnett, Colours in metal glasses, in metallic films, and in metallic solutions, Philos. Trans. R. Soc. Lond. A 205 (1905) 237–288
82. G. Mie, Ann. Phys. (Leipzig) 25 (1908) 377
83. D.A.G. Bruggeman, Ann. Phys. 24 (1935) 636

84. E. David, Interpretations of the anomalies in the optical constants of thin metal films, Z. Phys. 114 (1939) 389–406
85. H. Schopper, Z. Phys. 130 (1951) 565
86. D.M. Wood & N.W. Ashcroft, Effective medium theory of optical constants of small particle composites, Philos. Mag. 35(2) (1977) 269–280
87. T. Yamaguchi, M. Takiguchi, S. Fujioka & H. Takahashi, Optical absorption of submonolayer gold films: size dependence of ε_{bound} in small island particles, Surf. Sci. 138 (1984) 449–463
88. G.A. Niklasson & H.G. Craighead, Optical response and fabrication of regular arrays of ultrasmall gold particles, Thin Solid Films 125 (1985) 165–170
89. S. Norrman, T. Andersson & C.G. Granqvist, Optical absorption in discontinuous gold films, Solid State Commun. 23 (1977) 261–265
90. S. Norrman, T. Andersson & C.G. Granqvist, Optical properties of discontinuous gold films, Phys. Rev. B 18(2) (1978) 674
91. C.G. Granqvist & O. Hunderi, Optical properties of ultrafine gold particles, personal communication (1977)
92. L. Ward, The effective optical constants of thin metal films in island form, Br. J. Appl. Phys. (J. Phys. D) Ser. 2, 2 (1969) 123–125
93. J.P. Marton & M. Schlesinger, Optical constants of thin discontinuous nickel films, J. Appl. Phys. 40(11) (1969) 4529–4533
94. A.R. Vamdatt & Y.G. Naik, Application of Maxwell Garnett theory to antimony films, Thin Solid Films 8 (1971) R30–R32
95. F. Parmigiani, M. Scagliotti, G. Samoggia & G.P. Ferraris, Influence of the growth conditions on the optical constants of thin gold films, Thin Solid Films 125 (1985) 229–234
96. H. Kuwata, H. Tamaru, K. Esumi & K. Miyano, Resonant light scattering from metal nanoparticles: practical analysis beyond Rayleigh approximation, Appl. Phys. Lett. 83(22) (2003) 4625–4627
97. J.C.C. Fan & P.M. Zavracky, Selective black absorbers using MgO/Au cermet films, Appl. Phys. Lett. 29(8) (1976) 478–480
98. T. Yamaguchi, M. Sakai & N. Saito, Optical properties of well-defined granular metal systems, Phys. Rev. B 32(4) (1985) 2126–2130
99. M. Eichelbaum, B.E. Schmidt, H. Ibrahim & K. Rademann, Three-photon-induced luminescence of gold nanoparticles embedded in and located on the surface of glassy nanolayers, Nanotechnology 18 (2007) 355702
100. C.R. Sabanayagam & J.R. Lakowicz, Fluctuation correlation spectroscopy and photon histogram analysis of light scattered by gold nanospheres, Nanotechnology 18 (2007) 355402
101. R.C. Johnson, J. Li, J.T. Hupp & G.C. Schatz, Hyper-Rayleigh scattering studies of silver, copper, and platinum nanoparticle suspensions, Chem. Phys. Lett. 356 (2002) 534–540
102. D.D. Nolte, Optical scattering and absorption by metal nanoclusters in GaAs, J. Appl. Phys. 76(6) (1994) 3740–3745
103. R.H. Doremus, Optical properties of small silver particles, J. Chem. Phys. 42(1) (1965) 414–418
104. R.H. Doremus, Optical properties of small gold particles, J. Chem. Phys. 40 (1964) 2389–2396
105. R H. Doremus, Optical properties of thin metal films in island form, J. Appl. Phys. 37(7) (1966) 2775–2781

Chapter 6
Nanoparticle Fabrication

Y. Hayashi(✉) M. Inoue, H. Takizawa, and K. Suganuma

6.1 Introduction

Recently, a variety of high-value-added nanotechnologies have been developed. Nanomaterials involve the development and utilization of structures and devices with organizational features at the intermediate scale between individual atoms and under 100 nm, where novel properties occur as compared to bulk materials.

This implies the capability to build up tailored nanostructures and various properties for given functions by control at the molecular levels [1]. Nanoparticles make up one of the most important nanomaterials subgroups because nanoparticle manufacturing is an essential component of nanotechnology. Also, assembling of nanoparticles and related structures is the most generic route to generate nanostructured materials and to build up bulk nanomaterials.

Full-fledged metal nanoparticle research started in the 1970s. It was shown in this research that the melting point is decreased by reducing a metallic particle to the nanolevel [2]. This is theoretically shown for the proportion of the surface area in the entire volume to increase, and to become unstable by reducing it to the nanolevel. However, special nanoparticles have been used for much longer, as in the stained glass windows of a church [3]. This is the surface plasmonic effect of gold and silver nanoparticles, leading to red colors in gold nanoparticle dispersions glass and yellow colors for silver nanoparticle dispersions. That is, people have been using nanotechnology and nanoparticles since the age before quantum mechanics. The history of the metal nanoparticle is the oldest in nanotechnology.

Fabrication of the metal nanoparticle is easy, and is difficult. For example, the fabrication technique is very easy compared with ceramic nanoparticles. So, what is so difficult? The problem is the stabilization of the metallic nanoparticle. Oxidation and agglomeration are big problems. Since they have large surface areas, nanoparticles are easily oxidized, especially base metal nanoparticles, so a surface protection agent is necessary to prevent it. However, when the surface effect is used,

Y. Hayashi
Department of Applied Chemistry, Tohoku University, 6-6-07 Aoba Aramaki, Aoba-ku, Sendai 980-8579, Japan

the protection agent becomes a factor to disturb the nanocharacter. This problem is the most difficult one in metal nanoparticle fabrication.

This chapter explains methods of metal nanoparticle manufacture and control.

6.2 Metal Nanoparticle Fabrication Method

In general, metal nanoparticles are fabricated by breakdown or buildup methods. Figure 6.1 shows fabrication concepts of breakdown and buildup methods.

The breakdown method is a technique for crushing the bulk metal by mechanical grinding (MG) or mechanical milling (MM). It is very difficult to control the particle diameter at the nanolevel, though it is an easy technique. Moreover, there is a problem in that impurities are easily mixed by a vigorous and long-term milling, but the technique is considered unsuitable because plasticity is transformed in the case of soft metals.

The buildup method is a technique for assembling metallic atoms and has a lot of variations. This method is divided roughly into chemical and physical processings. A primary technique is introduced here for nanoparticle pastes for wiring.

6.2.1 Chemical Processing

Chemical processing especially covers a lot of different buildup techniques.

Basically, it is a technique for reducing ions in solution by a reducing agent and heating. Reduction techniques also include radiation and supersonic waves, etc.

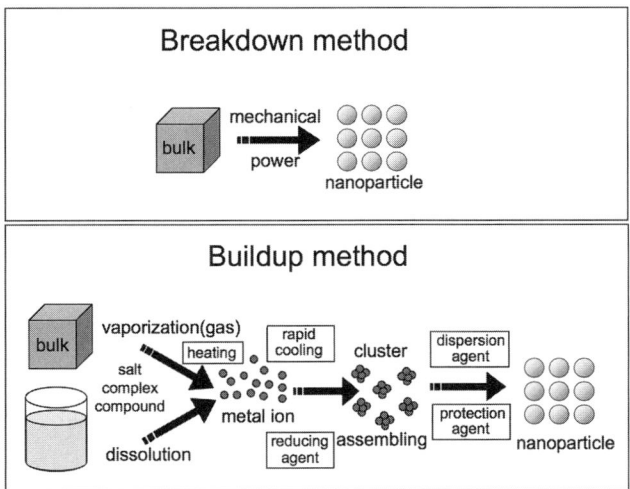

Fig. 6.1 Fabrication concepts of breakdown and buildup methods

Table 6.1 Popular reducing agent and energy for metal nanoparticles

Reducing materials	Alcohols, aldehydes, sugars hydroxy acids, hydroxylamines thiols, $NaBH_4$, B_2H_6, H_2, Sn^{2+}, Co^{2+}, etc.
Reducing energy (irradiation)	Heating (including special heating)photo ultrasound, etc.

Table 6.2 Popular metal sources for metal nanoparticle fabrication

Metal	Bulk, flake, powder, etc.
Metal salt	Nitrates, chlorides, hydrosulfates, cyanides
Organometal complex	carbonyls, fatty acids, alkoxides, etc.

The basic fabrication concept is the same as a traditional metallic reduction technique using metal salt solution and hydrazine, modified for the nanoparticle fabrication. It is the control of the reactive rate that differs from the older techniques. Supersaturation control is important in metal nanoparticle fabrication. When a metallic salt is used as the source in nanoparticle fabrication, the reducing agent is especially important. A mild reducing agent is suitable for nanoparticle fabrication, because the reduction speed is slowed and it is easy to control particle size. Particles grow large in the case of strong reducing agents such as hydrazine. With a fast reaction rate, it is not easy to control the nanoparticle size.

Popular reducing agents and energy used are shown in Table 6.1.

The amine-related compounds and organic acids (e.g., citric acid) act not only as the reducing agent but also as stabilization agents. In the reduction, the metallic source is also important.

The range of organic metallic compounds is infinite, and therefore the control of physical properties is relatively easy to control. For instance, synthesis is also possible of materials that act as protection agents for metal sources, such as metallic fatty acids, and are easily reduced to metal, like the carbonyl compounds.

Table 6.2 shows popular metal sources for metal nanoparticle fabrication.

In chemical processing, both choice and balance of a reducing agent and a metallic source are important, but heating is also an important element to control.

6.2.2 Physical Processing

Physical processing implies physical power and a phase reaction, as opposed to a chemical solute reaction. The breakdown method is a physical process and an example of physical power. Phase reactions are divided into gas and liquid phase methods (Fig. 6.2) [4–6].

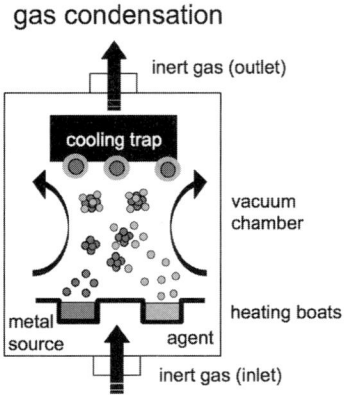

Fig. 6.2 Gas condensation and atomization methods

The gas condensation method is the most popular nanoparticle fabrication method amongst physical processes. In the gas condensation method, a metallic or inorganic material, or organometallic compound, is vaporized by thermal evaporation sources, such as Joule heated refractory crucibles, electron beam evaporation devices under low pressure, or inert gas. The metallic cluster is formed in the vicinity of the source by homogeneous nucleation in the gas phase. In this method, particle size depends on the residence time of the particles in the growth regime and can be influenced by the gas pressure and the kind of inert gas, evaporation rate, or vapor pressure. In general, the average particle size of the nanoparticles increases with gas pressure and vapor pressure.

Another technique is the atomization method [7], where the molten metal is rapidly cooled into droplets. Industry is very advanced in the methods of metallic particle fabrication at the micron and submicron levels.

6.3 Novel Processing Routes for Eco-Fabrication

It has been explained that the traditional fabrication processes of metal nanoparticles are basically either physical or chemical methods. Either way, to obtain nanoparticles, the reduction of metal ions normally requires high temperature, reducing agents, and supersaturated conditions. Here, we want to explain the problem for industrialization.

In industrialization, the large problems are cost and environmental problems, and these have been closely linked. Figure 6.3 shows the relationship of cost and environmental impact on metal nanoparticle preparation [8, 9].

In the physical method, it is easy to fabricate high purity nanoparticles because of their production in inert gas. However, high-temperature heating and a large-scale

6 Nanoparticle Fabrication

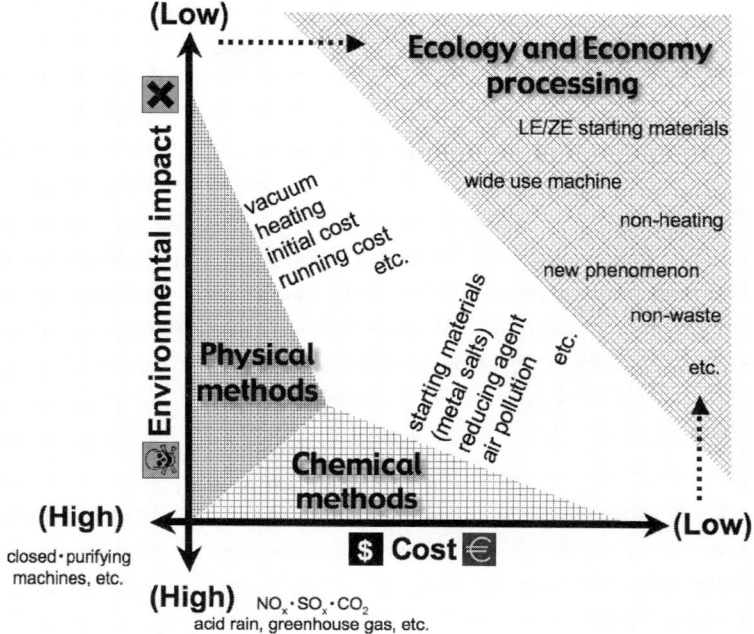

Fig. 6.3 Cost vs. environmental impact in metal nanoparticle fabrication

chamber filled with inert gas are necessary. And, yield is not so high either. Therefore, a large initial cost and high running costs are both necessary.

On the other hand, the chemical method is cheap because spontaneous chemical reduction reaction is used. However, there is a problem with the raw material (metal source and reduction agent). A metal salt is almost always used as a metal source if using a reducing agent, and many metallic salts contain the ions, NO^{3-}, SO_4^{2-}, Cl^-, etc., that produce an acid deposition problem, which necessitates a wet scrubber and washing of the nanoparticles to remove.

When an organometallic compound is used as the metallic source, 80 wt% or more might become organic waste. Naturally, it is necessary to remove this waste, so the cost barely changes between physical and chemical methods.

In industry, well-balanced fabrication is needed to control cost and to protect the environment. However, it is difficult to manage both low cost and low environmental impact in these methods. Therefore, innovative fabrication design is required to break through these problems now.

Environmental regulations are especially important in the electronic device field (e.g., nanopackaging), including waste electrical and electronic equipment (WEEE), restriction of hazardous substances (RoHS), and end-of-life vehicle (ELV) [10–12].

The coexistence of high performance and low environmental impact will become increasingly important in nanotechnologies containing nanopackaging. In addition, low product cost is also necessary to maintain an international competitive edge.

One example of metallic nanoparticle fabrication methods that achieve both ecological and economical goals is introduced as follows.

6.3.1 Liquid–Solid Sonochemical Reaction

Ecological and economical fabrication is becoming increasingly important in industry, as mentioned above. This synthesis can be achieved through a number of approaches. Rather than an expensive special-purpose device, an inexpensive general-purpose production device is desired. Also, when synthesizing, it is preferable for neither waste nor air pollution to be generated and for safe and nontoxic raw materials to be used. In addition, synthesis at the lowest temperature possible is desirable with respect to energy conservation (e.g., room temperature).

We developed a new metal nanoparticle synthesis method that provides these elements. This new synthesis method uses an ultrasonic cleaner as a general-purpose device and metal oxide (M_xO_y) and alcohol (C_xH_yOH) as raw materials. The metal oxide does not dissolve in ethanol, so the liquid–solid (alcohol–metal oxide) phase is ultrasonically agitated.

When a solid substance is subjected to pyrolytic treatment with a generating gas, the raw material is converted into fine powder [13, 14]. Noble metal oxides especially are simply decomposed by heating in air, i.e., without needing strongly reducing atmospheres [15]. This reduction is clean and ecological, because noble metal oxides are almost atoxic materials and only generate O_2 in decomposition. The starting materials are noble metal oxides and ethanol (EtOH), which are low emission (LE) materials.

The ultrasonic cleaner is an inexpensive home appliance, and metal oxide and alcohol are generally inexpensive and nontoxic. Ultrasonic processing as a chemical process is called a sonochemical process. In general, the properties of a specific energy source determine the course of the chemical reaction. The ultrasonic irradiation differs from traditional energy sources in duration, pressure, and energy per molecule and is unique in the interaction between energy and matter. The chemical effects of ultrasound do not come from a direct interaction with molecular species. Instead, they are derived principally from acoustic cavitation, which can produce temperatures as high as those on the surface of the sun and pressures as great as those at the bottom of the ocean [16]. Such a reactive place can be achieved with a home electronic ultrasonic cleaner.

We synthesized noble metal nanoparticles by ultrasound in a liquid–solid (EtOH–noble metal oxide) slurry. The synthesis technique is very simple. EtOH and the noble metal oxide powder are simply placed into a beaker and irradiated by ultrasound (Fig. 6.4). When the liquid–solid slurry is irradiated by ultrasound, it reduces to metal nanoparticles.

$$M_xO_y \xrightarrow[\text{in EtOH @ RT}]{} \text{Metal nanoparticle} + O_2 \uparrow \Delta G \qquad (6.1)$$

6 Nanoparticle Fabrication

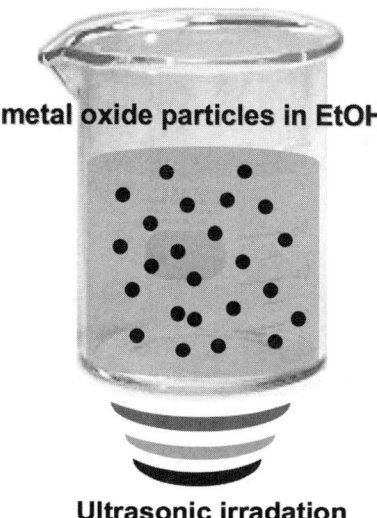

Fig. 6.4 Experimental procedure in liquid–solid sonochemical reaction

In this synthesis, ultrasonic irradiation is important. Cavitation in a liquid occurs due to the stresses induced in the liquid by the passing of a sound wave through the liquid [17–19]. These bubbles are now subjected to the stresses induced by the sound waves. The bubbles are filled with vapor and gas, and bubble implosions occur. These implosions are the remarkable part of sonochemical processes. Each of these imploding bubbles can be seen as a high-temperature hot spot having pressures of several hundred atmospheres. Hot spot reaction is considered to represent direct reduction, with the reduction of metal oxide induced by hot spots formed from ultrasonic cavitation and alcohol. EtOH is also important in this reduction. Ultrasonic reduction is accelerated by alcohol and protects metals from reoxidization. The nucleation of metal occurs at the hot spot in solution, followed by the growth and immobilization of the noble metal particles.

The Gibbs free energy (ΔG) is a thermodynamic potential, and is therefore a state function of a thermodynamic system [20, 21]. This is the most basic concept when thinking about the chemical reaction.

The metal oxide reduction for small ΔG happens easily in this reaction. The values of ΔG of base metal oxides are very large compared with those of noble metal oxides. In addition, the values of ΔG of noble metal oxides are extremely small compared with those of base metals. Among the noble metal oxides, ΔG is the smallest for gold oxide.

Figure 6.5 shows a TEM image of the products and the size distribution of the Ag particles formed in this synthesis. The Ag particles were conformed to be 1 nm in size. Ag particle sizes were changed by PVP concentration. UV absorption peaks were observed at approximately 400 nm. These absorption bands correspond to the

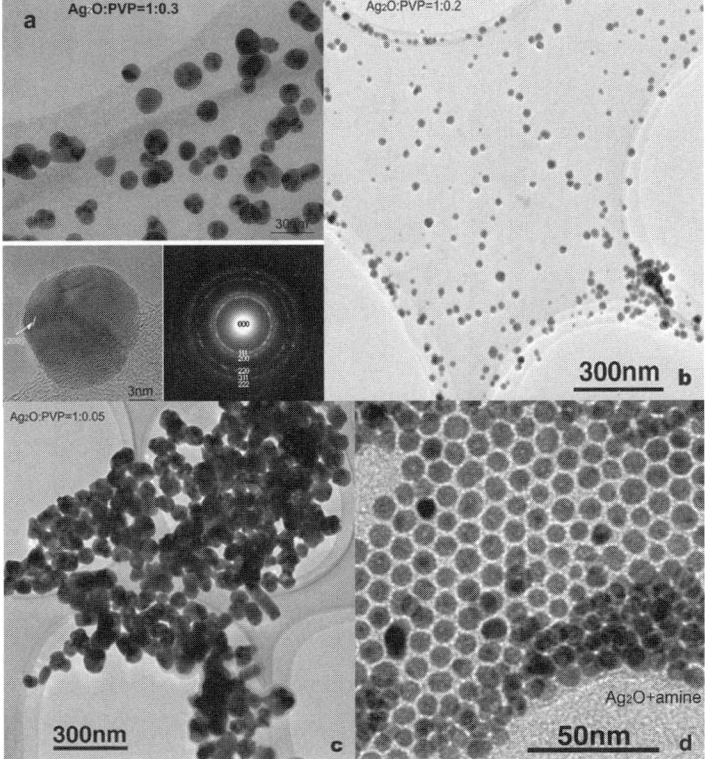

Fig. 6.5 Ag nanoparticle images by liquid–solid sonochemical fabrication

surface plasma resonance absorption of Ag nanosized metallic particles. The absorption peaks were shifted by changes in Ag particle size. Melting point is significantly lower for nanosized particles. These synthesized particles melt and sinter at approximately 100–150°C. If such a synthesized method is used, low-temperature nanowiring that combines low cost and low environmental impact can be achieved. Figure 6.6 shows a TEM image of Pt nanoparticles in sonochemical synthesis. Pt can synthesize the nanoparticles without the dispersing agent, which differs from Ag. Therefore, the surface effect can be expected in nanoparticles.

6.4 Techniques to Avoid Clustering

In general, metal particles immediately agglomerate in solution, so they may be hydrophobic. In the case of nanoparticles, this phenomenon is especially remarkable. To prevent agglomeration, various surface modifications are examined. Techniques to avoid clustering are similar.

6 Nanoparticle Fabrication

Fig. 6.6 Pt nanoparticle images by liquid–solid sonochemical fabrication

The technique that prevents the nanoparticle from clustering is very important. The controlling technique for single nanometer dimensions has essentially been established, and clustering can be prevented by using a surfactant agent. Moreover, if a uniform nanoparticle can be made by using a surfactant agent, it is also possible to fabricate self-assembled nanoparticle monolayers and superlattices.

The technique common to all these is to coat the surface of the particle and the cluster with the surfactant before particle agglomeration and large atom clustering. The surface modification varies according to the material, but the general concept is explained here. Generally, supersaturation is very important to make small particles and clusters.

For example, a raw material of a low concentration and a mild reactive condition (low temperature, mild reducing agent, etc.) are important in nanoparticle fabrication. However, the nanoparticle cannot be made only by the supersaturation. Because the surface of the cluster and the nanoparticle is very active, large clustering and particle agglomeration happen for stabilization. Surface modification of the cluster and the nanoparticle is necessary to prevent these.

Up to now, various materials and techniques have been used for surface modification and for fabricated colloids. One of the colloid fabrication techniques is to make the micelle by using a surface active agent and protective polymer.

A nanoparticle-dispersed solution can be achieved by adding a hydrophilic polymer with the protection effect during fabrication. In this case, the hydrophilic polymer becomes protective of the colloid. The hydrophilic polymer is a surfactant, which are usually organic compounds that are amphipathic, meaning they contain both hydrophobic groups and hydrophilic groups. Therefore, they are typically sparingly soluble in both organic solvents and water.

Micelles. For the solution at metal concentrations below the critical micellar concentration (CMC), the surfactant behaves as a powerful electrolyte above CMC. The monomers form spheroidal aggregates called micelles.

The particle size and morphology can change with increasing surfactant concentration [22, 23]. In terms of particle growth, some analogies between surfactant self-assemblies and natural media can be proposed. In both cases, the growth of particles needs a supersaturated media where the new nucleation takes place. In solution, surfactant molecules self-assemble to form aggregates. At low concentration, the aggregates are generally globular micelles, but these micelles can grow upon an increase of surfactant concentration. Micelles grow in agreement with theoretical prediction on micellization.

Reverse micelles. Reverse micelles are stabilized by a monolayer surfactant. The most famous useful surfactant is sodium bis sulfosuccinate (Na(AOT)) [23]. Besides being a surface active agent and protective polymer, it coats the inorganic matter as a silica excluding cluster, and there is also a technique for preventing clustering [3].

6.4.1 Surfactant of Metal Nanoparticle Paste

The surfactant choice for a metallic nanoparticle paste is amongst the most difficult in various nanoparticle fabrications.

For metal nanoparticles, surfactants are effective to prevent not only clustering and agglomeration but also oxidation of the surface. The metal nanoparticle surface is more active than that of a ceramic powder or micron-sized metal powder. Therefore, it tends to oxidize easily (other than gold nanoparticles).

Moreover, oxidation by ignition is a danger for metal nanoparticles, especially without surfactant. However, it is insufficient to only prevent agglomeration and oxidation in the case of paste.

The character that sintering and electroconduction are not obstructed is also desired in the nanoparticle paste. Concretely, surfactant agents are needed that resolve at low temperature, and obstruct sintering of metal nanoparticles.

Furthermore, the character to be able to distribute metal nanoparticles at high concentrations is requested. Thus, the demands upon the surfactant agent of metal nanoparticle paste point to the ultimate properties. The performance demanded as a surfactant is enumerated for metal nanoparticle pastes in the following:

1. Prevention of agglomeration and clustering
2. Prevention of oxidation
3. Low-temperature resolution
4. High metal nanoparticles concentration
5. Mass production

These factors are demanded in surfactants for metal nanoparticle pastes.

Various surfactants are used for the nanoparticle fabrication and reported in many papers. However, no surfactant that satisfies everything is reported. So, the surfactant ability is demonstrated by the interaction with the solvent, and it is not possible to discuss it only by surfactant effects. There are many trade secret cases without paper or patent.

Anyway, the know-how of this field is important, and metal nanoparticle paste is a necessity for the overall judgment by dispersion, sintering, and electronic properties.

6.5 Summary

Various metal nanoparticle fabrications were introduced in this chapter. The metal nanoparticles can be fabricated comparatively easily now. However, the hurdles for metal nanoparticle paste are high in the list (1)–(5) above.

Moreover, environmental problems are also important to the metal nanoparticle paste used with electronic equipment. Wiring technology that uses metal nanoparticles is being actively researched [24, 25].

The metal nanoparticle is effective for the nanoscale wiring and low printing temperature [26, 27]. In particular, inkjet wiring, which uses metal nanoparticles, has various advantages (e.g., waste fluid problem, many varieties, and small-quantity production on-demand) compared with past techniques. In chemical methods, metal nanoparticles are produced using ionic solutions that include nitrates, chlorides, carbonates, and cyanides, amongst others. These starting materials and reducing agents are toxic and generate pollution during fabrication. Moreover, toxic ions, such as NO_3^{2-}, Cl^-, and CN^-, remain in the preparation solutions after fabrication. These solutions require treatment, because these ions are the origin of environmental pollution problems such as acid rain and the generation of green house and toxic gases. Therefore, companies must treat these solutions with great care and must shoulder high treatment costs.

The cost is also important in practical use; not only high performance but also low production and low running cost technology are necessary in metal nanoparticle pastes.

Acknowledgment The contents of Sect. 6.3.1 were supported by the Industrial Technology Research Grant Program, 2005, through the New Energy and Industrial Technology Development Organization (NEDO) of Japan.

References

1. Pileni, MP (1993) Reverse micelles as microreactors. J. Phys. Chem., 97:6961–6973
2. Buffet, P, Borel, JP (1976) Size effect on the melting temperature of gold particles. Phys. Rev. A, 13(6):2287–2298
3. Mulvaney, P (2001) Not all that's gold does glitter. MRS Bull., 26(12):1009–1014
4. Gleiter, H (1990) Nanocrystalline materials. Prog. Mater. Sci., 33:233–315

5. Granqvist, C, Buhrman, R (1976) Ultrafine metal particles. J. Appl. Phys., 47:2200–2207
6. Hahn, HW et al. (1988) Ceramic powder science. In: Messing, GL et al. (eds) Ceramic Transaction, vol. 1, Part B. American Ceramic Society, Westerville, p 1115
7. Antony, LVM, Reddy, RG (2003) Processes for production of high-purity metal powders. JOM, 55(3):14–18
8. Hayashi, Y et al. (2005) Ecodesigns and applications for noble metal nanoparticles by ultrasound process. IEEE Trans. Electron. Packag. Manuf., 28:338–343
9. Hayashi, Y, Niihara, K (2004) Ceramics nanocomposite. Eng. Mater., 52:50–51
10. Directive 2002/96/EC of the European Parliament and of the Council of 27 January 2003 on waste electrical and electronic equipment (2003)
11. Directive 2002/95/EC of the European Parliament and of the Council of 27 January 2003 on the restriction of the use of certain hazardous substances in electrical and electronic equipment (2003)
12. Directive 2000/53/EC of the European Parliament and of the Council of 18 September 2000 on end-of-life vehicles (2000)
13. West, AR (1984) Basic Solid State Chemistry. Wiley, New York
14. Mizuta, S, Koumoto, K (1996) Materials Science for Ceramics. University of Tokyo Press, Tokyo
15. Hayashi, Y et al. (1999) Mechanical and electrical properties of ZnO/Ag nanocomposites. In: Singh JP et al. (eds) Advances in Ceramic Matrix Composites IV: Ceramic Transaction, vol. 96. American Ceramic Society, Westerville, pp 209–218
16. Crum, LA (1995) Bubbles Hotter than the Sun. New Sci., 146:36–40
17. Luce, JL (1994) Effect of ultrasound on heterogeneous systems. Ultrason. Sonochem., 1:S111–S118
18. Suslick, KS (1990) Sonochemistry. Science, 247:1439–1445
19. Suslick, KS, Price, GJ (1999) Applications of ultrasound to materials chemistry. Annu. Rev. Mater. Sci., 29:295–326
20. Perrot, P (1998) A to Z of Thermodynamics. Oxford University Press, London
21. Reiss, H (1965) Methods of Thermodynamics. Dover, New York
22. Bronstein, L, Antonietti, M, Valetsky, P (1998) Nanoparticles and Nanostructured Films. Wiley-VCH, Weinheim
23. Pileni, MP (1998) Nanoparticles and Nanostructured Films. Wiley-VCH, Weinheim
24. Suganuma, K (ed) (2006) Ink-Jet Wiring of Fine Pitch Circuits with Metallic Nano Particle Pastes. CMC, Tokyo
25. Nanoparticle Industry Review (2005) BCC Research, Stamford
26. Murata, K et al. (2005) Super-fine ink-jet printing: toward the minimal manufacturing system. Microsyst. Technol., 12:2–7
27. Fuller, SB et al. (2002) Ink-jet printed nanoparticle microelectromechanical systems. J. Microelectromech. Syst., 11:54–60

Chapter 7
Nanoparticle-Based High-k Dielectric Composites: Opportunities and Challenges

Jiongxin Lu and C.P. Wong(✉)

7.1 Introduction

The ever-increasing demands of miniaturization, increased functionality, better performance, and low cost for microelectronic products and packaging have been the driving forces for new and unique solutions in system integration such as system-on-chip (SOC) and system-in-package (SiP). Despite the high level of integration, the number of discrete passive components (resistors, capacitors, or inductors) remains very high. In a typical microelectronic product, about 80% of the electronic components are passive components which are unable to add gain or perform switching functions in circuit performance, but these surface-mounted discrete components occupy over 40% of the printed circuit/wiring board (PCB/PWB) surface area, account for up to 30% of solder joints and up to 90% of the component placements required in the manufacturing process. Embedded passives, an alternative to discrete passives, can address these issues associated with discrete parts, including substrate board space, cost, handling, assembly time and yield. Figure 7.1 schematically shows an example of realization of embedded passive technology by integrating resistor and capacitor films into the laminate substrates [1, 2].

By removing these discrete passive components from the substrate surface and embedding them into the inner layers of substrate board, embedded passives can not only provide the advantage of size and weight reduction but also have many other benefits such as increased reliability, improved performance, and reduced cost, which have driven a significant amount of effort during the past decade for this technology. For instance, NIST launched its Advanced Embedded Passives Technology (AEPT) project in 1999 with a group of industrial partners, focusing on developing the materials, design and processing technology for embedded passive devices in circuit board substrates.

C.P. Wong
School of Materials Science and Engineering, Georgia Institute of Technology, 771 Ferst Dr. NW, Atlanta, GA 30332, USA
cp.wong@mse.gatech.edu

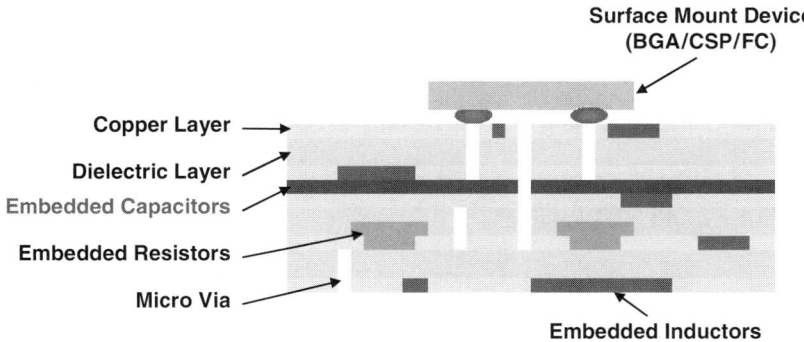

Fig. 7.1 Embedded passives integrated into the laminate substrate

However, embedded passive technology has still not been commercialized for electronic packages due to materials and process issues. Therefore, to enable embedded passive technology, it is necessary to develop materials that satisfy the requirements of fabrication as well as electrical and mechanical performances [3, 4]. High dielectric constant (k) and low dielectric loss are two most important prerequisites for these materials before any real applications are possible.

In this chapter, research and development on high-k polymer composites for embedded capacitor applications is reviewed and discussed. More specifically, current research efforts toward achieving high-k and low dielectric loss nanoparticle-based dielectric composites are presented.

7.2 Dielectric Mechanisms

7.2.1 Capacitance, Dielectric Constant, and Polarization

Capacitance (C) is used as the measure of how much electric charge can be stored in a capacitor. The relationship between capacitance C and dielectric constant ε_r (k) is given by the following equation:

$$C = \frac{\varepsilon_0 \varepsilon_r A}{t}, \qquad (7.1)$$

where ε_0 is the permittivity of free space (8.854×10^{-12} F m^{-1}), A is the area of the electrical conductor, t is the thickness of the dielectric layer, and ε_r is the relative dielectric constant of the dielectric layer. It is evident that the larger the dielectric constant, the larger the capacitance which can be realized in a given space. Therefore, materials of high dielectric constant are favored in practical design of embedded passives for miniaturization.

The ability of the dielectric materials to store energy is attributed to the polarization, i.e., electric field-induced separation and alignment of the electric charges. There are

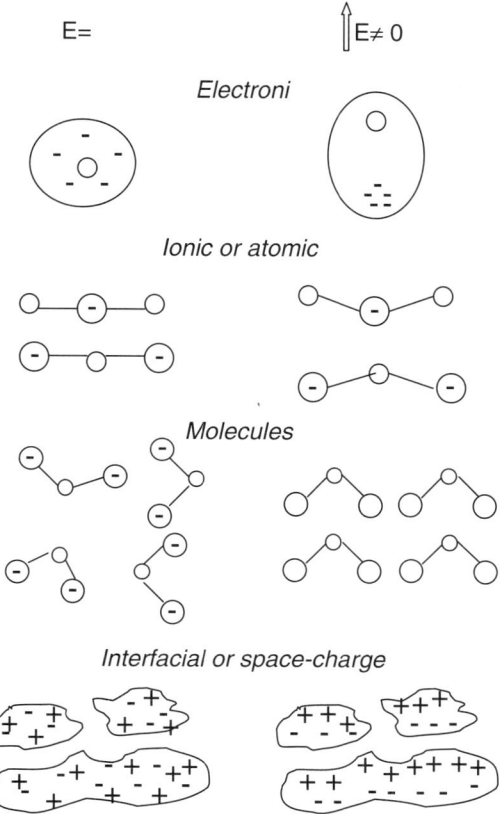

Fig. 7.2 Four major polarization mechanisms

several molecular mechanisms associated with this polarization (see Fig. 7.2). Ideally, the dielectric constant should be constant with regard to frequency, temperature, voltage, and time. However, each polarization mechanism has a characteristic relaxation frequency. Therefore, k values of most of the materials show a dependence on the frequency because the slower mechanism fails to respond and contribute to the dielectric storage when the frequency becomes larger. The k values of dielectric materials can also vary with temperature, bias, impurity, and crystal structure to different extents according to materials types [1, 5].

7.2.2 Dielectric Loss

Dielectric loss is a measure of energy loss in the dielectric during AC operation, which is a material property and does not depend on the geometry of a capacitor.

Dielectric loss is usually expressed as the loss tangent/factor (tan δ) or dissipation factor (Df), defined as

$$\tan\delta = \frac{\varepsilon''}{\varepsilon'} + \frac{\sigma}{2\pi f \varepsilon'}, \qquad (7.2)$$

where ε', ε'', and σ are the real and imaginary parts of the dielectric permittivity and the electrical conductivity of the materials, respectively, and f is the frequency.

In general, dielectric loss results from distortional, dipolar, interfacial, and conduction loss. Distortional loss is related to the electronic and ionic polarization mechanisms. Interfacial loss originates from the excessive polarized interface induced by the fillers and specifically the movement or rotation of the atoms or molecules in an alternating electric field. Conduction loss is attributed to the DC electrical conductivity of the materials, representing the flow of actual charge through the dielectric materials.

The energy dissipated in a dielectric material is proportional to the dielectric loss tangent, which can be determined by the following equation:

$$W \approx \pi\varepsilon' \xi^2 f \tan\delta, \qquad (7.3)$$

where ξ is the electric field strength and f is the frequency [6]. Therefore, a low dielectric loss is preferred to reduce the energy dissipation and signal losses, particularly for high frequency applications. Generally, a dissipation factor under 0.1% is considered to be quite low and 5% is high [1]. A very low dissipation factor is desired for RF applications to avoid signal losses, but much higher values can be tolerated for energy storage applications such as decoupling.

7.3 Materials Options for High-*k* Dielectrics

Dielectric theory suggests that high dielectric constant and low dielectric loss are the two most important parameters for dielectric materials to realize embedded capacitor applications. As such, to meet the stringent materials requirements, considerable attention has been devoted to the research and development of candidate high-*k* materials.

7.3.1 Ferroelectric Ceramic Materials

In the beginning, ferroelectric ceramic materials including barium titanate ($BaTiO_3$), barium strontium titanate ($BaSrTiO_3$), lead zirconate titanate ($PbZrTiO_3$), etc., have been used as dielectric materials for decoupling capacitors because of the high-*k* – up to thousands for this type of materials [1, 7]. However, the very high processing temperature (>600°C) required for sintering is unsuitable for the embedded capacitor applications in the low-cost organic PCB industry.

7.3.2 Ferroelectric Ceramic/Polymer Composites

Ferroelectric ceramic–polymer composites with high-k have also been actively explored as major material candidates. The methodology of this approach is to combine the advantages from the polymers, which meet the requirements for the low-cost organic substrate process, i.e., low-temperature processibility, mechanical flexibility, and low cost, with the advantages from the ferroelectric ceramic fillers, such as desirable dielectric properties [3, 4, 8–14]. However, some challenging issues in these polymer composites for high-k applications have been addressed such as limited dielectric constants, low adhesion strength, and poor processibility. A typical connectivity of polymer–ceramic composite is 0–3 composite, which indicates that ceramic fillers are dispersed without continuity in a three-dimensional connected continuous polymer matrix. Most of the k values of polymer–ceramic composites developed to date are below 100 at room temperature due to the low-k polymer matrix (usually in the range of 2–6). By employing the relatively high-k polymer matrix, the k values of polymer–ceramic composites can be effectively enhanced because the polymer matrix k has a very strong influence on the k of the final composites [4, 12]. For instance, poly(vinylidene fluoride–trifluoroethylene) (P(VDF–TrFE)) copolymer, a class of relaxor ferroelectric, can have a relatively high room temperature k (~40) after irradiation treatment [15]. Bai et al. [4] prepared $Pb(Mg_{1/3}Nb_{2/3})O_3$–$PbTiO_3$/P(VDF–TrFE) composites with k values above 200. Rao and Wong [16] reported a lead magnesium niobate–lead titanate (PMN–PT, 900 nm) + $BaTiO_3$ (50 nm)/high-k epoxy system (effective k: 6.4) composite with a k value about 150, for ceramic filler loading as high as 85% by volume. But the high filler loading of ceramic powders will lead to some technical barriers for real applications of polymer–ceramic composite in the organic substrate because it results in poor dispersion of the filler within the organic matrix, and almost no adhesion toward other layers in PCB as well due to the low polymer content.

7.3.3 Conductive Filler/Polymer Composites

Conductive filler/polymer composite is another approach toward ultrahigh-k materials for integral capacitor application of next-generation microelectronic packaging, which is a kind of conductor–insulator composite based on percolation theory [17]. For conductive filler/polymer composites, the effective electrical properties approaching the percolation threshold are determined by the scaling theory, which can be described as (7.4)–(7.6) [18]:

$$\sigma = \sigma_M (f - f_c)^t, \quad f > f_c, \tag{7.4}$$

$$\sigma = \sigma_D (f_c - f)^{-q}, \quad f < f_c, \tag{7.5}$$

$$\varepsilon = \varepsilon_D / |f - f_c|^q = \varepsilon_D (\sigma_M / \sigma_M)^s, \tag{7.6}$$

where σ_M and σ_D are the electrical conductivity of conductive filler and polymer, respectively; f and f_c are the concentration and the percolation threshold concentration of the conductive filler within the polymer matrix, respectively; ε_D is the dielectric constant of the polymer matrix; and q, s, and t are scaling constants, related to the material property, microstructure, and connectivity of the polymer-conductive filler interface [18]. Ultrahigh-k can be expected with conductive filler/polymer composites when the concentration of the conductive filler is close to (but does not exceed) the percolation threshold. Physically, this phenomenon can be interpreted in terms of a "supercapacitor network" with very large area and small thickness: when the concentration of the metal is close to the percolation threshold, large amount of conducting clusters are in proximity to each other but they are insulated by thin layers of dielectric material. Sometimes, the effective dielectric constant of the metal–insulator composite could be three or four orders higher than the dielectric constant of the insulating polymer matrix. And also, this percolation approach requires much lower volume concentration of the filler compared with traditional approaches of high dielectric constant particles in a polymer matrix. Therefore, this material option represents advantageous characteristics over the conventional ceramic/polymer composites, specifically, ultrahigh-k with balanced mechanical properties including the adhesion strength. Various conductive fillers, such as silver, aluminum, nickel, and carbon black, have been used to prepare the polymer-conductive filler composites or three-phase percolating composite systems [17, 19–24].

Although these composites were reported with high-k values at the percolation threshold, they still cannot be considered as effective materials for embedded capacitor applications due to the accompanied high dielectric loss tangent and conductivity. Some researchers use semiconductor fillers to achieve relatively low conductivity at the percolation threshold as compared with conductive fillers. Dang et al. [25] reported that LNO/PVDF (Li-doped NiO/polyvinylidene fluoride) composites were of effective k around 290 at 100 Hz at the percolation threshold $f_c = 0.10$. And the conductivity of the semiconductor fillers was also found to play an important role on the dielectric properties and the percolation threshold of the polymer–semiconductor composites. Other approaches to control dielectric loss will be discussed below.

7.4 Nanoparticle-Based Dielectric Materials

During recent years, great efforts have been made toward the synthesis and application of nanoparticles because of their unusual physical and chemical properties resulting from the nanosize and ultra-large surface area. Polymer composite materials based on nanoparticles provide a potential solution to meet present and future technological demands in terms of good processability and mechanical properties of polymers combined with the unique electrical, magnetic, or dielectric properties of nanoparticles [26]. Additionally, nanosized particles are preferred for high-k dielectric composite materials because they could help achieve thinner dielectric films leading to a higher capacitance density. Therefore, more nanoparticles of ceramic, metallic, or even organic semiconductor have been introduced to prepare high-k dielectric materials recently.

7.4.1 Ceramic Nanoparticle-Based Dielectric Composites

In the past decade, a great deal of effort has been devoted to the development of ceramic/polymer composites (0–3 composites), but most of the ceramic fillers used are in the micron size range. Although finer particle size is required to obtain a thin dielectric film and to increase the capacitance density, extremely fine ceramic particles may lead to the change of crystal structure from tetragonal, which results in the high permittivity, to cubic or pseudocubic. Generally speaking, the tetragonality and hence the permittivity of ceramic particles decreases with the particle size. Uchino et al. [27] and Leonard and Safari [28] found that the tetragonality of $BaTiO_3$ powders disappears finally when the particle size decreases to approximately 100 and 60–70 nm, respectively. Cho et al. [14] prepared $BaTiO_3$/epoxy composite embedded capacitor films (ECFs) with average particle sizes of 916 nm (P1) and 60 nm (P2); the k values of ECFs made of P1 were higher than those made of P2, so the coarser particle is more useful than the finer particle to obtain high-k of ECFs using unimodal powder. But by adopting bimodal fillers, fine nanoparticles can effectively enhance the k values by maximizing packing density and removing the voids and pores formed in the dielectric films. A dielectric constant of about 90 was obtained at a frequency of 100 kHz using these two differently sized $BaTiO_3$ powders.

7.4.2 Conductive or Semiconductive Nanoparticle-Based Dielectric Composites

The dramatic increase of dielectric constant observed in the conductor or semiconductor/insulator percolation systems close to the percolation threshold arouses interest in developing conductive or semiconductive filler/polymer composites as candidate materials for embedded capacitor applications. Especially, conductive filler/polymer nanocomposites have been identified as promising materials to fulfill the materials requirements for embedded capacitors. However, the dielectric loss of this type of materials is very difficult to control, because the highly conductive particles can easily form a conductive path in the composite as the filler concentration approaches the percolation threshold. Therefore, high dielectric loss and narrow processing windows have plagued metal/polymer composites in real applications. To solve the problems of the polymer-conductive filler composites, much work has been focused currently on the control of the dielectric loss of this system to overcome the above-mentioned drawbacks.

7.4.2.1 Effect of Dispersion

Uniform dispersion of nanoparticles in nanocomposite materials is required because multiparticle agglomerates inside the polymer matrix will lead to undesirable electrical or materials properties. Therefore, dispersion of nanofillers in composite materials is currently of great interest in both industry and academia. However, in many of the dielectric nanocomposite materials currently being produced, there is difficulty in obtaining both homogeneous materials and repeatable results where dielectric properties are dependent on nanofiller dispersion. Zhang et al. [29] selected CuPc oligomer, a class of organic semiconductor materials with k as high as 10^5, as the high-k filler dispersed in a P(VDF–TrFE) matrix. The composite showed a k of 225 and a loss factor of 0.4 at 1 Hz at low applied field. The high dielectric loss is due to the long-range intermolecular hopping of electrons. Wang et al. [30] further chemically modified CuPc to bond to the P(VDF–TrFE) backbone to improve the dispersion of CuPc in the polymer matrix. Compared to the simple blending method, the CuPc oligomer particulates in grafted sample are of relatively uniformly size in the 60–120 nm range, which is about five times smaller than that of the blended composite. Furthermore, dielectric loss was reduced and dielectric dispersion over frequency was weakened.

7.4.2.2 Effect of Surfactant Layer

The surfactant layer coated on nanoparticle surfaces during nanoparticle synthesis could serve as a barrier layer to prevent the formation of a conduction path to control the dielectric loss. A Ag/epoxy nanocomposite with 22 vol.% Ag possessing a high-k of 308 and a relatively low dielectric loss of 0.05 at a frequency of 1 kHz was reported by Qi et al. [19]. The 40-nm Ag nanoparticles coated with a thin layer of mercaptosuccinic acid were randomly distributed in the polymer matrix. As shown in Fig. 7.3, k and dielectric loss increase with the filler concentration up to 22 vol.%. The decrease of k after that point is not due to conduction, and this is attributed to the porosity caused by the absorbed surfactant layer and solvent residue, especially at a higher Ag content. In addition, no rapid increase of the dielectric loss tangent values was observed. Therefore, the observed highest k value was not considered as a real percolation threshold and the formation of a conducting filler network was prevented by the surfactant coating.

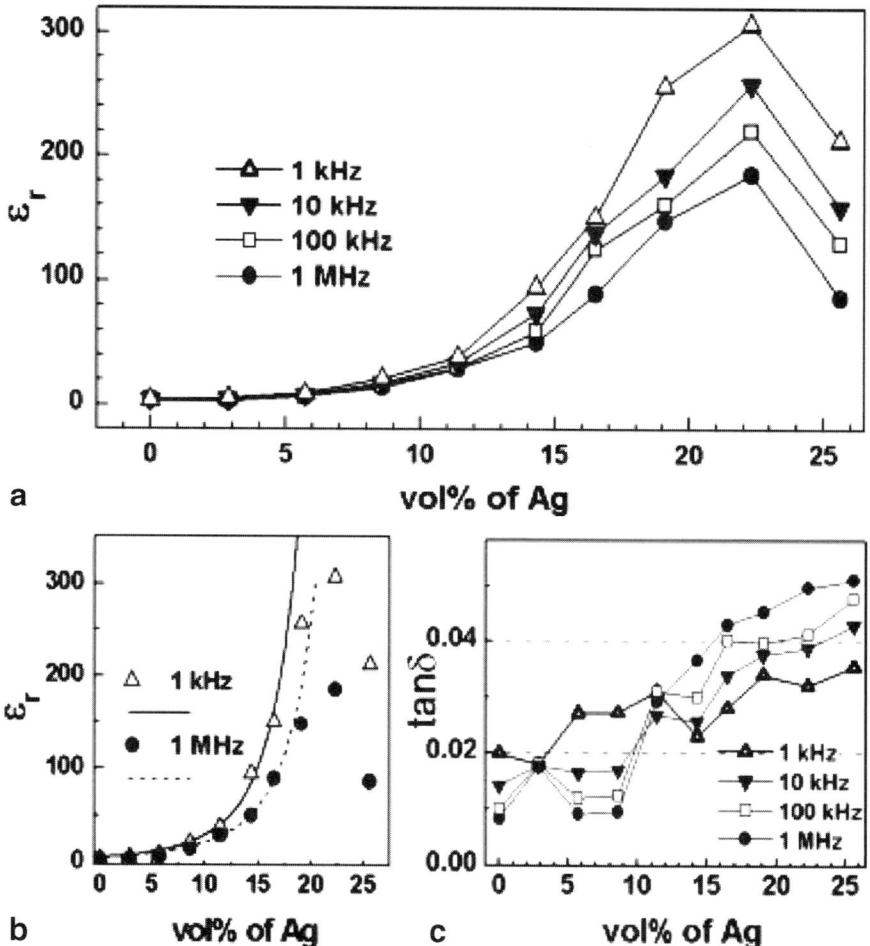

Fig. 7.3 k and loss tangent dependences on Ag volume fraction and frequency [19]

7.4.2.3 Coulomb Blockade (CB) Effect of Metal Nanoparticles

This novel approach is to take advantage of the unique properties of metal nanoparticles to control the dielectric loss of the conductive filler/polymer composite.

Ag nanoparticles were in situ synthesized in an epoxy resin matrix through the reduction of a silver precursor. The presence of the capping agent and its ratio with respect to the metal precursor were found to have great effect on the size and size distribution of the synthesized Ag nanoparticles in the nanocomposites. Nanoparticles of roughly two size ranges formed in all mixtures while the mixtures with higher concentrations of the capping agent showed the narrower size distribution as shown in Fig. 7.4.

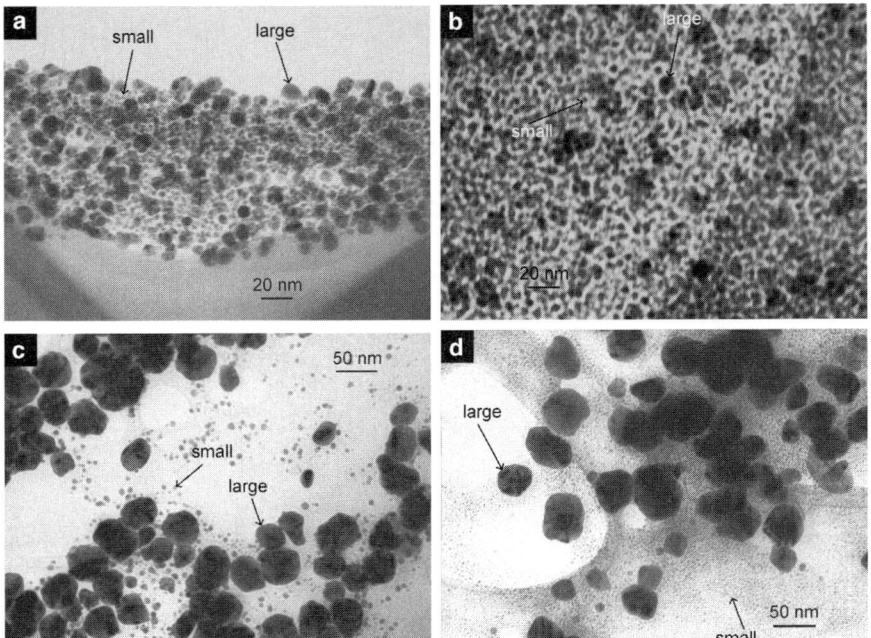

Fig. 7.4 TEM micrographs of Ag/epoxy composites in the presence of a capping agent (CA) with (**a**) [CA]/[silver] precursor ratio $R = 1$, (**b**) $R = 0.6$, (**c**) $R = 0.4$, and (**d**) $R = 0.2$ [20]

The Ag/CB/epoxy composite was prepared by mixing an in situ formed nano-Ag/epoxy composite and CB/epoxy composite. It can be seen from Fig. 7.5 that in situ formed Ag nanoparticles in the Ag/CB/epoxy composites increased the dielectric constant (k) value and decreased the dissipation factor (Df). The remarkably increased k of the nanocomposites was due to the piling of charges at the extended interface of the interfacial polarization-based composites. The reduced dielectric loss might be due to the CB effect of the Ag nanoparticles, a well-known quantum effect of metal nanoparticles. More specifically, the Ag nanoparticles of ultrafine size cause a high charging energy for the tunneling electrons and inhibit the charge transfer through the small metal island, reducing the conduction loss which represents the flow of charge through the dielectric materials.

The size, size distribution, and loading level of metal nanoparticles in the nanocomposite have significant influence on the dielectric properties of the nanocomposite system. Smaller size and narrower size distribution of Ag nanoparticles, obtained in the presence of larger amounts of a capping agent, resulted in more evident single-electron tunneling by Coulomb blockade effect and therefore reduction in conduction loss. Figure 7.6 illustrates the dielectric properties of the composites at different frequencies. The k values of composites containing Ag nanoparticles are larger over the whole frequency range than those of a control sample without Ag (Fig. 7.6a), while the decreased Df for nanocomposites containing

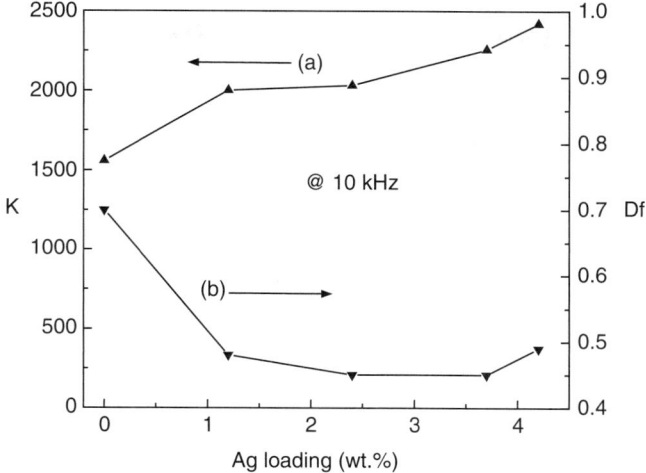

Fig. 7.5 Variation of k and Df at 10 kHz with different loading level of Ag nanoparticles [20]

Ag nanoparticles is observed in the low frequency range (10 and 100 kHz in Fig. 7.6b) only. This might be explained as the fact that the conduction loss contributes to the whole Df value less significantly as the frequency increases. Therefore, the effect of metal nanoparticles on suppressing dielectric loss is not obvious at higher frequency. Additionally, the contribution of interfacial loss is more evident in the high frequency range. Accordingly, the Df values of nanocomposites containing Ag nanoparticles are higher than those without Ag nanoparticles at higher frequencies (1 and 10 MHz in Fig. 7.6b).

7.4.2.4 Effect of High-k Silver Nanoparticles/Polymer Matrix Combined with Self-Passivated Conductive Particles

In this study, an in situ photochemical method was explored to prepare a metal nanoparticle–polymer composite as a high-k polymer matrix in which metal nanoparticles were generated by photochemical reduction of a metallic precursor within the polymer matrix. Compared with ex situ techniques, in situ techniques could facilitate a more uniform dispersion of nanoparticles in polymers and a photochemical approach provides the advantages of simplicity, reproducibility, versatility, selectivity, and ability of larger scale synthesis [31].

Figure 7.7 displays TEM micrographs of Ag nanoparticles synthesized via this method in an epoxy resin. Nanoparticle sizes ranged from 15 to 20 nm with smaller ones down to 3–5 nm. This demonstrated that ultrafine-sized, uniformly distributed and highly concentrated metal nanoparticles could be obtained in the polymer matrix via in situ photochemical reduction.

Self-passivated Al particles were then incorporated in the as-prepared Ag–epoxy nanocomposite to further improve the k of Al/epoxy composites while maintaining

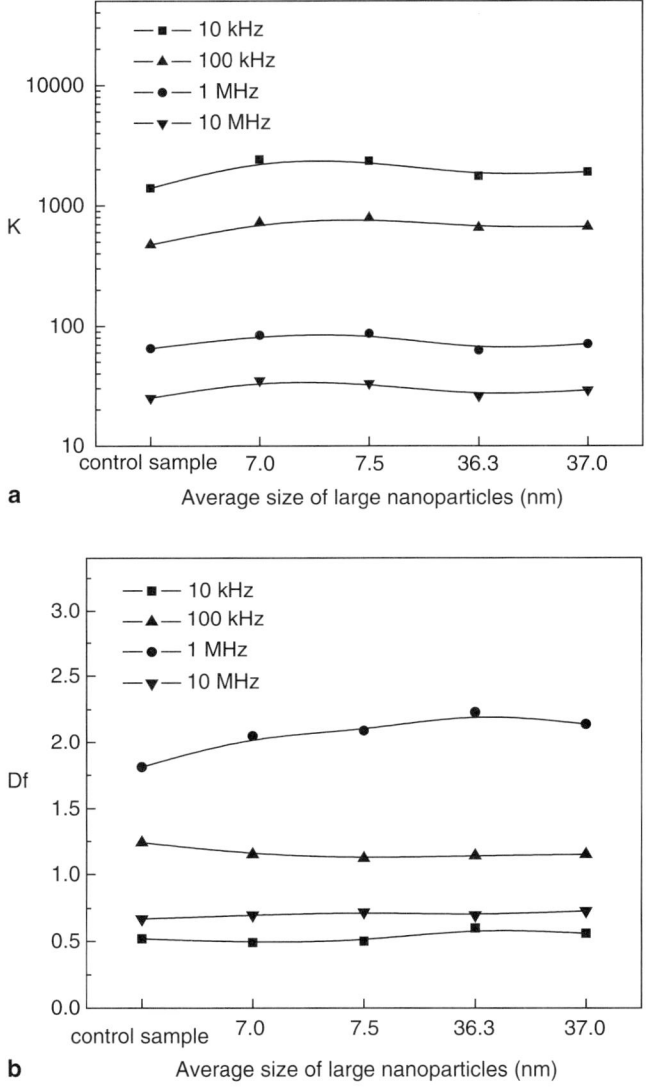

Fig. 7.6 Ag nanoparticle size effect on (**a**) k and (**b**) Df values of Ag/CB/epoxy composites at different frequencies [20]

the relatively low dielectric loss. Figure 7.8 displays the dielectric properties of Al/epoxy and Al/Ag–epoxy composites with different Al filler loading at a frequency of 10 kHz. The composites showed more than a 50% increase in k values compared with an Al/epoxy composite with the same filler loading of Al. Moreover, the dielectric loss was maintained below 0.05. The results suggested that the in situ formed Ag–polymer nanocomposites via photochemical approach can be employed as a high-k polymer matrix to host various fillers such as conductive metal or ferroelectric ceramic fillers to achieve both high-k and relatively low dielectric loss [31, 32].

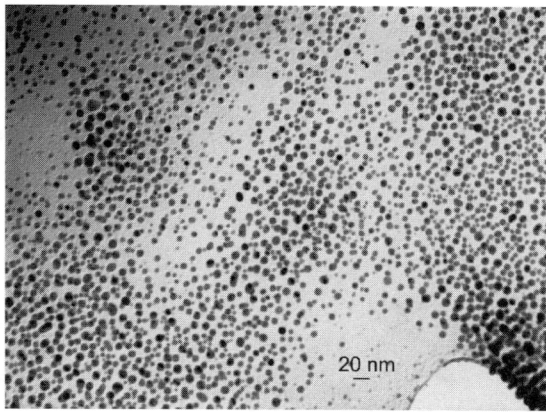

Fig. 7.7 TEM micrographs of Ag nanoparticles synthesized within epoxy resin via in situ photochemical reduction method [31]

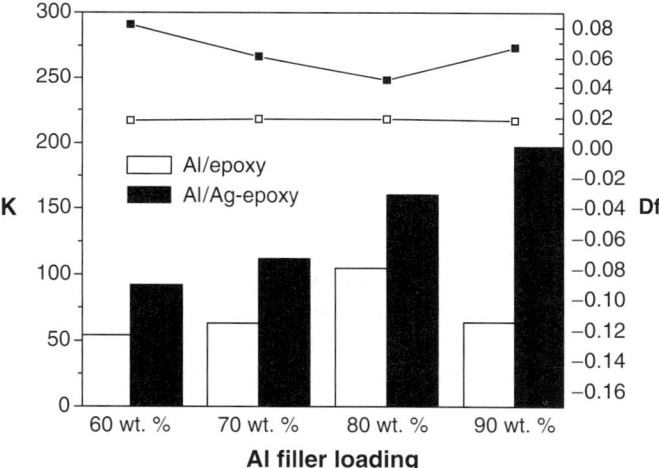

Fig. 7.8 Dielectric properties of Al/epoxy and Al/Ag–epoxy composites with different Al filler loading (at 10 kHz) [31]

7.4.2.5 Effect of Metal Nanoparticle Surface Modification

Surface modification of nanoparticles with organic molecules was employed to change the surface chemistry of nanoparticles and thus the interaction between nanoparticles and the polymer matrix. Full characterization of the surface-modified nanoparticle (SMN) showed that a thin layer was successfully coated on the nanoparticle surface via surface modification (Fig. 7.9).

The surface coating layer on the nanoparticles was demonstrated to be able to decrease the dielectric loss and enhance the dielectric breakdown strength, which can be attributed to the interparticle electrical barrier layer formed via surface

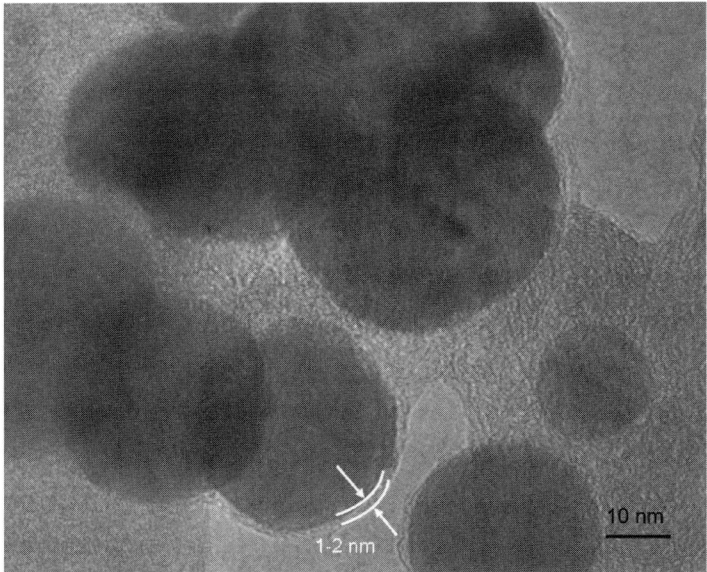

Fig. 7.9 High-resolution transmission electron microscopy (HRTEM) micrograph of SMN [33]

modification of nanoparticles preventing the metal cores from direct contact. Different surface modification conditions, such as surface modification agent type and concentration, solvent media, etc., may play complex roles in the degree of surface modification which impacts the changes of k and dielectric loss tangent values of SMN/polymer composites dramatically. Therefore, surface modification of nanoparticles is believed to be an effective approach to adjust the electrical features at the nanoparticle surface and the interface between the nanoparticle and the polymer matrix, and thus tailors the corresponding property of interest of nanocomposites [33, 34] (Fig. 7.10).

7.5 Summary

Generally speaking, high-k dielectric materials which can be realized for embedded passive applications are required to have high dielectric constant, low dissipation factor, high thermal stability, simple processibility, and good dielectric properties over broad frequency range. However, no such ideal materials that satisfy the above-mentioned prerequisites simultaneously have been realized till present. Nanocomposite materials based on nanosized particles have the potential to meet both present and future technological demands due to their unique properties and these materials have been studied extensively. Several techniques to further improve the overall dielectric properties of these candidate materials for real applications can be understood from theoretical predictions, which indicate that the dielectric

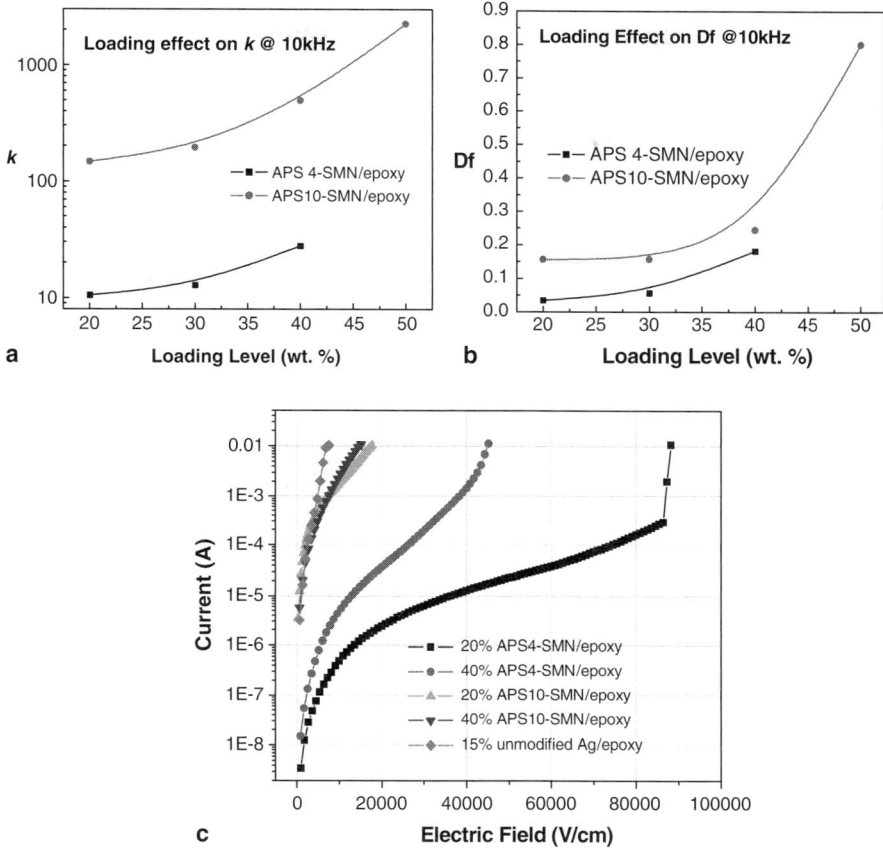

Fig. 7.10 Dielectric properties of SMN/epoxy composites with different SMN loadings and different modification degree [33]

constant of materials can be maximized and dielectric loss can be suppressed by following methods (1) optimized formulation of dielectric materials with high filler loading of high dielectric constant ceramics for ceramic–polymer nanocomposites and appropriate loading level of conductive fillers in the neighborhood of percolation threshold for conductive filler–polymer nanocomposites; (2) improvement in morphology of dielectric materials such as filler size and distribution, packing, and dispersion in the polymer matrix; (3) appropriate processing methods; and (4) modification of the filler interface to suppress the dielectric loss.

References

1. R.K. Ulrich and L.W. Schaper, Integrated Passive Component Technology, Wiley-IEEE, New York, 2003

2. J. Prymark, S. Bhattacharya, K. Paik, and R.R. Tummala, Fundamentals of Microsystems Packaging, McGraw-Hill, New York, 2001
3. R. Gregorio, M. Cestari, and F.E. Bernardino, Dielectric behavior of thin films of beta-PVDF/PZT and beta-PVDF/BaTiO3 composites, Journal of Materials Science, 31 (1996) 2925–2930
4. Y. Bai, Z.Y. Cheng, V. Bharti, H.S. Xu, and Q.M. Zhang, High-dielectric-constant ceramic-powder polymer composites, Applied Physics Letters, 76 (2000) 3804–3806
5. A.R. Blythe, Electrical Properties of Polymers, Cambridge University Press, Cambridge, 1979
6. J.P. Shaffer, A. Saxena, S.D. Antolovich, T.H.J. Sanders, and S.B. Warner, The Science and Design of Engineering Materials, McGraw-Hill, Boston, 1999
7. N. Tohge, S. Takahashi, and T. Minami, Preparation of PbZrO3–PbTiO3 ferroelectric thin films by the sol-gel process, Journal of the American Ceramic Society, 74(1) (1991) 67–71
8. K. Mazur, Polymer–Ferroelectric Ceramic Composites in Ferroelectric Polymers: Chemistry, Physics, and Applications, H.S. Nalwa, Ed., Dekker, New York, 1995
9. D.K. Dasgupta and K. Doughty, Polymer–ceramic composite materials with high dielectric constants, Thin Solid Films, 158 (1988) 93–105
10. S. Liang, S. Chong, and E. Giannelis, Barium titanate/epoxy composite dielectric materials for integrated thin film capacitors, Proceedings of the 48th Electronic Components and Technology Conference, pp. 171–175 (1998)
11. H. Windlass, P.M. Raj, D. Balaraman, S.K. Bhattacharya, and R.R. Tummala, Processing of polymer–ceramic nanocomposites for system-on-package applications, Proceedings of the 51st Electronic Components and Technology Conference, pp. 1201–1206 (2001)
12. Y. Rao, S. Ogitani, P. Kohl, and C.P. Wong, Novel polymer–ceramic nanocomposite based on high dielectric constant epoxy formula for embedded capacitor application, Journal of Applied Polymer Science, 83 (2002) 1084–1090
13. Z.M. Dang, Y.H. Lin, and C.W. Nan, Novel ferroelectric polymer composites with high dielectric constants, Advanced Materials, 15 (2003) 1625–1629
14. S.D. Cho, J.Y. Lee, J.G. Hyun, and K.W. Paik, Study on epoxy/BaTiO3 composite embedded capacitor films (ECFs) for organic substrate applications, Materials Science and Engineering B, 110(3) (2004) 233–239
15. Q.M. Zhang, V. Bharti, and X. Zhao, Giant electrostriction and relaxor ferroelectric behavior in electron-irradiated poly(vinylidene fluoride–trifluoroethylene) copolymer, Science, 280 (1998) 2101–2104
16. Y. Rao and C.P. Wong, Material characterization of a high-dielectric-constant polymer–ceramic composite for embedded capacitor for RF applications, Journal of Applied Polymer Science, 92 (2004) 2228–2231
17. Y. Rao, C.P. Wong, J. Xu, Ultra high k polymer metal composite for embedded capacitor application, US Patent 6864306 (2005)
18. C. Pecharroman and J.S. Moya, Experimental evidence of a giant capacitance in insulator–conductor composites at the percolation threshold, Advanced Materials, 12 (2000) 294–297
19. L. Qi, B.I. Lee, S. Chen, W.D. Samuels, and G.J. Exarhos, High-dielectric-constant silver–epoxy composites as embedded dielectrics, Advanced Materials, 17 (2005) 1777–1781
20. J. Lu, K.S. Moon, J. Xu, and C.P. Wong, Synthesis and dielectric properties of novel high-K polymer composites containing in situ formed silver nanoparticles for embedded capacitor applications, Journal of Materials Chemistry, 16(16) (2006) 1543–1548
21. J. Xu and C.P. Wong, Low loss percolative dielectric composite, Applied Physics Letters, 87 (2005) 082907
22. Z.M. Dang, Y. Shen, and C.W. Nan, Dielectric behavior of three-phase percolative Ni–BaTiO3/Polyvinylidene fluoride composites, Applied Physics Letters, 81 (2002) 4814–4816
23. H.W. Choi, Y.W. Heo, J.H. Lee, J.J. Kim, H.Y. Lee, E.T. Park, and Y.K. Chung, Effects of BaTiO3 on dielectric behavior of BaTiO3–Ni–polymethylmethacrylate composites, Applied Physics Letters, 89 (2006) 132910

24. J. Xu and C.P. Wong, Super high dielectric constant carbon black-filled polymer composites as integral capacitor dielectrics, Proceedings of the 54th IEEE Electronic Components and Technology Conference, Las Vegas, NV, pp. 536–541 (2004)
25. Z. Dang, C. Nan, D. Xie, Y. Zhang, and S.C. Tjong, Dielectric behavior and dependence of percolation threshold on the conductivity of fillers in polymer–semiconductor composites, Applied Physics Letters, 85(1) (2004) 97–99
26. L. Nicolais and G. Carotenuto, Metal–Polymer Nanocomposites, Wiley, Hoboken, NJ, 2005
27. K. Uchino, E. Sadanaga, and T. Hirose, Dependence of the crystal-structure on particle-size in BaTiO3, Journal of the American Ceramic Society, 72 (1989) 1555–1558
28. M.R. Leonard and A. Safari, Crystallite and grain size effects in $BaTiO_3$, Proceedings of the IEEE 10th International Symposium on Ferroelectric Applications, 2, pp. 1003–1005 (1996)
29. Q.M. Zhang, H.F. Li, M. Poh, F. Xia, Z.Y. Cheng, H.S. Xu, and C. Huang, An all-organic composite actuator material with a high dielectric constant, Nature, 419 (2002) 284–287
30. J. Wang, Q. Shen, C. Yang, and Q. Zhang, High dielectric constant composite of P(VDF–TrFE) with grafted copper phthalocyanine oligmer, Macromolecules, 37 (2004) 2294–2298
31. J. Lu, K.S. Moon, and C.P. Wong, Development of novel silver nanoparticles/polymer composites as high k polymer matrix by in situ photochemical method, IEEE Proceedings of the 56th Electronic Components and Technology Conference, San Diego, CA, pp. 1841–1846 (2006)
32. J. Lu, K.S. Moon, and C.P. Wong, "Silver/Polymer nanocomposites as high-k polymer matrix for dielectric composites with improved dielectric performance", Journal of Materials Chemistry, DOI: 10.1039/B807566B (2008)
33. J. Lu and C.P. Wong, Tailored dielectric properties of high-k polymer composites via nanoparticle surface modification for embedded passives applications, IEEE Proceedings of the 57th Electronic Components and Technology Conference, Reno, NV, pp. 1033–1039 (2007)
34. J. Lu and C.P. Wong, "Nano-scale Particle Surface Modification for Tailoring Dielectric Properties of Polymer Nanocomposites", Chemistry of Materials, to be submitted (2008)

Chapter 8
Nanostructured Resistor Materials

Fan Wu(✉) and James E. Morris

8.1 Introduction

This chapter focuses on nanostructured resistor materials in which there exists a macroscopic scale of inhomogeneity. In such a material, there are small, yet much larger than atomic, regions where macroscopic homogeneity prevails and where the foregoing macroscopic parameters suffice to characterize the physics, but different regions may have quite different values for those parameters. If we are interested in the physical properties only at scales that are much larger than those regions and at which the material appears to be homogeneous, then the macroscopic behavior can again be characterized by bulk effective values, σ and ε, of the conductivity and dielectric coefficient.

There is some literature [1–4] on the electrical properties and structural properties of the nanostructured resistor materials. However, there are few data, especially in a systematic way, on the relationship between the microstructure and the DC and AC electrical properties of nanostructured resistor materials. In this chapter, detailed studies of structural, compositional, and electrical properties of nanostructured resistor materials, such as $Cr_x(SiO)_{1-x}$, are presented. A detailed understanding of the charge transport in nanometallic particle systems such as three-dimensional cermets and two-dimensional discontinuous metal films is important to the achievement of a successful model for the conductivity of nanostructured resistor materials.

In this chapter, a brief overview of the nanostructured resistor materials is first given. Then, electrical conduction models including compositional, structural, and electrical (I–V, R–T) characterizations are discussed.

8.2 Nanostructured Resistor Material Overview

Nanostructured resistor materials are composite materials of conductor and dielectric. There are two types of nanostructured resistor materials. The first one includes films consisting of a physical or chemical mixture of metals and dielectrics and is

---

F. Wu
Zounds, Inc., 1910 S. Stapley Drive, Suite 202, Mesa, AZ85204, USA

generally classified under the general heading of cermets, or ceramic metal. Cermet films are prepared by a number of different methods such as evaporation or sputtering of an oxidizable metal in the presence of some oxygen [5, 6], coevaporation [7], cosputtering [8], simultaneous or alternating plasma polymerization and metal evaporation [9], and implantation of metal ions in polymers [10]. Another type is the discontinuous metal film which is formed during the initial stage of depositing thin metal films either by evaporation [11] or by sputtering [12, 13].

In the case of two-dimensional discontinuous films, these correspond to three growth stages. In the initial stages of growth, discrete nuclei are formed and these are generally stable once they consist of several atoms. The nuclei grow by capturing migrating surface adatoms or atoms direct from the vapor phase, and when the island separation is reduced to a few nm it is found that direct electrical current can pass through the film. With further island growth, the stage is reached where coalescence occurs. This is generally accompanied by a more rapid decrease in electrical resistance. Eventually, island coalescence leads to the formation of an interconnected network of capillaries that conduct like a normal metal. The important feature of this class of structure is that the overall film resistance is dominated not by the resistivity of the capillaries but by the manner in which they are connected. It is quite usual to observe very wide differences in resistance between films having nominally identical masses of metal per unit area and this is due to the distribution of material on the substrate. The third structural class, the continuous metal film, is formed when the holes between the capillaries are filled in and the film approximates to a plane-parallel slab, generally of polycrystalline metal. Surface and grain boundary scattering govern the resistivity of the film in this regime. Figure 8.1 shows atomic force microscopy (AFM) images of aluminum films with structures belonging to these three regimes [13].

Figure 8.2 shows the sheet resistance variation with the deposition time for the discontinuous aluminum films [13].

In the metallic regime, electronic conduction retains most of the properties of bulk metal. For example, granular Ni–SiO_2 films exhibit bulk ferromagnetism for $0.7 < x < 1$ [14], where x is the atomic percentage of Ni in the composite. However, properties which depend on electronic mean free paths are drastically modified due to strong electron scattering from dielectric particles and grain boundaries. For instance, the electrical conductivity decreases by orders of magnitude from its crystalline value, and the temperature coefficient of resistance (TCR), although positive, is very much smaller than in bulk metals [15].

In the dielectric regime which consists of isolated, nanometer size metal islands, two important physical quantities are substrate-assisted electron transport [16] and the thermally activated charging energy required to transfer an electron between two neutral islands [11].

Electrical conduction in the transition regime, which consists of a random interconnected array of metallic capillaries, is of particular interest because of its relevance to the mathematical topic of percolation theory. The electrical conductivity in this regime is due to percolation along the metallic capillaries and electron tunneling between isolated metal islands. Conduction in the island structure is, as

8 Nanostructured Resistor Materials

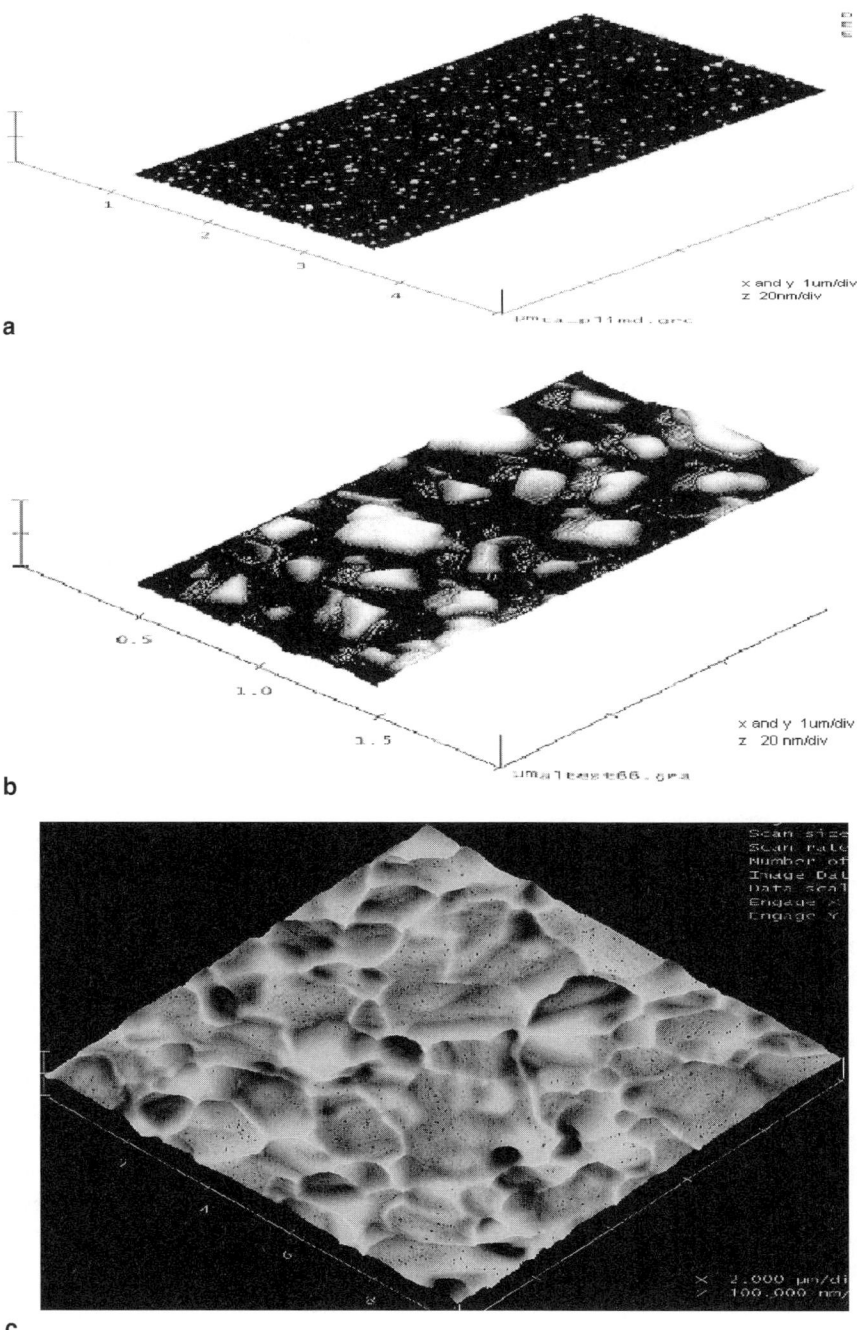

Fig. 8.1 AFM images of aluminum films with film structures of (**a**) nucleation stage, (**b**) coalescence stage, and (**c**) continuous film stage [13]

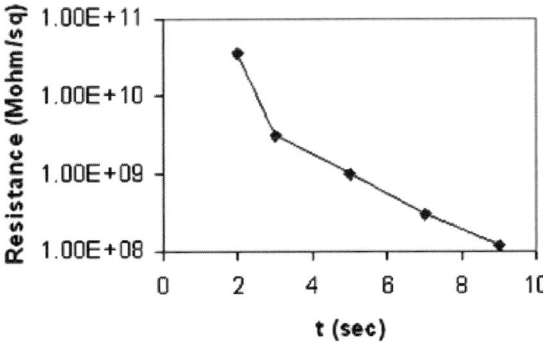

Fig. 8.2 Resistivity vs. deposition time [13]

stated earlier, an activated process which gives rise to a negative TCR whilst conduction in the capillaries is of a normal metallic type with a reduced positive TCR.

The transition from nonmetallic to metallic conduction in nanostructured resistor materials is governed by the volume or area fraction of the substrate covered by metal. Consider a substrate as an orthogonal lattice upon which metal nuclei can grow but only at the nodes (sites). As growth proceeds, the stage will be reached where adjacent nuclei actually come into contact and coalesce. Although in the high-temperature limit a single nearly spherical nucleus may be expected to be formed upon coalescence of two smaller nuclei, in general elongated capillaries tend to be formed. Thus, the nuclei constitute the sites, their size governs the strength of interaction and the capillaries are equivalent to the bonds. There are basically two problems in percolation theory. The "site" problem is concerned with the variation of physical properties in terms of the proportion of occupied sites. The "bond" problem is concerned with the variation in terms of bonds present. From the point of view of electrical conduction, it is of interest to determine the stage at which the transition from metallic regime to dielectric regime occurs. In fact, for a granular metal film, this will not be a sharp transition since there will be a range of structures in which conduction between islands will occur simultaneously with metallic capillary conduction, and initially the resistance contributions will be of similar values. Eventually, the tunnel contribution will be completely shorted by the capillaries. For a two-dimensional lattice, Scher and Zallen [17] got a critical surface fraction of 44% of the covered surface area when large-scale connected paths formed.

Analysis of the structural and electrical transport data yields detailed information on the granular metal film parameters: average metal particle size, average metal particle separation, and their distributions.

When an electron is transferred from one initially neutral island to another, a certain amount of work must be done against electrostatic forces. Therefore, the only electrons which could be transferred were those excited to a level of the order of $e^2/4\pi\varepsilon r$ above the Fermi level of the particular island of radius r [18], where e is the electronic charge and ε is the effective dielectric constant of the medium.

This energy can be supplied thermally at temperature T and, if Boltzmann statistics are obeyed, then the total number of charged islands n can be written as

$$n = N \exp(-\delta E / kT), \tag{8.1}$$

where N is the total number of islands and δE is the activation energy which must be supplied to overcome the electrostatic force, expressed by the Neugebauer and Webb formula:

$$\delta E = \frac{e}{4\pi\varepsilon_0 \varepsilon_r r}\left(\frac{r+s}{2r+s}\right) - Fs, \tag{8.2}$$

where s is the island separation and F is the applied electric field.

The activated tunneling conduction model expresses the conductivity as

$$\sigma = Ks^2 e^2 D \exp\left(-\frac{\delta E}{kT}\right), \tag{8.3}$$

where K is a geometrical constant and D is the interisland tunneling probability.

The Neugebauer and Webb theory and some other extended models proposed through the last several decades [19–22] have been able to account for several of the experimental observations and were in good qualitative agreement with experimental observation. Agreement between experimental and theoretical activation energy δE is sufficient to validate the fundamental electrostatic activated tunneling model(s), which is recognizable as the basis of the coulomb blockade. However, questions remained before the problem of the mechanism of transport in granular metal film systems could be totally resolved:

- The absolute conductances are found to be orders of magnitude greater than theory.
- Despite attempts to apply variable-range hopping concepts and percolation theory, the statistical effects of distributed r and s values, and island shapes, have yet to be included successfully.
- It is unclear whether the high or low frequency dielectric constant should be used in δE.
- Experimental results described in the next paragraphs are inconsistent with existing theories, and indicate that major modification of the basic thermally activated tunneling mechanism is needed.

Borziak et al. [23] reported three significant experiments. With previously deposited electrodes, they were able to fabricate discontinuous films with different island structures immediately adjacent to the contact, with smaller islands and/or wider gaps. These symmetrical "inhomogeneous" films showed that the voltage drop is always greater at the positive end of the film. The second result was the observation of stable and reproducible switching in such films, but the explanation of this effect is still elusive. In the third experiment, the inhomogeneous films were also made asymmetric, i.e., with different inhomogeneous structures at the two electrodes, whereupon the DC resistance became polarity dependent, i.e., a diode-like effect. These results cannot be explained on the basis of conduction models discussed in last paragraph, and indicate that the conduction mechanism must depend significantly upon the islands at the electrodes [24, 25].

The asymmetric film study has been extended to AC effects by Morris [25]. In the traditional model, the film is regarded as a matrix of identical island/gap elements, with the metal island resistance in series with the parallel combination of gap tunnel resistance R_g and capacitance C_g, where $\delta E = \frac{1}{2} e^2 / C_g$. C_g values determined by AC measurements on this model are universally orders of magnitude greater than those consistent with δE. With the asymmetrical inhomogeneous film, two corner frequencies appear, yielding two distinct values for both R_g and C_g, corresponding to the two electrodes. In addition, the C_g values match well to capacitances between the electrodes and film across a single gap width. At extreme asymmetries, a "pseudo-inductive" effect makes an appearance, as one contact resistance becomes very large, representing a time delay to establish steady-state conductance in the film by injection of the charge carriers which account for the higher conductances than predicted by a Boltzmann distribution of charged islands [26, 27]. These 2D effects are expected to be replicated in 3D cermet resistor films.

8.3 Physical Properties of $Cr_x(SiO)_{1-x}$ and $(Cr_xSi_{1-x})_{1-y}N_y$ Nanostructured Resistors

8.3.1 Microstructure and Composition

All the $Cr_x(SiO)_{1-x}$ samples were sputter deposited from Cr–SiO targets in an Ar ambient. All the $(Cr_xSi_{1-x})_{1-y}N_y$ films were deposited in Ar/N$_2$ ambients. C_{N2} is the nitrogen concentration in the Ar/N$_2$ mixtures.

8.3.1.1 Rutherford Backscattering Analysis

The chemical compositions of the samples were determined by Rutherford backscattering (RBS). We confirmed the consistency of these data by energy dispersive X-ray analysis (EDX) by means of a LINK AN10000 EDS system attached to a JSM-840 SEM. RBS is considered as likely to give the most accurate composition values since this method is quantitative from first principles and does not require elemental standards. The quantitative results of the EDX analysis are influenced by a series of effects; in particular, this method needs standards and various corrections.

Table 8.1 shows the atomic concentration data from RBS measurements. Figure 8.3 shows the depth profile generated from the theoretical fitting to the experimental data. The sample used in the RBS measurement is a 30-nm thick $Cr_x(SiO)_{1-x}$ film deposited from a target with [SiO]/[Cr] = 80/20.

8.3.1.2 X-Ray Photoelectron Spectroscopy/Electron Spectroscopy for Chemical Analysis (XPS/ESCA)

The purpose of the XPS analysis is to determine the chemical composition of annealed $Cr_x(SiO)_{1-x}$ films, evaluated at the surface and after ~10-nm sputter etch. The experimental details of the ESCA analysis are listed as follows.

The ESCA data of samples deposited from a target with [SiO]/[Cr] = 80/20 with different nitrogen concentrations C_{N2} are shown in Table 8.2. The survey spectra show the presence of Cr, Si, O, C, and N at the surface, as expected. The relative atomic percent of these species shown in the table matches the nitrogen partial pressure during the reactive sputtering very well. The carbon found in the films is from the target. No other species were detected.

Two peaks were identified in the region of Si. The low binding energy peak is suggestive of silicides or silicon, while the higher binding energy component is representative of silica type species. See Table 8.3 for relative amounts of Si species.

Table 8.1 Atomic concentration from RBS

Depth (nm)	N	O	Si	Cr	Si/Cr	Si/N
<32	5.03	45	31.0	16.5	1.88	0.59
32–274	–	67	33	–	–	–
>274	–	–	100	–	–	–

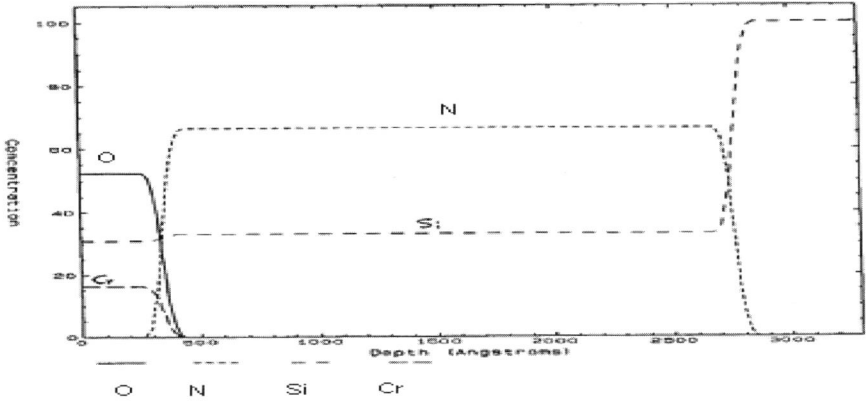

Fig. 8.3 Depth profile generated from the theoretical fitting to the experimental data

Table 8.2 Relative atomic percent determined from ESCA survey spectra

N$_2$ partial pressure (mTorr)	Anneal temperature (K)	at% O	at% C	at% Si	at% Cr	at% N
0.8	540	28.1	1.29	44.07	19.84	6.67
0.4	450	16.07	0.63	57.06	22.66	3.56
0.1	No	12.66	0.45	59.67	25.41	1.81
0.8	540	25.87	3.86	45.37	17.91	6.98
0.4	400	13.65	1.39	61.76	19.31	3.89
0.8	540	26.66	0.85	48.09	18.08	6.32
0.4	400	14.62	0.88	62.49	17.61	4.4
0.6	450	15.59	2.14	58.89	18.5	4.88
0.4	No	10.95	0.3	65.3	19.37	4.08

Table 8.3 Relative amounts of Si species

N$_2$ partial pressure (mTorr)	100 eV Silicon	103 eV Silica
0.8	44.71	55.29
0.4	24.86	75.14
0.1	23.47	76.53
0.8	40.23	59.77
0.4	16.40	83.60
0.8	12.71	87.29
0.4	33.42	66.58
0.4	13.69	86.31
0.4	5.12	94.88

8.3.1.3 Secondary Ion Mass Spectroscopy Analysis

The secondary ion mass spectroscopy (SIMS) analyses were performed on a PHI-6600 quadrupole mass spectrometer with oxygen primary ion bombardment and using positive secondary ion mass spectrometry.

The peaks in the mass spectra, shown in Fig. 8.4, have been assigned and indicate the presence of H, C, B, N, O, Cr, and Si in the sample. Figure 8.4 also indicates that there are several forms of Cr existing in the film: metallic Cr, CrB, CrO, and Cr silicide (CrSi).

8.3.1.4 Transmission Electron Microscopy Analysis

The microstructure of the samples was investigated by transmission electron microscopy (TEM). Samples for TEM studies were deposited on carbon-coated copper grids from Electron Microscopy Science or silicon nitride membrane windows™ grids from SPI.

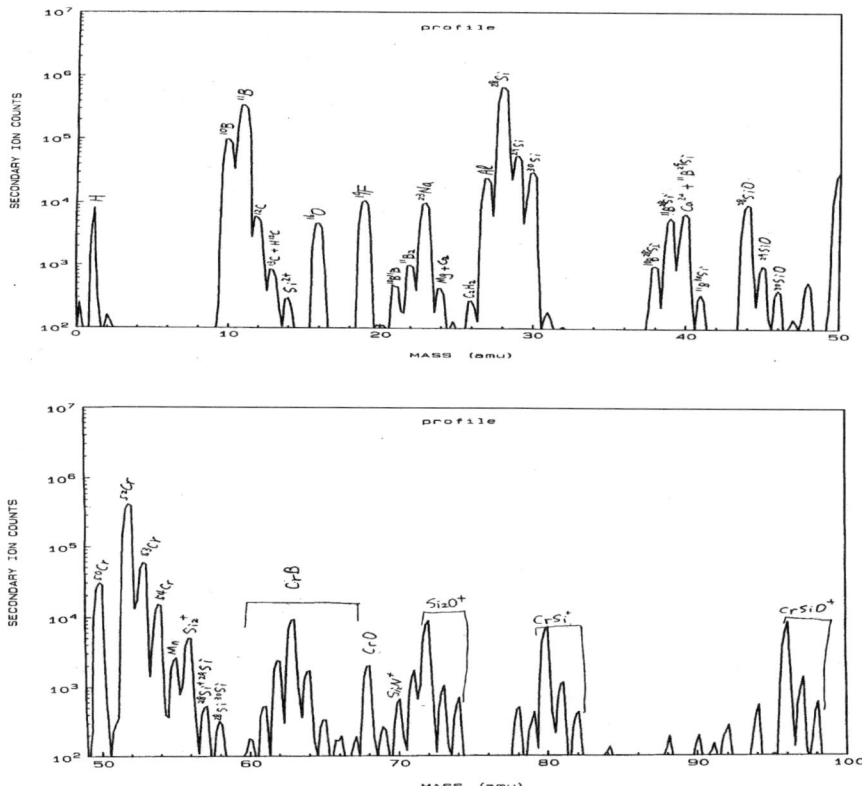

Fig. 8.4 SIMS mass spectra of $Cr_x(SiO)_{1-x}$ film

For electron microscope studies, films of about 20-nm thickness were deposited. Figure 8.5 shows electron micrographs for two film compositions. Figure 8.6 shows electron diffraction diagrams of the same samples. Figure 8.5a, c shows the structure observed as deposited at room temperature. In both cases, the films were found to be well amorphous with crystalline regions. The principal structures found in deposited films with [SiO]/[Cr] = 80/20 are a "cellular" microstructure of dark islands around 1–2 nm in size surrounded by medium of lighter contrast about 2–3 nm wide. It is likely that the dark islands contain the high Z component of the film, i.e., the Cr. EDX indicates that the film is amorphous or that there is only very short-range order in the film, which is indicated by the diffuseness of the diffraction patterns. Figure 8.5b, d shows the structures after annealing at 540°C in N_2 ambient for 0.5 h. Films with higher [SiO]/[Cr] ratios have finer metal island dispersions. Annealing has the same effect of islands growing coarser for both samples, but has the opposite effect of separation between islands for the samples. For the annealed samples, the microstructure remains similar to that of as deposited. However for the

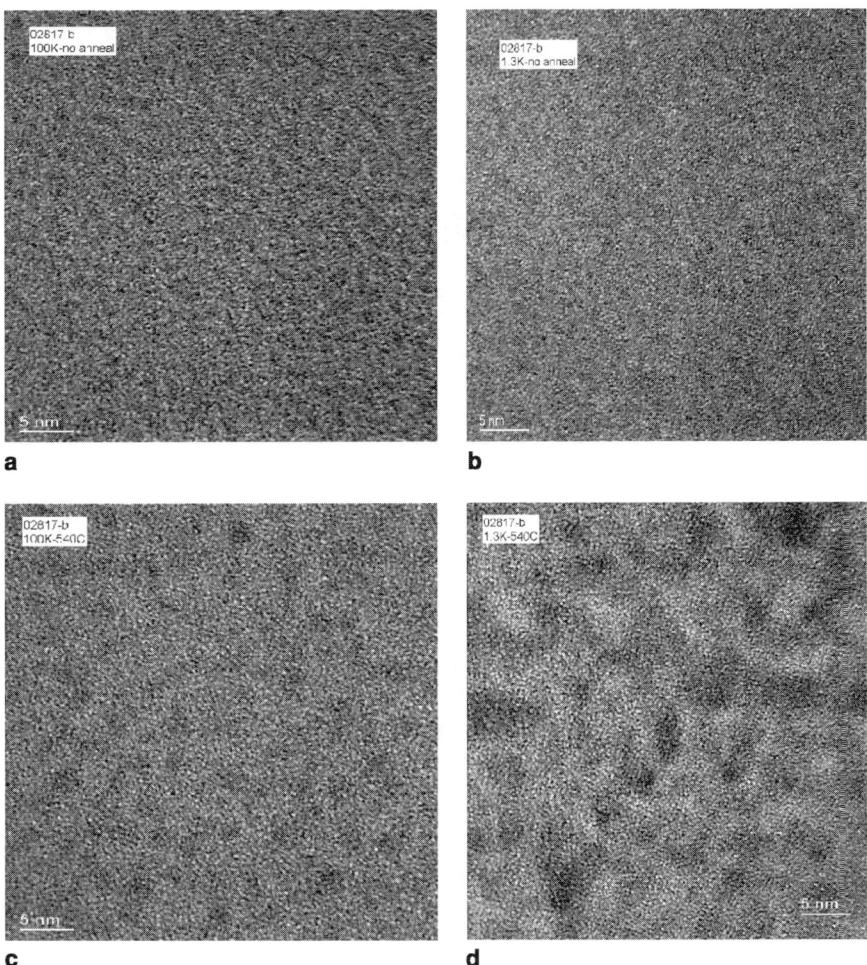

Fig. 8.5 Transmission electron microscopy images of $(Cr_xSi_{1-x})_{1-y}N_y$ films with $C_{N2} = 0.5\%$: (**a**) $x = 0.2$, unannealed; (**b**) $x = 0.2$, annealed at 540°C in N_2 ambient for 1.5 h; (**c**) $x = 0.6$, unannealed; and (**d**) $x = 0.6$, annealed at 540°C in N_2 ambient for 1.5 h

film with [SiO]/[Cr] = 80/20, the island size grows from ~1 to 2–4 nm and the spacing between islands increases from 2–3 to about 5 nm. The crystalline phase remains largely unchanged, especially as evidenced by little change seen in the diffraction pattern in Fig. 8.6b. For the sample deposited from a target with [SiO]/[Cr] = 40/60, the island size grows from 2–4 to 5–10 nm and the spacing between islands actually decreases to <1 nm or some of the islands are connected to form a metallic pathways. These structural changes are also reflected in electrical measurements on the samples prior and post-anneal. Based on the activated tunneling conduction model discussed above for granular metal, the film resistance R may be written approximately for effective tunneling barrier height Φ as

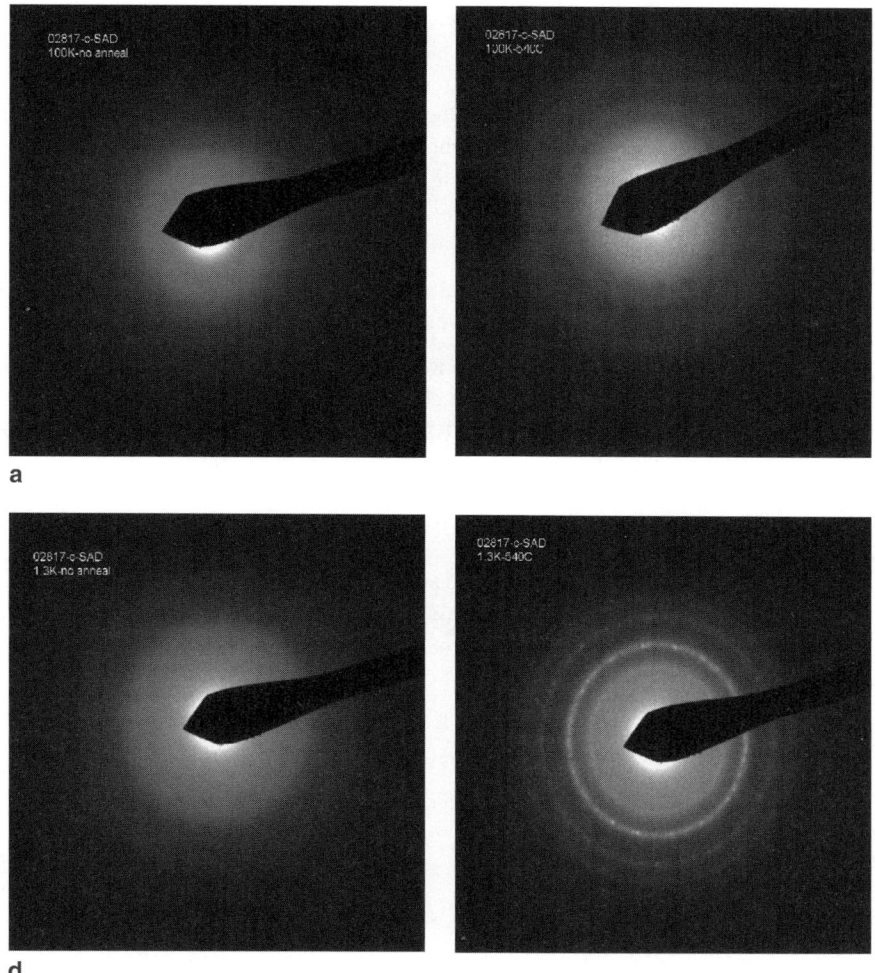

Fig. 8.6 Electron diffraction diagrams of $(Cr_xSi_{1-x})_{1-y}N_y$ films with $C_{N2} = 0.5\%$: (**a**) $x = 0.2$, unannealed; (**b**) $x = 0.2$, annealed at 540°C in N_2 ambient for 1.5 h; (**c**) $x = 0.6$, unannealed; and (**d**) $x = 0.6$, annealed at 540°C in N_2 ambient for 1.5 h

$$R = C \exp\left(A\phi^{1/2}s + \frac{\delta E}{kT}\right), \tag{8.4}$$

where A and C are constants and the TCR, α, is defined as

$$\alpha = \frac{1}{R}\frac{dR}{dT} = -\frac{\delta E}{k}\frac{1}{T^2}, \tag{8.5}$$

from which we see that the TCR becomes less negative as island radius increases. For the film with [SiO]/[Cr] = 80/20, the island radius r and spacing s both increase after annealing at 540°C.

Therefore, according to (8.4) and (8.5), the sheet resistance increases after annealing, and the TCR becomes less negative post-anneal. For the film with [SiO]/[Cr] = 40/60, the sheet resistance decreases after annealing, and the TCR changes sign from negative to positive post-annealed at 540°C. From Fig. 8.6d, it is observed that there are two possible charge transport paths in the annealed film with [SiO]/[Cr] = 40/60, these being:

1. Normal metallic conduction in the metallic capillaries
2. An activated conduction process between the metallic islands

So, there are two parallel-independent conduction processes in this annealed film, which is the reason behind the sign change of its TCR.

The diffraction pattern in Fig. 8.6d shows the presence of two new rings compared to the as-deposited sample. The change in the pattern indicates that a change is occurring in the film around this temperature. Crystalline grains of Cr and $CrSi_3$ were observed after 540°C anneal in N_2 ambient, which was also confirmed by the EDX analysis that will be presented later.

Figure 8.7 shows TEM images of four Cr_xSi_{1-x} films deposited from different Cr composition, 10, 30, 50, and 60%, respectively. Figure 8.8 shows their corresponding electron diffraction diagrams.

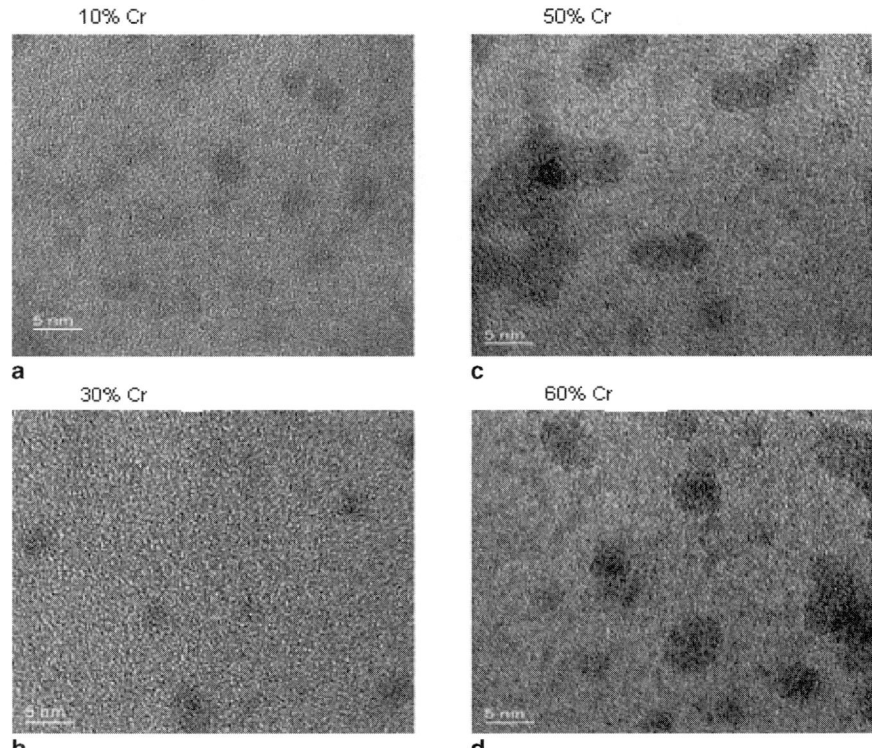

Fig. 8.7 High-resolution TEM (HRTEM) images of samples with different Cr compositions: (**a**) 10%, (**b**) 30%, (**c**) 50%, and (**d**) 60%

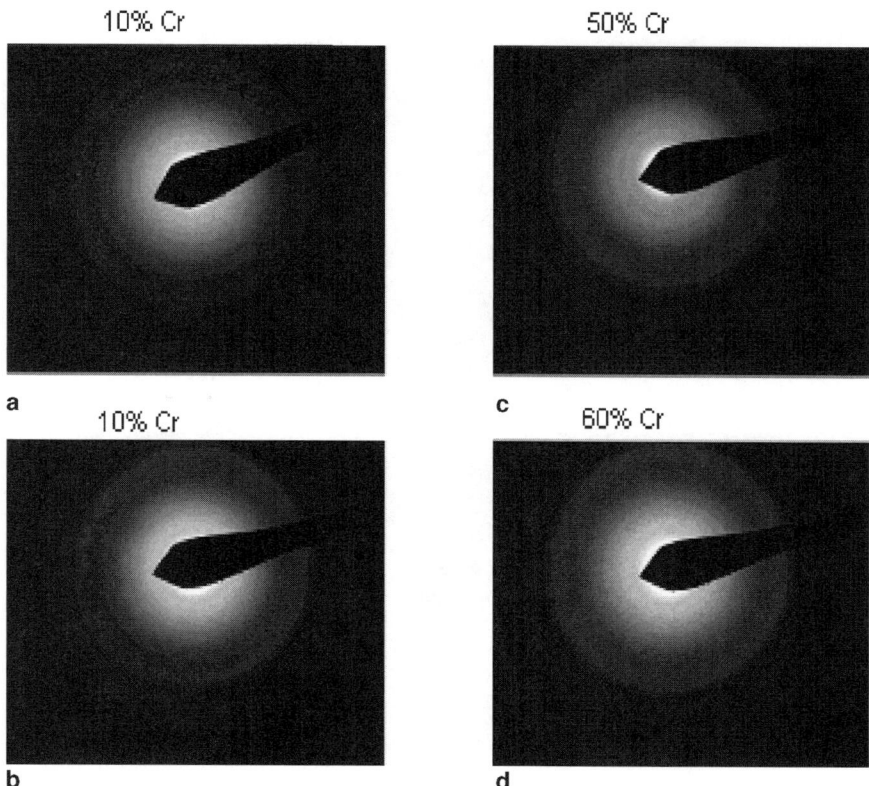

Fig. 8.8 Electron diffraction diagrams of samples with different Cr compositions: (a) 10%, (b) 30%, (c) 50%, and (d) 60%

Figure 8.9 a, b shows electron micrographs of an as-deposited $(Cr_xSi_{1-x})_yN_{1-y}$ film with $x = 0.4$, $y = 0$ and 0.1, respectively. The film thickness is 25 nm. The film consists of small Cr islands (dark areas), the mean diameter of which is around $d = 2$ nm, embedded in an amorphous Si matrix (light areas). Figure 8.9b shows that the structure of the reactively sputter-deposited film is similar to that of the film deposited by pure Ar sputtering.

Most of the Cr islands in Fig. 8.7 are disconnected by small SiO bridges, so the film is below the percolation threshold. In a schematic representation, Fig. 8.10 shows the structural properties of the Cr_xSi_{1-x} films, revealing both disconnected and connected Cr islands, simply given as circles in Fig. 8.10. As identified in Fig. 8.10, island diameters r and center-to-center distances R have been measured. From the corresponding mean value (r, R), the mean island separation $s = R - r$ was calculated.

In Table 8.4, x is the Cr volume fraction, t the film thickness, r the mean particle diameter, σ_r the standard deviation of r, R the mean center-to-center distance, σR the standard deviation of R, s the mean gap size between particles obtained from $s = R - r$, and σs is the standard deviation for s.

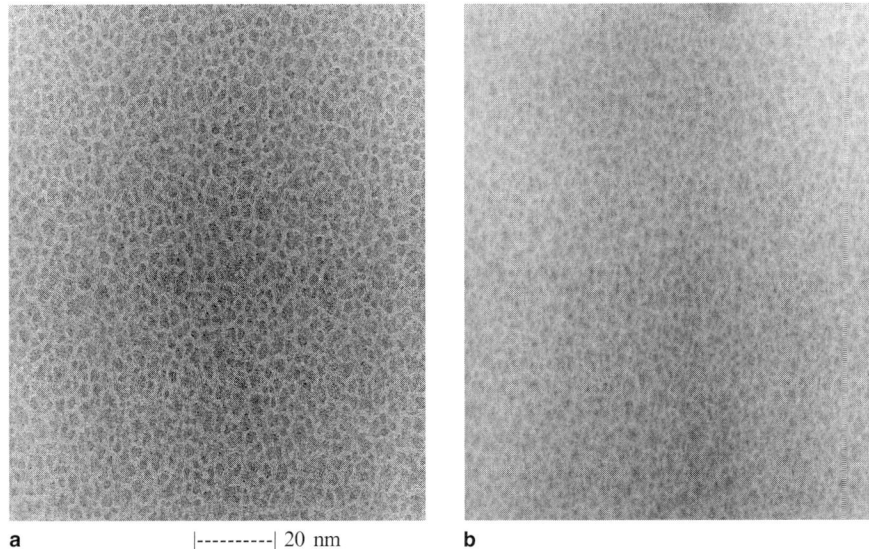

Fig. 8.9 Electron micrographs of $(Cr_xSi_{1-x})_{1-y}N_y$ films: (**a**) $x = 0.4$, $y = 0$ and (**b**) $x = 4$, $y = 0.1$

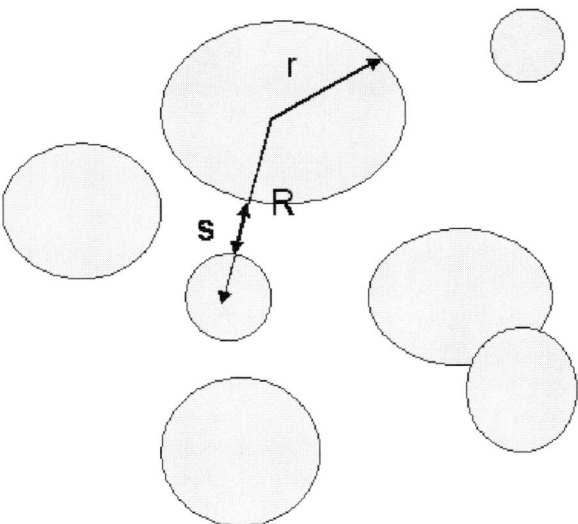

Fig. 8.10 Schematic representation of $Cr_x(SiO)_{1-x}$ film

Table 8.4 Measured (unannealed) $Cr_x(SiO)_{1-x}$ film parameters (defined in Fig. 8.10)

Film	t (nm)	r (nm)	σr (nm)	R (nm)	σR (nm)	s (nm)	σs (nm)
1	20	4.3	1.61	4.65	1.54	1.69	0.59
2 ($y = 0.1$)	21	2.6	0.79	4.03	0.91	2.43	0.71

Cr/SiO = 60/40

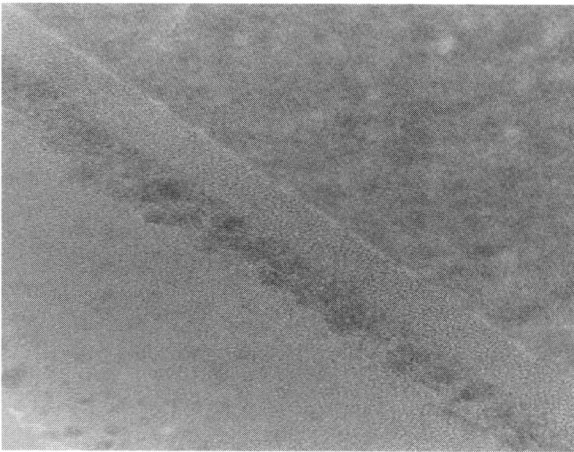

Fig. 8.11 Cross-sectional HRTEM image of a $(Cr_xSi_{1-x})_{1-y}N_y$ film with $x = 0.2$ and $y = 0$, unannealed

Cross-sectional HRTEM is also performed on $Cr_x(SiO)_{1-x}$ films as deposited, as shown in Fig. 8.11. No magnification markers are shown on the micrographs since the lattice fringes serve as scale markers (with the main spacing of the fringes corresponding to the Si(111) planes, i.e., 0.313 nm). The specimens were tilted to the Si[110] zone axis to ensure that the $Cr_x(SiO)_{1-x}$ layer was viewed directly edge on. The images show the Si substrate, the oxide layer, the CrSi layer, and finally part of the passivation layer. A combination of high-resolution and diffraction contrast imaging was used to enhance contrast of the layers, while still providing lattice images from which to accurately determine the size of metallic islands in the $Cr_x(SiO)_{1-x}$ layer and the layer thickness. While the boundary between the oxide and the $Cr_x(SiO)_{1-x}$ layer is quite well defined, the top part of the $Cr_x(SiO)_{1-x}$ layer is somewhat diffuse, suggesting that the density of the $Cr_x(SiO)_{1-x}$ film varies along the growth direction. Diffraction contrast imaging reveals small dark regions, 1.5–2 nm in size, in the $Cr_x(SiO)_{1-x}$ layer. No lattice fringes are evident in this layer in any of the images. This matches the observations from top TEM images and indicates that there is no long-range crystalline order within the $Cr_x(SiO)_{1-x}$ layer as deposited.

8.3.2 Temperature Dependence of Resistivity

Two systems were investigated in this study:

1. $Cr_x(SiO)_{1-x}$ with various x values
2. $(Cr_xSi_{1-x})_{1-y}N_y$ with fixed x and various y values

It can be seen that the metal–insulator transition occurs by choosing proper x or y values. There are lots of data available on the first system but this study was the first investigation on the $(Cr_xSi_{1-x})_{1-y}N_y$ system.

The conductivity of the Cr–Si–N(O) films depends on temperature, Cr concentration, and on the deposition conditions. In this system, the concentration x of the Cr is increased from zero; at a certain value of x, the film structure changes from the dielectric regime to the metallic regime. It can also be explained that the states at the Fermi level become delocalized and a transition to the metallic state occurs.

Three regions (metallic, semiconducting, and transition) can be distinguished in the temperature dependence of the conductivity. The results for the Cr–Si–N system with [SiO]/[Cr] = 40/60 for various nitrogen concentrations C_N are shown in a $R(T)/R_0$ vs. log T plot in Fig. 8.12a and $\log(R(T))$ vs. $T^{-1/2}$ in Fig. 8.12b: sample with nitrogen concentration $C_N = 2.6\%$ is metallic, samples with $C_N = 26$ and 40% are semiconducting, and the sample with $C_N = 12.6\%$ should be classified as in the transition region. The results for the Cr–Si–N system with [SiO]/[Cr] = 80/20 for various C_N are shown in a $R(T)/R_0$ vs. log T plot in Fig. 8.13a and $R(T)$ vs. log $T^{1/2}$ plot in Fig. 8.13b: samples with $C_N = 2.6, 12.6, 26,$ and 40% are all to be classified as semiconducting.

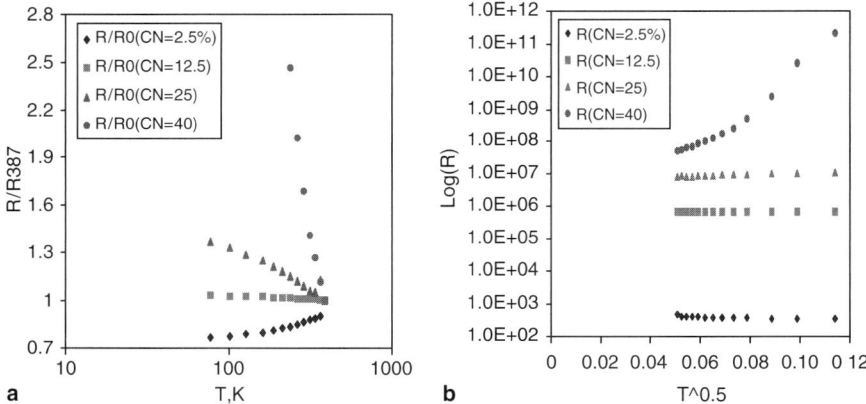

Fig. 8.12 Cr–Si–N system with [SiO]/[Cr] = 40/60: (**a**) $R(T)/R_0$ vs. log Tt and (**b**) $\log(R(T))$ vs. $T^{-1/2}$

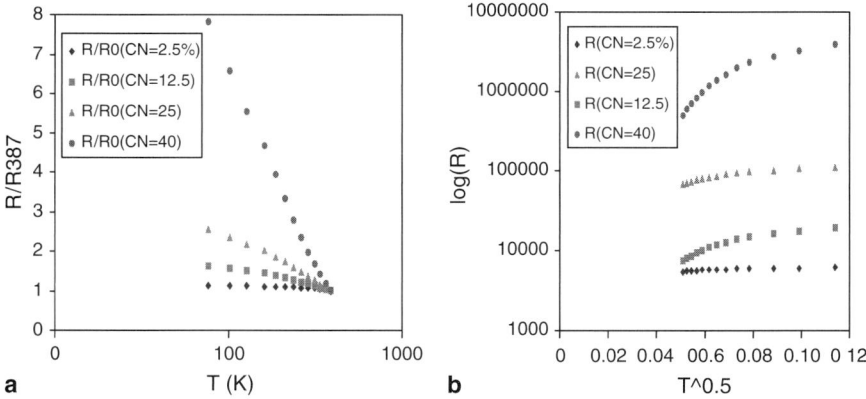

Fig. 8.13 Cr–Si–N system with [SiO]/[Cr] = 80/20: (**a**) $R(T)/R_0$ vs. log Tt and (**b**) $\log(R(T))$ vs. $T^{-1/2}$

8 Nanostructured Resistor Materials

The conductivity is shown as a log R vs. $T^{-1/2}$ plots in Figs. 8.12b and 8.13b. The semiconducting samples are well described by

$$\sigma(T) = \sigma_1 \exp\left\{-\left(\frac{T_0}{T}\right)^{1/2}\right\} \tag{8.6}$$

(where σ_1 and T_0 are constants) over only certain temperature regions, indicating again that the variable-range hopping theory is not applicable to all cermet systems. It can be seen that the results are qualitatively similar for both systems, but the semiconducting behavior for the silicon-to-chromium ratio of 40/60 sets in at much higher nitrogen concentrations than for a silicon-to-chromium ratio of 80/20. The decrease in σ_1 with increasing nitrogen concentration and increasing T_0 should be noticed in Figs. 8.12b and 8.13b. This is one important difference from the behavior of the binary system Cr_xSi_{1-x}, in which σ_1 in the semiconducting region is the same for all as-deposited samples. Annealing has opposite effects on σ_1 between the binary systems with silicon-to-chromium ratio of 80/20 and 40/60. Annealing the $Si_{1-x}Cr_x$ samples leads to a decrease in σ_1 for the [SiO]/[Cr] = 40/60 system but an increase in σ_1 for the [SiO]/[Cr] = 80/20 system. The change of σ_1 can be interpreted as a consequence of the formation and growth of metallic clusters. In this interpretation, the size of metallic clusters in Cr–Si–N films should increase with increasing C_N. The increase in T_0 with C_N implies an increase in the distance between the metallic clusters. Thus, the dependence of the parameters T_0 and σ_1 of (8.6) on C_N can be explained by an increase in both the average size and the separation of the metallic clusters with C_N.

Note that Neugebauer [28] demonstrated consistency in resistivity variations with composition between M–SiO and M–SiO$_2$ cermets (where M represents a variety of metals) if the M–SiO data are reinterpreted as M/Si–SiO$_2$, i.e., if the excess Si is regarded as adding to the metallic content. Cr cermets were included in the study, but are significantly more complex than the noble metal cermets, due to the formation of Cr silicides and oxides.

8.3.3 I–V Characteristics

There are two kinds of test structures, lateral and vertical, on which I–V characteristics were measured. Figure 8.14 shows these two structures schematically. Figure 8.14a shows the lateral test structure and Fig. 8.14b shows the vertical one. For the lateral structure, two Al electrodes 50 × 100 µm and 110-nm thick were first deposited onto a Si substrate through a mask. A $Cr_x(SiO)_{1-x}$ film was then deposited by the cosputtering of SiO and Cr. The volume fraction of Cr was controlled by varying the number of Cr strips placed on a Si target during the cosputtering. The thickness of the film was varied from 10 to 15 nm. The substrate was kept at ambient temperature during the sputtering. For the vertical test structure, a bottom electrode is deposited first. Then, the $Cr_x(SiO)_{1-x}$ film is deposited.

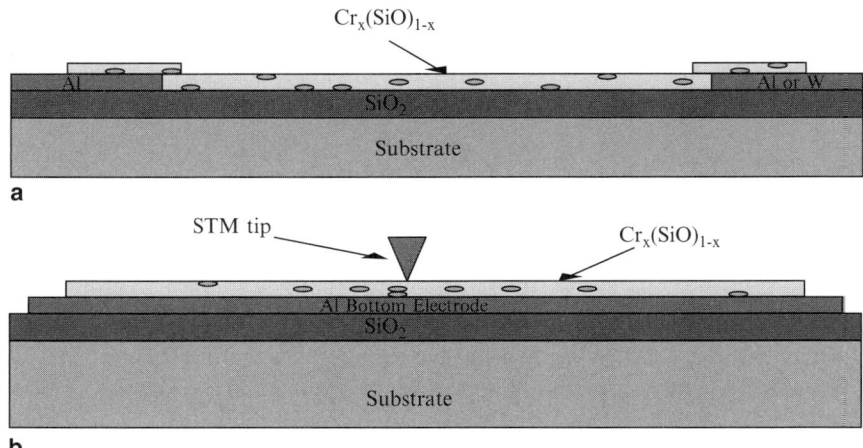

Fig. 8.14 (a) Lateral and (b) vertical test structures for *I–V* characteristics

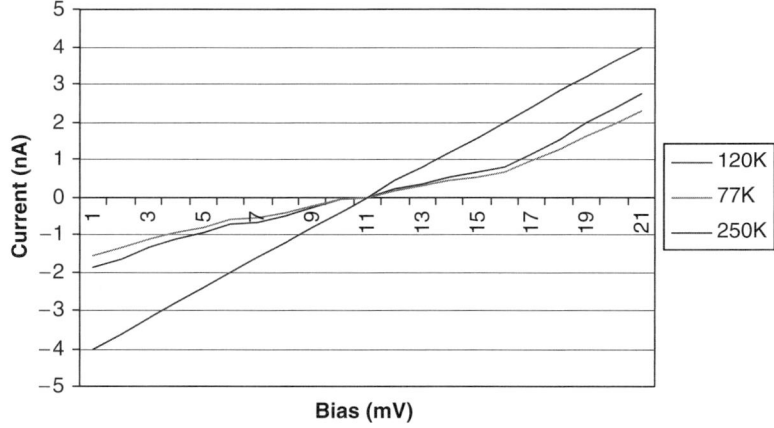

Fig. 8.15 *I–V* characteristics of a vertical $Cr_x(SiO)_{1-x}$ granular film test structure at different test temperatures

Figure 8.15 shows typical *I–V* characteristics of a vertical $Cr_x(SiO)_{1-x}$ granular film test structure shown in Fig. 8.14b. The thickness of the $Cr_x(SiO)_{1-x}$ film is about 10 nm. As the temperature decreases, step-like structures appear in the *I–V* curve. Similar step-like *I–V* curves have been reported for the system containing nanoparticles [29–31]. The observed step structures can be explained by the Coulomb blockade (CB) and Coulomb staircase (CS). Electrons transport between metal Cr islands by thermal-activated tunneling. When tunneling occurs, the charge on the island suddenly changes by the quantized amount *e*. The associated change in the Coulomb energy is conveniently expressed by the approximate form of (8.2)

$$\delta E = \frac{e^2}{4\pi\varepsilon_0\varepsilon_r r}, \tag{8.7}$$

where $\varepsilon_r = 3.8$ is the relative dielectric constant of the SiO_2. From the TEM images, it can be seen that a typical island radius $r = 2$ nm. The charging energy is ~0.59 eV, which is easily resolvable even at room temperature. However, in present samples, the electron transport across the film occurs via Cr islands. Since the device size is very large (50 × 50 μm), a very large number of parallel local current paths exist between the two electrodes. In each current path, there are only a few metal islands, assuming evenly distributed Cr islands with diameter = 2 nm and gap sizes between islands = 1 nm, $n < 3$. The local transport property of each path will be controlled by the charging energy of the islands contained in it, and the CB and CS structures will appear on the I–V curve of each local path. Such local transport properties of granular metal films can be studied by scanning tunneling microscopy (STM), and the observation of CB and CS structures has already been reported [32–34].

In contrast to the previous local transport measurements by STM, the I–V curves obtained from the lateral test structure shown in Fig. 8.16 reflect the macroscopic conductance of the whole system. If each local path in a large device has a similar conductance, the CB and CS structures observed for many paths will be smeared out by averaging the conductance of many paths. On the other hand, if the conductance of the local paths is distributed over many orders of magnitude, the macroscopic conductance of the films under low bias voltage is considered to be determined by a special path with the largest conductance. In this case, the CB and CS structures are expected to be observable even for a macroscopic size device.

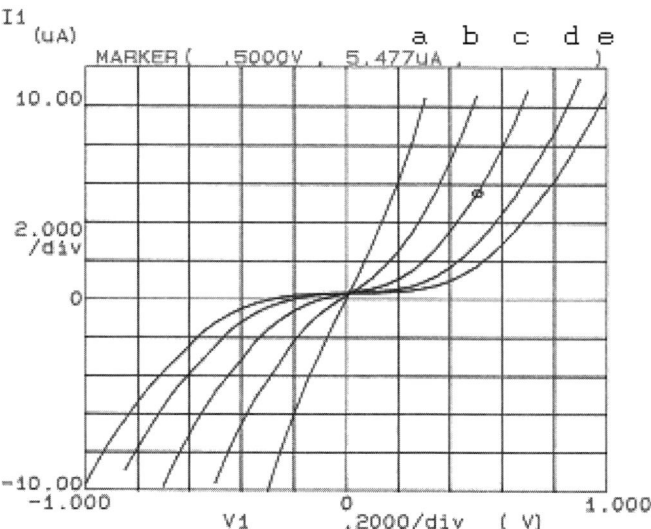

Fig. 8.16 I–V characteristics of a lateral $Cr_x(SiO)_{1-x}$ granular film test structure at different test temperatures: (**a**) 293 K, (**b**) 250 K, (**c**) 180 K, (**d**) 120 K, and (**e**) 77 K

The current–voltage characteristics obtained in this study yield a power-law dependence over a wide temperature range. The steady-state current may be expressed as a function of bias voltage V and temperature T:

$$I = f(V,T). \tag{8.8}$$

The total derivative is

$$dI = \left.\frac{\partial I}{\partial V}\right|_T dV + \left.\frac{\partial I}{\partial T}\right|_V dT. \tag{8.9}$$

The term

$$g_T = \left.\frac{\partial I}{\partial V}\right|_T$$

is the small signal conductance at constant temperature. From Figs. 8.13 and 8.16, for constant bias voltages,

$$I|_V = m_1 T + I_0, \tag{8.10}$$

$$g|_V = m_2 T + g_{0V}. \tag{8.11}$$

From Fig. 8.16, for constant temperature,

$$g|_T = m_3 V + g_{0T}, \tag{8.12}$$

where m_1, m_2, m_3, I_0, g_{0V}, and g_{0T} are small signal constants. Substituting in (8.9) yields

$$dI = (m_3 V + g_{0T})dV + m_1 dT \tag{8.13}$$

assuming $g_{0T} = 0$ and hence,

$$I = m_3 \frac{V^2}{2} + m_1[T(V) - T(0)]. \tag{8.14}$$

This equation would be expected to hold over the linear regions of the above results. However, it would not be expected to hold for bias voltages approaching 0 V, and for higher temperature. Equation (8.14) can also be rewritten as

$$I = AV + BV^2, \tag{8.15}$$

where A and B are temperature-dependent constants. This dependence suggests that the total current consists of two components: one ohmic in nature and the other more akin to a space charge-limited flow.

The tunneling process between islands can be expressed by rewriting (8.8) as following:

$$I(V,T) \propto V \exp(-\delta E / kT) \qquad (8.16)$$

which corresponds to the ohmic term in (8.15). Now, if there is a capacitance C_g associated with the two metal particles, then a space charge will exist in the dielectric given by

$$Q = C_g V, \qquad (8.17)$$

where V is the voltage drop between the islands. Assuming that this charge can move in the insulator by a tunneling process between traps, then the space charge contribution to the current is given by

$$I_s = C_g V / \tau, \qquad (8.18)$$

where τ is the transition time. τ is inversely proportional to the transmission probability D, which is directly proportional to V. Thus, we arrive at the relationship

$$I \propto V^2. \qquad (8.19)$$

Equations (8.16) and (8.19) can be combined to give an overall I–V characteristic of the form suggested by (8.15). The observed temperature dependence above can be explained well by the tunneling theory by taking into account the electrostatic charging energy of metal islands.

The CB suppression of current around the zero voltage region is clearly observed from the current–voltage (I–V) characteristics in Figs. 8.15 and 8.16, especially at the low-temperature measurements at 77 K. It is also found that the observed structures in I–V curves vary from one sample to another, even for samples prepared under the same conditions. This suggests that the transport in metal granular films is a highly selective process and that the transport is dominated by only a few special paths with the largest conductance. In Fig. 8.15, I–V curves are shown for the vertical sample. A CB and the CS step structure with a periodic step width of about 10.4 mV are clearly seen. Each step corresponds to the incremental charging of a nanoparticle by a single electron. The step width of 104 mV (kT) corresponds to a capacitance of 1.6×10^{-18} F. Since the tunneling resistance is very sensitive to the barrier width, the current path with the largest conductance is expected to contain a particle whose size is in the large-size tail of the particle size distribution. The diameter of the largest particle that was found from the HRTEM image of Fig. 8.17a (plane view) and Fig. 8.17b (cross-section view) is about 5 nm. Taking the

Fig. 8.17 HRTEM micrographs of $Cr_x(SiO)_{1-x}$ granular films deposited on (100) Si wafer with 400-nm thermal oxide: (**a**) plane view and (**b**) cross section

dielectric constant of the SiO_2 barrier as 3.9, the capacitance of a Cr nanoparticle with respect to the Al electrode is estimated to be 1.5×10^{-18} F by the image charge method, which is in good agreement with the experimental value. The agreement supports that the macroscopic conductance of sample 1 is governed by a single path containing a serial connection of two Cr nanoparticles of about 5-nm diameter.

References

1. C.J. Adkins, Conduction in granular metals – variable-range hopping in a Coulomb gap?, Journal of Physics: Condensed Matter, 1 (1989) 1253–1259
2. B. Abeles, H.L. Pinch, and J.I. Gittleman, Percolation conductivity in $W-Al_2O_3$ granular metal films, Physical Review Letters, 35 (1975) 247–250
3. B. Abeles, P. Sheng, M.D. Coutts, and Y. Arie, Advances in Physics, 24 (1975) 407
4. E.B. Priestley, B. Abeles, and R.W. Cohen, Surface plasmons in granular $Ag–SiO_2$ films, Physical Review B, 12 (1975) 2121–2124

5. L.G. Feinstein and R.D. Huttemann, Annealing and phase stability of tantalum films sputtered in Ar–O$_2$, Thin Solid Films, 20 (1974) 103–114
6. B. Abeles, R.W. Cohen, and W. Stowell, Critical magnetic fields of granular superconductors, Physical Review Letters, 18 (1967) 902–905
7. B. Long, K. Li, Y. Qin, Z. Chen, and L. Zhang, Direct current electrical conductivity of a Ge–Au composite thin film near the critical threshold, Journal of Physics: Condensed Matter, 9 (1997) 4175–4183
8. M. Fujii, T. Nagareda, S. Hayashi, and K. Yamamoto, Japanese Journal of Applied Physics, 61 (1992) 754
9. A. Heilmann, U. Kreibig, A. Kiesow, and M. Gruner, Optical and electrical properties of embedded silver nanoparticles at low temperatures, Thin Solid Films, 343–344 (1999) 175–178
10. R.I. Khaibullin, Y.N. Osin, A.L. Stepanov, and I.B. Khaibullin, Synthesis of metal/polymer composite films by implantation of Fe and Ag ions in viscous and solid state silicone substrates, Nuclear Instruments and Methods in Physics Research B, 148 (1999) 1023–1028
11. C.A. Neugebauer and M.B. Webb, Electrical conduction mechanism in ultrathin, evaporated metal films, Journal of Applied Physics, 33 (1962) 74–82
12. S.K. Mandal, A. Gangopadhyay, S. Chaudhuri, and A.K. Pal, Electron transport process in discontinuous silver film, Vacuum, 52 (1999) 485–490
13. F. Wu and J.E. Morris, Morphology and electrical characteristics of thin aluminum film grown by DC magnetron sputtering onto SiO2 on Si(100) substrate, International Spring Seminar on Electronics 2002, Prague, Czech, May 2002
14. J.I. Gittleman, Y. Goldstein, and S. Bozowski, Magnetic properties of granular nickel films, Physical Review B, 5 (1972) 3609–3621
15. T.J. Coutts, Conduction in thin cermet films, Thin Solid Films, 4 (1969) 429–443
16. G. Dittermer, Electrical conduction and electron emission of discontinuous thin films, Thin Solid Films, 9 (1972) 317–328
17. H. Scher and R. Zallen, Critical density in percolation processes, Journal of Chemical Physics, 53 (1970) 3759–3761
18. E. Darmois, Journal of Applied Physics, 17 (1956) 211
19. J.E. Morris and T.J. Coutts, Electrical conduction in discontinuous metal films: a discussion, Thin Solid Films, 47 (1977) 3–65
20. R.M. Hill, Electrical conduction in ultra thin metal films. I. Theoretical, Proceedings of the Royal Society of London. Series A, Mathematical and Physical Sciences, 309 (1969) 377–395
21. J.E. Morris, Contact angle contribution to the negative TCR of discontinuous metal films, Thin Solid Films, 29 (1976) L9–L12
22. B. Abeles, P. Sheng, M.D. Coutts, and A. Arie, Structural and electrical properties of granular metal films, Advances in Physics, 24 (1975) 407–459
23. P. Borziak, V. Diukov, A. Kostenko, Yu. Kulyupin, and S. Nepijko, Electrical conductivity in structurally inhomogeneous discontinuous metal films, Thin Solid Films, 36 (1976) 21–24
24. J.E. Morris, A. Mello, and C.J. Adkins, In: G.D. Cody, T.H. Beballe, and P. Sheng (Eds.), Physical Phenomena in Granular Materials, Materials Research Society Proceedings 195, MRS, Pittsburgh, pp. 181–186, 1990
25. J.E. Morris, AC effects in asymmetric discontinuous metal films, Thin Solid Films, 193/194 (1990) 110–116
26. F. Wu and J.E. Morris, Modeling conduction in asymmetrical discontinuous thin metal films, Thin Solid Films, 317 (1998) 178–182
27. J.E. Morris, Recent progress in discontinuous thin metal film devices, Vacuum, 50(1–2) (1998) 107–113
28. C.A. Neugebauer, Resistivity of cermet films containing oxides of silicon, Thin Solid Films, 6 (1970) 443–447
29. L.R.C. Fonseca, A.N. Korotkov, K.K. Likharev, and A.A. Odintsov, A numerical study of the dynamics and statistics of single electron system, Journal of Applied Physics, 78 (1995) 3238–3251

30. C. Wasshuber, About single electron circuits and devices, Ph.D. Thesis, University of Glasgow, 1994
31. E. Bar-Sadeh, Y. Goldstein, C. Zhang, H. Deng, and B. Abeles, Single-electron tunneling effects in granular metal films, Physical Review B, 50 (1994) 8961–8964
32. P. Radojkovic, M. Schwartzkopff, T. Gabriel, and E. Hartmann, Metallic nanoparticles for compact nanostructure fabrication and observation of single-electron phenomena at room temperature, Solid-State Electronics, 42 (1998) 1287–1292
33. R.P. Andres, T. Bein, M. Dorogi, S. Feng, J.I. Henderson, C.P. Kubiak, W. Mahoney, R.G. Osifchin, and R. Reifenberger, Science, 272 (1996) 1323
34. F. Wu and J.E. Morris, Electrical and structural characterization of CrSi resistive films, 26th International Electronic Components and Technology Conference, New Orlean, LA, May, 2003

Chapter 9
Nanogranular Magnetic Core Inductors: Design, Fabrication, and Packaging

Gopal C. Jha, Swapan K. Bhattacharya(✉), and Rao R. Tummala

9.1 Introduction

Technological advances often lead to a transition when the vision of the past no longer remains a vision and becomes the demand of the present. Miniaturized, cost-effective, and mega functional devices are the present day's demand. With Moore's law driving the miniaturization of active devices, passive components have been left behind. As the trend continues toward miniaturization, electronic industries are experiencing an immense demand for miniaturized and more efficient passive components.

Presently, high-end electronic devices are composed of almost 90% passive components, taking up almost 70% of the total board area [1, 2]. Inductors are important passive elements readily used in military, communication, automotive, computer, and other portable devices. They are of utmost importance for high frequency applications, especially in wireless/radio communication devices. "Inductor" is a generic term for components having specific inductance that can be used to store energy in the form of a magnetic field [3]. They cover a wide range of applications including DC–DC converters, voltage-controlled oscillators, clocks, filters, power amplifiers, low noise amplifiers, voltage regulator modules, phase-locked loops, and point of load converters.

Recently, significant attention has been given to achieve high-quality factor (Q-factor) inductors with significant size and cost reduction. Various works have been carried out by different research groups to bring out an efficient way for more compact packaging. Efficient designs have been made and studied to increase the quality factor and inductance of the devices, including innovative designs reported by Patranabis et al. [4], Park and Allen [5–7], and Chuang et al. [8]. Yamaguchi et al. studied and reported microslits [9, 10] and surface planarization [11]. Allen [12] studied microelectromechanical system (MEMS)-based inductors. However, the hunt for high-quality factor and high inductance has never ceased.

This chapter presents a comprehensive review of inductor research, including design, fabrication, material, characterization, and packaging, with special attention

---

K. Bhattacharya
Henkel Loctite (China) Co., Ltd, 90 Zhujiang Road, Yantai ETDZ, Shandong, China 264006
daniel.lu@ieec.org

to recent advances in nanogranular magnetic materials for high performance inductive cores. Air core inductors are generally preferred over magnetic core inductors, where loss is of significant concern. For magnetic cores, however, high permeability to achieve inductance, high saturation magnetization to avoid dramatic decrease in the inductance, low coercivity, high resistance to avoid eddy current loss, and high frequency characteristics have been identified as key parameters that can significantly decrease the number of windings needed to achieve required inductances, and thus could catalyze the effective miniaturization of devices. Low dimensionality, small length scale leading to enhanced exchange coupling, and ease of tailoring properties through additional degrees of freedom make nanogranular magnetic materials an attractive choice in achieving a high degree of compactness for such high performance devices [13]. The enhanced properties at nanoscales, however, do not fit well with the classical theory of magnetism. Significant works on modeling of local anisotropy and exchange interactions between constituents in nanostructures by Herzer et al. are important in this regard [14, 15]. In this chapter, we have attempted to give an insight to the basic concepts of evolution of novel magnetic properties in nanomaterials. Furthermore, the translation of research into the mainframe product can only be possible if a cost-effective and well-defined fabrication process can be designed. Demands for nanostructured materials call for advanced sample growth techniques such as sputtering, plasma-enhanced deposition processes, and electron-induced deposition as well as patterning techniques to achieve feature sizes of the order of nanometers, beyond the limits of conventional photolithography [13]. Compatible fabrication processes are also discussed briefly in this chapter.

9.2 Inductor Design

Performance of a device depends upon two key factors – design and inherent material properties. Design is an important parameter that takes care of electrical, mechanical, and reliability issues. A novel design with careful packaging can dramatically enhance the performance of the device, which would not be possible with just a simple design backed up by inherent material properties. Number of turns, width of metal traces, spacing between them, inner and outer diameters are some of the important design parameters. Various designs have been proposed to achieve high-quality factor. This section reviews some of the important designs.

9.2.1 Spiral Inductor

Spiral inductors are the most widely studied, which can be attributed to their high efficiency and simple fabrication process. This design renders a very high-quality factor due to the ease in attaining very large core cross-sectional areas. Air core and magnetic core spiral inductor designs have been studied by various researchers. Yamaguchi et al. proposed variations in the design, including closed magnetic circuit type spiral inductors [10], sandwiched spiral inductors [16], on-top type ferromagnetic spiral inductors [17], microslit spiral inductors [9, 10], and surface planarized spiral inductors [11]. They also reported various patterned spiral inductors for better performance [10]. A number of magnetic cores have been studied with spiral inductors, including Co–Zr–Nb [18], Co–Fe–B [19], Co–Fe–B–N [20], and Fe–Hf–N [21]. A quality factor of more than 10 was reported for all these inductors. Spiral inductors are designed in different shapes including square, hexagonal, octagonal, and circular [22]. Figure 9.1 illustrates schematics of some of them. Polygonal spiral inductors have been reported to render higher quality factor and performance.

In recent years, RF technology has identified CMOS as the enabling technology. There are, however, concerns due to conducting substrates which limit the use of inductors. Hizon et al. [23] have proposed several polygonal designs of monolithic

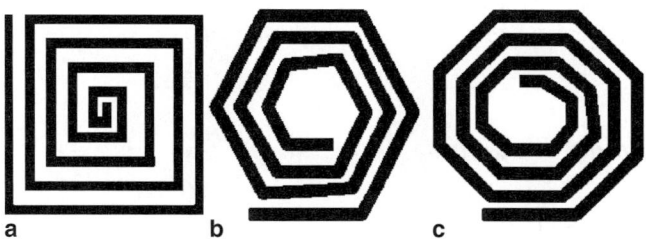

Fig. 9.1 Polygonal spiral inductors: (**a**) square, (**b**) hexagonal, and (**c**) octagonal (© IEEE 1999) [22]

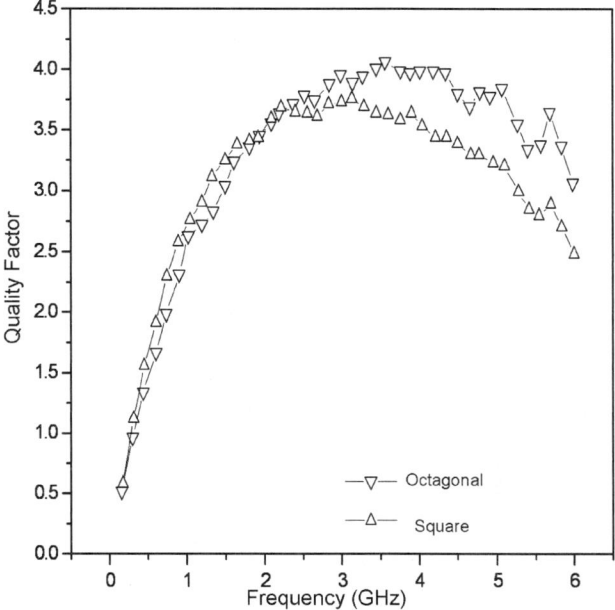

Fig. 9.2 Comparative extracted quality factors of 3-nH square and octagonal spiral inductors (© IEEE 2005) [23]

spiral inductors for very high frequency applications. Figure 9.2 illustrates the comparison between the performances of square and octagonal inductors. As is evident, the quality factor of octagonal spirals is higher than that of square spirals due to small series resistance that is extracted in octagonal spiral inductors. Octagonal structures have also been reported to have lower shunt capacitance because of their more hollow structure.

Inductance (L) of a spiral inductor depends on several parameters including the number of turns (N), and fill ratio (R_{Fill}) that is defined as the ratio of the difference between outer and inner diameters to the average diameter (D_{Avg}) [22], as evident from

$$L = \rho_L N D_{Avg} \ln(8R_{Fill}), \tag{9.1}$$

where ρ_L is a material-dependent constant. Therefore, an optimization of all the above-mentioned parameters is necessary to insure high performance of the device.

9.2.2 Toroidal Inductors

Liakopoulos and Ahn [24] proposed a novel design for toroidal inductors for high current and power electronic applications as can be seen in Fig. 9.3. Some of the important design parameters for such kind of inductors include number of turns (N), coil length (L_{coil}), length of air gap (L_{gap}), relative permeability of core materials (μ_{rel}), effective magnetic path area (A_{mag}), and effective gap area (A_{gap}) [24]. A theoretical expression for effective inductance can be given by [24]

$$L = \frac{0.4 \times 10^{-8} \pi N^2}{\dfrac{L_{coil}}{\mu_{rel} A_{mag}} + \dfrac{L_{gap}}{A_{gap}}}. \tag{9.2}$$

Evidently, the inductance is expected to increase with increase in the number of coil turns. Also, it should be noted that increase in air gap and effective magnetic path area can enhance the inductance of the device.

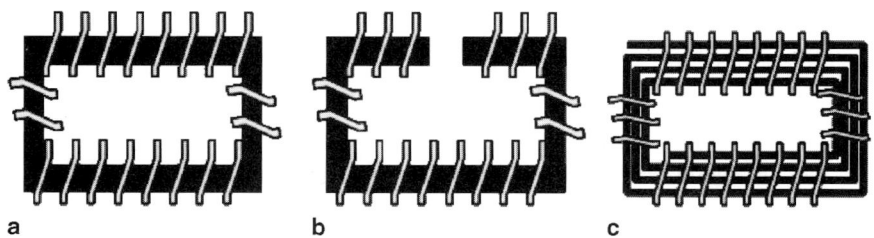

Fig. 9.3 Different designs of toroidal inductors: (**a**) core without air gap, (**b**) core with an air gap, and (**c**) core in a spiral shape (© IEEE 2005) [24]

Inductors with planar permalloy cores with no air gaps were reported to have inductance values of more than 10 µH and a DC resistance as low as 1–4 Ω at low frequency (<2 kHz), thus making them suitable for power applications [24]. However, the inductance decreased drastically at higher frequency. High frequency performance could, however, be enhanced by introducing an air gap in the core and also by introducing spiral cores instead of planar cores. Both of the above designs introduce air gaps in the magnetic flux path and thus reduce the eddy current loss at high frequencies.

9.2.3 Solenoid Inductors

Most discrete inductors are of the solenoid type. Such inductors offer high inductance, high-quality factor, and low DC resistance [25]. However, inefficient packaging and high leakage current at high frequency make them less attractive. In addition, parasitic effects resulting from conductor line spacing and dielectric constant of the substrate are of major concerns. Edelstein and Burghartz [26] reported comparison between magnetic performances of spiral and solenoidal inductors. They reported that the magnetic fluxes in the solenoid type inductors are mostly associated with the coil itself and thus they are less prone to eddy current loss as compared with spiral inductors. However, the substrate losses are reported to be comparable with those in spiral inductors.

9.2.4 Inductors with Microslits

At high frequency applications, as in wireless and communications, losses are key parameters. Due to material considerations, air cores are preferred over ferromagnetic cores. However, a ferromagnetic core with high permeability can achieve high inductance and thereby reduce the number of windings needed for a specific application. Novel designs have been proposed to minimize the losses of a magnetic core including introduction of microslits in the core material along the easy axis. Microslits introduce attributes that artificially control the shape anisotropy and magnetostatic energy to avoid ferromagnetic resonance (FMR) at high frequency [9]. Such designs, in addition, reduce the eddy current loss and leakage flux to a significant extent by ensuring fully closed magnetic circuits [27]. Reduction of leakage flux is vital as the leakage flux interferes with the electronic circuits and does not contribute to the effective inductance. There are various parameters that dictate the design of microslits and assure high-quality factors. These include compatibility of microslit fabrication with existing technology, thermal stresses and stability, easy axis orientation associated with demagnetization and domain configuration, geometrical design, leakage flux, and stray capacitance [10].

Microslit inductors are fabricated on ferromagnetic film by patterning. The simplest pattern (shown in Fig. 9.4a) consists of microslits along the easy axis direction.

Width and spacing of the microslits and judicial selection of the ferromagnetic core are key parameters that determine the anisotropy field, effective demagnetization field (N_d), FMR, inductance, and quality factor (Q) [9].

Various research groups have demonstrated enhanced performance with microslited inductor cores. Yamaguchi et al. [9] reported micropatterned sputter-deposited CoNbZr ferromagnetic films of 200-nm thickness with an enhanced anisotropy field of 70 Oe as compared with the intrinsic value of 10 Oe. Similarly, the FMR was reported as high as 2.5 GHz with microslit of thickness 200 nm, width 20 μm, and 4-μm spacing, as a result of the microslits' introduction. Also, the quality factor was improved. Similar work on micropatterned FeAlO films showed an increase in the quality factor (from 5.6 to 7.7) at 1 GHz for a 2,000-nm slit due to effective reduction in the resistance [17].

The aforementioned unidirectionally patterned film, however, utilizes only half of the total area available on the film. Performance of the device can be increased further by an improved design that can make use of more area. Such design consists of preferential micropatterning to form bidirectional microwire arrays as shown in Fig. 9.4b [28]. Preferential micropatterning helps split the easy axis into two different directions making use of the high shape anisotropy of the film. This arrangement makes the full area of the film active through the excitation of the hard axis. In addition, it further raises the FMR frequency.

Typical fabrication flow for such micropatterns involves deposition of thin magnetic film followed by ion milling or photolithography for micropatterning. Yamaguchi et al. [28] reported such bidirectional microslit fabrication on 100–300 nm thick amorphous $Co_{85}Nb_{12}Zr_3$ film, deposited by radio frequency magnetron sputtering, using ion milling. Such a patterned structure was reported to render coercivity as low as 0.4–0.7 Oe. Moreover, the effective anisotropy field was improved; for 100-nm film, it was noted as high as 30 Oe. The frequency profile also improved. Real permeability did not show any degradation at higher frequency as compared to vertically aligned inductors with no slit. Almost 11% increase in the inductance (7.5 nH) over that with an air core inductor was also noted with a 5-nm thick underlayer

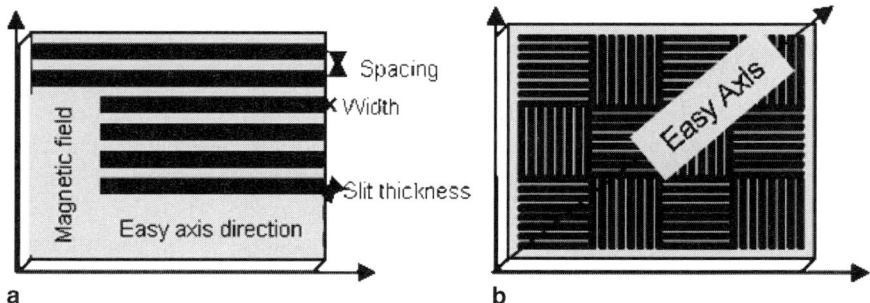

Fig. 9.4 Inductor core with patterned microslits: (**a**) unidirectional slits along the easy axis direction [9] and (**b**) bidirectional micropatterns (© IEEE 2000) [28]

of Ti. The quality factor (6.3) was comparable to that of an air core. Baba et al. [29] reported even larger inductance of 8.2 nH in a bidirectional patterned film with a 10-μm line and 1-μm space, which was twice as large as that in a similar inductor with a uniaxial magnetic film. It should be noted that this is attributed to the variation in slit width and other parameters. As described before, slit width is one of the parameters that dictates the quality factor and the inductance of the pattern [28]. Figure 9.5a, b illustrates the frequency-dependent quality factor and inductance of such bidirectional microslits at various slit widths. Also, it compares the values with air core inductors. As evident from Fig. 9.5a, a magnetic core has a higher inductance than an air core at all frequency ranges of interest. Introduction of microslits, however, witnesses a decrease in inductance in the low frequency range. The amount of drop in inductance increases with increase in the slit width. Nevertheless, a significant enhancement in the quality factor is observed with micropatterns (Fig. 9.5b). Increase in slit width further increases the quality factor.

Introduction of orthogonal bar slit patterns or cross-slit patterns, instead of parallel slits, can also increase the available magnetic film area utilization [11]. Schematics of such designs are shown in Fig. 9.6a and b, respectively.

Yamaguchi et al. [11] studied such designs on conventional Al–Si spiral coils with $Co_{85}Nb_{12}Zr_3$ film cores. Figure 9.7a, b illustrates the magnetic performances of such designs. As evident from Fig. 9.7a, the inductance with no-slit and cross-slit patterning is 21% larger than that of air core inductors. Peak inductance is achieved at 1.3 GHz. Also, it can be noted that cross-slit patterns have the highest inductance as compared with orthogonal bar slit and parallel slit patterns. The FMR frequency is shifted toward higher frequency range through an artificially controlled demagnetizing field and magnetostatic energy. A decrease in the eddy current losses was also noted. However, the quality factor witnessed a decrease due to large resistance (Fig. 9.7b).

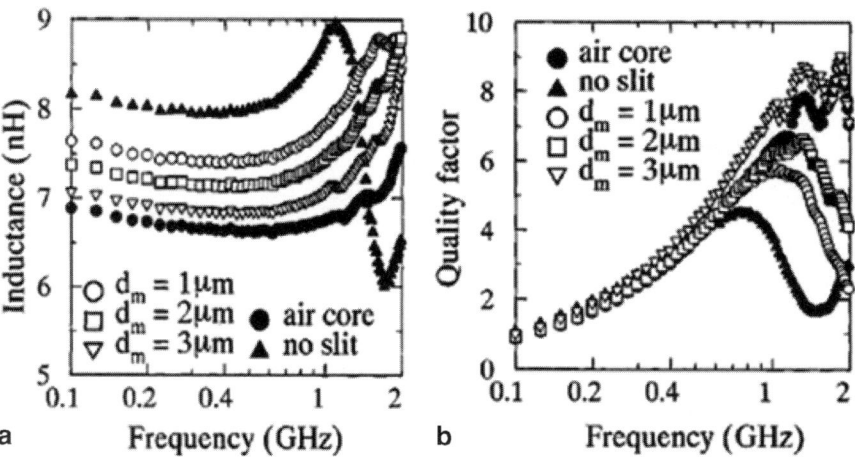

Fig. 9.5 Magnetic characteristics of bidirectional micropatterned CoNbZr film: (**a**) inductance and (**b**) quality factor (© IEEE 2000) [28]

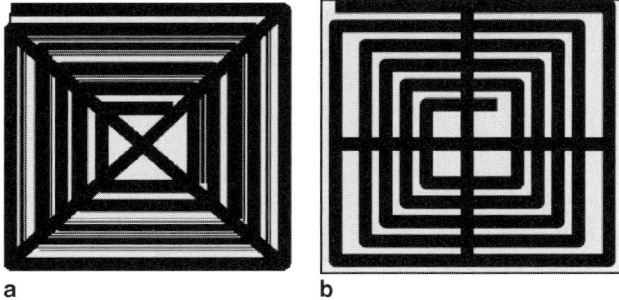

Fig. 9.6 Micropatterned inductors: (**a**) with orthogonal bar slits and (**b**) with cross-slits (© IEEE 2000) [11]

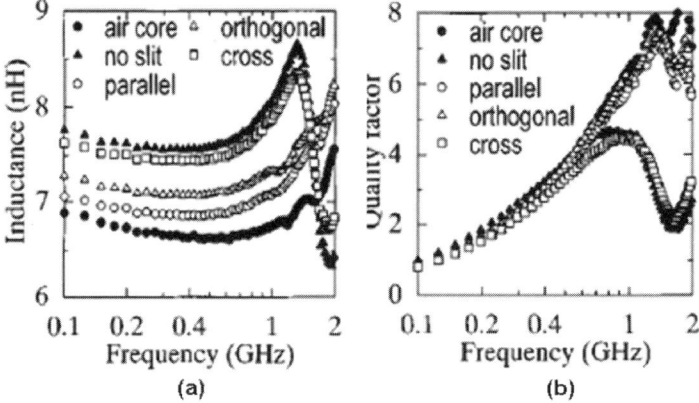

Fig. 9.7 Magnetic characteristics of inductors with various slit patterns: (**a**) inductance as a function of frequency and (**b**) quality factor as a function of frequency (© IEEE 2000) [11]

9.2.5 Surface Planarization

Surface roughness on the inductor coil deteriorates the magnetic performance of the inductor. Surface planarization was found to be effective in raising the inductance and the quality factor simultaneously [11]. Surface planarization reduces the surface roughness and thus makes the gap between coil and magnetic film nearly constant and small. Yamaguchi et al. [11] achieved surface roughness as low as 1.2 nm using surface planarization. A significant enhancement in the inductance was reported up to 2 GHz. Planarized inductors were measured to have inductance values as high as 8.26 nH at 1 GHz (22% larger than that of the air core and 2% more than the non-planarized inductor). This increase in the inductance is attributed to the enhanced magnetic field near the magnetic film due to the decrease in the coil to film gap. The quality factor also showed a 14% increase as a result of planarization.

9.2.6 Closed Magnetic Circuit Type Inductor

Yamaguchi et al. [10] reported that the performance of a magnetic core spiral inductor can be improved by an optimized arrangement of patterned magnetic films against the spiral legs. They studied different arrangements of patterned films and compared the results with nonpatterned films. Figure 9.8 shows various arrangements of patterned films.

Plain film inductors are provided with no slit patterned on the magnetic film and nonterminated top and bottom magnetic film edges. The aligned type faces the leg of the spiral. The shift type faces the gaps of each leg, and the closed magnetic circuit type is equivalent to the aligned type with terminated top and bottom magnetic film edges. In-plane eddy current loss becomes prominent in the shift type arrangement, as the middle of the coil leg, where the leakage flux attains a maximum, faces the gap between the magnetic films. Similarly, stray capacitance is dependent on the film arrangement. Any voltage difference between the coil turns causes a displacement current. Plain type and shift type arrangements shunt the displacement current caused by voltage difference, which might result in a low self-resonant frequency [10]. The comparative performance of each arrangement is shown in Fig. 9.9. As is evident, the plain, closed, aligned, and shifted type arrangements show inductance in decreasing order. Due to overlap between the two ferromagnetic layers, the closed type has a lower FMR frequency than that observed in either shift or aligned inductors. Nevertheless, it is higher than that of plain film inductors. The air core inductor resonates at 15 GHz, whereas the self-resonant frequency of the ferromagnetic inductor was found to be 10 GHz, with the FMR frequency in the range of 1–3 GHz due to stray capacitance and inductance. The resonant frequency values are in increasing order with plain film inductor, shifted, closed, aligned, and air core inductors. The resonant frequency is dictated by the inherent conductivity of the film. The

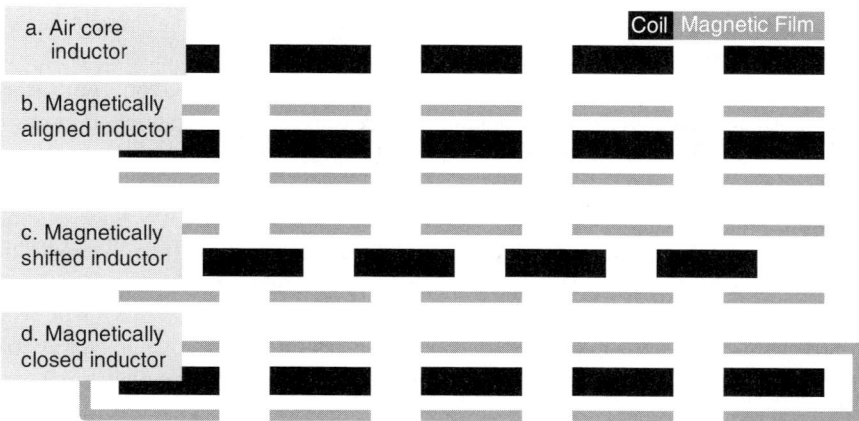

Fig. 9.8 Cross-sectional view of relative position of magnetic film with respect to the spiral leg [10]

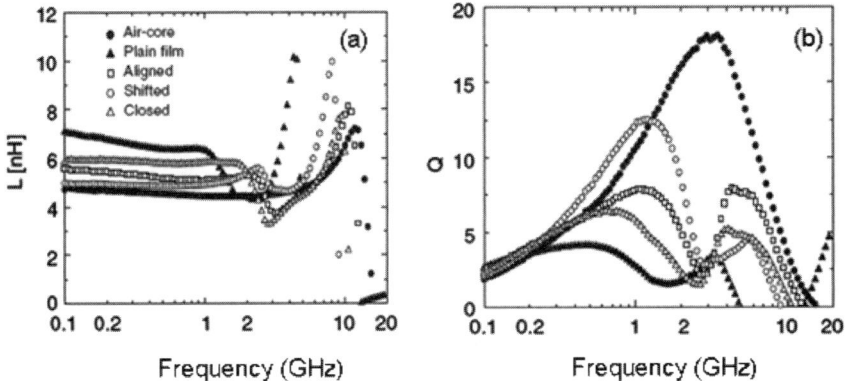

Fig. 9.9 Comparative evaluation of magnetic characteristics of various inductors as a function of frequency [10]

highest air core resonant frequency is attributed to the absence of any conductive film. An additional terminating structure between two ferromagnetic films in the aligned type introduces added stray capacitance, and thus they resonate at lower values than closed and aligned types. Plain inductors have the largest electrodes, resulting in the highest stray capacitance, and thus the lowest resonant frequency value.

DC resistance was recorded in decreasing order with closed, aligned, shifted, and air core. As the resistance is inversely related to the quality factor, the shifted type showed higher quality factor (as high as 13) than others as evident from Fig. 9.9b. However, a localized decrease in the quality factor at 3 GHz was noted due to FMR.

9.2.7 Composite Planar Inductor

The need for miniaturized inductors with low DC resistance and high rating current has stimulated research in thin film composite inductors. Yamaguchi et al. [16] proposed sandwich structures to increase the quality factor of the inductor. The Vishay Group [30] makes high-quality inductors by pressing low permeability iron–resin composites and magnetowire coil together. The iron powder, however, has been reported to be very prone to oxidation and thus requires a passive layer. Also, the inductor is very thick (3 mm), which limits the size of the device [30]. Kowase et al. [31] proposed a composite magnetic core of Mn–Zn ferrite/polyimide to avoid this problem. This film was reported to have very large saturation magnetization (M_s) of about 2 kG and was still not saturated even at large magnetic fields up to 2.5 kOe. The proposed structure consisted of two designs: type 1 with an inner square spiral coil between a top and bottom composite magnetic film core and type 2 a planar inductor with the same spiral core between a top composite magnetic core and a bottom 1-mm thick Ni–Zn ferrite substrate. The design is schematically illustrated in Fig. 9.10.

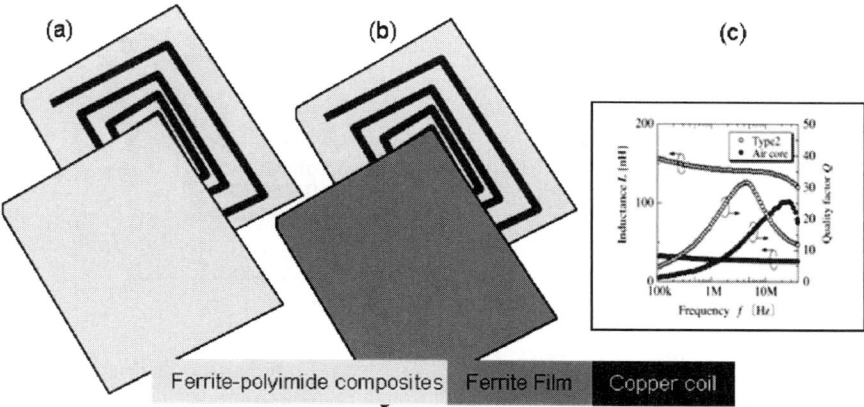

Fig. 9.10 Composite magnetic core inductor: (**a**) an inner square spiral coil between a top and bottom composite magnetic film core, (**b**) a planar inductor with the same spiral core between a top composite magnetic core and a bottom 1-mm thick Ni–Zn ferrite substrate, and (**c**) comparison of magnetic performances of type 2 inductor with air core inductor (© IEEE 2005) [31]

It was found that the inductance increased with the film thickness. For type 1, the saturation was reached at 300 μm whereas the type 2 planar inductor still recorded increase even in thicker conditions. This difference is attributed to the high permeability ferrite substrate layer. They also reported larger than threefold increase in inductance in type 2 as compared to the air core inductor as evident from Fig. 9.10c. The quality factor also improved (Fig. 9.10c).

9.2.8 Suspended Air Core Design

Losses and parasitics are prime packaging design issues. These issues have been addressed by using suspended air core inductors. Air gaps reduce the winding capacitance, which results from the spatially separated coils from the central lead as in spiral coils. In addition, the DC handling capacity is improved before saturation takes place. A schematic of such design is shown in Fig. 9.11. Goldfarb and Tripathi [32] reported such suspended spiral inductor in integration with a transistor using air bridge technology. A 3-μm gap between the inductor and the substrate was achieved.

Park and Allen [7] reported an improved design for a suspended inductor with air core using surface micromachining. The air gap between the inductor and the substrate was made as high as 60 μm and a thick electroplated conductor line of copper was also deposited. Thick conductors improved the quality factor. They also demonstrated LC filters based on these inductors. The advantage of this design over the air bridge technique is that a large air gap can be maintained with the help of plated copper. Moreover, thick copper plating increases the cross-sectional area and thus

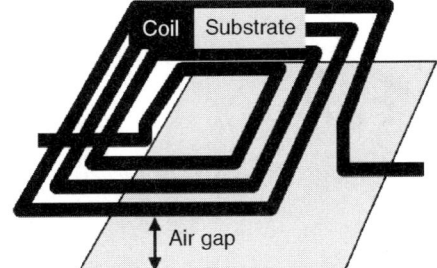

Fig. 9.11 Suspended air core inductor (© IEEE 1999) [7]

decreases the DC resistance. Chang et al. [33] also reported suspended inductors on silicon, where the air gap was introduced by selective etching of silicon under the inductor coil to minimize the effect of lossy silicon. Another air gap spiral inductor structure was reported, which used glass microbump bonding (GMBB), by Chuang et al. [8]. Kim and Allen [34] also investigated the effect of air gaps in solenoid inductors. The air gap was introduced using micromachining and electroplating.

9.3 Fabrication Techniques

A cost-effective and well-defined fabrication process is needed to translate the research into useable product. Demand for nanostructured materials calls for advanced sample growth techniques such as sputtering, plasma-enhanced deposition processes, and electron-induced deposition as well as patterning techniques to achieve feature sizes of the order of nanometers, beyond the limits of conventional photolithography [13]. The overall fabrication process of an inductor with a nanogranular magnetic core consists of thin film deposition, patterning, surface treatment whenever necessary, and annealing (magnetic field and/or temperature annealing). Various thin film deposition processes including radio frequency magnetron sputtering (RFMS), electroplating, and plasma-enhanced chemical vapor deposition (PECVD) have been studied for the deposition of magnetic thin films. Electroplating, nonetheless, is the most popular choice – due to its low cost, simplicity, and low temperature process. Electroplating involves deposition of materials on a conducting surface under the influence of an electric current. This is particularly useful for post-lithography processing. Electrodeposition, however, is not very suitable for generating complex stoichiometry. For complex stoichiometric magnetic thin films, RF sputtering has been studied the most. Spin coating and sol-gel processes have also been reported for the deposition of polymeric composite materials as the magnetic core. Recently, spin-sprayed magnetic thin films have drawn significant attention [35–39]. Patterning is a very important process for the design of conductors as well as the magnetic films. Patterned magnetic films increase the high frequency characteristics of inductors as described in Sect. 9.2. The shape anisotropy and DC resistance can be artificially enhanced by suitable patterning. This enables the designer to design tunable inductance profiles in a single film [13, 40]. Moreover, closure domain structures at the edges of patterned film can also be suppressed [40]. A suitable design with very small feature size is only possible with advanced lithographic techniques. Martin et al. [13] have done a comprehensive review on lithographic processes. Several modifications have been reported to make lithography more compatible with materials and technologies. For fabrication of multilayer 3D microinductor components, a new lithography technique (UV LIGA) has been developed [24]. Other patterning processes include micromachining [7] and ion milling [10]. Surface micromachining is a low cost fabrication process compatible with integrated passive technology. Micromachining has been readily studied to fabricate suspended inductors. Predeposition of a sacrificial layer (e.g., polyimide) is another technique to fabricate suspended cores. Advances in MEMS technology have been extended to the fabrication of inductive components. Recently, a new class of fabrication process has been reported based on MEMS technology [12].

RF inductors usually require a domain structure along the length of the wire. Various kinds of annealing treatment are done to align the domain including non-magnetic field annealing, static magnetic field annealing, and rotational magnetic field annealing.

9.4 Nanogranular Magnetic Core Materials

High efficiency, greater compactness, and multimega functionality are the driving needs of the electronic industries. Inductors are represented as bulky components unless realized as embedded thin films. Reduction of size in the inductors without distorting performance is possible with the discovery of new high performance materials as magnetic cores [41]. High permeability and enhanced magnetic flux bring down the overall size by reducing the number of turns. Resulting reduction in the coil length decreases the coil loss. Size reduction improves the resonant frequency also. Stray capacitance decreases as a result of low coil length. Also, the eddy current loss could be minimized as the magnetic flux is mostly associated with the core. This brings down the required thickness of insulator between coil and substrate even for low resistivity substrates. The limitations of microgranular magnetic cores, however, are not able to meet the growing demand of electronic devices especially at high frequencies where losses become very prominent due to the inherent conductivity of the magnetic materials. In addition, eddy current losses decrease the permeability and hence, the inductance of the device. One possible remedy is the use of high resistivity material. High resistance of inductor, however, decreases the quality factor of the inductor [42]. So, there is a growing need to develop new materials that could satisfy the present needs. The issue has been addressed by the introduction of ultrasoft nanocrystalline thin films with ultrahigh frequency permeability [43]. It has been found that nanostructures improve the magnetostatic and magnetodynamic properties of the material. The use of nanogranular ultrathin magnetic films to improve the magnetic performance seems to defy the classical theory [44]. According to the classical physics, the effective coercive field strength (H_{CK}) is equal to the summation of intrinsic coercive strength caused by magnetostrictive residual stresses, nonmagnetic inclusion, and the high energy distorted region of the grain boundaries [45–47]. Equations (9.3) and (9.4), respectively, show the theoretical values of grain boundary component of coercivity, and initial relative permeability (μ_r), as determined by the classical theory [44, 48, 49]:

$$H_C \approx p_C \frac{\gamma_W}{J_s D}, \tag{9.3}$$

$$\mu_r \approx p_\mu \frac{J_s^2}{\mu_0 \sqrt{AK_1}} D, \tag{9.4}$$

where p_w is the boundary wall energy, which depends upon the anisotropic constant (K_1), lattice parameter, and curie temperature; J_s is the saturation polarization; D is the grain diameter; μ_0 is the permeability; A is the exchange stiffness; and p_C and $p\mu$ are material-dependent constants. Evidently, from the above equations, the coercive field is expected to increase and the permeability to decrease with increase in the grain boundary area, i.e., with decrease in the grain size.

Anisotropy and coercivity modeling by Herzer [49] for nanostructured materials, however, found that the classical rule does not hold good at the nanoscale. Critical phenomena at the nanoscale result in attractive properties. It has been found that the classical rule is valid only when the grain size is greater than the ferromagnetic exchange length, where the magnetocrystalline anisotropy is not suppressed due to the ferromagnetic exchange interaction [50].

Due to easy magnetization alignment along the easy axis and in-grain domain formation, magnetization in large grains is a function of magnetocrystalline anisotropy, which is measured by the difference in the B–H hysteresis loop along the easy and hard axes. At the nanoscale, however, it is determined by the simultaneous occurrence of magnetic anisotropy energy and the ferromagnetic exchange energy. Ferromagnetic exchange interaction forces the magnetic moment to align parallel to it and thus restricts the alignment along the easy axis. When the grain size is smaller than the effective exchange interaction length L_{ex} (9.5), the coercivity and the permeability are given by (9.5) and (9.6), respectively [49], as follows:

$$L_{ex} = \sqrt{\frac{A}{\langle k \rangle}}, \tag{9.5}$$

$$H_C = p_C \frac{K_1^4 D^6}{J_s A^3}, \tag{9.6}$$

$$\mu_r = \rho_\mu \frac{J_s^2 A^3}{\mu_0 K_1^4 D^6}, \tag{9.7}$$

where $\langle k \rangle$ is the anisotropy density that is given by the mean fluctuation amplitude of anisotropy energy and other constants are as discussed above [49].

Therefore, materials with intrinsically low anisotropy, e.g., permalloy, can exhibit very low coercivity with very high permeability, if the grain size is restricted below the effective exchange coupling length as can be seen in Fig. 9.12. As evident, coercivity is maximum, and permeability is minimum when the grain size equals L_{eff}. Dramatic changes in the coercivity and anisotropy are interpreted in terms of the smoothing part of the exchange interaction, averaging out locally fluctuating anisotropy, so that there is only a small net anisotropy effect on the magnetization process.

Nanostructures do not merely decrease the coercivity and increase the permeability, but they also limit the losses due to high frequency. Various loss mechanisms including eddy current loss, FMR loss, and Landau–Lifshitz (LL) damping loss limit the applicability of magnetic films at high frequencies [43]. Eddy current losses, e.g., resistive, capacitive coupling through space layers, etc., are determined by the conductance, shape, and thickness of the film. The cutoff frequency for infinitely wide films due to eddy current loss can be given by

$$f_{\text{cutoff}} = \frac{4\rho}{\pi \mu_0 \mu_i d^2}, \tag{9.8}$$

where ρ is the resistivity, μ_i is the intrinsic initial permeability, and d is the thickness of the film [43].

Thin coating with foreign or native oxides and increase in the DC resistance are prevalent methods to decrease the eddy current loss. However, it should be noted that the former does not improve frequency response and the latter decreases the quality

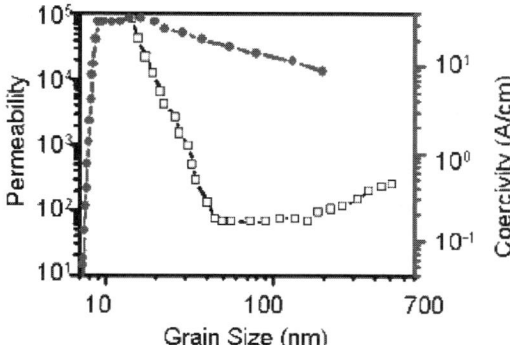

Fig. 9.12 Coercivity and initial permeability as functions of the average grain size D (© IEEE 1990) [49]

factor [40]. Advanced engineered nanostructural materials can render an optimized value of high resistivity (low eddy current loss) as well as high-quality factor.

Losses due to FMR and LL damping occur when the applied field frequency matches with the characteristic frequency of the materials [43]. LL damping is caused by structural factors that include demagnetization fields, magnetization dispersion, and domain structure-ripple fields. The FMR frequency for a thin film with a unidirectional demagnetizing field is given by

$$f_{\text{FMR}} = 2\pi\gamma(H_k\mu_0 M_s)^{1/2}, \tag{9.9}$$

where γ represents the gyromagnetic constant and H_k is the in-plane anisotropy field. The critical phenomena as seen in nanogranular thin films have also been reported to occur in sandwich structures when the layer thickness is restricted below the exchange coupling length. Dirne et al. [51] reported such critical behavior in Fe/CoNbZr multilayers.

Above discussions reinforce the attractive properties of nanogranular magnetic materials. Such specially designed materials can render simultaneous low anisotropy and low magnetostriction, e.g., addition of silicon to low anisotropic nanomaterials can decrease the magnetostriction also. Various kinds of nanostructures based on nanogranular thin films, nanocomposite thin films, and sandwich structures with nanometer-scale layers have been studied. In general, popular ferromagnetic materials can be divided into three broad categories – iron based, cobalt based, and iron–cobalt based. Polymeric nanocomposites and advanced novel materials are other attractive choices.

9.4.1 Iron-Based Nanostructural Cores

Permalloy (Fe–Ni) is the most popular among iron-based alloys for inductor applications [40, 52–60]. Magnetic properties are best in the composition range of 30–80% Ni [48]. Good anisotropy is observed due to short-range atomic pairing [61, 62]. As discussed, the magnetic properties are functions of crystalline anisotropy and magnetostriction. Magnetostriction of an inductor is defined as the mechanical response to magnetization (i.e., the change in the shape of inductor under changing magnetic field). Low magnetostriction helps to achieve high permeability, avoids mechanical stress-induced magnetoelastic anisotropy, and reduces noise (e.g., in switched mode converters used in voltage regulator modules and toroidal inductors used in hi-fi amplifiers). Magnetostriction can be decreased in these alloys by addition of nonferrous elements such as copper, silicon, and molybdenum. Yoshizawa et al. [63] reported low magnetostriction in ultrafine (~10 nm) $Fe_{73.5}Cu_1Nb_3Si_{13.5}B_9$ alloys due to addition of copper and silicon. Herzer [49] reported a similar effect. A variation of the above-mentioned composition ($Fe_{73}Cu_1Nb_3Si_{16}B_7$) is commercially available under the name of Vitroperm-800 [64]. Microstructural studies of these alloys revealed homogeneously distributed ultrafine grains of body-centered cubic Fe–Si [14]. Addition of nitrogen and aluminum, similarly, increases the saturation magnetization (M_s). Fe–Al–N has been reported to have $M_s > 20$ kG [65, 66]. Such attractive properties are, however, limited to small compositional range. In comparison, Fe–Al–O gives better properties and large compositional freedom. This alloy is reported to have large M_s (>12 kG), DC resistivity (~500 μΩ cm), and high resonant frequency (~2 GHz), making it suitable for high frequency applications [9, 17]. Various designs have been proposed that use Fe–Al–O nanogranular inductor cores. Yamaguchi et al. [9, 41] reported very high inductance in such designs (~8 nH at 1 GHz for 100-nm thin film). However, a flux saturation reduces the inductance (~7.2 nH) [17]. Low eddy current loss has also been reported in two phase heteroamorphous/granular Fe-based alloys [21, 67, 68]. However, such structure has limited application due to low anisotropy field.

9.4.2 Cobalt-Based Nanostructural Cores

Magnetostriction can further be decreased with Co-based alloys. Masumoto et al. [69, 70] have reported almost zero magnetostriction. Studies of various oxide alloys reveal that only Co–Al–O and Co–Zr–O are suitable for the low coercivity applications as can be seen in Fig. 9.13 [71].

Nanostructural films of Co–Al–O have been reported [71, 72]. Nanometer particle size (53 nm) $Co_{85}Al_{15}$-based oxides, with uniformly distributed FCC phase of α-cobalt oriented along the lowest anisotropy field plane, were reported to have

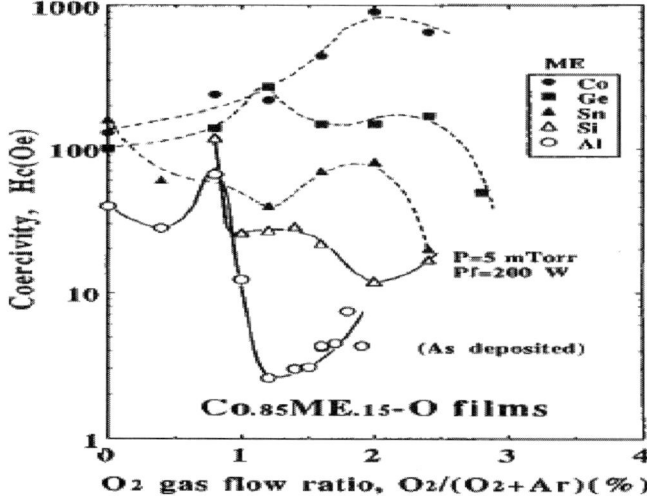

Fig. 9.13 Coercivities of various compositions of Co–X–O based alloys [71]

very low coercivity (~5 Oe) due to an incoherent magnetization and low anisotropic energy orientation [71]. Magnetic closure-domain formation due to nanogranular dipole moment also significantly decreases the coercivity [50, 73]. Eddy loss is also minimal in these films due to high resistance. Energy dispersive X-ray analysis and electron energy loss spectroscopy studies reveal aluminum- and oxygen-rich interparticle regions [71]. This results in preferential oxidation of the interparticle region and thus, enhanced DC resistance and reduced eddy current loss. The concentration of oxygen, however, should be optimized for an optimum combination of various magnetic parameters. The optimum value reported for $Co_{85}Al_{15}$-based oxide films includes resistivity of 500 μΩ cm (as high as Fe-based alloys), M_s ~ 10 kG, H_c < 5 Oe, and H_k ~ 70 Oe at 1% oxygen concentration [71]. Such optimal properties are, however, possible at high temperatures.

In contrast, Co–Zr–O alloys reveal optimal properties even on water-cooled substrates, making it more suitable for CMOS technology [74]. Studies of the magnetic behavior of these films show that unlike Co–Al–O, Co–Zr–O alloys do not exhibit superparamagnetism at 55% cobalt. Moreover, least coercivity is visible at more than one composition (Co – 55 and 70%). Anisotropy as high as 150 Oe is reported in $Co_{60}Zr_{10}O_{30}$ films. A very low value of H_c is also noted when the percolation threshold is approached due to formation of virtual multidomains. These films have been reported to show excellent high frequency properties. The real component of permeability has been reported to remain almost constant up to 1 GHz. The FMR frequency is equal to 3 GHz. However, the absolute value of permeability is much less (~60).

Studies on nanostructural evolution in such alloys reveal initial formation of fine-dispersed ZrO as a result of preferential oxidation. Resulting stress in the matrix results in nanostructure formation. Additional stress and large concentration of Co in the matrix are the primary known factors that affect the final coercivity, anisotropy, and resistivity of the matrix [74].

Nanostructural nitrides of Co alloys exhibit more compositional freedom than oxides [71]. Structural studies reveal amorphous interparticulate region surrounding microcrystalline nanoparticles. Nitrides, however, exhibit low anisotropy (~35 Oe). Another kind of Co-based magnetic nanocrystalline alloys is Co–Zr–Nb [9, 10, 41, 75]. These alloys exhibit resonant frequency less than 1 GHz. Various designs have been studied using such core. Details of these studies could be found in [9, 10, 41, 75].

9.4.3 Fe–Co Based Nanostructural Cores

Fe-based and Co-based nanogranular alloys can yield very high resistance resulting in low eddy current loss. They, however, exhibit low permeability and saturation magnetization at high frequencies. This issue can be addressed using Fe–Co based oxides and nitrides. According to the Pauling–Slater curve, Fe–Co alloys have the highest magnetization (>20 kG) among the iron alloys [75]. However, high magnetostriction in such alloys allows neither high anisotropy nor low H_c. Composite structures, nevertheless, can partly address this issue. Addition of nonmagnetic elements, e.g., silicon and nitrogen, has also been useful to decrease the magnetostriction and thus coercivity. Addition of nickel has also been reported to decrease the coercivity (~1.2 Oe) [76]. Saturation magnetization (M_s) higher than that achieved in permalloy is also noted due to nickel addition. However, Osaka et al. [76] reported very low DC resistivity (~21 µΩ cm) in such alloys making them unattractive for high frequency applications. Addition of aluminum in Co–Fe nitrides has also been found to decrease the coercivity to 1 Oe as compared with 5 Oe in unadulterated alloys [77, 78]. Low saturation magnetization in such alloys can also be enhanced to a value as high as 17.6 kG [79]. Anisotropy coupling between magnetic nanograins and electric field-induced atomic ordering also results in high magnetocrystalline anisotropy (~45.6 Oe) [80]. Addition of boron also increases the saturation magnetization of Fe–Co [19]. Minor et al. [81] report saturation magnetization as high as 24 kG in Fe–Co–B thin films.

Fe–Co–Zr–O have also been found to exhibit excellent properties for inductor cores [82, 83]. Randomly oriented nanogranular alloy (~10 nm) can decrease the coercivity (~1.9 Oe). The averaging effect of randomly orientated exchange coupling, however, decreased the net magnetocrystalline anisotropy. Such film rendered permeability as high as 400 up to 1 GHz. Also, such films showed much lower DC resistivity (~36 µΩ cm) than expected in nanogranular structures. This is attributed to electron percolation due to connected conducting grains. Resistivity can, however, be increased by engineering the microstructure. Lee et al. [84] achieved very high resistivity by ensuring fine and isolated α-Fe(Co) grains in $Co_{17.08}Fe_{49.76}Zr_{16.24}O_{16.91}$ film on a silicon substrate. However, such engineered microstructure increases the volume fraction of oxide phases and thus can have detrimental effect on saturation magnetization. Therefore, it is imperative to control the process parameter to ensure low coercivity, high resistance, and high saturation magnetization. Magnetostatic and magnetodynamic property analysis of thin films showed simultaneous occurrence of coercivity of 0.3 Oe, anisotropy of 44.9 Oe, saturation magnetization of 16.8 kG, and resistivity of 462.8 µΩ cm in random grains (average size 10 nm). Better high frequency performance has also been reported. The effective permeability as high as 800 was reported up to 2 GHz.

Fe–Co–Al–O nanogranular thin films have also been found to be attractive. RF reactive magnetron sputter-deposited films (50–1,000 nm) have been reported to have larger M_s (~22 kG) [43] than 15 kG as achieved in Fe-based and Co-based (~15 kG) [13], and 16.8 kG as could be achieved in the Fe–Co–Zr–O alloys [84]. Ultrahigh DC resistivity (900 µΩ cm) was also achieved. The resistivity of the films increased as the extent of oxides increased at the grain boundary as well as with the increment of refinement of the CoFe grains, completely separated by the oxide matrix (Al_2O_3 or FeCo oxide) and nanogranular structure. Moreover, high anisotropy as well as permeability was observed.

9.4.4 Permalloy/Fe–Co Alloy Thin Film Sandwich

As discussed in earlier sections, conventional nanomaterials are not very promising for high frequency applications. Exotic structures such as composite/sandwich structures have been studied to address this issue [85]. According to the LL equation, high saturation magnetization, high resistance, and high anisotropy are necessary for a magnetic material for high frequency applications [86]. Such sandwich structures can render saturation magnetization and permeability as high as observed in Fe–Co alloys. In addition, this can exhibit high resistivity and high anisotropy as observed in permalloy [10]. Wang et al. [77, 87, 88] reported a sandwich structure of $Ni_{0.81}Fe_{0.19}$ (5 nm)/$Fe_{0.7}(Co_{0.3})_{0.95}N_{0.05}$ (100 nm)/$Ni_{0.81}Fe_{0.19}$ (5 nm). Reportedly, even a small volume of permalloy could dramatically reduce coercivity (~0.6 Oe). Ultrahigh and almost constant permittivity (~1,000) up to 1.2 GHz was also achieved. The resistivity was, however, very small (50 μΩ cm). Katada et al. [89] reported such structures with saturation magnetization of 24 kG and DC coercivity of 3 Oe.

Native oxides, similarly, have been found to decrease the eddy losses. High resistance of native oxide in combination with the high saturation magnetization flux and high anisotropy of Fe–Co–N film makes sandwich structures of native oxide and Co–Fe–N suitable for high frequency applications [86]. Kakazei et al. [90] reported the magnetic behavior of CoFeN sandwiched between native and deposited nonmetallic oxides. Hai et al. studied a sandwich structure of $Fe_{70}Co_{30}N$ (45 nm)/$Ni_{55}Fe_{45}$ [91]. The composite structure was fabricated using RF reactive sputtering. The resistivity of the film increased with increase in the nitrogen content. A single layer was isotropic and significantly higher coercivity was achieved (~80 Oe). However, a seed layer of NiFe of 1-nm thickness significantly reduced the coercivity (~6 Oe) and well-defined uniaxial anisotropy was achieved due to effective magnetic coupling between the magnetic layers. It can also be attributed to partial cancellation of crystalline and magnetostatic anisotropy due to epitaxially grown FeCoN nanograins on NiFe seed layer.

Decrease in the coercivity is also witnessed with nonmagnetic seed layers [85, 87, 92]. Ha et al. [75] reported another sandwiched structure consisting of discontinuous CoFeN thin films and native oxide. The structure was fabricated using RF reactive magnetron sputtering ($Co_{30}Fe_{70}$ target). The CoFeN film thickness was maintained at 2 nm. The native oxide layers were formed by exposing the film to an oxygen atmosphere. Coercivity in such multilayers was found to be dependent on the exchange coupling of metallic magnetic nanograins through magnetic oxides. Ha et al. also reported linear dependence of the anisotropy field with external applied field during deposition. H_k as high as 67 Oe was achieved for an external applied field of 120 Oe. The coercivity was as low as 0.32 in the hard axis. They also studied the effect of resistivity on thickness. It was found that the resistivity decreases with increase in the thickness in the 1.6–5 nm range. Saturation magnetization flux was also reported to be high. This was attributed to the formation of magnetic oxides ($CoFe_2O_4$ or Fe_3O_4) during oxidation [93]. Moreover, this structure showed excellent high frequency characteristics. Permeability as high as 1,100 was achieved up to 1 GHz.

9.4.5 Novel Materials

Low weight, low cost, and high performance materials are needed for new generations of electronics. Mobile devices are the most important driving force. This problem has partially been addressed by low temperature cofired ceramics (LTCC) [42]. However, the high processing temperature is not compatible with polymeric substrate technology. Moreover, high cost also makes LTCC unattractive for consumer products [94]. Recently, various researches have been carried out to develop compatible polymer matrix composites for inductor cores [95–99]. Liu et al. [42] proposed an alternative solution to low inductive magnetic powder-filled composites. They reported a low temperature process-capable amorphous nickel–zinc ferrite-filled nanocomposite, and a three-layer structure with a planar gold coil was fabricated on a glass wafer. The composite was deposited on the coil and heated at different temperatures. The filler was prepared by sol-gel and the particle size was less than 200 nm. They reported a relative magnetic permeability of 150–200 for 20% filler by volume, which is six to eight times the reported value for polymer matrix composites with 90% filler.

9.5 Packaging Issues

The ultimate goal of packaging is to achieve high-quality factor, high density, high reliability, small size, and cost-effective inductors. Studies on on-chip integrated inductors have shown that the thick porous silicon substrate (through the wafer) is a better choice to achieve high-quality factor in RF integrated inductors [100]. The quality factor may also be increased by using a grating metal structure with low-K benzocyclobutene (BCB) and electroplated copper [101]. Almost 15% increase in the quality factor was noted with this structure, which was attributed to the reduction in series resistance due to enhanced effective current area. The possibilities and limits of ultrathin inductors have been explored by Waffenschmidt [102]. There still exists a need to research more to get high packaging efficiency, high-quality factor, and highly miniaturized inductors to meet today's requirements.

9.6 Conclusion

In this chapter, design, packaging, and materials aspects of nanogranular inductor cores have been addressed. Various kinds of inductor designs were reviewed, and recent developments in this regard were addressed. Packaging issues were also reviewed and it was concluded that there is a need for new materials and advanced designs for effective miniaturization of the device. Novel nanogranular magnetic material was identified as one of the key elements for such applications. For magnetic cores, however, high permeability to achieve inductance, high saturation magnetization to avoid dramatic decrease in the inductance, low coercivity, high resistance to avoid eddy current loss, and high frequency characteristics have been identified as key parameters that can significantly decrease the number of windings needed to achieve required inductances, and thus could catalyze the effective miniaturization of the devices. There is still a need to improve the magnetic properties to meet the high frequency requirements, which calls for further research and development in nanogranular magnetic materials.

References

1. Tummala, R.R., ed., *Fundamentals of Microsystems Packaging*, 1st ed. McGraw-Hill, New York, 2001
2. Yang, R., Atsushi, T., and Wong, C.P. Di-block copolymer surfactant study to optimize filler dispersion in high dielectric constant polymer–ceramic composite. Compos. Part A: Appl. Sci. Manuf., 34A(11) (2003) 1113–1116
3. http://whatis.techtarget.com/definition/0,,sid9_gci212339,00.html
4. Patranabis, D., Tripathi, M.P. and Roy, S.B. A new approach for lossless floating inductor realization. IEEE Trans. Circuits Syst., CAS-26(10) (1979) 892–893
5. Park, J.Y. and Allen, M.G. High Q spiral-type microinductors on silicon substrates. IEEE Trans. Magn., 35(5) (1999) 3544–3546
6. Park, J.Y. and Allen, M.G. New micromachined inductors on silicon substrates. IEEE Trans. Magn., 35(5) (1999) 3547–3549
7. Park, J.Y. and Allen, M.G. Packaging-compatible high Q microinductors and microfilters for wireless applications. IEEE Trans. Adv. Packag., 22(2) (1999) 207–213
8. Chuang, J., et al., Low loss air-gap spiral inductors for MMICs using glass microbump bonding technique. IEEE MTT-S Int. Microw. Symp. Digest, 1 (1998) 131–135
9. Yamaguchi, M., , Magnetic thin-film inductors for RF-integrated circuits. J. Magn. Magn. Mater., 215–216 (2000) 807–810
10. Yamaguchi, M., Ki Hyeon, K., and Ikedaa, S. Soft magnetic materials application in the RF range. J. Magn. Magn. Mater., 304(2) (2006) 208–213
11. Yamaguchi, M., et al., Improved RF integrated magnetic thin-film inductors by means of micro slits and surface planarization techniques. IEEE Trans. Magn., 36(5) (2000) 3495–3498
12. Allen, M.G., MEMS technology for the fabrication of RF magnetic components. IEEE Trans. Magn., 39(5) (2003) 3073–3078
13. Martin, J.I., et al., Ordered magnetic nanostructures: fabrication and properties. J. Magn. Magn. Mater., 256(1–3) (2003) 449–501
14. Flohrer, S., , Interplay of uniform and random anisotropy in nanocrystalline soft magnetic alloys. Acta Mater., 53(10) (2005) 2937–2942
15. Herzer, G., Anisotropies in soft magnetic nanocrystalline alloys. J. Magn. Magn. Mater., 294(2) (2005) 99–106
16. Yamaguchi, M., Baba, M. and Arai, K.I. Sandwich-type ferromagnetic RF integrated inductor. IEEE Trans. Microw. Theory Tech., 49(12) (2001) 2331–2335
17. Yamaguchi, M., et al., Microfabrication and characteristics of magnetic thin-film inductors in the ultrahigh frequency region. J. Appl. Phys., 85(11) (1999) 7919–7922
18. Shimada, Y., Amorphous Co-metal films produced by sputtering. Physica Status Solidi A, 83(1) (1984) 255–261
19. Munakata, M., et al., Magnetic properties and frequency characteristics of $(CoFeB)_x$–$(SiO_{1.9})_{1-x}$ and CoFeB films for RF application. Trans. Magn. Soc. Jpn., 2(5) (2002) 388–393
20. Kim, I., et al., High frequency characteristics and soft magnetic properties of FeCoBN nanocrystalline films. Physica Status Solidi A, 201(8) (2004) 1777–1780

21. Viala, B., et al., Bidirectional ferromagnetic spiral inductors using single deposition. IEEE Trans. Magn., 41(10) (2005) 3544–3549
22. Mohan, S.S., et al., Simple accurate expressions for planar spiral inductances. IEEE J. Solid State Circuits, 34(10) (1999) 1419–1424
23. Hizon, J.R.E., et al., Integrating spiral inductors on 0.25 μm epitaxial CMOS process. Asia-Pacific Microw. Conf. Proc., 1 (2005) 1606314
24. Liakopoulos, T.M. and Ahn, C.H. 3-D microfabricated toroidal planar inductors with different magnetic core schemes for MEMS and power electronic applications. IEEE Trans. Magn., 35(5) (1999) 3679–3681
25. Bhattacharya, S.K. and Tummala, R.R. Next generation integral passives: materials, processes, and integration of resistors and capacitors on PWB substrates. J. Mater. Sci.: Mater. Electron., 11(3) (2000) 253–268
26. Edelstein, D.C. and Burghartz, J.N. Spiral and solenoidal inductor structures on silicon using Cu-damascene interconnects. In International Interconnect Technology Conference, 1998, IEEE, San Francisco, CA
27. Park, J.Y. and Allen, M.G. A comparison of micromachined inductors with different magnetic core materials. In Proceedings of the 46th Electronic Components and Technology Conference, 1996, IEEE, Orlando, FL
28. Yamaguchi, M., et al., Application of bi-directional thin-film micro wire array to RF integrated spiral inductors. IEEE Trans. Magn., 36(5) (2000) 3514–3517
29. Baba, M., et al., RF integrated inductor using a bidirectional micro-patterned magnetic thin film. J. Magn. Soc. Jpn., 25(4) (2001) 1091–1094
30. http://www.vishay.com
31. Kowase, I., et al., A planar inductor using Mn–Zn ferrite/polyimide composite thick film for low-voltage and large-current DC–DC converter. IEEE Trans. Magn., 41(10) (2005) 3991–3993
32. Goldfarb, M.E., and Tripathi, V.K. The effect of air bridge height on the propagation characteristics of microstrip. IEEE Microw. Guided Wave Lett., 1(10) (1991) 273–274
33. Chang, J.Y.C., Abidi, A.A. and Gaitan, M. Large suspended inductors on silicon and their use in a 2 μm CMOS RF amplifier. IEEE Electron Device Lett., 14(5) (1993) 246–248
34. Kim, Y.-J. and Allen, M.G. Surface micromachined solenoid inductors for high frequency applications. IEEE Trans. Compon. Packag. Manuf. Technol. Part C, 21(1) (1998) 26–33
35. Matsushita, N., et al., Ni–Zn ferrite films with high permeability ($\mu' = \sim 30$) at 1 GHz prepared at 90°C. J. Appl. Phys., 91(10) (2002) 7376–7378
36. Kondo, K., et al., FMR study on spin-sprayed Ni–Zn–Co ferrite films with high permeability usable for GHz noise suppressors. IEEE Trans. Magn., 41(10) (2005) 3463–3465
37. Shimada, Y., et al., Study on initial permeability of Ni–Zn ferrite films prepared by the spin spray method. J. Magn. Magn. Mater., 278(1–2) (2004) 256–262
38. Kondo, K., et al., FMR study on spin-sprayed Ni–Zn–Co ferrite films with high permeability usable for GHz noise suppressors. In Digest of the IEEE International Magnetics Conference, 2005, pp. 901–902
39. Fu, C.M., et al., High frequency conductivity of spin-spray plated Ni–Zn ferrite thin films. In Intermag Europe 2002 Digest of Technical Papers. 2002 IEEE International Magnetics Conference, 2002, p. FD11
40. Shimada, Y., et al., Granular thin films with high RF permeability. IEEE Trans. Magn., 39(5) (2003) 3052–3056
41. Yamaguchi, M., et al., Magnetic RF integrated thin-film inductors. MTT-S Int. Microw. Symp. Digest, 1 (2000) 205–208
42. Liu, C.K., et al., Development of low temperature processable core material for embedded inductor. In 2nd International IEEE Conference on Polymers and Adhesives in Microelectronics and Photonics (POLYTRONIC), 2002, IEEE, Zalaegerszeg, Hungary
43. Ha, N.D., et al., High frequency permeability of soft magnetic CoFeAlO films with high resistivity. J. Magn. Magn. Mater., 290–291 (2005) 1571–1575
44. Mager, A., The influence of grain size on coercive force. Annalen der Physik, 11(1) (1952) 15–16

45. Adler, E. and Pfeiffer, H. The influence of grain size and impurities on the magnetic properties of the soft magnetic alloy 47.5% NiFe. IEEE Trans. Magn., 10(2) (1974) 172–174
46. Pfeifer, F. and Kunz, W. The influence of grain structure and non-magnetic particles on the magnetic properties of high-permeability Ni–Fe alloys. J. Magn. Magn. Mater., 4(1–4) (1977) 214–219
47. Kunz, W. and Pfeifer, F. The influence of grain structure and non-magnetic inclusions on the magnetic properties of high permeability Fe–Ni alloys. AIP Conf. Proc., 34 (1976) 63–65
48. Pfeifer, F. and Radeloff, C. Soft magnetic Ni–Fe and Co–Fe alloys – some physical and metallurgical aspects. J. Magn. Magn. Mater., 19(1–3) (1980) 190–207
49. Herzer, G., Grain size dependence of coercivity and permeability in nanocrystalline ferromagnets. IEEE Trans. Magn., 26(5) (1990) 1397–1402
50. Herzer, G., Grain structure and magnetism of nanocrystalline ferromagnets. IEEE Trans. Magn., 25(5) (1989) 3327–3329
51. Dirne, F.W.A., et al., Soft-magnetic properties and structure of Fe/CoNbZr multilayers. Appl. Phys. Lett., 53(24) (1988) 2386–2388
52. Jing, Z., et al., A novel fabrication and properties investigation of permalloy-SiO2 granular films with induced anisotropy. Mater. Lett., 61(2) (2007) 491–495
53. Weiping, N., Jinsook, K. and Kan, E.C. Permalloy patterning effects on RF inductors. IEEE Trans. Magn., 42(10) (2006) 2827–2829
54. Salvia, J., J.A. Bain, and C.P. Yue, Tunable on-chip inductors up to 5 GHz using patterned permalloy laminations. In 2005 International Electron Devices Meeting Technical Digest, IEDM, 2005, pp. 943–946
55. Gao, X.-y., et al., Application of different magnetic core materials in microinductor. Semicond. Technol., 30(10) (2005) 58–61
56. Wieserman, W.R., Schwarze, G.E. and Niedra, J.M. Magnetic and electrical characteristics of cobalt-based amorphous materials and comparison to a permalloy type polycrystalline material. In Collection of Technical Papers – 3rd International Energy Conversion Engineering Conference, American Institute of Aeronautics and Astronautics, San Francisco, CA, 2005
57. Zhao, J., et al., Radio-frequency planar integrated inductor with permalloy-SiO2 granular films. IEEE Trans. Magn., 41(8) (2005) 2334–2338
58. Kim, J., Ni, W. and Kan, E.C. Integrated on-chip planar solenoid inductors with patterned permalloy cores for high frequency applications. In Materials Research Society Symposium Proceedings, Materials Research Society, Warrendale, PA, 2005
59. Zhuang, Y., et al., Magnetic properties of electroplated nano/microgranular NiFe thin films for rf application. J. Appl. Phys., 97(10) (2005) 10N305-1–10N305-3
60. Kim, J., W. Ni, and E.C. Kan, Integrated on-chip planar solenoid inductors with patterned permalloy cores for high frequency applications. In Materials, Integration and Packaging Issues for High-Frequency Devices II, 2004, pp. 135–140
61. Neel, L., Magnetic surface anisotropy and superlattice formation by orientation. J. Physique Radium, 15 (1954) 225–239
62. Chikazumi, S. and Oomura, T. On the origin of magnetic anisotropy induced by magnetic annealing. J. Phys. Soc. Jpn., 10(10) (1955) 842–848
63. Yoshizawa, Y., Oguma, S. and Yamauchi, K. New Fe-based soft magnetic alloys composed of ultrafine grain structure. J. Appl. Phys., 64(10) (1988) 6044–6046
64. Petzold, J., Advantages of soft magnetic nanocrystalline materials for modern electronic applications. J. Magn. Magn. Mater., 242–245 (2002) 84–89
65. Terada, N., et al., Synthesis of iron–nitride films by means of ion beam deposition. IEEE Trans. Magn., MAG-20(5) (1984) 1451–1453
66. Wang, S., Guzman, J.I. and Kryder, M.H. High moment soft amorphous CoFeZrRe thin-film materials. J. Appl. Phys., 67(9) (1990) 5114–5116
67. Jiang, H., et al., High moment FeRhN/NiFe laminated thin films for write head applications. J. Appl. Phys., 91(10) (2002) 6821–6823
68. Karamon, H., A new type of high-resistive soft magnetic amorphous films utilized for a very high-frequency range. J. Appl. Phys., 63(8) (1988) 4306–4308

69. Ohnuma, S. and Masumoto, T. Amorphous magnetic alloys (Fe, Co, Ni)–(Si, B) with high permeability and its thermal stability. In *Rapidly Quenched Metals III*, ed. B. Cantor. Vol. II. The Metals Society, Brighton, UK, 1978, pp. pp. 197–204
70. Fujimori, H. and Masumoto, T. Magnetic properties of an Fe-13P-7C amorphous ferromagnet – the effects of stress, stress-annealing and magnetic-field-annealing. Trans. Jpn. Inst. Metals, 17(4) (1976) 175–180
71. Ohnuma, S., et al., High-frequency magnetic properties in metal–nonmetal granular films. J. Appl. Phys., 79(8) (1996) 5130–5135
72. Ohnuma, M., et al., Microstructure of Co–Al–O granular thin films. J. Appl. Phys., 82(11) (1997) 5646–5652
73. Hoffmann, H., Quantitative calculation of the magnetic ripple of uniaxial thin permalloy films. J. Appl. Phys., 35(6) (1964) 1790–1798
74. Ohnuma, S., et al., Co–Zr–O nano-granular thin films with improved high frequency soft magnetic properties. IEEE Trans. Magn., 37(4) (2001) 2251–2254
75. Ha, N.D., et al., High frequency characteristics and magnetic properties of CoFeN/native-oxide multilayer films. J. Magn. Magn. Mater., 286 (2005) 267–270
76. Osaka, T., et al., A soft magnetic CoNiFe film with high saturation magnetic flux density and low coercivity. Nature, 392 (1998) 796–798
77. Wang, S.X., et al., Properties of a new soft magnetic material. Nature, 407 (2000) 150–151
78. Iwasaki, H., Akashi, R. and Ohsawa, Y. Soft magnetic properties of Co–Fe–Al–N films. J. Appl. Phys., 73(12) (1993) 8441–8446
79. Ha, N.D., et al., Soft magnetic properties of CoFeAlN thin films. J. Magn. Magn. Mater., 290–291 (2005) 1469–1471
80. Li, W.D., Kitakami, O. and Shimada, Y. Study on the in-plane uniaxial anisotropy of high permeability granular films. J. Appl. Phys., 83(11) (1998) 6661–6663
81. Minor, M.K., et al., Stress dependence of soft, high moment and nanocrystalline FeCoB films. J. Appl. Phys., 91(10) (2002) 8453–8455
82. Xiong, X.Y., et al., Microstructure of soft magnetic FeCo–O(–Zr) films with high saturation magnetization. J. Magn. Magn. Mater., 265(1) (2003) 83–93
83. Ohnuma, S., et al., FeCo–Zr–O nanogranular soft-magnetic thin films with a high magnetic flux density. Appl. Phys. Lett., 82(6) (2003) 946–948
84. Lee, K.E., et al., Microstructure and soft magnetic properties of CoFeZrO thin films. J. Magn. Magn. Mater., 304(1) (2006) e192–e194
85. Jiang, H., Chen, Y. and Lian, G. Sputtered FeCoN soft magnetic thin films with high resistivity. IEEE Trans. Magn., 39(6) (2003) 3559–3562
86. Van de Riet, E., Klaassens, W. and Roozeboom, F. On the origin of the uniaxial anisotropy in nanocrystalline soft-magnetic materials. J. Appl. Phys., 81(2) (1997) 806–814
87. Sun, N.X. and Wang, S.X. Soft high saturation magnetization (Fe0.7Co0.3)1−xNx thin films for inductive write heads. IEEE Trans. Magn., 36(5) (2000) 2506–2508
88. Sun, N.X. and Wang, S.X. Anisotropy dispersion effects on the high frequency behavior of soft magnetic Fe–Co–N thin films. J. Appl. Phys., 93(10) (2003) 6468–6470
89. Katada, H., et al., Soft magnetic properties and microstructure of NiFe/FeCo/NiFe thin films with large saturation magnetization. J. Magn. Soc. Jpn., 26(4) (2002) 505–508
90. Kakazei, G.N., et al., Eur. Phys. J. B, 25(177) (2002) 189
91. Hai, J., Sin, K. and Yingjian, C. High moment soft FeCoN/NiFe laminated thin films. IEEE Trans. Magn., 41(10) (2005) 2896–2898
92. Jung, H.S., et al., Soft anisotropic high magnetization Cu/FeCo films. Appl. Phys. Lett., 81(13) (2002) 2415–2417
93. Sahoo, S., et al., Magnetic states of discontinuous Co80Fe20–Al2O3 multilayers. J. Magn. Magn. Mater., 240(1–3) (2002) 433–435
94. Park, J.Y., Lagorce, L.K. and Allen, M.G. Ferrite-based integrated planar inductors and transformers fabricated at low temperature. IEEE Trans. Magn., 33(5) (1997) 3322–3324
95. Park, J.Y. and M.G. Allen, Low temperature fabrication and characterization of integrated packaging-compatible, ferrite-core magnetic devices. In APEC'97. 12th Annual Applied Power Electronics Conference and Exposition 1, 1997, pp. 361–367

96. Park, J.Y. and Allen, M.G. Packaging compatible micromagnetic devices using screen printed polymer/ferrite composites. Int. J. Microcircuits Electron. Packag., 21(3) (1998) 243–252
97. Park, J.Y. and Allen, M.G. Integrated electroplated micromachined magnetic devices using low temperature fabrication processes. IEEE Trans. Electron. Packag. Manuf., 23(1) (2000) 48–55
98. Tang, S.C., Hui, S.Y.R. and Chung, H.S.H. A low-profile power converter using printed-circuit board (PCB) power transformer with ferrite polymer composite. IEEE Trans. Power Electron., 16(4) (2001) 493–498
99. Arshak, K.I., Ajina, A. and Egan, D. Development of screen-printed polymer thick film planar transformer using Mn–Zn ferrite as core material. Microelectron. J., 32(2) (2001) 113–116
100. Yang, L., et al., Backside growth thick porous silicon layers for high Q on-chip RF integrated inductors. Rare Metal Mater. Eng., 35(6) (2006) 966–969
101. Yeo, S.-K., Shin, S.-H. and Kwon, Y.-S. Grating metal structure with low-K benzocyclobutene and electroplated copper for high-Q spiral inductors. Jpn. J. Appl. Phys., 45(4B) (2006) 2997–3001
102. Waffenschmidt, E., Performance limits of ultra-thin printed circuit board inductors. In IEEE Power Electronics Specialists Conference, 2006, p. 7

Chapter 10
Nanoconductive Adhesives

Daoqiang Daniel Lu(✉), Yi Grace Li, and C.-P. Wong

10.1 Introduction

Electrically conductive adhesives (ECAs) are composites of polymeric matrices and electrically conductive fillers. Polymeric matrices have excellent dielectric properties and thus are electrical insulators. The conductive fillers provide the electrical properties and the polymeric matrix provides mechanical properties. Therefore, electrical and mechanical properties are provided by different components, which is different from metallic solders that provide both the electrical and mechanical properties. ECAs have been with us for some time. Metal-filled thermoset polymers were first patented as ECAs in the 1950s [1–3]. Recently, ECA materials have been identified as one of the major alternatives for lead-containing solders for microelectronic packaging applications. There are two types of conductive adhesives: isotropically conductive adhesives (ICAs) and anisotropically conductive adhesives/films (ACAs/ACFs).

ICAs, also known as "polymer solder," are conductive in all directions. The conductive fillers provide the composite with electrical conductivity through contact between the conductive particles. With increasing filler concentrations, the electrical properties of an ICA transform it from an insulator to a conductor. Percolation theory has been used to explain the electrical properties of ICA composites. At low filler concentrations, the resistivities of ICAs decrease gradually with increasing filler concentration. However, the resistivity drops dramatically above a critical filler concentration V_c called the *percolation threshold* (Fig. 10.1). It is believed that at this concentration, all the conductive particles contact each other and form a three-dimensional network. The resistivity decreases only slightly with further increases in the filler concentrations [4–6]. To achieve conductivity, the volume fraction of conductive filler in an ICA must be equal to or slightly higher than the critical volume fraction. Similar to solders, ICAs provide the dual functions of electrical connection and mechanical bond in an interconnection joint. In an ICA joint (Fig. 10.2), the polymer resin provides mechanical stability and the

---

D.D. Lu
Henkel Loctite (China) Co., Ltd, 90 Zhujiang Road, Yantai, ETDZ, Shandong, China 264006
Daniel.d.lu@ieee.org

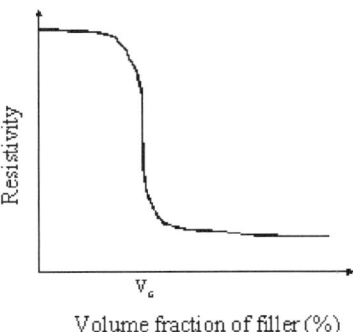

Fig. 10.1 Effect of filler volume fraction on the resistivity of ICA systems

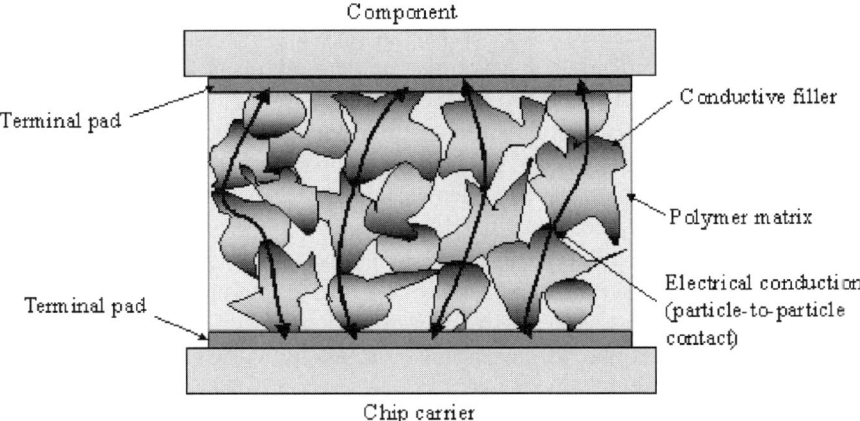

Fig. 10.2 ICA particle contacts establish conduction from component to chip carrier

conductive filler provides electrical conductivity. Filler loading levels that are too high cause the mechanical integrity of adhesive joints to deteriorate. Therefore, the challenge in formulating an ICA is to maximize conductive filler content to achieve a high electrical conductivity without adversely affecting the mechanical properties. In a typical ICA formulation, the volume fraction of the conductive filler is about 25–30% [7, 8].

Recently, ACAs/ACFs are becoming popular as promising candidates for lead-free interconnection solutions in microelectronic packaging applications due to their technical advantages such as fine pitch capability (<40-μm pitch), low-temperature processing ability, low cost and environmentally friendly materials and processing, etc. ACAs/ACFs consist of conducting particles (typically 5–10 μm in diameter) and polymer matrix, which provide both attachment and electrical interconnection between electrodes [9–11]. In particular, ACFs are widely used for high-density interconnection between liquid–crystal display (LCD) panels and tape carrier packages (TCPs) to replace the traditional soldering or rubber connectors. In LCD applications, traditional soldering may not be as effective as ACFs in interconnecting materials between indium tin oxide (ITO) electrodes and TCP. ACFs have also been used as an alternative to soldering for interconnecting TCP input lead bonding to printed circuit boards (PCBs). ACAs/ACFs provide unidirectional electrical conductivity in the vertical or Z-axis. This directional conductivity is achieved by using a relatively low volume loading of conductive filler (5–20 vol.%). The low volume loading is insufficient for interparticle contact and prevents conductivity in the X–Y plane of the adhesive. The ACA/ACF is interposed between the two surfaces to be connected. Heat and pressure are simultaneously applied to this stack-up until the conductive particles bridge the two conductor surfaces. Figure 10.3 shows the configuration of a component and a substrate bonded with ACA. Once the electrical continuity is produced, the polymer matrix is hardened by thermally initiated chemical reaction (for thermosets) or by cooling (for thermoplastics). The hardened dielectric polymer matrix holds the two components together and helps maintain the pressure contact between component surfaces and conductive particles. Because of the anisotropy, ACA/ACF may be deposited over the entire contact region, greatly facilitating materials application. Also, an ultrafine pitch interconnection (<40 μm) could be achieved easily. The fine pitch capability of ACA/ACF would be limited by the particle size of the conductive filler, which can be a few microns or a few nanometers in diameter.

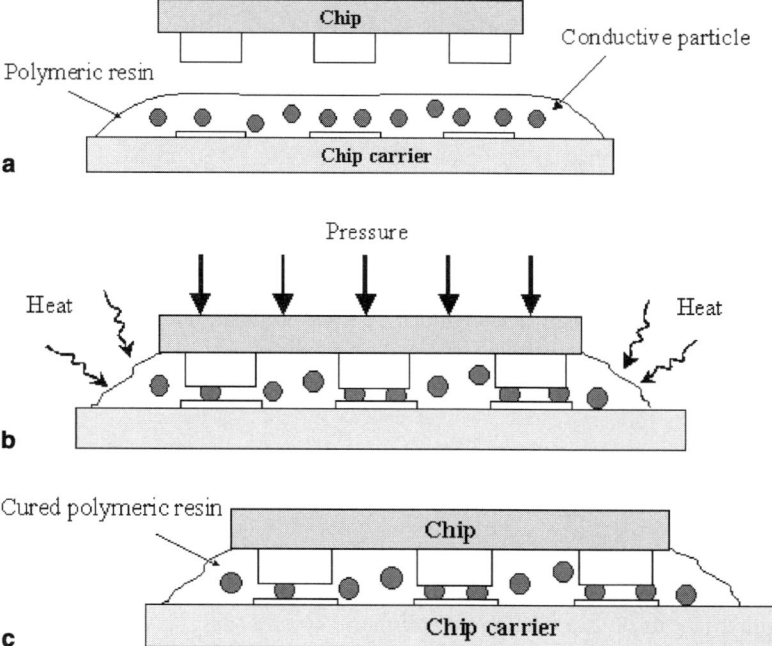

Fig. 10.3 A series of schematics illustrating the steps in forming an ACA joint. (**a**) Component parts: a bumped die and mating carrier with ACA spread over the surface. (**b**) Die is mounted with the carrier and held in place when cured. (**c**) Side view of the completed assembly

To meet the requirements for future fine pitch and high performance interconnects in advanced packaging, ECAs with nanomaterial or nanotechnology attract more and more interests due to the specific electrical, mechanical, optical, magnetic, and chemical properties. There has been extensive research on nanoconductive adhesives which contain nanofiller such as nanoparticles, nanowires, or carbon nanotubes. This chapter provides a comprehensive review of most recent research results on nanoconductive adhesives.

10.2 Recent Advances on Nano-Isotropic Conductive Adhesive (Nano-ICA)

10.2.1 ICAs with Silver (Ag) Nanowires

Wu et al. [12] developed an ICA filled with nanosilver wires and compared the electrical and mechanical properties of this nano-ICA with two other ICAs filled with micrometer-sized (roughly 1 µm and 100 nm, respectively) silver particles. The nanowires had a diameter of roughly 30 nm and a length up to 1.5 um, and the nanowires are polycrystalline in nature. It was found that, at a low filler loading (e.g., 56 wt%), the bulk resistivity of ICA filled with the Ag nanowires was significantly lower than the ICAs filled with 1-µm or 100-nm silver particles. The better electrical conductivity of the ICA-filled nanowires was attributed to the lower contact resistance between nanowires and more significant contribution from the tunneling effects among the nanowires [12].

It was also found that, at the same filler loading (e.g., 56 wt%), the ICAs filled with Ag nanowires showed similar shear strength to those of the ICAs filled with the 1-µm and 100-nm silver particles. However, to achieve the same level of electrical conductivity, the filler loading must be increased to at least 75 wt% for the ICA filled with micrometer-sized Ag particles, and the shear strength of these ICAs is then decreased (lower than that of the ICA filled with 56 wt% nanowires) due to the higher filler loading.

10.2.2 Effect of Nanosized Ag Particles on ICA Conductivities

Lee et al. [13] studied the effects of nanosized filler on the conductivity of conductive adhesives by substituting nanosized Ag colloids for microsized Ag particles with either in part or as a whole to a polymeric system (polyvinyl acetate – PVAc). Electrical resistivity was then measured as a function of silver volume fraction.

It was found that, when nanosized silver particles were added into the system at 2.5 wt% each increment, the resistivity increased in almost all cases, except when the quantity of microsized silver was slightly lower than the threshold value. At that point, the addition of the about 2.5 wt% brought significant decrease in resistivity. Near the percolation threshold, when the microsized silver particles are still not connected, the addition of a small amount of nanosized silver particles helps to build the conductive network and lowers the resistivity of the composite. However, when the filler loading is above the percolation threshold and all the microsized particles are connected, the addition of nanoparticles seems only to increase the relative contribution of contact resistance between the particles. Due to its small size, for a fixed amount of addition, the nanosized silver colloid contains a larger number of particles when compared with microsized particles. This large number of particles should be beneficial to the interconnection between particles. However, it also inevitably increases the contact resistance. As a result, the overall effect is an increase in resistivity upon the addition of nanosized silver colloids.

Ye et al. [14] also reported a similar phenomenon, i.e., the addition of nanoparticles showed a negative effect on electrical conductivity. They proposed two types of contact resistance, i.e., restriction resistance due to small contact area and tunneling resistance when nanoparticles are included in the system. It was believed that the conductivity of microsized Ag particle-filled adhesives is dominated by constriction resistance, while the nanoparticle-containing conductive adhesives are controlled by tunneling and even thermionic emission. Fan et al. [15] also observed the similar phenomenon (adding nanosize particles reduced both electrical and thermal conductivities).

Lee et al. [13] also studied the effects of temperature on the conductivity of ICAs. Heating the composite to a higher temperature can reduce the resistivity quite significantly. This is likely due to the high activity of nanosized particles. For microsized paste, this temperature effect was considered negligible. The interdiffusion of silver atoms among nanosized particles helped to reduce the contact resistance quite significantly and the resistivity reached 5×10^{-5} Ω cm after treatment at 190°C for 30 min. Jiang et al. [16] showed that, when suitable surfactant was used in the nano-Ag-incorporated ICA, the dispersion and interdiffusion of silver atoms among nanosized particles could be facilitated and the resistivity of ICA could be reduced to 5×10^{-6} Ω cm.

10.2.3 ICA Filled with Aggregates of Nanosized Ag Particles

To improve the mechanical properties under thermal cycling conditions while still maintaining an acceptably high level of electric conductivity, Kotthaus et al. [17] studied an ICA material system filled with aggregates of nanosized Ag particles. The idea was to develop a new filler material which did not deteriorate the mechanical property of the polymer matrix to such a great extent. A highly porous Ag powder was attempted to fulfill these requirements. The Ag powder was produced by the inert gas condensation method (IGC). The powders consist of sintered networks of ultrafine particles in the size range 50–150 nm. The mean diameter of these aggregates could be adjusted down to some microns. The as-sieved powders were characterized by low level of impurity content, an internal porosity of about 60%, and a good ability for resin infiltration.

Using Ag IGC instead of Ag flakes is more likely to retain the properties of the resin matrix because of the infiltration of the resin into the pores. Measurements of the shear stress–strain behavior indicated that the thermomechanical properties of bonded joints may be improved by a factor up to 2, independent of the chosen resin matrix.

Resistance measurements on filled adhesives were performed within a temperature range from 10 to 325 K. The specific resistance of an Ag IGC-filled adhesive is about 10^{-2} Ω cm and did not achieve the typical value of commercially available adhesives of about 10^{-4} Ω cm. The reason may be that Ag IGC particles are more or less spherical whereas Ag flakes are flat. So, the decrease of the percolation threshold because of the porosity of Ag IGC is overcompensated by the disadvantageous shape and the intrinsically lower specific conductivity. For certain applications where mechanical stress plays an important role, this conductivity may be sufficient and therefore the porous Ag could be suitable as a new filler material for conductive adhesives.

10.2.4 Nano-Ni Particle-Filled ICA

It is generally known that metal powders present the properties that are different from those of bulk metals when their particle sizes are made as small as nanometer size. Powders are classified into particles, microparticles, and nanoparticles according to size. Although the classification criteria are not clear, particles with diameter smaller than 100 nm are generally called *nanoparticles*. This classification is based on the fact that, when its particle size is smaller than 100 nm, the particle has properties that are not found in the microparticles larger than 100 nm. For example, when the particle diameter of such magnetic materials as iron and nickel nears to 100 nm, their magnetic domains are changed from multiple to single, and their magnetic properties also change [18]. Majima et al. [18] reported an application example of metal nanoparticles to conductive pastes, focusing on the properties of a new conductive adhesive that were not found in conventional ICAs.

Sumitomo Electric Industries, Ltd. (SEI) has developed a new liquid-phase deposition process using plating technology. This new nanoparticle fabrication process achieves purity greater than 99.9% and allows easy control of particle diameter and shape. The particles' crystallite size calculated from the result of X-ray diffraction measurement is 1.7 nm, which leads to an assumption that the particle size of primary particles is extremely small. When the particle size of nickel and other magnetic metals becomes smaller than 100 nm, they are changed from the multidomain particles to the single-domain particles, and their magnetic properties change. That is, if the diameter of nickel particles is around 50 nm, each particle acts like a magnet that has only a pair of magnetism, and magnetically connects with each other to form chain-like clusters. When the chain-like clusters are applied to conductive paste, electrical conduction of the paste is expected to be better than the existing paste. The developed chain-like nickel particles were mixed with predefined amount of polyvinylidene fluoride (PVdF) that acts as an adhesive.

Then, N-methyl-2-pyrrolidone was added to this mixture to make conductive paste. This paste was applied on a polyimide (PI) film and then dried to make a conductive sheet. Specific volume resistivity of the fabricated conductive sheet was measured by the quadrupole method. The same measurement was also conducted on the conductive sheet that uses paste made of conventional spherical nickel particles. Measurement of the sheet resistance immediately after paste application defined that the developed chain-like nickel powders had low resistance of about one eighth of that of the conventionally available spherical nickel particles. This result showed that, when the newly developed chain-like nickel particles were applied as conductive paste, high conductivity can be achieved without pressing the sheet. SEI tested and developed the metal nanoparticles and investigated the possibility of application to conductive paste.

10.2.5 Nano-ICAs for via Filling in Organic Substrates

Das et al. [19] have developed conductive adhesives using controlled-sized particles, ranging from nanometer scale to micrometer scale, and use them to fill small diameter holes to fabricate Z-axis interconnections in laminates for interconnect applications (see Chap. 11 for a detailed treatment).

10.2.6 Nano-ICAs Filled with Carbon Nanotubes (CNTs)

10.2.6.1 Electrical and Mechanical Characterization of CNT-Filled ICAs

The density of commercially available silver-filled conductive adhesive is around 4.5 g cm^{-3} after cure. Metal-filled ECAs offer an alternative to typical lead–tin soldering with the advantages of being simple to process at lower temperatures without toxic lead or corrosive flux. The disadvantage of conventional metal-filled conductive adhesives is that high loading of filler decreases the mechanical impact strength, while decreasing filler loading results in poor electrical properties. CNTs are a new form of carbon, which was first identified in 1991 by Iijima [20] of NEC, Japan. Nanotubes are sheets of graphite rolled into seamless cylinders. Besides growing single wall nanotubes (SWNTs), nanotubes can also have multiple walls (MWNTs) – cylinders inside the other cylinders. The CNT can be 1–50 nm in diameter and 10–100 μm or up to a few millimeters in length, with each end "capped" with half of a fullerene dome consisting of five and six member rings. Along the sidewalls and cap, additional molecules can be attached to functionalize the nanotube to adjust its properties. CNTs are chiral structures with a degree of twist in the way that the graphite rings join into cylinders. The chirality determines whether a nanotube will conduct in a metallic or semiconducting manner. CNTs possess many unique and remarkable properties. The measured electrical conductivity of metallic CNTs is in the order of 104 S cm^{-1} [21]. The thermal conductivity of CNTs at room temperature can be as high as 6,600 W mK^{-1} [22]. The Young's modulus of a CNT is about 1 TPa. The maximum CNT tensile strength is close to 30 GPa, with some reported at TPa [23]. The density of MWNTs is 2.6 g cm^{-3} and the density of SWNTs ranges from 1.33 to 1.40 g cm^{-3} depending on the chirality [24]. Since CNTs have very low density and long aspect ratios, they have the potential of reaching the percolation threshold at very low weight percent loading in the polymer matrix.

Li and Lumpp [25] developed new epoxy-based conductive adhesives filled with MWNTs. Preparation and processing methods for the new conductive adhesives were developed. It was found that ultrasonic mixing process helped disperse CNTs in the epoxy more uniformly and made them contact better, and thus lower electrical resistance was achieved [25]. The contact resistance and volume resistivity of the conductive adhesive decreased with increasing CNT loading. The percolation threshold for the MWNTs used in Li's experiments is less than 3 wt%. With 3 wt% loading, the average contact resistance was comparable with solder joints. It was also found that the performance of CNT-filled conductive adhesives was comparable with solder joints at high frequency. By replacing metal particle fillers with CNTs in the conductive adhesive, a higher percentage of mechanical strength was retained. For example, with 0.8 wt% of CNT content, 80% of the shear strength of the polymer matrix was retained, while conventional metal-filled conductive adhesives only retain less than 28% of the shear strength of the polymer matrix.

Experiments conducted by Qian et al. [26] show 36–42% and 25% increases in elastic modulus and tensile strength, respectively, in polystyrene (PS)/CNT composites. The TEM observations in their experiments showed that cracks propagated along weak CNT–polymer interfaces or relatively low CNT density regions and caused failure. If the outer layer of MWNTs can be functionalized to form strong chemical bonds with the polymer matrix, the CNT/polymer composites can be further reinforced in mechanical strength and have controllable thermal and electrical properties.

10.2.6.2 Effect of Adding CNT to the Electrical Properties of ICAs

Lin and Lin [27] studied the effect of adding CNT to the electrical conductivity of silver-filled conductive adhesive which had various filled loadings. It was found that the CNT could enhance the electrical conductivity of the conductive adhesives greatly when the silver filler loading was still below percolation threshold. For example, the 66.5 wt% filled silver conductive adhesive without CNT had a resistivity of 10^4 Ω cm, but showed a resistivity of 10^{-3} Ω cm after adding 0.27 wt% CNT. Therefore, it is possible to achieve the same level of electrical conductivity by adding a small amount of CNT to replace the silver fillers.

10.2.6.3 Composites Filled with Surface-Treated CNTs

Although CNTs have exceptional physical properties, incorporating them into other materials has been inhibited by the surface chemistry of carbon. Problems such as phase separation, aggregation, poor dispersion within a matrix, and poor adhesion to the host must be overcome. Zyvex claimed that they have overcome these restrictions by developing a new surface treatment technology that optimizes the interaction between CNTs and the host matrix [28]. A multifunctional bridge was created between the CNT sidewalls and the host material or solvent. The power of this bridge was demonstrated by comparing the fracture behavior of the composites filled with untreated and surface-treated nanotubes. It was observed that the untreated nanotubes interacted poorly with the polymer matrix, and thus left behind voids in the matrix after fracture. However, for composite filled with treated nanotubes, the nanotube remained in the matrix even after the fracture, indicating strong interaction with the matrix. Due to their superior dispersion in the polymer matrix, the treated nanotubes achieved the same level of electrical conductivity at much lower loadings than the untreated nanotubes [28].

10.2.7 Inkjet-Printable Nano-ICAs and Inks

Areas for printing very fine pitch matrix (e.g., very fine pitch paths, antennas) are very attractive. But there are special requirements for inkjet printing materials, namely the most important ones are low viscosity and very homogeneous structure (like a molecular fluid) to avoid sedimentation and separation during the process. Additionally, for electrical conductivity of printed structures, the liquid has to contain conductive particles, with nanosized dimensions to avoid blocking the printing nozzle and to prevent sedimentation phenomenon. Nanosized silver seems to be one of the best candidates for this purpose, especially when its particle size dimensions will be less than 10 nm (see also Chap. 12).

Inkjet is an accepted technology for dispensing small volumes of material (50–500 pL). Currently, traditional metal-filled conductive adhesives cannot be processed by inkjetting (due to their relatively high viscosity and the size of filler material particles). The smallest droplet size achievable by traditional dispensing techniques is in the range of 150 μm, yielding proportionally larger adhesive dots on the substrate. Electrically conductive inks are available on the market with metal particles (gold or silver) < 20 nm suspended in a solvent at 30–50 wt%. After deposition, the solvent is eliminated and electrical conductivity is enabled by a high metal ratio in the residue. Some applications include a sintering step. However, these traditional nanofilled inks do not offer an adhesive function [29, 30].

There are many requirements for an inkjettable, Ag particle-filled conductive adhesive. The silver particles must not exceed a maximum size determined by the diameter of the injection needle used. At room temperature, the adhesive should resist sedimentation for at least 8, preferably 24 h. A further requirement by the end user on the adhesive's properties was a two-stage curing mechanism. In the first curing step, the adhesive surface is dried and remains meltable. In this state, the product may be stored for several weeks. The second curing step involves glueing the components with the previously applied adhesive. By heating and applying pressure, the adhesive is remelted and cured. Thus, the processing operation is similar to that required for soldering. A conductivity in the range of 10^{-4} Ω cm in the bulk material is required. An adhesive less prone to sedimentation was formulated by using suitable additives. Furthermore, the formation of filler agglomerations during deflocculation and storage was reduced. This effect was achieved by making the additives adhere to the filler particle surfaces. This requires a very sensitive balance. If the insulation between individual silver particles becomes too strong, overall electrical conductivity is significantly reduced.

Kolbe et al. [31] and Mosicki et al. [32] both demonstrated feasibility of an inkjettable, ICA in the form of a silver-loaded resin with a two-step curing mechanism. In the first step, the adhesive was dispensed (jetted) and precured leaving a "dry" surface. The second step consisted of assembly (wetting of the second part) and final curing. The attainable droplet sizes were in the range of 130 μm, but could be further reduced by using smaller (such as 50 μm) and more advanced nozzle shapes.

See Chap. 12 for inkjet-deposited interconnections.

10.3 Recent Advances of Nano-ACA/ACF

10.3.1 Low-Temperature Sintering of Nano-Ag-Filled ACA/ACF

One of the concerns for ACA/ACF is the higher joint resistance since interconnection using ACA/ACF relies on mechanical contact, unlike the metal bonding of soldering. An approach to minimize the joint resistance of ACA/ACF is to make the conductive fillers fuse to each other and form metallic joints such as metal solder joints. However, fusing metal fillers in polymers does not appear feasible, since a typical organic PCB (T_g ~ 125°C), on which the metal-filled polymer is applied, cannot withstand such a high temperature; the melting temperature (T_m) of Ag, for example, is around 960°C. Research showed that T_m and sintering temperatures of materials could be dramatically reduced by decreasing the size of the materials [33–35]. It has been reported that the surface premelting and sintering processes are a primary mechanism of the T_m depression of the fine nanoparticles (<100 nm) [34]. For nanosized particles, sintering behavior could occur at much lower temperatures and, as such, the use of the fine metal particles in ACAs would be promising for fabricating high electrical performance ACA joints through eliminating the interface between metal fillers. The application of nanosized particles can also increase the number of conductive fillers on each bond pad and result in more contact area between fillers and bond pads. Figure 10.4 shows the SEM photographs of nano-Ag

Fig. 10.4 SEM photographs of 20-nm-sized Ag particles annealed at different temperatures for 30 min: (**a**) room temperature (no annealing), (**b**) annealed at 100°C, (**c**) annealed at 150°C, (**d**) annealed at 200°C, and (**e**) annealed at 250°C (*markers*: (**a**) 3 μm, (**b**) 2 μm, (**c**) 2 μm, (**d**) 3 μm, and (**e**) 3 μm) [33]

particles annealed at various temperatures. Although very fine particles (20 nm) were observed for as-synthesized (Fig. 10.4a) and 100°C treated particles (Fig. 10.4b), dramatically larger particles were observed after heat treatment at 150°C and above. With increasing temperatures, the particles became larger and appeared as solid matter rather than porous particles or agglomerates. The particles shown in Fig. 10.4c–e were fused through their surface and many dumbbell type particles could be found. The morphology was similar to a typical morphology of the initial stage in the typical sintering process of ceramic, metal, and polymer powders. This low-temperature sintering behavior of the nanoparticles is attributed to the extremely high interdiffusivity of the nanoparticle surface atoms, due to the significantly energetically unstable surface status of the nanosized particles with large proportion of the surface area to the entire particle volume.

For the sintering reaction in a certain material system, temperature and duration are the most important parameters, in particular, the sintering temperature. Current–resistance (I–R) relationship of the nano-Ag-filled ACA is shown in Fig. 10.5. As can be seen from the figure, with increasing curing temperatures, the resistance of the ACA joints decreased significantly, from 10^{-3} to 5×10^{-5}. Also, higher curing temperature ACA samples exhibited higher current-carrying capability than the

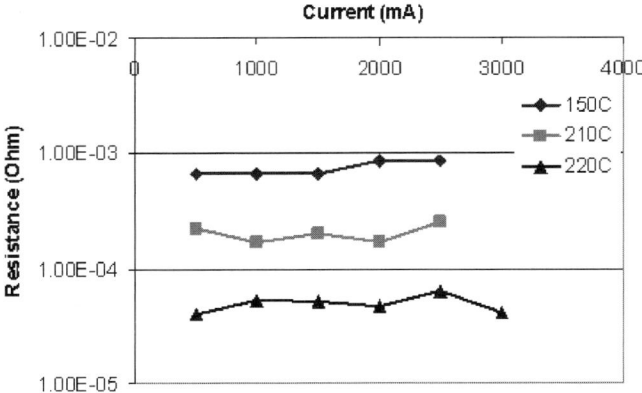

Fig. 10.5 Current–resistance (*I–R*) relationship of nano-Ag-filled ACAs with different curing temperatures [36]

low-temperature samples. This phenomenon suggested that more sintering of nano-Ag particles and subsequently superior interfacial properties between fillers and metal bond pads were achieved at higher temperatures [36], yet the *X–Y* direction of the ACA maintains an excellent dielectric property for electrical insulation.

10.3.2 Self-Assembled Molecular Wires for Nano-ACA/ACF

To enhance the electrical performance of ACA/ACF materials, self-assembled molecular wires (SAMs) have been introduced into the interface between metal fillers and metal-finished bond pad of ACAs [37, 38]. These organic molecules adhere to the metal surface and form physico-chemical bonds, which allow electrons to flow. As such, they reduce electrical resistance and enable a high current flow. The unique electrical properties are due to the tuning of metal work functions by those organic monolayers. The metal surfaces can be chemically modified by the organic monolayers and the reduced work functions can be achieved by using suitable organic monolayer coatings. An important consideration when examining the advantages of organic monolayers pertains to the affinity of organic compounds to specific metal surfaces. Table 10.1 gives the examples of molecules preferred for maximum interactions with specific metal finishes; although only molecules with symmetrical functionalities for both head and tail groups are shown, molecules and derivatives with different head and tail functional groups are possible for interfaces concerning different metal surfaces.

Different organic molecular wires, dicarboxylic acid, and dithiol have been introduced into ACA/ACF joints. For SAM-incorporated ACA with micron-sized gold/polymer or gold/nickel fillers, lower joint resistance and higher maximum allowable current (highest current applied without inducing joint failure) were achieved for low-temperature curable ACA (<100°C). For high curing temperature

Table 10.1 Potential organic monolayer interfacial modifiers for metal finishes

Formula	Compound	Metal finish
H–S–R–S–H	Dithiols	Au, Ag, Sn, Zn
N≡C–R–C≡N	Dicyanides	Cu, Ni, Au
O=C=N–R–N=C=O	Diisocyanates	Pt, Pd, Rh, Ru
HOOC–R–COOH (structure)	Dicarboxylates	Fe, Co, Ni, Al, Ag
Imidazole (structure)	Imidazole and derivatives	Cu
	Organosilicone derivatives	SiO_2, Al_2O_3, quartz, glass, mica, ZnSe, GeO_2, Au
R–SiOH		

R denotes alkyl and aromatic groups

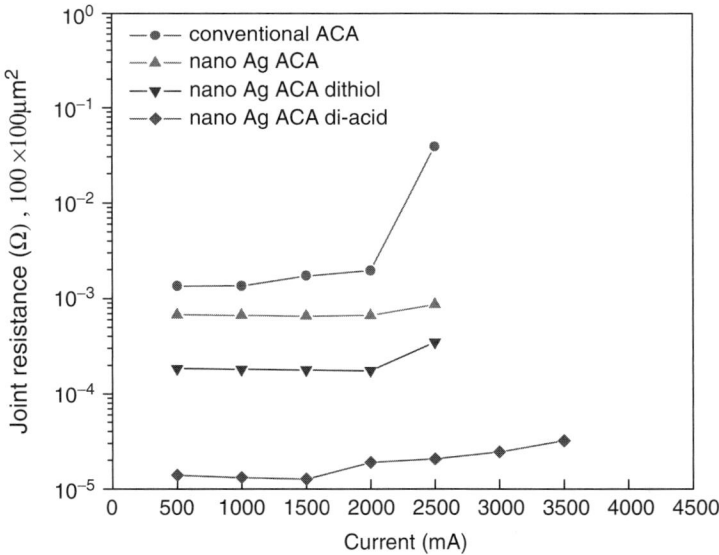

Fig. 10.6 Electrical properties of nano-Ag ACA with dithiol or dicarboxylic acid [40]

ACA (150°C), however, the improvement was not as significant as low curing temperature ACAs, due to the partial desorption/degradation of organic monolayer coating at the relatively high temperature [39]. However, when dicarboxylic acid or dithiol was introduced into the interface of nanosilver-filled ACAs, significantly improved electrical properties could be achieved for a high-temperature curable ACA/ACF, suggesting that the coated molecular wires did not suffer degradation on silver nanoparticles at the curing temperature (Fig. 10.6). The enhanced bonding could be attributed to the larger surface area and higher surface energy of nanoparticles, which enabled the monolayers to be more readily coated and relatively thermally stable on the metal surfaces [40].

10.3.3 Silver Migration Control in Nanosilver-Filled ACA

Silver is the most widely used conductive fillers in ICAs and exhibits exciting potentials in nano-ACA/ACF due to many unique advantages of silver. Silver has the highest room temperature electrical and thermal conductivity among all the conductive metals. Silver is also unique among all the cost-effective metals by nature of its conductive oxide (Ag_2O). In addition, silver nanoparticles are relatively easily formed into different sizes (a few nanometers to 100 nm) and shapes (such as spheres, rods, wires, disks, flakes, etc.) and well dispersed in a variety of polymeric matrix materials. Also, the low-temperature sintering and high surface energy make silver one of the promising candidates for conductive filler in nano-ACA/ACF. However, silver migration has long been a reliability concern in the electronics industry. Metal migration is an electrochemical process, whereby, metal (e.g., silver), in contact with an insulating material, in a humid environment and under an applied electric field, leaves its initial location in ionic form and deposits at another location [41]. It is considered that a threshold voltage exists above which the migration starts. Such migration may lead to a reduction in electrical spacing or cause a short circuit between interconnections. The migration process begins when a thin continuous film of water forms on an insulating material between oppositely charged electrodes. When a potential is applied across the electrodes, a chemical reaction takes place at the positively biased electrode where positive metal ions are formed. These ions, through ionic conduction, migrate toward the negatively charged cathode and over time, they accumulate to form metallic dendrites. As the dendrite growth increases, a reduction of electrical spacing occurs. Eventually, the dendrite silver growth reaches the anode and creates a metal bridge between the electrodes, resulting in an electrical short circuit [42].

Although other metals may also migrate under specific environment, silver is more susceptible to migration, mainly due to the high solubility of silver ion, low activation energy for silver migration, high tendency to form dendrite shape, and low possibility to form stable passivation oxide layer [43–45]. The rate of silver migration is increased by (1) an increase in the applied potential, (2) an increase in the time of the applied potentials, (3) an increase in the level of relative humidity, (4) an increase in the presence of ionic and hydroscopic contaminants on the surface of the substrate, and (5) a decrease in the distance between electrodes of the opposite polarity.

To reduce silver migration and improve the reliability, several methods have been reported. The methods include (1) alloying the silver with an anodically stable metal such as palladium [42] or platinum [46] or even tin [47]; (2) using hydrophobic coating over the PWB to shield its surface from humidity and ionic contamination

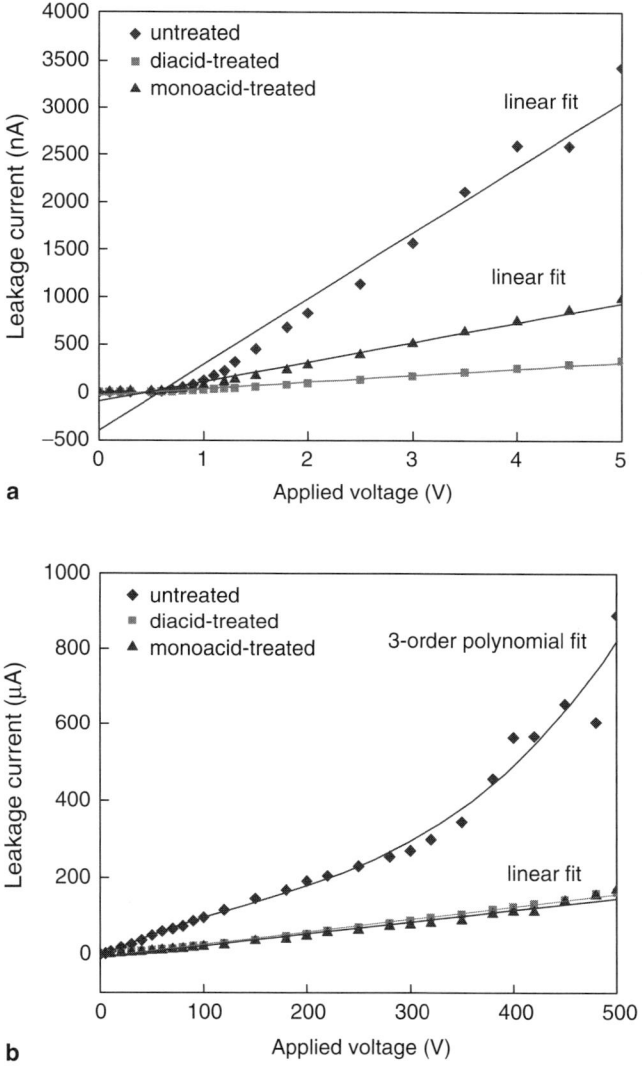

Fig. 10.7 Leakage current–voltage relationship of nano-Ag conductive adhesives at (**a**) low voltages and (**b**) high voltages [52]

[48], since water and contaminates can act as a transport medium and increase the rate of migration; (3) plating of silver with metals such as tin, nickel, or gold, to protect the silver fillers and reduce migration; (4) coating the substrate with polymer [49]; (5) applying benzotriazole (BTA) and its derivatives in the environment [50]; (6) employing siloxane epoxy polymers as diffusion barriers due to the excellent adhesion of siloxane epoxy polymers to conductive metals [51]; and (7) chelating silver fillers in ECAs with molecular monolayers [52]. As an example shown in Fig. 10.7 [53], with carboxylic acids and forming chelating compounds with silver

ions, the silver migration behavior (leakage current) could be significantly reduced and controlled.

10.3.4 ACF with Straight-Chain-Like Nickel Nanoparticles

cles with a straight-chain-like structure as conductive fillers [54]. They applied the formulated straight-chain-like nickel nanoparticles and solvent in a mixture of epoxy resin on a substrate film. Then, the particles were made to orient toward the vertical direction of the film surface and fixed in resin by evaporating the solvent. In the estimation using 30-μm pitch IC chips and glass substrates (the area of Au bumps was 2,000 μm^2 and the distance of space between neighboring bumps was 10 μm), the new ACF showed excellent reliability of electrical connection after high-temperature, high-humidity (60°C/90%RH) test and thermal cycle test (between −40 and 85°C). The samples were also exposed to high-temperature, high-humidity (60°C/90%RH) for insulation ability estimation. Although the distance between two electrodes was only 10 μm, ion migration did not occur and insulation resistance was maintained at over 1 GΩ for 500 h. This result showed that the new ACF has superior insulation reliability. This indicates that this new ACF has potential to be applied in very fine interconnections.

10.3.5 Nanowire ACF for Ultrafine Pitch Flip Chip Interconnection

To satisfy the reduced I/O pitch and avoid electric shorting, a possible solution is to use high-aspect-ratio metal posts. Nanowires exhibit high possibilities due to the small size and extremely high-aspect-ratio. In the literature, nanowires could be applied in FET sensors for gas detection, magnetic hard disk, nanoelectrodes for electrochemical sensor, thermal electric device for thermal dissipation and temperature control, etc. [55–57]. To prepare nanowires, it is important to define nanostructures on photoresist. Many expensive methods such as e-beam, X-ray, or scanning probe lithography have been used but micrometer length nanowires cannot be achieved. Another less expensive alternative is electrodeposition of metal into a nanoporous template such as anodic aluminum oxide (AAO) [58] or block copolymer self-assembly template [59]. The disadvantages of a block copolymer template include thin thickness (that means short nanowires), nonuniform distribution, and poor parallelism of nanopores. However, AAO has benefits of higher thickness (>10 μm), uniform pore size and density, larger size and very parallel pores. Lin et al. [60] developed a new ACF with nanowires. They used AAO templates to obtain silver and cobalt nanowire arrays by electrodeposition. And then, low viscous PI was spread over and filled into the gaps of nanowire arrays after surface treatment. The bimetallic Ag/Co nanowires could keep parallel during fabrication by magnetic interaction between the cobalt and the applied magnetic field. The silver and cobalt nanowires/polyimide composite films could be obtained with nanowire diameter of about 200 nm and maximum film thickness up to 50 μm. The X–Y insulation resistance is about 4–6 GΩ and Z-direction resistance including the trace resistance (3-mm length) is less than 0.2 Ω. They also demonstrated the evaluation of this nanowire composite film by stress simulation. They found that the most important factor for designing nanowire ACF was the volume ratio of nanowires. But actually, the ratio of nanowires cannot be too small to influence the electric conductance. They concluded that it is important to get a balance between electric conductance and thermal–mechanical performance by increasing film thickness or decreasing the modulus of the polymer matrix instead of reducing the ratio of nanowires.

10.3.6 In Situ Formation of Nanoconductive Fillers in ACA/ACF

One of the challenging issues in the formation of nanofiller ACA/ACF is the dispersion of nanoconductive fillers in ACA/ACF. A lot of research has been going on in recent years to address the dispersion issue of nanocomposites because nanofillers tend to agglomerate. For the fine pitch electronic interconnects using nano-ACA/ACF, the dispersion issues need to be solved. The efforts usually include the physical approaches such as sonication and chemical approaches such as surfactants. Recently, a novel ACA/ACF incorporated with in situ formed nanoconductive particles was proposed for next-generation high performance fine pitch electronic packaging applications [61, 62]. This novel interconnect adhesive combines the electrical conduction along the Z-direction (ACA-like) and the ultrafine pitch (<100 nm) capability. Instead of adding the nanoconductive fillers in the resin, the nanoparticles can be in situ formed during the curing/assembly process. By using in situ formation of nanoparticles, during the polymer curing process, the filler concentration and dispersion could be better controlled and the drawback of surface oxidation of the nanofillers could be easily overcome.

References

1. Wolfson H, Elliot G (1956) Electrically conducting cements containing epoxy resins and silver. US Patent 2,774,747
2. Matz KR (1958) Electrically conductive cement and brush shunt containing the same. US Patent 2,849,631
3. Beck DP (1958) Printed electrical resistors. US Patent 2,866,057
4. Jana PB, Chaudhuri S, Pal AK, De SK (1992) Electrical conductivity of short carbon fiber-reinforced carbon polychloroprene rubber and mechanism of conduction. Polymer Engineering and Science 32:448–456
5. Malliaris A, Tumer DT (1971) Influence of particle size on the electrical resistivity of compacted mixtures of polymers and metallic powders. Journal of Applied Physics 42:614–618

6. Ruschau GR, Yoshikawa S, Newnham RE (1992) Resistivities of conductive composites. Journal of Applied Physics 73(3):953–959
7. Gilleo K (1995) Assembly with conductive adhesives. Soldering & Surface Mount Technology 19:12–17
8. Hariss PG (1995) Conductive adhesives: a critical review of progress to date. Soldering & Surface Mount Technology 20:9–21
9. Li Y, Moon K, Wong CP (2005) Electronics without lead. Science 308:1419–1420
10. Li Y, Wong CP (2006) Recent advances of conductive adhesives as a lead-free alternative in electronic packaging: materials, processing, reliability and applications. Materials Science & Engineering R: Reports 51:1–35
11. Lau J, Wong CP, Lee NC, Lee SWR (2002) Electronics Manufacturing: with Lead-Free, Halogen-Free, and Conductive-Adhesive Materials. McGraw-Hill, New York, NY
12. Wu H, Wu X, Liu J, Zhang G, Wang Y, Zeng Y, Jing J (2006) Development of a novel isotropic conductive adhesive filled with silver nanowires. Journal of Composite Materials 40(21):1961–1968
13. Lee HS, Chou KS, Shih ZW (2005) Effect of nano-sized silver particles on the resistivity of polymeric conductive adhesives. International Journal of Adhesion and Adhesives 25:437–441
14. Ye L, Lai Z, Liu J, Tholen A (1999) Effect of Ag particle size on electrical conductivity of isotropically conductive adhesives. IEEE Transactions on Electronics Packaging Manufacturing 22(4):299–302
15. Fan L, Su B, Qu J, Wong CP (2004) Electrical and thermal conductivities of polymer composites containing nano-sized particles. In: Proceedings of Electronic Components and Technology Conference, Las Vegas, NV, pp. 148–154
16. Jiang HJ, Moon K, Li Y, Wong CP (2006) Surface functionalized silver nanoparticles for ultrahigh conductive polymer composites. Chemistry of Materials 18(13):2969–2973
17. Kotthaus S, Günther BH, Haug R, Schafer H (1997) Study of isotropically conductive bondings filled with aggregates of nano-sized Ag-particles. IEEE Transactions on Components, Packaging, and Manufacturing Technology: Part A 20(1):15–20
18. Majima M, Koyama K, Tani Y, Toshioka H, Osoegawa M, Kashihara H, Inazawa S (2002) Development of conductive material using metal nano particles. SEI Technical Review 54:25–27
19. Das RN, Lauffer JM, Egitto FD (2006) Electrical conductivity and reliability of nano- and micro-filled conducting adhesives for Z-axis interconnections. In: Proceedings of Electronic Components and Technology Conference, San Diego, CA, pp. 112–118
20. Iijima S (1991) Helical microtubules of graphitic carbon. Nature 354:56
21. Thess A, Lee R, Nikolaev P, Dai H, Petit P, Robert J, Xu C, Lee YH, Kim SG, Rinzler AG, Colbert DT, Scuseria G, Tománek D, Fischer JE, Smalley RE (1996) Crystalline ropes of metallic carbon nanotubes. Science 273:483
22. Berber S, Kwon YK, Tomànek D (2000) Unusually high thermal conductivity of carbon nanotubes. Physical Review Letters 84(20):4613–4616
23. Yu MF, Files BS, Arepalli S, Ruoff RS (2000) Tensile loading of ropes of single-wall carbon nanotubes and their mechanical properties. Physical Review Letters 84(24):5552–5555
24. Gao G, Cagin T, Goddard WAIII (1998) Energetics, structure, mechanical and vibrational properties of single walled carbon nanotubes (SWNT). Nanotechnology 9:184–191
25. Li J, Lumpp JK (2006) Electrical and mechanical characterization of carbon nanotube filled conductive adhesive. In: Proceedings of Aerospace Conference, Manhattan, CA, pp. 1–6
26. Qian D, Dickey EC, Andrews R, Rantell T (2000) Load transfer and deformation mechanisms in carbon nanotube-polystyrene composites. Applied Physics Letters 76:2868
27. Lin XC, Lin F (2004) Improvement on the properties of silver-containing conductive adhesives by the addition of carbon nanotube. In: Proceedings of High Density Microsystem Design and Packaging, Shanghai, China, pp. 382–384
28. Rutkofsky M, Banash M, Rajagopal R, Chen J (2006) Using a carbon nanotube additive to make electrically conductive commercial polymer composites. Zyvex Corporation, Application Note 9709. http://www.zyvex.com/Documents/9709.PDF, 28

29. Kamyshny A, Ben-Moshe M, Aviezer S, Magdassi S (2005) Ink-jet printing of metallic nanoparticles and microemulsions. Macromolecular Rapid Communications 26:281–288
30. Cibis D, Currle U (2005) Inkjet printing of conductive silver paths. In: 2nd International Workshop on Inkjet Printing of Functional Polymers and Materials, Eindhoven, The Netherlands
31. Kolbe J, Arp A, Calderone F, Meyer EM, Meyer W, Schaefer H, Stuve M (2005) Inkjettable conductive adhesive for use in microelectronics and microsystems technology. In: Proceedings of IEEE Polytronic 2005 Conference, Wroclaw, Poland, pp. 1–4
32. Moscicki A, Felba J, Sobierajski T, Kudzia J, Arp A, Meyer W (2005) Electrically conductive formulations filled nano size silver filler for ink-jet technology. In: Proceedings of IEEE Polytronic 2005 Conference, Wroclaw, Poland, pp. 40–44
33. Moon K, Dong H, Maric R, Pothukuchi S, Hunt A, Li Y, Wong CP (2005) Journal of Electronic Materials 34:132–139
34. Matsuba Y (2003) Erekutoronikusu Jisso Gakkaishi 6(2):130–135
35. Efremov MY, Schiettekatte F, Zhang M, Olson EA, Kwan AT, Berry RS, Allen LH (2000) Physical Review Letters 85:3560–3563
36. Li Y, Moon K, Wong CP (2006) Enhancement of electrical properties of anisotropically conductive adhesive (ACA) joints via low temperature sintering. Journal of Applied Polymer Science 99(4):1665–1673
37. Li Y, Moon K, Wong CP (2004) In: Proceedings of 54th IEEE Electronic Components and Technology Conference, Las Vegas, NV, pp. 1968–1974
38. Li Y, Wong CP (2005) In: Proceedings of 55th IEEE Electronic Components and Technology Conference, Lake Buena Vista, FL, pp. 1147–1154
39. Li Y, Moon K, Wong CP (2005) Journal of Electronic Materials 34(3):266–271
40. Li Y, Moon K, Wong CP (2006) Journal of Electronic Materials 34(12):1573–1578
41. Davies G, Sandstrom J (1976) Circuits Manufacturing 56–62
42. Harsanyi G, Ripka G (1985) Electrocomponent Science and Technology 11:281–290
43. Giacomo GA (1992) In: J McHardy and F (Eds) Ludwig Electrochemistry of Semiconductors and Electronics: Processes and Devices. Noyes, Park Ridge, NJ, pp. 255–295
44. Manepalli R, Stepniak F, Bidstrup-Allen SA, Kohl PA (1999) IEEE Transactions on Advanced Packaging 22:4–8
45. Giacomo D (1997) Reliability of Electronic Packages and Semiconductor Devices. McGraw-Hill, New York (Chap. 9)
46. Wassink R (1987) Hybrid Circuits 13:9–13
47. Shirai Y, Komagata M, Suzuki K (2001) In: 1st International IEEE Conference on Polymers and Adhesives in Microelectronics and Photonics, Potsdam, Germany, pp. 79–83
48. Marderosian D, Raytheon Co. Equipment Division, Equipment Development Laboratories, pp. 134–141
49. Schonhorn H, Sharpe LH (1983) Prevention of surface mass migration by a polymeric surface coating. US Patent 4,377,619
50. Brusic V, Frankel GS, Roldan J, Saraf R (1995) Journal of the Electrochemical Society 142:2591–2594
51. Wang PI, Lu TM, Murarka SP, Ghoshal R (2005) US Pending Patent (No. 20050236711)
52. Li Y, Wong CP (2005) US Pending Patent
53. Li Y, Wong CP (2006) Monolayer protection for electrochemical migration control in silver nanocomposite. Applied Physics Letters 81:112
54. Toshioka H, Kobayashi M, Koyama K, Nakatsugi K, Kuwabara T, Yamamoto M, Kashihara H (2006) SEI Technical Review 62:58–61
55. Lieber CM (2001) Nanowire nanosensors for high sensitive and selective detection of biological and chemical species. Science 293:1289–1292
56. Prinz GA (1998) Science 282:1660
57. Martin CR, Menon VP (1995) Fabrication and evaluation of nanoelectrode ensembles. Analytical Chemistry 67:1920–1928
58. Xu JM (2001) Fabrication of highly ordered metallic nanowire arrays by electrodeposition. Applied Physics Letters 79:1039–1041

59. Russell TP (2000) Ultra-high density nanowire array grown in self-assembled di-block copolymer template. Science 290:2126–2129
60. Lin R-J, Hsu Y-Y, Chen Y-C, Cheng S-Y, Uang R-H (2005) In: Proceedings of 55th IEEE Electronic Components and Technology Conference, Orlando, FL, pp. 66–70
61. Li Y, Moon K, Wong CP (2006) In: Proceedings of 56th IEEE Electronic Components and Technology Conference, IEEE, NJ, pp. 1239–1245
62. Li Y, Zhang Z, Moon K, Wong CP (2006) Ultra-fine pitch wafer level ACF (anisotropic conductive film) interconnect by in situ formation of nano fillers with high current carrying capability. US Pending Patent

Chapter 11
Nanoparticles in Microvias

Rabindra N. Das(✉) and Frank D. Egitto

11.1 Introduction

Electronic packaging provides for mounting and physical support of electronic components, removal of heat from devices (e.g., integrated circuit chips), protection of devices from the environment, and electrical interconnection of components. This electrical interconnection enables distribution of both electronic signals and power throughout the package by means of multiple layers of metal circuit traces. Electrical interconnection between layers (vertically) is typically made with drilled and plated holes.

The demand for high-performance, lightweight, portable computing power is driving the microelectronics industry toward miniaturization of many electronic products and the components that comprise them [1]. Greater I/O density of IC chips and more demanding performance requirements necessitate greater wiring density and a concomitant reduction in feature sizes for electronic packages. To incorporate a greater degree of electronic function into a smaller volume, circuit traces and the holes used to connect them must have smaller physical dimensions [2–6]. The term "microvia" has been coined to describe small holes used for layer-to-layer electrical interconnection. Somewhat arbitrarily, they are generally accepted to have diameters in the order of 150 μm, or smaller.

Today, the high end of the semiconductor market appears to be standard Application-Specific Integrated Circuits (ASICs), structured ASICs, and Field-Programmable Gate Arrays (FPGAs). These devices continue to need an increasing number of signal, power, and ground die pads. A corresponding decrease in pad pitch is required to maintain reasonable die sizes. The combination of these two needs is pushing complex semiconductor packaging designs. This packaging challenge is especially critical in flip chip ball grid array (BGA) packages where the need for density has to coexist with good electrical, thermal, and reliability performance. Migration from wirebond to flip chip packages has been driven by the

R.N. Das
Endicott Interconnect Technologies, Inc., Endicott, NY, USA

need to combine electrical and thermal performance advantages with density, without compromising component reliability. In flip chip designs, the backs of the die are exposed, so direct access is available for heat sinking as opposed to a wirebond die-up package. Further, flip chip packaging provides improvements in electrical performance, such as reduced power supply noise due to reduction in inductance as well as increases in power and ground connections. Flip chip packaging also provides significant increases in the number of rows of signal die pads that can be interconnected. However, typical flip chip packages increase the mechanical coupling of the die to the printed wiring board (PWB). This normally results in lower reliability performance, either at the die bumps, as in the case of organic BGAs, or at the PWB-to-package interface, as is the case for ceramic BGA packages. Various interconnection methods have been employed to compensate for this increased coupling. Although a typical plastic flip chip package normally exhibits good BGA reliability, the coefficient of thermal expansion (CTE) mismatch between die and package typically limits the size of the device that can be used. Techniques such as BGA ball depopulation under the die, at the corners of the die, or at the die periphery reduce mechanical coupling and therefore allow increased device sizes. Unfortunately, this reduces the improvement in signal I/O. Ceramic flip chip packages have good CTE match to silicon. However, these packages have a large CTE mismatch to the PWB. This constrains the package size and, in turn, constrains the package I/O. There are new materials that increase the CTE of the ceramic package to provide increased second level reliability performance at the expense of additional CTE mismatch to the die. Replacement of the BGA ball by a solder column is also helpful.

Specific organic package material sets have been investigated and developed to eliminate the deficiencies of both typical plastic and ceramic packages. A highly compliant poly(tetrafluoroethylene) (PTFE) material has been proven [7] to meet all these organic flip chip reliability challenges. But each new generation of device technology places further significant pressures on package density.

There are a number of traditional approaches to reducing size and increasing density in packages. These include reduction in the width and spacing of metal traces and the addition of wiring layers. Formation of circuit features with reduced dimensions by subtractive techniques is facilitated by the use of thinner metals (typically copper), but this sometimes compromises electrical performance by way of higher resistance of the thinner, narrower circuit lines. This shortcoming may be mitigated somewhat by the use of semiadditive (or pattern plating) processes to form lines having a greater aspect ratio (ratio of line height to line width). Adding wiring layers is a straightforward means of providing greater density of function in the package. However, added layers invariably translate to added cost. It is therefore imperative to make the most efficient use of real estate used for wiring to keep the number of wiring layers to a minimum. For interconnection with traditional plated through-hole (PTH) technology, two PTHs are required to complete a circuit trace. PTHs block channels that could otherwise be used for wiring (Fig. 11.1a). Packaging designs that are most effective in optimizing the use of available wiring space incorporate blind and buried vias. As opposed to PTHs, blind and buried vias

11 Nanoparticles in Microvias

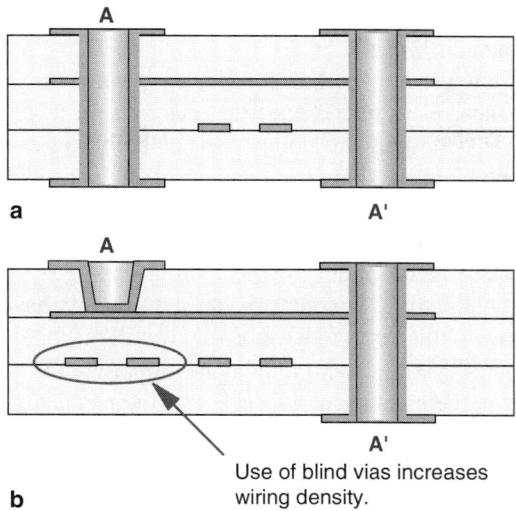

Fig. 11.1 Blind via use increases wiring density in circuit layers below the via

do not span the full thickness of the substrate. A blind via is formed in a manner whereby it terminates on a previously formed metal feature internal to the substrate, and is subsequently metallized. A buried via is formed as a through hole in a thin core. Following metallization and formation of circuit traces, the thin core is laminated with other subcomposites into a thicker substrate. In this chapter, the term microvia is used to describe both blind and buried vias with small dimensions. Blind and buried vias provide for vertical electrical interconnection that can be terminated at any wiring plane(s), at any depth, within the packaging structure. Replacement of conventional plated through holes with such vertically terminated vias opens up additional wiring channels on layers above and below the terminated vias.

As via diameters decrease to accommodate more dense designs, plating of the vias becomes more of a challenge. This problem is alleviated to a degree by use of thinner, laser-friendly [8–12] dielectric materials. Although the use of blind vias frees up wiring space, its utility is limited by the challenge of plating blind vias with aspect ratios (depth to diameter) greater than 1:1. As the distance between the planes to be interconnected increases, it is necessary to either increase the size of the blind vias, or sequentially add several layers of dielectric material, each interconnected from layer to layer with blind vias. However, such a fabrication technique can be costly owing to the cumulative yield loss incurred when adding layers sequentially.

One method of extending wiring density beyond the limits imposed by these approaches, while fabricating structures with vertically terminated vias of arbitrary depth, is a strategy that allows for metal-to-metal Z-axis interconnection of subcomposites during lamination to form a composite structure [13–19]. Interconnection is made using an electrically conducting adhesive (ECA). The adhesive is generally in the form of a conductive paste.

During the past few years, there has been an increasing interest in using electrically conductive adhesives as interconnecting materials in the electronics industry [20–25]. Conductive adhesives are composites of polymer resin and conductive fillers. Metal-to-metal bonding between conductive fillers provides electrical conductivity [22–29], while the polymer resin provides favorable processing attributes and mechanical robustness [22–25, 30]. Several investigators have reported on the electrical resistivity of metal-filled polymer composites [31–38]. In an effort to increase electrical conductivity, conductive adhesives usually have a high degree of filler loading. However, this tends to weaken the overall mechanical strength of the adhesive. For layer-to-layer electrical interconnection, reliability of the conductive joint formed between the conductive adhesive and the metal surface to which it is mated is of prime importance.

The area of nanotechnology, encompassing the synthesis of nanoscale materials, understanding and utilization of their exotic physicochemical and electronic properties, and organization of nanoscale structures into predefined superstructures, promises to play an increasingly important role in many key technologies of the new millennium [39–51]. As far as high-density electronic packaging is concerned, there is an ever-growing need to achieve high-density PWBs and laminate chip carriers through Z-axis electrical interconnection using paste-filled microvias [13–19]. Current trends indicate that via diameters for PWBs are approaching 100 μm, whereas those in laminate chip carriers are in the order of 50 μm or less. Conductive adhesives can have broad particle size distributions, and larger particles can be a problem when filling smaller holes (e.g., diameter of 60 μm or less), resulting in voids. Consequently, researchers in the field of materials processing have been looking at nanoparticle approaches for the development of conductive adhesives that can fill smaller diameter holes.

Nanoparticles (1–100 nm) have been investigated for more than a decade using a rather wide range of experimental methods [52, 53]. Various investigations of their chemical, mechanical, electrical, magnetic, and optical behavior have already demonstrated the possibility of controlling the properties of nanoparticles through control of the sizes of their constituent clusters or powders, and the manner in which the constituents are assembled. The microstructural features of importance include particle/grain size, distribution, and morphology. Nanoparticles exhibit a variety of considerably improved properties with respect to coarse-grained particles. These include increased hardness/strength, enhanced diffusivity, improved ductility and toughness, lower electrical resistivity, higher thermal expansion coefficient, lower thermal conductivity, and soft magnetic properties. Recently, conducting nanoparticles, nanotubes, and nanowires are getting significant attention in the microelectronics industry for miniaturization. Nanomaterials are flexible to use as conductive adhesives and inks and can be screen- or injection- printed to produce fine conducting features. Low-temperature sintering to achieve high conductivity is one of the major advantages of nanosystems. However, only limited literature is available on the electrical properties of nanoparticles within the microvia.

In recognition of the importance and issues of nanoparticle-based conductive adhesives, there has been a worldwide research and development effort directed toward high-performance adhesives in recent years. The scope of this chapter is to summarize some recent activities and advances in the area of electrically conductive adhesives formulated using controlled-size particles, ranging from nanometer scale to micrometer scale, and used to fill small diameter holes (microvias) for Z-axis electrical interconnection applications. A purpose of this review is to provide a better understanding of the nature of nanoparticle-based conductive adhesives, as well as to highlight the significant progress made on microvia filling for Z-axis electrical interconnections. In particular, a strategy that allows for metal-to-metal Z-axis (vertical) interconnection using ECA-filled microvias is described. This chapter discusses nanoparticles and nanoparticle-based conducting adhesives in microvias for these Z-axis interconnections. Recent work on adhesives formulated using controlled-size particles to fill small-diameter microvias is highlighted, particularly with respect to their integration into organic laminate chip carrier substrates. The mechanical strength and reliability of the electrically conductive joints formed between the adhesive and metal surfaces are addressed. A variety of conductive adhesives with particle sizes ranging from nanometer scale to submicron and micron scale were investigated. The review also describes the microstructures, conducting mechanism, volume resistivity, mechanical strength, and reliability of adhesives formulated with a variety of metals and alloys, such as Cu, Ag, and low melting point (LMP) alloy, used as conductive filler particles in the ECA. As a case study, an example of a silver-filled conductive adhesive used to fill microvias for Z-axis interconnection in a flip chip plastic ball grid array package having a 150-μm die pad pitch is given. The processes and materials used to achieve smaller feature dimensions, satisfy stringent registration requirements, and achieve robust electrical interconnections are discussed.

11.2 Electrically Conductive Adhesives/Inks for Microvias

A conductive adhesive is a composite material consisting of a nonconductive polymer binder and conductive filler particles. When the filler content is high enough, the system is transformed into a good electrical conductor. Conductive-adhesive-filled microvias can provide the conductive path required to achieve connection from one circuit element to another. Electrical connection is achieved primarily by interparticle conduction. For electrical conduction, particles should make intimate contact (physical and/or tunneling) and form a network (conductive chain), which helps in the transfer of electrons. This conductive path is formed at a threshold volume fraction of conductive filler, which can be calculated using the percolation theory of spherical particles. Conductive ink composed of metal particles can also be used in microvias. Inks are typically made of a dispersion of nanoparticles in solution. Metal-to-metal sintering between conductive filler particles provides electrical conductivity.

The electrically conductive material used for via filling must be corrosion-resistant, easily adaptable to the existing manufacturing process (e.g., screen printing), demonstrate low shrinkage after processing, possess high mechanical strength, exhibit good adhesion to PCB materials, and be available at a low cost. Figure 11.2 shows a typical example of a conductive adhesive used in the micovia. Here conducting particles provide low resistance, and the polymer component of the adhesive produces reliable bonding within the joints. Conductive adhesives used for via filling generally need to be isotropic conductive adhesives (ICAs), that is, exhibit good conductivity in all directions. Figure 11.2 shows land-to-land (z-direction) and inner plane connections (x, y direction). Either thermosetting or thermoplastic materials may be used as the polymer matrix in the adhesive. Epoxy, cyanate ester, silicone, and polyurethane are all widely used thermosets. Phenolic epoxy, maleimide, acrylic, and preimidized polyimide are commonly used thermoplastics. Thermoset epoxies are by far the most common binders. They provide excellent adhesive strength, good chemical and corrosion resistance, and low cost. Thermoplastics are usually added to allow softening and rework under moderate heat. An attractive advantage of thermoplastic ICAs is that they are reworkable, that is, they can easily be repaired. A major drawback of thermoplastic ICAs, however, is the degradation of adhesion at high temperature between the ICA and the substrate to which it is bonded. A drawback of polyimide-based ICAs is that they generally contain solvents. During heating, voids are formed when the solvent evaporates. Most commercial ICAs are based on thermosetting resins.

Conductive fillers provide the composite with electrical conductivity through contact between the conductive particles. Conductive filler materials include silver (Ag), gold (Au), nickel (Ni), copper (Cu), and carbon (C) particles in various sizes and shapes. Among the different metal particles, Ag flakes are the most commonly used conductive fillers for current commercial ICAs because of their high conductivity, simple paste formulation process, and excellent contact among the flakes. In addition, Ag is unique among all the cost-effective metals by nature of its conductive oxide. Oxides of most common metals are good electrical insulators. Copper powder, for example, becomes a poor conductor after aging. Nickel- and

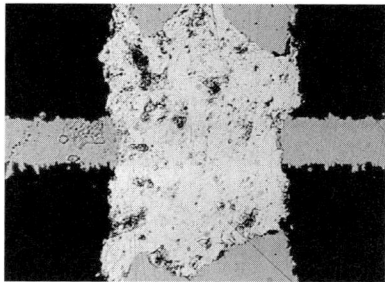

Fig. 11.2 Photograph of a cross section showing a conductive adhesive in a microvia. In this instance, interconnection is made vertically between copper pads, and laterally to a copper inner plane

Table 11.1 Resistivities of various materials

Materials	Resistivity (Ω cm)
Cured unfilled epoxy	10^{15}
Diamond	10^{14}
Glass	10^{10}
Undoped ICP	10^{10}
Silicon	10^{5}
Doped polyaniline	10^{-1}
Best ICP	10^{-4}
Nanotubes	10^{-4}
Silver-filled epoxy	10^{-4}
Solder	10^{-5}
Silver	10^{-6}

copper-based conductive adhesives generally do not have good resistance stability, because both nickel and copper are easily oxidized. Even with antioxidants, copper-based conductive adhesives show an increase in bulk resistivity after aging, especially under high-temperature and high-humidity conditions. Table 11.1 summarizes bulk resistivities of some pertinent materials.

11.3 Nanoparticle-Based Conductive Adhesives in Microvias

11.3.1 Material Categories

The term "nanoparticle" generally refers to the class of ultrafine metal particles with a physical structure or crystalline form that measures less than 100 nanometers (nm) in size. Nanoparticles can be 3D (block), 2D (plate), or 1D (tube or wire) structures [54, 55]. In general, nanoparticle-filled conductive adhesives are defined as containing at least some percentage of nanostructures (1D, 2D, and/or 3D) that enhance the overall electrical conductivity or sintering behavior of the adhesives. Figure 11.3 represents a theoretical comparative model for a variety of possible structures based on powder filling in a microvia. In this model, the volume of the microvia is constant for all six cases. Conductivity is achieved through metal–metal bonding. Increasing the number density of particles increases the probability of metal–metal contact. Each contact spot possesses a contact resistance. For microparticles, the number density of particles will be much less than for nanoparticles. Therefore, microparticle-filled vias will tend to have a lower contact resistance (fewer contacts required over a given length), although the probability of particle–particle contact will be less. In the case of a nano–micro mixture, the microscale particles could maintain a low contact resistance, whereas the nanoscale particles can increase the number of particle contacts. Nano- and microparticle mixtures could be nanoparticle–microparticle, nanoplate (2D)–microparticle, nanotube (1D)–microparticle, or any combination of these three cases.

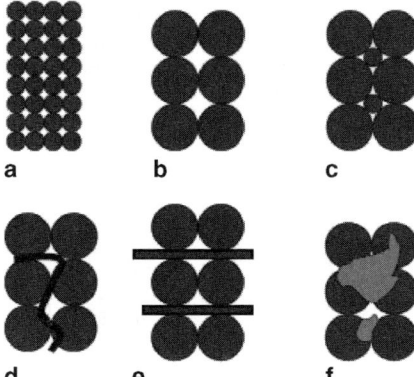

Fig. 11.3 A variety of adhesive-filled microvia structures. The adhesive consists of a polymer and (**a**) nanoparticles, (**b**) controlled-size microparticles, (**c**) nanomicroparticle mixture, (**d**) nanotube/wire–microparticle mixture, (**e**) microparticle–sheet/flake (2D), and (**f**) micro/nano–low melting point (LMP) particles

Another possibility is the use of a LMP filler. The LMP filler melts and reduces interparticle resistance. Hence, nanoparticle-based conductive adhesives can be categorized as nano, controlled micro, nano–micro, or LMP-based systems. Finally, use of a conducting polymer [56, 57] instead of the high-resistance polymer binder has been reported to enhance the overall conductivity of adhesives.

11.3.1.1 Nano-Based Adhesives (Nbas)

Metal nanoparticles with particle size less than 100 nm are useful for developing a variety of ink or paste formulations that can be used to generate small conducting features. Larger particles can be a problem when filling small micovias, resulting in voids. Such problems can be solved by employing nanometer-sized metal particles. However, aggregation of the particles driven by van der Waal forces and/or high surface energies must be overcome. Homogeneous dispersion of fine nanoparticles in the organic matrix is a critical step to achieve highly conducting adhesives that sinter at low temperatures. Organic modification of the surface of the particles can alter the surface energy of the particles such that aggregate-free dispersions can be obtained [58]. A silane modifier can influence dispersibility but may also alter the curing behavior (e.g., cross-link density) of the polymeric matrix [59, 60]. Organic protective agents [61] such as polyvinyl pyrrolidone (PVP) protect nanoparticles from agglomerations and disperse easily in an organic matrix. In the case of inks, nanoparticle inks can be used to fill microvias using an ink-jet process [61–67].

Nanoparticles can self-sinter and form a continuous conduction path. The high surface area possessed by nanoparticles necessitates an excess amount of solvent to

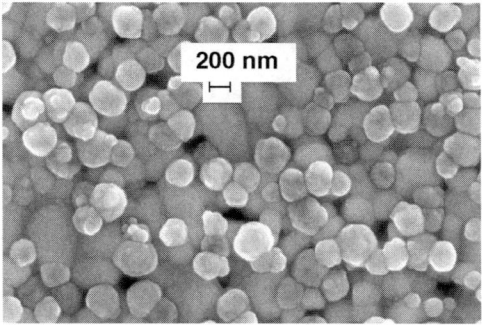

Fig. 11.4 SEM micrograph of a nanoparticle-based conductive adhesive

make a highly loaded silver paste or ink. Figure 11.4 shows the microstructure of a nano silver-filled adhesive. Nanofilled adhesives achieve electrical conductivity through sintering during the polymer curing and solvent evaporation process. However, solvent evaporation may cause paste shrinkage. Adhesives with high cure shrinkage generally exhibit voiding leading to resistive opens. Therefore, it is important to use as little solvent as possible in pastes used for via filling. Nanoparticles, nanotubes, and nanowires can be used as fillers for pastes used in via fill applications. For example, a novel ICA has been developed by using Ag nanowires as conductive fillers. The electrical and mechanical properties of this adhesive were compared with that of conventional ICAs filled with micrometer-sized Ag particles, or nanometer-sized Ag particles [68]. It was found that at a lower filler content, the ICA filled with Ag nanowires exhibited lower bulk resistivity and higher shear strength than the ICA filled with micrometer-sized Ag particles or nanometer-sized Ag particles. Carbon nanotube-based adhesives are another possible option for use in microvias [69, 70].

11.3.1.2 Controlled-Size Submicroparticles and Microparticles

As particle size increases from nano to micro, the sum of the surface areas of the particles within a given volume decreases. Microparticles need higher loading to achieve percolation. Particle size and distribution of particles is important for via fill applications. Larger particles could plug a microvia during the filling operation, and generate voids and open circuits. Recently, controlled-size particles in the range of a few microns have been given interest for via fill applications [13]. Figure 11.5 shows Cu- and Ag-based microparticle-filled conductive adhesives. In the silver adhesive, the average filler diameter is in the range of 5 µm. Filler loading was high and adjacent particles united mutually, and necking phenomena between fillers occurred; namely, a conduction path was achieved [26], as shown in Fig. 11.5a. A similar result was observed for an adhesive filled with 4-µm Cu particles (Fig. 11.5b). A mixture of different shapes and sizes of microparticles

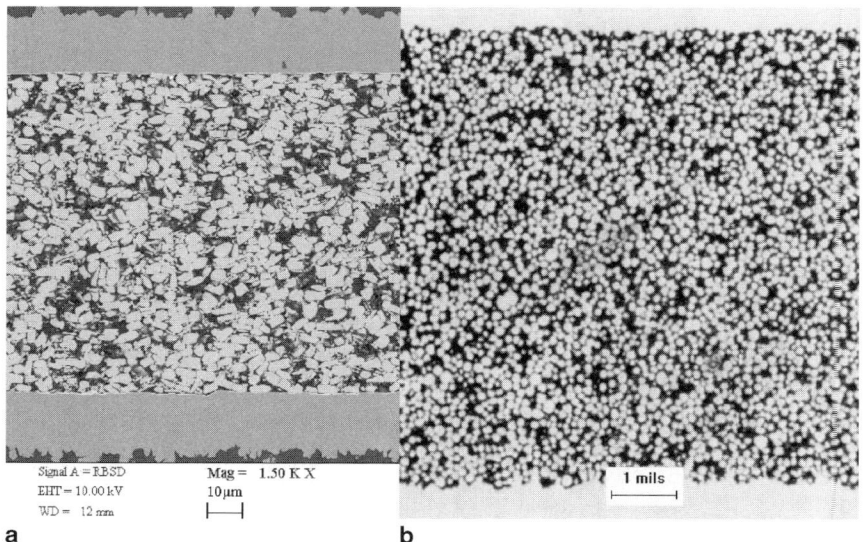

Fig. 11.5 Micrographs for the cross-sectional views of adhesives made with (**a**) silver microparticles, and (**b**) Cu microparticles

can be used for highly conductive adhesives. Silver particles consisting of a mixture of micro (2.0–3.5 μm) granular particles, submicro (0.6–1.5 μm) granular particles, and flake (0.5–5.0 μm) particles have been used to achieve very low resistivity [71]. Silver flakes sometimes show thickness in the submicron to nanometer range. Low-cost submicrometer nickel particles can also be used for conductive adhesives [72]. Metal nanocoated microparticle-based adhesives are also used in microvia applications. One of the common examples is the use of gold- or silver-coated copper particles [73]. Typically, the gold or silver nanocoating protects copper surfaces from oxidation.

11.3.1.3 Nano–Micro Mixture

Nanosized conductive particles have been proposed as conductive fillers in ICAs for fine pitch interconnects. Although the nanosilver fillers in ICAs can reduce the percolation threshold, there has been concern that incorporation of nanosized fillers may introduce more contact spots due to their high surface area, and consequently induce higher resistivity compared with micron-sized fillers. In nano–micro mixtures, nanoparticles occupy interstitial positions to improve particle–particle contact for conductivity [13]. Figure 11.6 represents microstructures of nano–micro silver-filled adhesives. It can be seen that individual nanoparticles connect the larger microparticles.

Fig. 11.6 SEM micrograph showing a cross-sectional view of an adhesive with a mixture of silver nanoparticles and microparticles

Addition of 2% (wt/wt) nanoparticles into microparticles dispersed in epoxy was reported to improve electrical conductivity [74]. Another study [75] showed that addition of nanoparticle content to conventional conductive adhesives increased electrical resistance. Addition of nanosized silver colloids to microsized Ag flakes usually increased resistivity, probably due to increased contact resistance. Only near the percolation threshold would the addition of nanosized silver particles decrease resistivity by helping to form a conductive path. It was also reported [76] that addition of nanosized particles to microparticles reduced the percolation threshold from 60 to 50 wt%, but increased overall resistivity. Nanosized silver particles can fill the gaps between silver flakes of conductive adhesives and help electron transport at lower metal loading. However, due to the small particle size and high surface area (and consequently, high contact resistance) of nanoparticles, the measured resistivity of the adhesives was higher than that of the Ag-flake-filled adhesives. So, it is clear that resistivity of nano–micro system will depend on their particle sizes (nano/micro) and concentration.

11.3.1.4 LMP Fillers

Another interesting approach for improving electrical conductivity is to incorporate LMP filler into the epoxy matrix [77–79]. Solder is the best-known example of a LMP material. A LMP/polymer composite paste material can be developed by mixing solder powder particles, thermoplastic polymer resin in a volatile solvent, and a fluxing agent [77]. Upon reflow, an oxide-free, partially coalesced LMP/solder connection is obtained, which is polymer-strengthened and reworkable at a low reflow temperature or in the presence of an organic solvent. It is also possible to use a hybrid of solder and metal powder with high melting point to exploit the advantages of both [78]. This conductive adhesive is a mixture of a LMP powder, a metal powder of high melting point (such as copper), a fluxing agent, a polymer resin, and other additives. The low-melting-alloy filler melts when its melting point

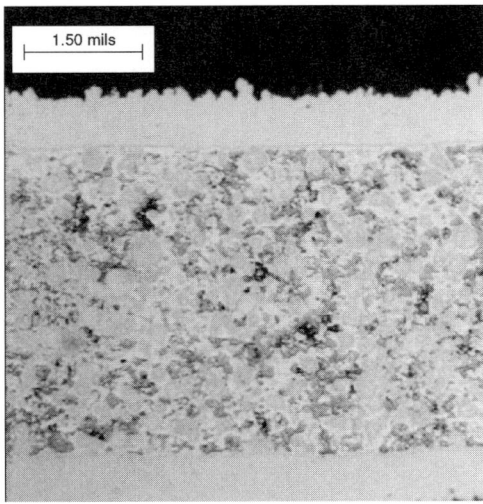

Fig. 11.7 Micrograph showing the cross-sectional view of adhesives made with LMP fillers

is achieved during the cure of the polymer matrix. The liquid phase dissolves the high melting point particles. The liquid exists only for a short period of time and then forms an alloy and solidifies. Electrical conduction is established through a plurality of metallurgical connections formed in situ from these two powders in a polymer binder. Figure 11.7 shows a cross section of a LMP-based adhesive. LMP melts and produces a continuous metallic network. Here, the electrical connection is established through transient liquid phase sintering (TLPS) among metal and LMP powders. High electrical conductivity can be achieved using this method [79]. One critical limitation of this technology is that the number of combinations of low-melting and high-melting fillers is limited. Only certain combinations of two metallic fillers, which are mutually soluble, exist to form this type of metallurgical interconnections.

11.3.2 Nanoparticle Sintering

In general, nanoparticles in conductive adhesives can reduce the percolation threshold and introduce more contact spots due to the high surface area, and consequently induce higher resistivity compared with micron-sized fillers. A recent study showed that nanosilver particles could exhibit sintering behavior at the curing temperature of adhesives [80]. It is well established that sintering temperature increases with increasing particle size. Addition of nanoparticles into microfilled adhesives will reduce sintering temperature significantly. The number of contact spots of nanoparticles is greater than with microparticles, for the same volume, as shown in Fig. 11.3a, b.

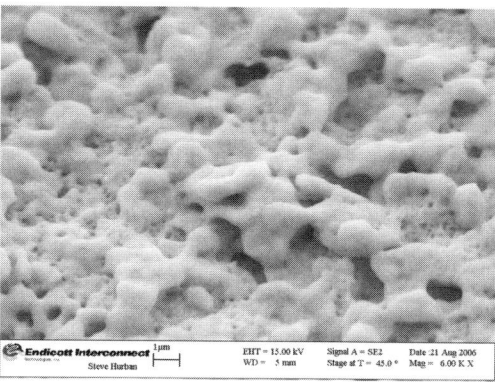

Fig. 11.8 SEM micrograph of an electrically conducting adhesive (ECA) fabricated using a mixture of silver nanoparticles and microparticles, and laminated at 275°C. Sintering is observed

The overall resistance of the adhesive-based inerconnection is the sum of the resistances of the individual fillers, the resistance between fillers (interparticle resistance), and the resistance between the filler and pads. To decrease the overall resistance, it is desirable to reduce the contact resistance between filler particles, and rely more on bulk metal conductance. If nanoparticles are sintered together, then the number of contacts between filler particles is reduced. This leads to lower contact resistance (Fig. 11.8). By using effective surfactants for the dispersion, and effectively capping those nanosized silver fillers in ECAs, obvious sintering behavior of the nanofillers can be achieved. As a result, an improved electrical conductivity of nanosilver-filled ICAs can be achieved at a lower loading level than that of microfilled ICAs with a filler loading of 80 wt% or higher. Sintering of silver adhesive was further evaluated using high-temperature curing. Here a mixture of nano and micro silver particles was used. Figure 11.8 shows a SEM micrograph of an ECA fabricated using a mixture of silver nanoparticles and microparticles, and cured at 275°C. Sintering is observed, and conductivity is achieved through a continuous metallic network. Comparable sintering was not observed at this temperature for ECAs fabricated without the addition of nanoparticles. It appears that at a sufficiently high concentration, nanoparticles are more prone to immediate particle–particle contact, facilitating sintering.

11.3.3 Conductivity Requirements in Microvias

Obviously, low resistance is a desirable attribute of joints formed by conductive adhesive-filled microvias. Typical volume resistivity of the conductive adhesives is in the range of 10^{-3} to 10^{-6} Ω cm. Volume resistivity of Cu, LMP, and silver-filled pastes has been reported [13] to be 5×10^{-4}, 5×10^{-5}, and 2×10^{-5} Ω cm, respectively, for conductive adhesives cured at ~190°C for 2 h. All composites fabricated

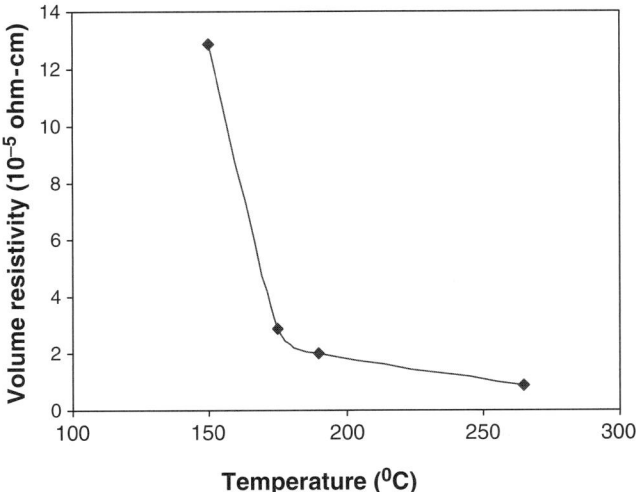

Fig. 11.9 Volume resistivity of silver adhesive as a function of curing temperature

from LMP and Cu–epoxy nanocomposites showed a resistivity of about 10^{-4} to 10^{-5} Ω cm, whereas silver composites showed resistivity of about 10^{-4} to 10^{-6} Ω cm. Silver nanoparticles showed volume resistivity in the range of 10^{-4} Ω cm, and the resistivity decreased to 10^{-5} Ω cm when nano–micro mixtures were used. Resistivity was lowest for silver-filled paste. Volume resistivity decreases with increasing curing temperature due to sintering of metal particles. Figure 11.9 shows volume resistivity of silver–epoxy adhesive as a function of curing temperature. There is a significant resistivity drop with increasing curing temperature from 150 to 175°C.

Exposure of ECAs to ambient conditions before curing can affect paste properties of the final product. Figure 11.10 presents volume resistivity values for the same adhesive, cured at various temperatures, as a function of aging times [13] at room temperature prior to thermal curing. After 72 h, curing of the ECA at 50, 190, and 265°C resulted in values of 50×10^{-5}, 32×10^{-5}, and 2×10^{-5} Ω cm, respectively. Change in resistivity with aging was significant when cured below 200°C, but it was not significant when cured at or above 250°C. A sharp increase in resistivity is observed up to 24 hours of aging, and thereafter it increases slowly.

Figure 11.11 shows viscosity as a function of exposure time at room temperature for a silver-filled epoxy ECA. Viscosity measurement was done using a 50-Pa stress under N_2. Adhesive viscosity increased by about 30% after 40 h, and doubled after 70 h. Differential Scanning Calorimetry (DSC) measurements indicated that this change was the result of polymer cross-linking such that the adhesive was partially cured at room temperature. When subsequently cured at 200°C, the resistivity of this adhesive is greater than that of cured adhesive that was not stored at room temperature for any significant time. For curing above 250°C, particle sintering plays an important role in maintaining low volume resistivity or high conductivity.

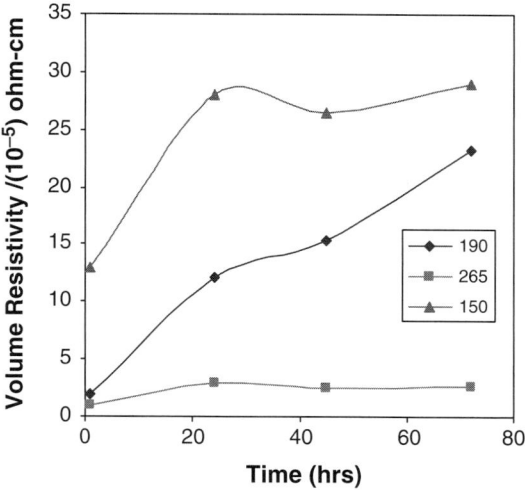

Fig. 11.10 Volume resistivity of silver adhesive with aging as a function of curing temperature

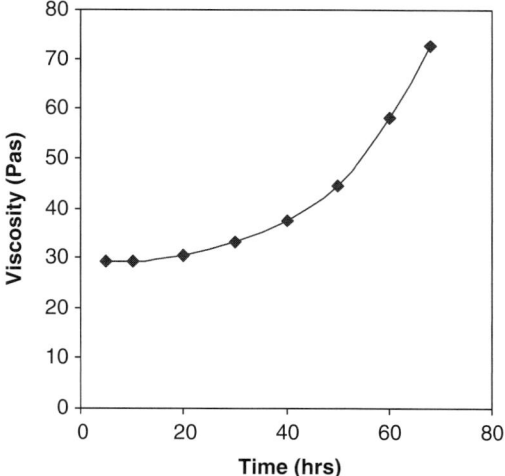

Fig. 11.11 Viscosity as a function of exposure time at room temperature for a silver–epoxy ECA

In general, ICA pastes exhibit high resistivity before cure, but the conductivity increases dramatically after curing. ICAs achieve electrical conductivity during the polymer curing process as a result of shrinkage of the polymer binder. Accordingly, ICAs with high cure shrinkage generally exhibit better conductivity. With increasing cross-linking density of ECAs, the shrinkage of the polymer matrix increases, and subsequently decreased resistivity is observed. For epoxy-based ICAs, a small amount of a multifunctional epoxy resin can be added into an ICA formulation to increase cross-linking density, shrinkage, and thus increase electrical conductivity.

Jeong et al. [81] reported the effect of curing behaviors, solvent evaporation, and shrink on conductivity of adhesives. As curing time increases, the silver particles in the polymer are concentrated due to the incremental solvent evaporation rate and the shrink rate. As a result, the silver particles in the polymer form an electric path. These results reveal that the increased shrink rate and solvent evaporation rate in conductive adhesives during the curing process improve their conductivity.

With the addition of only a small amount of short-chain dicarboxylic acid, the conductivity of ICA can be improved significantly due to the easier electronic tunneling/transportation between Ag flakes, and subsequently the intimate flake–flake contact [82]. The conductivity of silver oxides formed at the surface of silver flakes is inferior to that of the metal (silver) itself. Incorporation of reducing agents such as aldehydes further improves the electrical conductivity of ICAs due to a reduction reaction between aldehydes and silver oxides that generates pure metal silver in ECAs during the curing process [83].

Electrical resistivity of the specimens also varied significantly depending on the subsequent annealing process. However, the electrical resistivity achieved after annealing at temperatures above the curing temperatures clearly depended on the particular curing temperature that was used. The characteristics of the polymer structure in the adhesive binder vary with curing temperature, and this affects the electrical properties of the ICA. That is, the characteristics of the polymer structure obtained during the curing process affect the electrical resistance of the ICA, even after subsequent annealing processes.

11.3.4 Adhesion of Conductive Joints

Adhesion between the adhesive and the substrate to which it is mated is critical to the reliability of the semiconductor package. There are two types of adhesion mechanisms, physical bonding and chemical bonding, which contribute to the overall adhesion strength of a polymer on a surface [84]. Chemical bonding involves the formation of covalent or ionic bonds to link the polymer and the substrate [84]. Physical bonding involves mechanical interlocking or physical adsorption between the polymer and the surface of substrate. In mechanical interlocking, polymer and substrate interact on a more macroscopic level, where the polymer flows into the crevices and the pores of the substrate surface to establish adhesion [84]. Therefore, a polymer is expected to have better adhesion on a rougher surface because there is more surface area and "anchors" to allow for interlocking between the polymer and the substrate.

Bond strength of adhesive joints was evaluated using 90° peel test and tensile strength measurements. Peel strengths as high as 2.75 lbs/in. were measured for silver-filled pastes on roughened Cu foils, while LMP-filled pastes yielded peel strengths as low as 1.00 lb/in. Temperature cycle tests were run using a thermal shock chamber, cycling between −55 and 125°C, with exposure times of 10 min at each temperature. Tensile strength was measured before the test, and after 1,000 cycles, using an MTS tensile testing machine at a pulling rate of 0.025 in./min, and measuring until the joint

Table 11.2 Tensile strength and failure modes for a variety of ECA formulations

Adhesive	90° Peel strength (lb/in.)	Tensile strength (PSI)	Failure mode
Low melting point (LMP) alloy	1	600	Cohesive
Copper (Cu)	1.77	2,056	Cohesive
Silver (Ag)	2.75	3,370	Cohesive

ruptured. All pastes were stable after 1,000 cycles and maintained similar (within 10%) peel and tensile strength, even after 1,000 cycles. For all tensile strength test samples, cohesive failure within the paste was the observed failure mode. Table 11.2 summarizes 90° peel strength and tensile strength for various conductive adhesives. Silver-filled paste yielded the maximum mechanical strength.

Plasma cleaning of surfaces has been considered as one of the effective approaches to enhance the adhesion strength of conductive adhesives [85]. During the plasma etching process, the plasma radicals react with contaminants, and long-chain organic molecules can be broken down into small gaseous ones (mostly gaseous water and carbon-oxide conjunctions). These particles can be removed during the plasma cleaning process.

Another approach to improve adhesion is by using coupling agents [86]. Coupling agents are organofunctional compounds based on silicon, titanium, or zirconium. A coupling agent consists of two parts and acts as an intermediary to "couple" the inorganic substrate and polymer. For example, silane has different type organic chains that interact with the polymer and the substrate. Gianelis et al. reported various silane coupling agents [87]. Roughening of surfaces, for example, by sand blasting, chemical etching, plasma treatment, or anodization to specific morphologies has been employed to enhance the adhesion strength and provide structural durability in humid or corrosion environments [88, 89].

A further approach is to lower the elastic modulus of adhesive resins. By using low elastic modulus resins, the thermal stress at the adhesion interface can be reduced, improving the adhesion strength [90, 91]. However, a modulus value that is too low deteriorates the cohesive force and thus, decreases the adhesion strength. Therefore, the elastic modulus needs to be optimized to improve the adhesion properties.

In addition to the methods listed earlier, other factors such as curing conditions and IC packaging structure may also affect the adhesion strength of conductive adhesives.

11.3.5 Reliability of Conductive Joints

Conductive adhesives are of little value in electronic packaging unless they can survive the rigors of testing, which modules or boards receive. To test the reliability of joints formed using conductive adhesives, a film of adhesive about 100-μm thick was laminated between two copper substrates. The adhesive film was surrounded by a 100-μm thick layer of glass cloth-reinforced dielectric (prepreg). The surrounding 100-μm thick dielectric helped to maintain proper adhesive thickness during lamination. Reliability

Table 11.3 Reliability test results

Tests	Silver adhesive	Cu adhesive	LMP adhesive
IR reflow (3X reflows at 225°C)	Passed	Passed	Passed
PCT (4 h) (121°C/100%RH)	Passed	Passed	Passed
Solder shock (15 s dip at 260°C)	Passed	Passed	Passed
IR reflow + PCT + Solder shock (3X reflows + 121°C/100%RH + 15 s solder dip at 260°C)	Passed	Passed	Passed
IR reflow + solder (3X reflows + 15 s solder dip at 260°C)	Passed	Passed	Passed

of the laminate was ascertained by pressure cooker test (PCT), solder shock, and IR reflow. For PCT, samples were exposed to 100% humidity with a constant pressure of 19 PSI at 121°C. Table 11.3 summarizes test results. Samples were stable after reliability test, and there was no delamination after PCT, solder shock, and IR reflow. Laminates were also exposed to PCT (4 h) followed by 15 s of solder dip at 260°C.

11.4 Microvia Hole-Fill Study

Reliable metal–epoxy adhesives were used for microvia fill applications to fabricate Z-axis interconnections in laminates. Figure 11.12 shows LMP, Cu, and Ag adhesive-filled microvias as representative examples. Holes having a diameter of roughly 55 μm, with an aspect ratio of about 3:1, were filled with different pastes. All pastes had continuous connection from top to bottom. LMP melted and grew as a big grain and separate organics (black regions). Cu and Ag both maintained their particle–particle connection mechanism and also maintained paste uniformity in the holes. Thus, it is possible to make a wide variety of conductive adhesives that can be used for Z-axis electrical interconnection in electronic packages.

Reliability of conductive joints in the test vehicle was further examined by 1,000 cycles of deep thermal cycling (DTC), IR reflow (3X, 225°C), PCT, and solder shock. No intrinsic failure mechanisms were observed. There was no delamination. Conductive joints are stable even after multiple IR reflow (3X) followed by multiple (3X, 15 s) solder dips.

11.5 Case Study: Test Vehicle with Filled Microvias for Z-Axis Interconnection

11.5.1 Core Fabrication

Integral to the methodology described in this chapter is the use of core building blocks that can be laminated in a manner such that electrical interconnection between adjacent cores is achieved. The cores can be structured to contain a variety

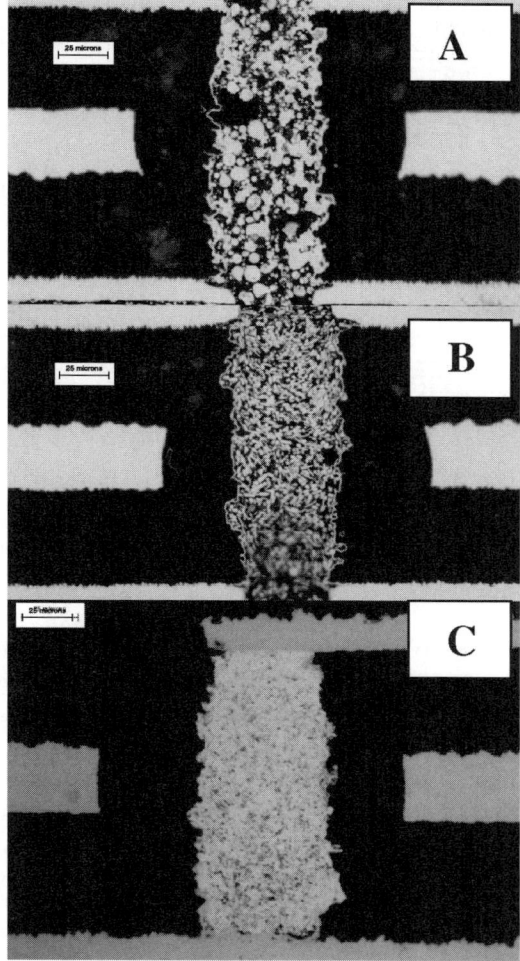

Fig. 11.12 Adhesive-filled joining cores (**a**) LMP, (**b**) Cu, and (**c**) silver

of arrangements of signal, voltage, and ground planes. In addition, signal, voltage, and ground features can reside on the same plane. As a case study, this z-interconnection methodology was used to fabricate a package for a flip chip device having a pad pitch of 150 μm. Two basic building blocks are used for this case study (Fig. 11.13). One is a 2S/1P core. The power plane (P), a 35-μm thick copper foil, is sandwiched between two layers of a PTFE-based dielectric. The PTFE is filled (60% by weight) with silica particles to achieve a reduced CTE for the dielectric material, more closely matched to that of copper. PTFE is used because of its favorable electrical, mechanical, and thermal properties. The dielectric constant and loss tangent of the silica-filled PTFE at 10 GHz are 2.7 and 0.003, respectively.

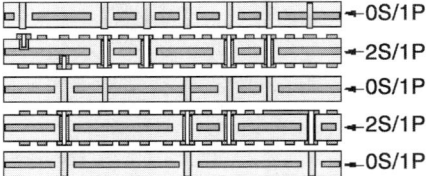

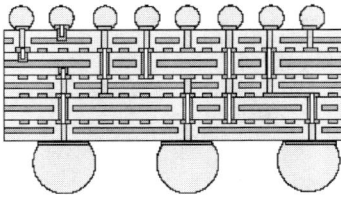

Fig. 11.13 Parallel lamination of subcomposites (cores) to form a laminate chip carrier having four signal wiring planes with a stripline transmission line structure

The signal (S) layers comprise copper features generated using a semiadditive (pattern plating) process. A line thickness of 12 µm was achieved with minimum dimensions for line width and space of 25 µm each. Minimum land-to-line spacing was also 25 µm. Laser-drilled through vias had a diameter of 40 µm while blind vias were laser-drilled with a 50-µm diameter. The latter diameter was selected to avoid the need to plate blind vias having an aspect ratio greater than 1:1. The diameter of plated pads around the through and blind vias was 75 µm. These dimensions enabled wiring designs having one line per channel in the most densely populated areas of the chip site.

The second building block in this case study is a 0S/1P core, or "joining core." This core is constructed using a copper power plane, 35-µm thick, sandwiched between layers of a dielectric material composed of a silica-filled allylated polyphenylene ether (APPE) polymer. Dielectric constant of this material is 3.23 at 1 GHz and the dielectric loss tangent is 0.003 at 1 MHz. Through holes in the core are filled with an electrically conductive adhesive (as seen in Fig. 11.12). Figure 11.14 shows a process flow chart for fabrication of the 0S/1P cores.

11.5.2 Composite Lamination

By alternating 2S/1P and 0S/1P cores in the lay-up prior to lamination, the conductive paste electrically connects copper pads on the 2S/1P cores that reside on either side of the 0S/1P core. Two signal layers are added to the composite structure each time one adds an additional 2S/1P core and an additional 0S/1P core. A structure

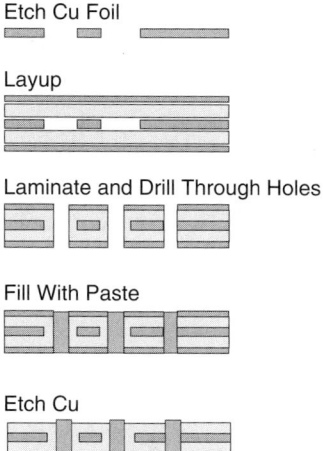

Fig. 11.14 Schematic process flow for fabrication of a 0S/1P joining core

with four signal layers composed of five subcomposites (two 2S/1P cores and three 0S/1P cores) is shown schematically in Fig. 11.13. Although this particular construction comprises alternating 2S/1P and 0S/1P cores, it is possible to place multiple 0S/1P cores adjacent to each other in the stack.

The adhesive-filled joining cores were laminated with circuitized subcomposites to produce a composite structure. High temperature/pressure lamination was used to cure the adhesive in the composite and provide Z-interconnection among the circuitized subcomposites. Figure 11.15 shows optical photographs and SEM micrographs, taken prior to composite lamination, of a joining core having paste-filled holes with a diameter of 55 μm. It can be seen that the conductive adhesive height is higher than that of the surrounding dielectric (see Fig. 11.15d). This excess height helps to produce robust conductive joint between two 2S/1P cores during lamination process. Photographs of a composite laminate structure are shown in cross section in Fig. 11.16.

Special processing of the metal surfaces on 2S/1P cores and the conductive paste in the 0S/1P cores is required to provide robust and reliable joints with the conductive paste. Figure 11.17a shows photos of a cross section taken from a laminate for which metal surfaces were not treated to provide robust mating with the paste. Following exposure of the laminate to four cycles of conditions that simulated solder reflow temperatures, the paste separates from the opposing metal surface. Figure 11.17b illustrates the result after similar temperature cycling for metal surfaces that have been treated. These joints are very robust.

Figure 11.18 (*top*) shows a SEM micrograph of a portion of a via stack, in cross section, from which the dielectric surrounding the plated via in the 2S/1P core and the paste column in the 0S/1P core had been removed using a CO_2 laser. The bottom micrograph in Fig. 11.18 shows the metal pad, 75-μm diameter, after having been removed from the laminated composite. The conductive paste column in contact with this land was fractured after removal. The fracture occurred within the paste,

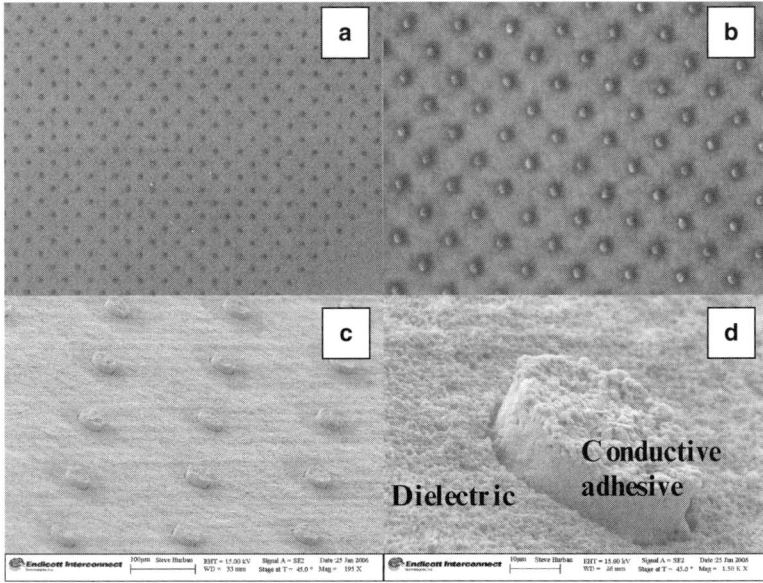

Fig. 11.15 Photographs of adhesive-filled joining cores: (**a**) large-area optical photograph, (**b**) higher magnification optical photograph, (**c**) corresponding low-magnification SEM micrograph, and (**d**) higher magnification SEM micrograph

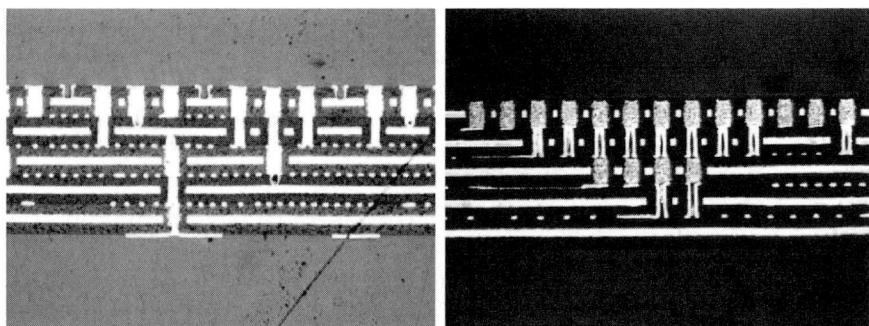

Fig. 11.16 Photograph of z-interconnect laminates shown in cross section

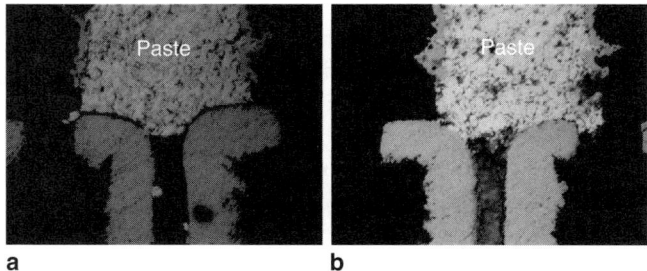

Fig. 11.17 Photographs of cross sections taken from temperature-cycled metal–paste joints for which the metal mating surfaces were untreated (**a**) and treated (**b**)

11 Nanoparticles in Microvias

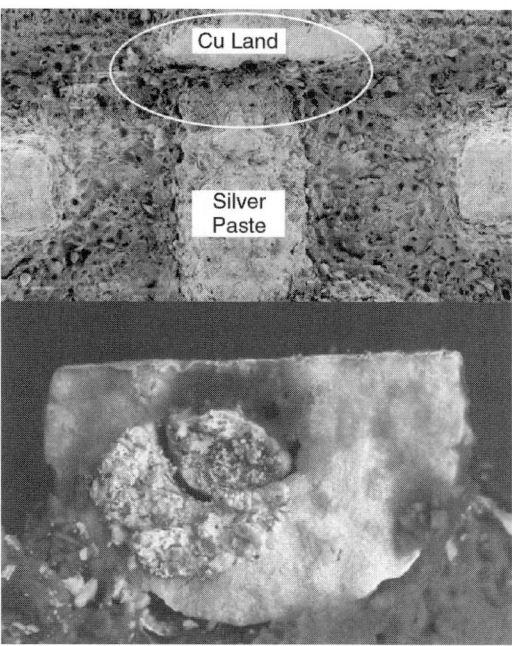

Fig. 11.18 SEM micrograph showing cross section of paste-to-land joint in via stack (*top*). The dielectric surrounding the stack had been removed from the cross section using a CO_2 laser. A SEM micrograph of the joint circled in the *top* photo after removal of the via stack from the sample, and fracture of the paste column is shown in the *bottom* photo. Silver paste remained on the land after fracture

indicating that the joint between the metal and the paste had extremely good mechanical integrity.

The *z*-interconnect package technology in tkhis study uses a high-performance material set to yield excellent reliability and electrical performance [47, 92, 93]. It also offers excellent escape density and wireability. In addition to making wiring channels available, this *z*-interconnection technology reduces losses for high-speed signals. PTHs have been replaced with blind and buried vias, reducing or eliminating stubs (Fig. 11.19).

11.5.3 Reliability of Paste-Filled Microvias in the Final Package

For the laminate in the earlier example, the average CTE of the composite is 18.3 ppm/°C. This is comparable to that of copper, 17 ppm/°C, whereas the CTE of the silica-filled PTFE is 25 ppm/°C, and that of silica-filled APPE is 41 ppm/°C. It is apparent that the CTE of the laminate structure is dominated by that of the copper planes in the composite cross section.

The test vehicle was a chip carrier having a flip chip die pad pitch of 150 μm. The die size was 9.3 mm^2 owing to the limitation of the BGA I/O (pitch and substrate

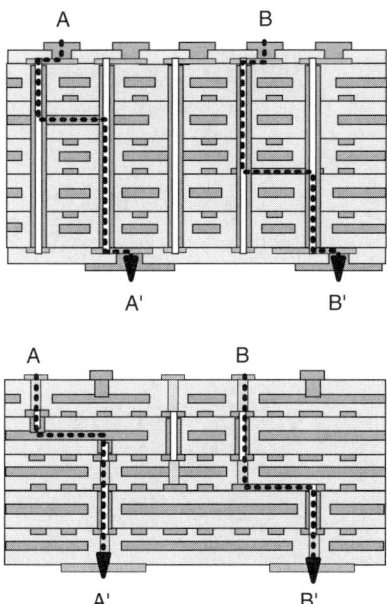

Fig. 11.19 Ability to terminate vias at any internal layer (*bottom*) provides additional channels for wiring and reduces or eliminates stubs associated with PTHs (*top*)

body size) and the die pad pitch. The body size (package outer dimensions) was 52.5 mm² with a 1-mm BGA pad pitch.

Assembled components (chip on composite) were subjected to JEDEC level 4 preconditioning per the following conditions:

Five cycles from −40 to +60°C
Twenty-four hours at 125°C
Ninety-six hours at 30°C and 60% RH
3X reflows, 225°C peak

Components were then subjected to environmental stress testing using the tests and conditions outlined in Table 11.4. No intrinsic failure mechanisms were observed. There was no die cracking, underfill delamination, BGA ball fatigue, dielectric cracking, or delamination.

11.6 Conclusion

Nanoparticles and nanoparticle-based adhesives have shown remarkable advantages and are attractive for use in microvia fill applications. A variety of nano- and microfilled Cu, silver, and LMP-based conducting adhesives can be used in microvias

Table 11.4 Stress testing conditions

Test	Conditions	Duration
Accelerated thermal cycling (component on card with heat sink)	0–100°C BLR	3,600 Cycles
Deep thermal cycling (DTC)	−40 to 125°C	1,000 Cycles
High-temperature storage	150°C	1,000 h
Pressure cooker test (PCT)	121°C/100%RH/2 atm	96 h
Wet thermal shock	−55 to 125°C	1,000 Cycles

for Z-axis interconnection. High aspect ratio, small diameter (~55 μm) holes have been successfully filled. Reports show that silver-filled adhesives are electrically and mechanically better than Cu- and LMP-filled adhesives. Nanoparticle-based adhesives exhibit sintering at lower temperatures than are required for sintering of microparticle-based adhesives, resulting in higher electrical conductivity. All adhesives maintain high tensile strength even after 1,000 cycles of DTC testing. Conductive joints were stable after 3X IR reflow, 1,000 cycles of DTC, PCT, and solder shock. Adhesive-filled joining cores were laminated with circuitized subcomposites to produce a composite structure. High temperature/pressure lamination was used to cure the adhesive in the composite and provide stable, reliable nanoparticle-filled microvia-based z-interconnections among the circuitized subcomposites.

A high-performance z-interconnect package can be provided, which meets or exceeds JEDEC level requirements if specific materials, design, and manufacturing process requirements are met. Proper lamination process settings, core metallurgy treatment, and selection of proper joining metallurgy result in an excellent package that can be used in single and multichip applications. By designing an organic package without electrical stubs and without through holes, high wiring density and excellent electrical performance can be achieved. Novel means of providing vertical electrical interconnection in organic substrates can help semiconductor packaging keep pace with the needs of the semiconductor marketplace.

References

1. Atluri VP, Mahajan RV, Patel PR, Mallik D, Tang J, Wakharkar VS, Chrysler GM, Chiu C-P, Choksi GN, Viswanath RS (2003) Critical aspects of high-performance microprocessor packaging. MRS Bulletin, 28(1):21–34.
2. McDermott BJ, Tryzbiak S (1997) The practical application of photo-defined micro-via technology. SMI Proceedings, San Jose, CA, pp. 199–207.
3. Singer AT (1999) Microvia cost modeling. Proceedings from IPC Works, Washington, DC, p. S-14–2.
4. Numakura DK, Dean SE, McKenney DJ, DiPalermo JA (1999) Micro hole generation processes for HDI flex circuit. HDI EXPO'99, Mesa, AZ, pp. 443–450.
5. Reboredo L (1999) Microvias: a challenge for wet processes. IPC Expo 99, Long Beach, CA, p. S12–1.

6. Schmidt W (1999) High performance microvia PWB and MCM applications. IPC Expo 99, Long Beach, CA, p. S17–5.
7. Alcoe D, Blackwell K, Youngs T (2000) Qualification results of hyper-BGA, IBM's high performance flip chip organic BGA chip carrier. Semicon West Conference, San Jose, CA.
8. Illyefalvi-Vitez Z, Ruszinko M, Pinkola J (1998) Recent advancements in MCM-L imaging and via generation by laser direct writing. Proceedings of 48th Electronic Components and Technology Conference, Seattle, WA, pp. 144–150.
9. Moser D (1997) Sights set on small holes? How to get there with lasers. Printed Circuit Fabrication, 20–22.
10. Owen M, Roelants E, Van Puymbroeck J (1997) Laser drilling of blind holes in FR4/glass. Circuit World, 24:45.
11. Schaeffer RD (1998) Laser microvia drilling: recent advances. CircuiTree, 12:38.
12. Young T, Polakovic F (1999) Thermal reliability of laser ablated microvias and standard through-hole technologies. IPC Expo 99, Long Beach, CA, p. S17.
13. Das RN, Lauffer JM, Egitto FD (2006) Electrical conductivity and reliability of nano- and micro-filled conducting adhesives for Z-axis interconnections. Proceedings of 56th Electronic Components and Technology Conference, IEEE, Piscataway, NJ, pp. 112–118.
14. Egitto FD, Krasniak SR, Blackwell KJ, Rosser SG (2005) Z-axis interconnection for enhanced wiring in organic laminate electronic packages. Proceedings of 55th Electronic Components and Technology Conference, IEEE, Piscataway, NJ, pp. 1132–1138.
15. Kang SK, Buchwalter SL, LaBianca NC, Gelorrme J, Purushothaman S, Papathomas K, Poliks M (2001) Development of conductive adhesive materials for via fill applications. IEEE Transactions on Component and Packaging Technologies, 24:431–435.
16. Curcio BE, Egitto FD, Japp RM, Miller TR, Nguyen M-QT, Powell DO (2004) Method and structure for small electrical Z-axis electrical interconnections. US Patent 6,790,305.
17. Egitto FD, Farquhar DS, Markovich VR, Poliks MD, Powell DO (2004) Multilayered interconnect structure using liquid crystalline polymer dielectric. US Patent 6,826,830.
18. Gonzalez CG, Wessel RA, Padlewski SA (1999) Epoxy-based aqueous processable photo dielectric dry film and conductive viaplug for PCB build-up and IC packaging. IEEE Transactions on Advanced Packaging, 22(3):385–390.
19. Lasky R (1998) New PCB technologies emerge for high density interconnect. Electronic Packaging of Products, 75.
20. Liu J (1999) Conductive adhesives for electronics packaging. Electrochemical Publications, British Isles, pp. 317–320.
21. Liu J, Rorgren R, Ljungkrona L (1995) High volume electronics manufacturing using conductive adhesives for surface mounting. Journal of Surface Mount Technology, 8:30–41.
22. Murray CT, Rudman RL, Sabade MB, Pocius AV (2003) Conductive adhesives for electronic assemblies. MRS Bulletin, 28:449–454.
23. Inoue M, Suganuma K (2006) Effect of curing conditions on the electrical properties of isotropic conductive adhesives composed of an epoxy-based binder. Soldering & Surface Mount Technology, 18:40–45.
24. Li Y, Wong CP (2006) Recent advances of conductive adhesives as a lead-free alternative in electronic packaging: materials, processing, reliability and applications. Materials Science and Engineering Reports, 51:1–35.
25. Coughlan FM, Lewis HJ (2006) A study of electrically conductive adhesives as a manufacturing solder alternative. Journal of Electronic Materials, 35(5):912–921.
26. Ye L, Lai Z, Liu J, Tholen A (1999) Effect of Ag particle size on electrical conductivity of isotropically conductive adhesives. IEEE Transactions on Electronics Packaging and Manufacturing, 22:299–302.
27. Yasuda K, Kim JM, Rito M, Fujimoto K (2003) Joining mechanism and joint property by polymer adhesive with low melting alloy filler. International Conference on Electronic Packaging, Tokyo, Japan, pp. 149–154.

28. Yasuda K, Kim JM, Yasuda M, Fujimoto K (2003) New process of self-organized interconnection in packaging by conductive adhesive with low melting point filler. International Conference on Solid State Devices and Materials, Tokyo, Japan, pp. 390–391.
29. Yasuda K, Kim JM, Fujimoto K (2003) Adhesive joining process and joint property with low melting point filler. Third International IEEE Conference on Polymer and Adhesives in Microelectronics and Photonics, Montreux, Switzerland, pp. 5–10.
30. Yao Q, Qu J (2002) Interfacial versus cohesive failure on polymer–metal interfaces in electronic packaging—effects of interface roughness. ASME Jouranl of Electronic Packagaging, 124:127–134.
31. Bueche F (1972) Electrical resistivity of conducting particles in an insulating matrix. Journal of Applied Physics, 43:4837–4838.
32. Springet BE (1973) Conductivity of system of metallic particles dispersed in an insulating medium. Journal of Applied Physics, 44:2925–2926.
33. Nicodemo L, Nicolais L, Romeo G (1978) Temperature effect on electrical resistivity of metal polymer composites. Polymer Engineering and Science, 18:293–298.
34. Bhattacharya SK, Chaklader ACD (1982) Review on metal-filled plastics and electrical conductivity. Polymer–Plastics Technology and Engineering, 19:21–51.
35. Lyons AM (1991) Electrical conductivities and effect of particle composition and size distribution. Polymer Engineering and Science, 31:445–450.
36. Lee BL (1992) Electrically conductive polymer composites and blends. Polymer Engineering and Science, 32:36–42.
37. Guerrero C, Aleman C, Garza R (1997) Conductive polymer composites. Journal of Polymer Engineering, 17:95–110.
38. Weber M, Kamal MR (1996) Estimation of the volume resistivity of electrically conductive composites. Polymer Composites, 18:711–725.
39. Bachtold A (2001) Logic circuits with carbon nanotube transistors. Science, 294:1317.
40. Duesberg GS (2003) Growth of isolated carbon nanotubes with lithographically defined diameter and location. Nano Letters, 3:257.
41. Ramanathan K, Bangar MA, Yun MH, Chen WF, Mulchandani A, Myung NV (2004) Individually addressable conducting polymer nanowires array. Nano Letters, 4:1237–1239.
42. Chung HJ, Jung HH, Cho YS, Lee S, Ha JH, Choi JH, Kuk Y (2005) Cobalt-polypyrrole-cobalt nanowire field-effect transistors. Applied Physics Letters, 86:213113.
43. Zhang Y, Gu C (2006) Optical trapping and light-induced agglomeration of gold nanoparticle aggregates. Physical Review B, 73:165405.
44. Tseng RJ, Ouyang J, Chu CW (2006) Nanoparticle-induced negative differential resistance and memory effect in polymer bistable light-emitting device. Applied Physics Letters, 88:123506.
45. Hu MS, Chen HL, Shen CH (2006) Photosensitive gold-nanoparticle-embedded dielectric nanowires. Nature Materials, 5:102–106.
46. ShenharR, Norsten TB, Rotello VM (2005) Polymer-mediated nanoparticle assembly: structural control and applications. Advanced Materials, 17:657–669.
47. Chou YI, Chen CM, Liu WC (2005) A new Pd-InP Schottky hydrogen sensor fabricated by electrophoretic deposition with Pd nanoparticles. IEEE Electron Device Letters, 26:62–65.
48. Tondelier D, Lmimouni K, Vuillaume D (2004) Metal/organic/metal bistable memory devices. Applied Physics Letters, 85:5763–5765.
49. Wu JH, Ma LP, Yang Y (2004) Single-band Hubbard model for the transport properties in bistable organic/metal nanoparticle/organic devices. Physical Review B, 69:115321.
50. Quidant R, Girard C, Weeber JT (2004) Tailoring the transmittance of integrated optical waveguides with short metallic nanoparticle chains. Physical Review B, 69:085407.
51. Hutter E, Cha S, Liu JF (2001) Role of substrate metal in gold nanoparticle enhanced surface plasmon resonance imaging. Journal of Physical Chemistry B, 105:8–12.
52. Whitesides GM, Mathias JP, Setu CT (1991) Science, 254:1312.
53. Das RN (2005) Preparation of free-standing PZT and gold nanoparticles. Materials Research Society Symposium Proceedings, 879:Z10.19.

54. Siegel RW (1991) Cluster-assembled nanophase materials. Annual Review of Materials Science, 21:559–578.
55. Siegel RW (1990) Nanophase materials assembled from atomic clusters. MRS Bulletin, 15:60–67.
56. Blanchet GB, Loo Y-L, Rogers JA, Gao F, FincherCR (2003) Large area, high resolution, dry printing of conducting polymers. Applied Physics Letters, 82:463–465.
57. Lui CJ, Oshima K, Shimomura M, Miyauchi S (2005) All polymer PTC devices: temperature-conductivity characteristics of polyisothianaphthene and poly(3-hexylthiophene) blends. Journal of Applied Polymer Science, 97:1848–1854.
58. Ramesh S, Huang C, Liang S, Giannelis EP (1999) Integral thin film capacitors: interfacial control and implications for their fabrication and performance. Proceedings of 49th Electronic Components and Technolnogy Conference (ECTC'99), San Diego, CA, pp. 99–104.
59. Banhegyi G (1986) Comparison of electrical mixture rules for composites. Colloid and Polymer Science, 26:1030–1050.
60. Lestriez B, Maazouz A, Gerard JF, Sautereau H, Boiteux G, Seytre G, Kranbuehl DE (1998) Is the Maxwell–Sillars–Wagner model reliable for describing the dielectric properties of a core-shell particleepoxy system? Polymer, 39:6733–6742.
61. Lee HH, Chou K-S, Huang K-C (2005) Inkjet printing of nanosized silver colloids. Nanotechnology, 16:2436–2441.
62. Fuller S, Wilhelm EJ, Jacobson JM (2002) Ink-jet printed nanoparticle microelectromechanical systems. Journal of Microelectromechical Systems, 11:54–60.
63. Bieri NR, Chung J, Haferl SE, Poulikakos D, Grigoropoulos CP (2003) Microstructuring by printing and laser curing of nanoparticle solutions. Applied Physics Letters, 82:3529–3531.
64. Chung J, Ko S, Bieri NR, Grigoropoulos CP, Poulikakos D (2004) Microconductors by combining laser curing and printing of gold nanoparticle inks. Applied Physics Letters, 84:801.
65. Bieri NR, Chung J, Poulikakos D, Grigoropoulos CP (2004) Manufacturing of nanoscale thickness gold lines by laser curing of a discretely deposited nanoparticle suspension. Superlattices and Microstructures, 35:437.
66. Szczech JB, Megaridis CM, Zhang J, Gamota DR (2004) Ink jet processing of metallic nanoparticle suspensions for electronic circuitry fabrication. Microscale Thermophysical Engineering, 8:327–339.
67. Szczech JB, Megaridis CM, Gamota DR, Zhang J (2002) Fine-line conductor manufacturing using advanced drop-on-demand PZT printing technology. IEEE Transactions on Electronic Packaginging and Manufacturing, 25:26–33.
68. Haiping W, Xijun W, Lui J, Zhang G, Wang Y, Zeng Y, Jing J (2006) Development of a novel isotropic conductive adhesive filled with silver nanowires. Journal of Composite Materials, 40:1961–1969.
69. Zhao Y, Tong T, Delzeit L, Kashani A, Meyyappan M, Majumdar A (2006) Interfacial energy and strength of multiwalled-carbon-nanotube-based dry adhesive. Journal of Vacuum Science & Technology B, 24:331.
70. Vigolo B, Coulon C, Maugey M, (2005) An experimental approach to the percolation of sticky nanotubes. Science, 309(5736):920–923.
71. Cheng WT, Chin YW, Lin CW (2005) Formulation and characterization of UV-light curable electrically conducting pastes. Journal of Adhesion Science and Technology, 19:511–525.
72. Goh CF, Yu H, Yong SS, Mhaisalkar SG, Boey FY, Teo PS (2006) The effect of annealing on the morphologies and conductivities of sub-micrometer sized nickel particles used for electrically conductive adhesive. Thin Solid Films, 504(1–2):416–420.
73. Iwasa Y (1997) Conductive adhesive for surface mount devices. Electronic Packagaging and Production, 11:93.
74. Jeong WJ, Nishikawa H, Itou D, Takemoto T (2005) Electrical characteristics of a new class of conductive adhesive. Materials Transactions, 46(10):2276–2281.
75. Lee HH, Chou KS, Shih ZW (2005) Effect of nano-sized silver particles on the resistivity of polymeric conductive adhesives. International Journal of Adhesion and Adhesives, 25(5):437–441.

76. Chiang H-W, Chung C-L, Chen L-C, Li Y, Wang CP, Fu S-L (2005) Processing and shape effects on silver paste electrically conductive adhesives (ECAs). Journal of Adhesion Science and Technology, 19:565–578.
77. Huang WS, Khandros I, Saraf R, Shi L (1991) Solder/polymer composite paste and method. US Patent 5,062,896.
78. Gallagher C, Matijasevic G, Capote M (1995) Transient liquid phase sintering conductive adhesives. Proceedings of Surface Mount Techolnology International Conference, San Jose, CA, p. 568.
79. Gallagher C, Matijasevic G, Maguire JF (1997) Transient liquid phase sintering conductive adhesives as solder replacements. Proceedings of 47th Electronic Components and Technology Conference, San Jose, CA, p. 554.
80. Jiang HJ, Moon KS, Lu JX, Wong CP (2005) Conductivity enhancement of nano silver-filled conductive adhesives by particle surface functionalization. Journal of Electronic Materials, 34(11):1432–1439.
81. Jeong WJ, Nishikawa H, Gotoh H, Takemoto T (2005) Effect of solvent evaporation and shrink on conductivity of conductive adhesive. Materials Transactions, 46:704–708.
82. Li Y, Moon K, Li H, Wong CP (2004) Conductivity improvement of isotropic conductive adhesives with short-chain dicarboxylic acids, Las Vegas, NV, pp. 1959–1964.
83. Li Y, Whitman A, Moon K, Wong CP (2005) High performance electrically conductive (ECAs) modified with novel aldehydes, Lake Buena Vista, FL, pp. 1648–1652.
84. Gent AN, Hamed GR, Kroschwitz JI, Mark HF, Bikales NM, Overberger CJ, Menges G (Eds.) (1985) Encyclopedia of Polymer Science and Technology, Vol. 1. Wiley, New York.
85. Morris JE, Probsthain S (2000) Investigations of Plasma Cleaning on the Reliability of Electrically Conductive Adhesives, Espoo, Finland, pp. 41–45.
86. Liong S, Wong CP, Burgoyne WF (2005) IEEE Transactions on Components Packagaging and Technology, 28(2):327–336.
87. Ramesh S, Shutzberg BA, Haung C, Gao J, Giannelis EP (2003) Dielectric nanocomposites for integral thin film capacitors: materials design, fabrication, and integration issues. IEEE Transactions on Advanced Packaging, 26(1):17–24.
88. Venables JD (1984) Adhesion and durability of metal-polymer bonds. Journal of Materials Science, 19:2431–2453.
89. Davis GD, Venables JD (1983) Surface and interfacial analysis. In: Kinloch AJ (ed.), Durability of Structural Adhesives. Applied Science, Essex, UK, p. 43.
90. Nagai A, Takemura K, Isaka K, Watanabe O, Kojima K, Matsuda I, Watanabe K (1998) Anisotropic conductive adhesive films for flip-chip interconnection onto organic substrates. Proceedings of the Second IEMT/IMC Symposium, Tokyo, Japan, pp. 353–357.
91. Watanabe I, Fujinawa T, Arifuku M, Fujii M, Gotoh Y (2004) Recent advances of interconnection technologies using anisotropic conductive films in flat panel applications: Processes, Properties and Interfaces, Atlanta, GA, pp. 11–16.
92. McBride R, Rosser SG, Nowak RP (2003) Modeling and simulation of 12.5 Gb/s on a hyper-BGA Package. IEEE/CPMT/SEMI International Electronics Manufacturing Technology Symposium, San Jose, CA, pp. 1–5.
93. Budell T, Audet J, Kent D, Libous J, O'Connor D, Rosser SG, Tremble E (2001) Comparison of multilayer organic and ceramic package simultaneous switching noise measurements using a 0.16 μm CMOS test chip. Proceedings of 50th Electronic Components and Technology Conference, Las Vegas, Nevada, pp. 1087–1094.

Chapter 12
Materials and Technology for Conductive Microstructures

Jan Felba (✉) and Helmut Schaefer

12.1 Conductive Nanosized Particles for Microelectronics

Packaging of today's miniaturized electronics is based on conductive microstructures and contacts with dimensions in the range of tens of micrometers. Such lines, patterns, or arrays of dots are possible under the condition that both special technologies and materials are applied. Since the late eighties of the last century, digital injection technology has been widely explored by the electronic industry in developing new manufacturing processes. The technology is favored mainly due to the higher processing precision as compared with traditional processes. Ink-jet technology needs a special liquid, usually termed ink, which should satisfy at least the following three requirements: very low viscosity, can be treated as a *true solvent* without component separation during high acceleration, and is able to make electrically conductive structures. If a suspension is used as the fluid for printing, the *conductivity* condition requires applying electrically conductive particles, and the true solvent demands particles with dimensions as small as possible, not higher than tens of nanometers. Ink for printing conductive microstructures is typically based on noble metals of nanosized dimensions because of their chemical inertness in ambient atmosphere and good electrical conductivity – mainly Ag, although nano-Au is also in use.

There are many methods of producing nano-Ag particles. The methods can be briefly described as *metal dissipation in plasma process*, *chemical reaction process*, *electrochemical process*, *thermal decomposition process*, *vapor condensation process*, and others. The final product must be protected from agglomeration. Because of this, only methods that yield singular nanoparticles separated from others can be used for production of nanosized particles as filler for printed formulations.

Among methods mentioned earlier and described in the literature, only a few are worth taking into consideration, because they actually are applied in nanometals manufacturing/production. One of them is the gas evaporation process. This process

---

J. Felba
Faculty of Microsystem Electronics and Photonics, Wroclaw University of Technology, Wroclaw Poland

has some advantages, such as contamination-free, narrow size distribution and broad selectivity of metals. Au particle size distribution includes dimensions from ~4.2 to ~7.0 nm with a maximum at 5 nm [1]. Similar dimensions of particles are obtained for Ag, about 7 nm [2]. The particles are stabilized and protected by the dispersing agent. They act almost like a liquid due to the stable dispersion at room temperature. As each nano particle is covered with the dispersing agent, even dispersions with a high metal content do not show any cohesion at room temperature.

One of the known ways of obtaining Ag powder with particles in the atomic size range is the release of metallic Ag from Ag salts of fatty acids during their thermal decomposition in an oxygen-free atmosphere. To obtain highly divided powders (to the size of several nanometers), it is necessary to moderate their coagulation during the production process, e.g., by a protective coating of fatty acid. This recognized fact has been applied in the technology of obtaining Ag salts of fatty acids covered with excess acid, which moderates the coagulation of the released silver during further thermal processing. During studies of thermal decomposition of silver salts/ fatty acids, it was found that depending on the manner of conducting the process the decomposition end product contained various volumes of fractions, which were insoluble in nonpolar solvents. Those materials might contain, for instance, particles of too large dimensions or not fully decomposed Ag salts of fatty acid.

Experiments conducted regarding the fabrication of nano-Ag and analysis of that process lead to conclusions regarding the determination of decomposition conditions [3]. Finally, reaction conditions have the main influence for end results of nano-Ag with average dimension of about 6 nm [3, 4]. As a standard result, very homogeneous and pure products are obtained with constant reaction parameters.

Metal nanoparticles can be prepared by laser ablation from a metal target. Such technology was applied for nano-Ag production [5]. The cleaned Ag target was placed on the bottom of a glass vessel filled with ethanol, deionized water, or acetone, which were used as the liquid environment. The liquid and its type play very important roles in the technological process. High-polar molecules provide a strong surrounding electrical double layer, which prevents growth, aggregation, and precipitation. As a result, changing the nature of the liquid environment is an easy and flexible way to control the size distribution and stability of Ag colloidal nanoparticles. Experiment with a pulsed Nd:YAG laser showed that Ag nanoparticles in acetone have a narrow size distribution with a mean size of 5 nm. In deionized water, a rather narrow size distribution with a mean size of 13 nm was observed.

Besides traditional methods of nanometal production, new methods have been developed in recent years. The first group is based on chemical reduction of Ag salts using various reducing agents, such as reduction with borohydrate in aqueous solution, reduction of Ag acetylacetonate with dimethylamine borane in the presence of a fluorinated surfactant in supercritical carbon dioxide, or reduction of Ag iodide with alkali metal in ammonia. Ag nanoparticles have also been produced using different electrochemical methods, microwave irradiation, and sonochemical synthesis. Some methods published recently are described as *new* (e.g., Ag polarization in nonaqueous solution of sodium nitrate in ethanol was investigated by means of cyclic voltammetry and chronoamperometry [6]). Methods may differ

Table 12.1 Specifications of inks for printing

Metal	Production method	Average particle size (nm)	Particle size distribution (nm)	Carrier media	References
Ag	Laser ablation	5	3–10	Acetone	[5]
Ag	Laser ablation	13	6–27	Water	[5]
Ag	Laser ablation	22	10–50	Ethanol	[5]
Au	Gas evaporation	5	3–7		[1]
Ag	Electrochemical synthesis	20		Ethanol	[6]
Ag	Gas evaporation	7			[2]
Ag	Thermal decomposition	6	4–10		[3, 4]
Ag	Thermolysis decomposition	8.8	6–22		[7]
Ag (86%)–Pd alloy		6.5			[7]

regarding efficiency, costs, environmental protection, etc., but the aim of all of those methods is to receive pure nanometal, which can be used for injection technology. Some of the results are listed in Table 12.1. It is worth noting that only dimensions of the particles obtained are commonly presented by producers. There is a lack of information about their shape or structure. Nevertheless, it is possible to detect these. By X-ray diffraction patterning, it was stated that in the case of Ag nanoparticles prepared by the controlled thermolysis of Ag carboxylate, the core metal has a face-centered-cubic (FCC) structure [7].

12.2 Nanomaterials for Printing Technologies

In the majority of work, the technology of nanofluid dispensing without direct contact between dispenser and substrate is termed "ink-jet printing." Because of this, the fluids are usually described as inks, but in fact, they are homogeneous suspensions containing metallic nanoparticles with viscosity no higher than tens of mPa s. Some producers (e.g., [2]) named their product also as nanopaste, even though its viscosity is relatively low. The name *nanoparticle paste* is legitimate when materials with metallic particles of nano dimension are used for screen printing. Such technology can also be applied for making fine pitch patterns with 30-μm resolution [7]. If a nanofluid was used for ink-jet printing, then only the term "ink" is used in this chapter, independently of the author's original terminology.

As mentioned earlier, an ink with low viscosity containing nanometals ought to behave as a true solvent. And it should be stable at room temperature for weeks without any sedimentation. Besides the two basic components, nanometal and solvent, there are also many additives, which are usually not disclosed by manufacturers. Therefore, examples of ink specifications in Table 12.2 contain only basic information.

Table 12.2 Specifications of inks for printing

Symbol	NPS-J	NPS-J-HTB	NPFS	NPG-J
Metal	Ag	Ag	Ag	Au
Particle size (nm)	3–7	3–7	1÷10	3–7
Metal content (wt%)	57–62	53–58	30	46–52
Color	Dark blue			Dark red
Diluent	Tetradecane	Tetradecane	Toluene	AF[a]
Viscosity (mPa s)	5–10	8–13	1÷2	5–10
Specific gravity[c] (g/cm^3)	1.6–2.0	1.5–1.8		1.5–1.8
Surface tension				
Sintering conditions	230°C; 60 min	350–500°C; 30–50 min	300°C; 15 min	250°C; 60 min
Resistivity (Ω cm)	3 × 10^{-6}	3 × 10^{-6}	3.5 × 10^{-5}	7 × 10^{-6}
Reference	[1]	[2] Harima	[8]	[1]

Symbol	AX NJP-6F	AG-IJ-G-100-S1	Metalon JS-011	
Metal	Ag	Ag	Ag	
Particle size (nm)	4–8		D_{50} 200; D_{95} 400	
Metal content	40–60%	20 wt%	20 wt%	
Color	Dark brown			
Diluent			Water	
Viscosity (mPa s)	4.4–15.5[b]	14		
Specific gravity[c] (g/cm^3)	1.3–1.6	1.24	1.4	
Surface tension	28.5–32.5 mN m^{-1}	31 dynes/cm		
Sintering conditions	250°C; 60 min	100–350°C; 1–30 min	100°C	Xenon lamp 300 μs
Resistivity (Ω cm)	(1÷3) × 10^{-5}	4–32 × 10^{-6}	26 × 10^{-6}	9 × 10^{-6}
Reference/Prod.	[9] Amepox	[10] Cabot	[11] NovaCentrix	

Symbol	DGP-45-LT	DGP-45-HT/ DGP-45-HTG	DGH-55-LT	DGH-55-HT/ DGH-55-HTG
Metal	Ag	Ag	Ag	Ag
Particle size (nm)	5–11	5–11	5–11	5–11
Metal content (wt%)	45	45	>50	>50
Color	Dark brown	Dark brown	Dark brown	Dark brown
Diluent	Polar solvent	Polar solvent	Nonpolar hydrocarbon solvent	
Viscosity				
Specific gravity[c]				
Surface tension				
Sintering conditions	150–300°C; 30 min	150–300°C; 30 min	150–300°C; 30 min	150–300°C; 30 min
Resistivity (Ω cm)	3.3–15 × 10^{-6}	3.3–15 × 10^{-6}	2.6 × 10^{-2} to 2.3 × 10^{-6}	2.6 × 10^{-2} to 2.3 × 10^{-6}
Reference	[12] ANP	[12] ANP	[12] ANP	[12] ANP

Symbol	DGH-(T)-50LT 25°C			
Metal	Au			
Particle size				

(continued)

Table 12.2 (continued)

Metal content	50 wt%
Color	
Diluent	
Viscosity	3 mPa s
Specific gravity[c]	
Surface tension	
Sintering conditions	250°C; 60 min
Resistivity	
Reference	[12] ANP

[a] Mixture of petroleum hydrocarbons
[b] Brookfield LVDVII + CP; 100 rpm; 20°C
[c] After sintering

For users, the further parameters of inks are important: viscosity, surface tension, resistivity of printed structures, and sintering conditions. The first two features determine the dimensions and droplet stability, which influence the pitch, while the last two set the electrical properties of printer structures.

Viscosity is a measure of the resistance of a fluid to deform under shear stress. It is commonly perceived as *thickness*, or resistance to flow. Viscosity describes a fluid's internal resistance to flow and may be thought of as a measure of fluid friction. The SI physical unit of dynamic viscosity is the Pascal second (Pa s). Viscosity μ tends to fall as temperature increases exponentially:

$$\mu(T) = \mu_0 \exp(-bT), \qquad (12.1)$$

where T is temperature and μ_0 and b are coefficients.

Surface tension is an effect within the surface layer of a liquid that causes that layer to behave as an elastic sheet. Surface tension is caused by the attraction between the molecules of the liquid by various intermolecular forces. In the bulk of the liquid, each molecule is pulled equally in all directions by neighboring liquid molecules, resulting in a net force of zero. Surface tension is measured in Newtons per meter (N m^{-1}).

By practice, it is required that ink viscosity should not be higher than tens of mPa s with its surface tension in the order of tens of mN m^{-1}.

The resistivity ρ of the printed structures, usually expressed in Ω cm, is calculated from the simple formula:

$$\rho = \frac{Rdh}{l}, \qquad (12.2)$$

where R is measured resistance, and d, h, and l are the width, thickness, and length of measured lines, respectively. Resistance is usually measured with a four-point probe, and d and l by optical microscopy. Printed structures are very thin (in the order of less than 1 μm) and additionally may have not stable structures. Because of this, the accuracy of h value measurement can be the source of significant error when resistivity is estimated. Because of this, and the current levels that may be needed, there are no absolute resistivity requirements. Nevertheless one aspires to achieve values lower than 10^{-4} Ω cm.

The low resistivity is obtained after thermal processing. It is desired to keep both the temperature and operation time as low as possible. Unfortunately, inks in use today usually need heating times of tens of minutes at temperatures higher than 200°C (Table 12.1). There is a dependence of time vs. temperature for successful thermal processes. Nevertheless, it is impossible to reach acceptable resistivity below some temperature threshold, even with very long heating times (Fig. 12.1).

It is reported [13] that there are three key steps for the nano-Ag ink preparation. In the first step organic components such as dispersant, thinner, and binder are mixed with the nano-Ag particles along with a sufficient amount of organic solvent (such as alcohol or acetone) for complete dispersion of Ag particles. An important consideration during this step is the selection of organic components with low burnout temperatures to achieve low-temperature sintering. Dispersion of the nano-Ag particles was obtained by both mechanical stirring and ultrasonic vibration. During the third step, the solvent was eventually evaporated out leaving a viscous nano-Ag paste behind. The viscosity of the formulation could be adjusted by varying the amounts of the organic components.

It is also possible to use particle-free conductive inks for ink-jet technology. A novel aqueous solution consists of Ag nitrate and additives, demonstrates excellent adherence to glass and polymers, and has an electrical resistivity only 2.9 times that of bulk Ag after curing [14]. Ag nitrate decomposes to Ag at temperatures from 440 to 500°C. Ag neodecanoate dissolved in toluene starts to decompose at 175°C, and completes the decomposition process at 250°C. The Ag trace further has to be annealed at 580°C [14]. The electrical resistivity of the final consolidated trace produced with this kind of ink is very close to that of bulk Ag,

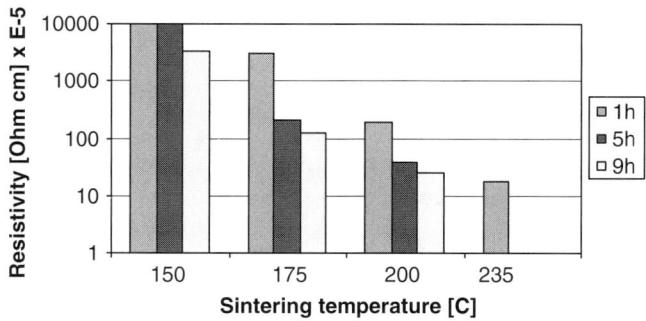

Fig. 12.1 The resistivity of printed structures vs. temperature (150, 175, 200, and 235°C) and time (1, 5, and 9 h) of sintering. For these structures the resistivity of 3×10^{-5} Ω cm was obtained after 250°C, 1 h [4]

only 1.6–2.0 times higher. In addition, the traces exhibit excellent wear and fracture resistance.

Not only Ag, but also Au microstructures can be achieved by ink decomposition. The ink containing 31 wt% Au mercaptoproprionylglycine with viscosity of 7.5 mPa s was used as the liquid for ink-jet printing, and after firing in air at a temperature close to 500°C, the Au pattern was obtained. The printed line width and thickness were estimated as 190 and 1.4 µm, respectively. The lowest resistivity is about ten times higher than that of bulk pure Au [15].

A breakthrough in application of Ag nanoink for consumer electronics needs a new formulation with a much lower temperature thermal/chemical process, which is necessary to obtain relatively high conductivity. Such requirements are met by Ag carboxylate inks containing 30–50 wt% of Ag [16]. The compounds can be obtained as solid powders after purification. This powder could be easily dissolved in typical solvents (e.g., polyvinyl pyrrolidine) or distilled water to form an ink. The decomposition temperature of the inks can be effectively controlled between 110 and 170°C. Quite good electrical resistivity of 9.0×10^{-5} Ω cm can be obtained at 160°C after 30 min of curing (on a glass substrate).

Current inks do not offer an adhesive function. Such properties would be very desirable for microelectronic packaging, especially for flip-chip technologies. There has been work on material with both functions, adhesive and conductive [17]. For the time being, the ink is saturated by micro-Ag particles of 4.2 µm average value at a filler content of 70 wt%. The acrylate-methacrylate-epoxy adhesive shows shear thinning behavior. Therefore, at shear rates below 1 s^{-1} the viscosity is higher than 1 Pa s. This supports the sedimentation stability. During the printing process, the shear rates are very high. The estimated shear rate is on the order of 10,000 s^{-1}. At a shear rate of 2,500 s^{-1} the formulation developed for ink-jet technology has a viscosity of 30 mPa s. As a result of printing, droplets of 130 µm can be produced (such as presented in Fig. 12.18 ahead). After curing, the resistivity of printed structures was measured as 6×10^{-4} Ω cm.

One can also mention that it is possible to eject molten metal drops through the nozzle of a jet printing system. The driving mechanism is similar to that in micro ink-jet printing. In an experiment [18], a fusible alloy (Bi-Pb-Sn-Cd-In alloy) with a melting point of 47°C was used. Nevertheless as the diameters of hemispherical metal dots adhered on a base plate were about 400 µm, this technique cannot be included as a nanotechnology.

It is worth knowing that adding nano-Ag to a paste actually in use (containing 3–5 µm Ag particles) improves its electrical and thermal conductivities significantly. It is reported [19] that the hybrid paste containing both silver microparticles and nanoparticles with a size of 3–7 nm in diameter shows very low electrical resistivity of 6×10^{-6} Ω cm, after thermal processing below 200°C. It was also stated that the thermal conductivity of such paste increases about 2.5 times in comparison with a conventional silver paste.

12.3 The Principle and Equipment for Ink-Jet Printing

Generally, making conductive microstructures and connections of complicated configurations at the microscale requires that dispensed dots have diameters in the range of less than 1 µm to not higher than 200 µm, depending on the application. The larger value equates to a point volume of less than 1 nl [20], while the small one is a few pl. A spherical drop with volume of 2 pl has diameter of 16 µm [21].

Microdispensing systems have been well known for many years and are used for applying liquids and adhesives. The printing method principle depends mainly on the viscosity of the printed medium. For relatively high to medium viscosity, application systems can be generally divided into mechanical displacement (rotational screw, piston, peristaltic, etc.) and compressed-air displacement systems [20]. Such equipment is mainly devoted to printing relatively large dots at dimensions of tens of micrometers. A peristaltic microdispensing system dedicated to adhesive printing makes reproducible dots with diameters of 190 µm and heights of 30–40 µm on a substrate [20]. An air pressure head for jetting of underfill materials can produce dots of SMD adhesive of ~225 µm or about 3.5-nl volume [22].

For low-viscosity printed liquids, the ink-jet method is mainly in use. Parameters of liquid drops depend on the droplet generator (also called a microdispenser). There are a few basic methods of liquid doses dispensing by ink-jet printing technology. One of the simplest seems to be deposition by valve technology. Liquid from a pressurized reservoir is pressed into the dispenser head, which consists of a microvalve and a nozzle. By opening the valve, liquid flows through the nozzle and forms a well-defined jet. The integrated microvalve switches the jet on and off (Fig. 12.2a). The special nozzle geometry forms a laminar jet, which is stable over a wide range before breaking off into droplets, and in such systems volumes down to 200 nl can be dispensed [23]. The maximum throughput is estimated as 2 ml/s provided that liquid viscosity is not higher than 50 mPa s.

The much more popular system uses a piezo actuator, which surrounds a glass capillary. The end of the capillary is formed to a nozzle with diameter in the range of a few dozen of micrometers. Applying a voltage pulse, the piezo actuator contracts and creates a pressure wave, which propagates through the glass into the liquid. In the nozzle region the pressure wave accelerates the liquid (Fig. 12.2b) with

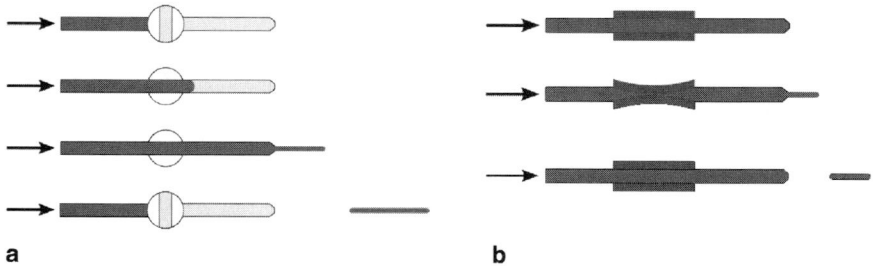

Fig. 12.2 The principle of material deposition by (**a**) valve technology and (**b**) piezo drive system [23] (courtesy of Microdrop Technologies GmbH)

up to 10^5 g [24]. A small liquid column leaves the nozzle, breaks off, and forms a droplet, which flies freely through the air (Fig. 12.3). Depending on the nozzle sizes (30–100 μm), volumes of 25 up to 500 pl (corresponding to drop diameters of 35–100 μm) are generated with drop rates up to 2,000/s and maximum throughput of 1 μl/s. Such printing efficiency is possible under the condition that liquid viscosity is in the range of 1–100 mPa s [23] for unheated systems. When the reservoir of liquid and the printing head are heated, it is possible to print ink with initial higher viscosity as it tends to diminish with increasing temperature (see 12.1).

It was calculated by numerical analysis and experimentally observed that droplet formation depends strongly on the shape of the electrical pulse applied to the piezo actuator. It was stated [25] that there are two ways to obtain smaller ejected ink droplets: by lowering the operating voltage or by varying the discharge time of ending the singular impulse, when the voltage is changing its value from maximum to zero.

Surface tension plays an important role in droplet formation. The smaller the volume, the higher the force that keeps small drops together. Liquid acceleration inside and outside of the droplet generator nozzle makes these phenomena much more complicated. Three-dimensional simulations of the droplet formation during the injection printing process show that formation of the meniscus and its shape are the results of both pressure and the surface tension, which significantly influences the generation process [26].

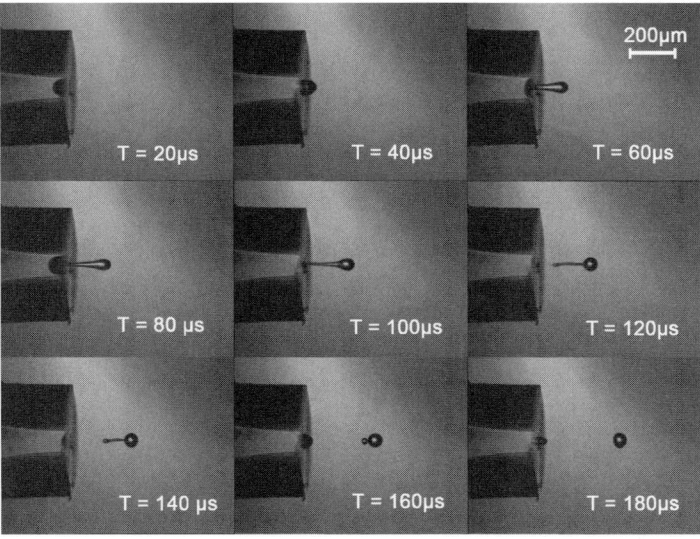

Fig. 12.3 The steps of droplet forming in ink-jet technology with the use of a piezo actuator [23] (courtesy of Microdrop Technologies GmbH)

As an example, a commonly used droplet generator can be presented. It consists of a glass capillary tube with a single 60-μm circular orifice, surrounded by a radially polarized piezoelectric ceramic crystal, and may be cited as the system used for fine-line conductor manufacturing [8]. Such a generator ejects droplets of diameter of 50–70 μm with velocity of 1–3 m s^{-1}.

The drop volume is affected by the dispenser performance and the properties of the liquid. The dispenser performance is controlled by the nozzle diameter and the driving parameters: voltage, pulse length, and – in case of a heated dispenser – also temperature. Increasing the voltage pulse on the piezo actuator causes an increase in drop speed that means an increase of the kinetic energy of the system leading to a larger drop volume. The dependence is almost linear and with changing voltage from 60–120 V, the drop speed increases more than 3.3 times [24].

There are two basic technologies of ink-jet printing, which have been used in the laboratory and industry: drop-on-demand (DOD) and continuous ink-jet (CIJ) [27]. The principles of both technologies are presented in Figs. 12.4 and 12.5. In a DOD system, one single drop is ejected through an orifice (nozzle of printing head) at a specific point in time. As a substrate moves opposite the drop's source, the combination of the speed and direction of substrate movement, as well as the frequency of *shots* or breaks, makes obtaining the required printed pattern shape possible.

Continuous ink-jet systems produce a pressurized fluid stream that is broken into droplets using a piezoelectric element. By stimulating the piezoelectric element at high frequencies (in the range of 20 Hz to 80 kHz), capillary waves are generated within the fluid domain, and the fluid stream forms continuous and consistent droplets with uniform size and spacing. The main difference in comparison to former technologies is that droplets are selectively charged as they pass through an electrode channel that is incorporated into a print head. These charged droplets can then

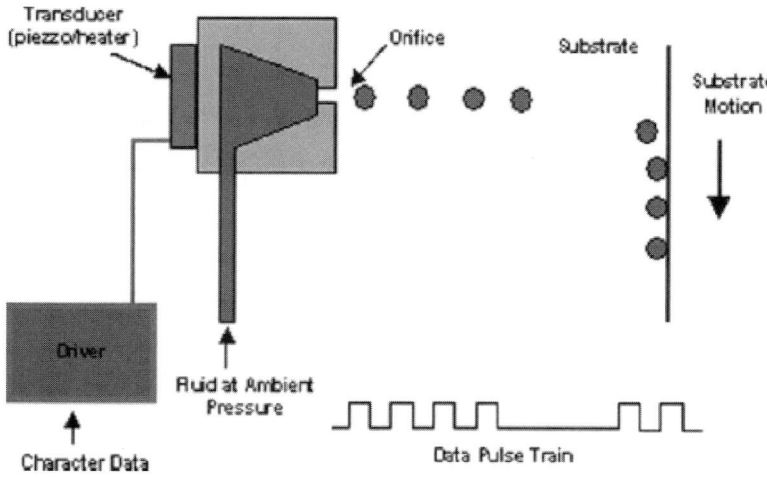

Fig. 12.4 Drop-on-demand (DOD) ink-jet printing: single droplets are ejected through an orifice at a specific point of time [14] (with permission)

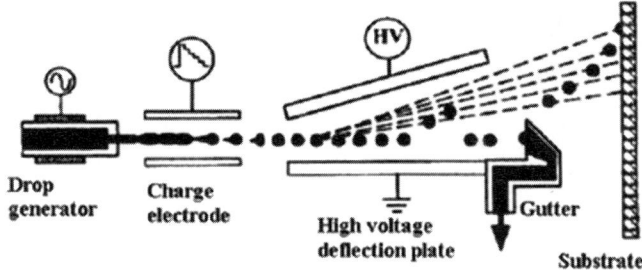

Fig. 12.5 Continuous ink-jet printing (CIJ): the drop generator works continuously and a pattern on a substrate depends on the deflection system [14] (with permission)

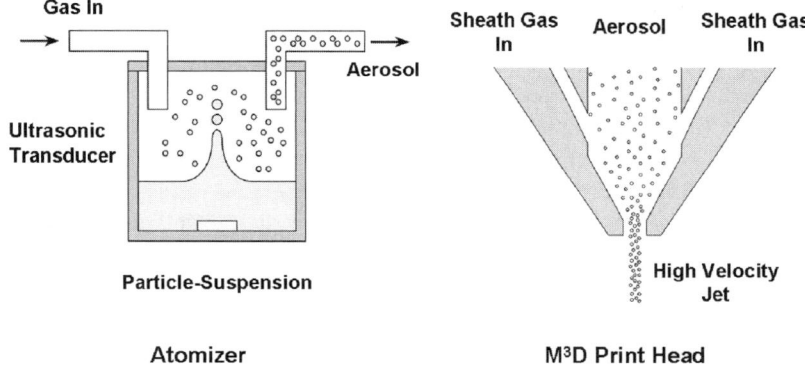

Fig. 12.6 M³D printing principle [28]

be deflected by means of high-voltage deflector plates to form various patterns on a substrate. The lack of deflector activity plays the same role as switching-off the drop generator; uncharged droplets are captured by a gutter and recirculated through the system, and there is no action on a substrate.

M³D® (Maskless Mesoscale Material Deposition) technology represents another continuously working printing technique [28]. The source material (fluid or suspension with a viscosity from ca. 1 to 1,000 mPa s) is atomized in a mist generator. The resulting aerosol stream is aerodynamically focused by an annular flow of sheath gas to as small as less than a tenth of the size of the nozzle orifice (typically 100 µm), Fig. 12.6. Together with a relatively large standoff distance (>5 mm) to the substrate this technology allows the deposition of fluids and nanomaterial formulations on nonplanar substrates even for high aspect ratio structures. After deposition the material may undergo a post treatment, e.g., with a laser process to enhance its final properties, such as electrical or thermal conductivity, adhesion to the substrate, etc. The resulting layer thickness may vary from ca. 10 nm to several micrometers, the line width from ca. 5 to 150 µm.

For 3D ink-jet printing, the continuous system seems to be better [14]. In the drop-on-demand technique, there is no external backpressure in the fluid since surface tension is used to hold the fluid inside the orifice. Without external pressure, the DOD process has a limited throw distance (less than 10 mm), which limits its ability to control the trajectory of droplets. In the case of continuous ink-jet technology, the ink droplets travel a relatively long distance (100 mm) before being deposited onto the substrate. Additionally, through the use of electrostatic forces, CIJ systems are inherently designed to control the movement of droplets once they leave the printing head nozzle, which allows greater flexibility when printing in 3D space.

Besides the two basic technologies presented earlier, there are new systems developed, which allow arrangements of dots with a smaller size, higher repeatability, and better quality of printing. There is a system that is able to print dots in the range of 1 μm [29].

The electrical field, without the deflection system (as in the CIJ system), may also be helpful for microline printing. When a liquid is supplied to a nozzle and the interface between air and the liquid is charged to a sufficiently high electrical potential (~kV), the liquid meniscus takes the form of a stable cone, whose summit emits a microscopic jet. This is referred to as the cone-jet mode in electrospray. In the system, the diameter of the nozzle used (above 100 μm) can be larger than that in *standard* ink-jet printing. The use of a larger nozzle prevents its clogging and allows easier processing of a viscous suspension containing a high level of solid particles [30]. Using an ink containing nano-Ag with the average diameter of 3–7 nm, lines as fine as 32 μm in width and 0.3 μm in thickness were printed. The line resistivity was about 13×10^{-6} Ω cm.

Active alignment control makes printing more precise. The system consists of nozzles, image sensors, and a target tracking system. Each nozzle is integrated with a microlens for an image sensor and a pair of electrodes to control ink jets according to the output of the image processing. The performance of the target detecting system was accurate enough to obtain 5-μm precision alignment. The fabricated nozzles achieved the patterning of Ag nanoparticles, which can be used as conductive wires after sintering. The width of the pattern was less than 20 μm with errors less than 3 μm [31].

12.4 Physical Processes for Increasing the Conductivity of Printed Microstructures

Microstructures made with the use of inks containing nanoparticles of metals are electrically nonconductive just after printing, because during the production process, all particles are safely dispersed without possibility of agglomeration. For this, each particle is protected by a special kind of layer, which is electrically insulating. The protection material occurs in the range of several percent of total metal mass (in the case of Ag particles with average diameter of 6 nm – not higher than 4% [3]).

As a result, this protective layer on single nano particles makes DC conduction impossible (at relatively low voltage).

To obtain good electrical conductivity, each type of initial product needs additional energy – mostly thermal by heating. The processes depend on the form of the initial products and may differ in their stages. As an example, in Fig. 12.7 the changes of the resistivity of printed lines made by ink with nano-Ag are presented with heating time [32]. The dependence was measured after the short preheating process (at about 110°C) during which solvent evaporated out of the ink formulation. In the first step of the heating, the thin protective layer is removed by evaporation. The protective layer should be very effective at low and medium temperature, but it has to be easily removed by temperatures close to the *full cure temperature*.

Parallel investigation reveals the loss of the carbon presence from EDX spectra during the thermal process. It is estimated that the conductive layer's carbon content falls from about 10% just after printing to less than 4% after the full thermal process. It seems to be clear that removing the protective layer makes printed structures as conductive as almost pure metal. Such a conclusion can be supported by the result of impedance measurement at different frequencies. Figure 12.8 presents the absolute impedances of layers after different times of heating at 250°C. The straight lines after 50 min of heating indicate the removal of the protective layer.

SEM observation of the specimen surface reveals the nucleation of pure Ag particles in the form of spheroids of diameter about 0.1 μm [4]. Probably this is due to the second step of the heating – the sintering of nanoparticles. The sintering or other

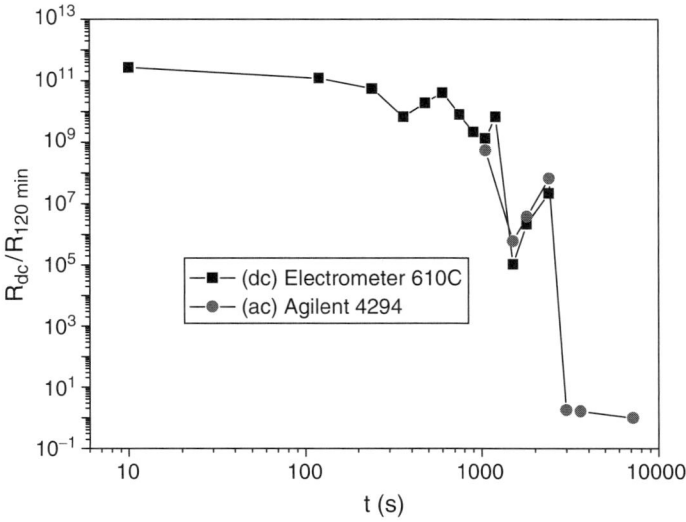

Fig. 12.7 The relative resistivity of printed structures vs. time of heating at 250°C (measured by impedance analyzer Agilent 4292A and analogue electrometer Keithley 610°C) [33]

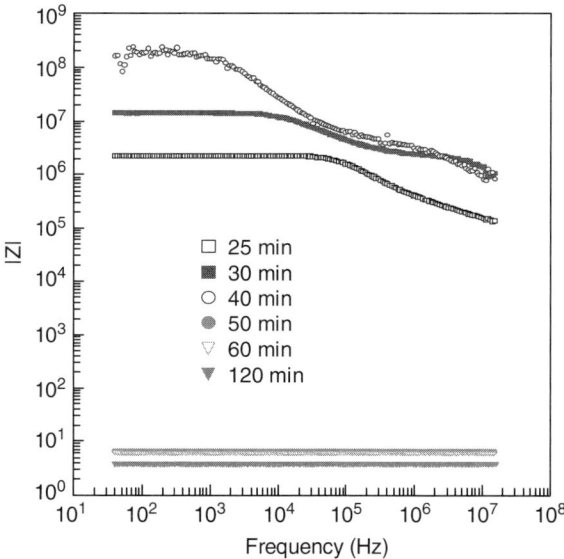

Fig. 12.8 Impedance absolute value of printed layer at different times of heating process at 250°C in the air atmosphere [33]

thermal phenomena are key to understanding the relationship between microstructural evolution, resistivity, and substrate adhesion.

In some works presenting the use of formulations containing nanosized particles for ink-jet technology, the process of achieving the high conductivity is based on the large depression of melting point below particle size of 5 nm (Fig. 12.9) [34]. By utilizing this phenomenon, the bulk metal structure has to be spontaneously obtained by heating at the temperature much lower than the bulk melting point.

Although conductive inks contain nanoparticles, the average diameters are usually much higher than 2.5–3 nm, and melting point depression cannot always be observed.

Figure 12.10 presents the microstructure of a Ag-containing formulation sintered at 280°C (the temperature ramp 20 min, sintering 10 min, and dwell 10 min) [35]. It can be seen that the structure changed into a dense network of Ag with micropores. EDX analysis shows that almost all the organic components have burned off after the thermal process [13]. It is worth noting that the values of the density, specific heat, thermal diffusivity, and thermal conductivity of the sintered Ag change in comparison with the bulk silver. The data are listed in Table 12 3.

The sintering of silver nanoparticles below 200°C is difficult. Thus, the most pertinent curing temperature range is from 220 to 250°C, as confirmed by resistivity measurement. The resistivity gradually decreases down to 1×10^{-5} Ω cm with increasing curing time at 220 and 250°C, and maintains a constant value after 30 min

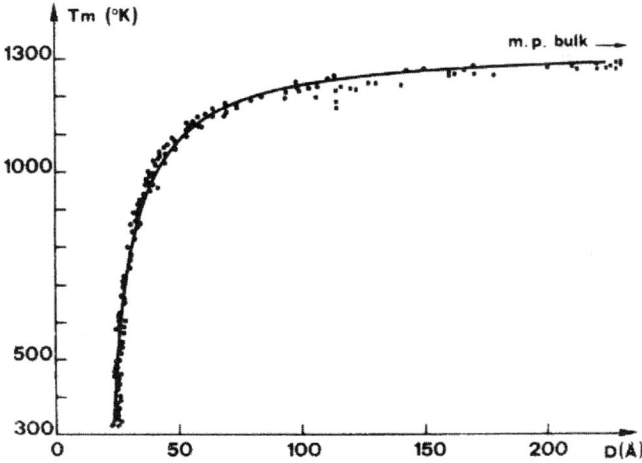

Fig. 12.9 Theoretical and experimental values of the melting point temperatures of Au particles vs. their size, in 10^{-1} nm [34] (with permission)

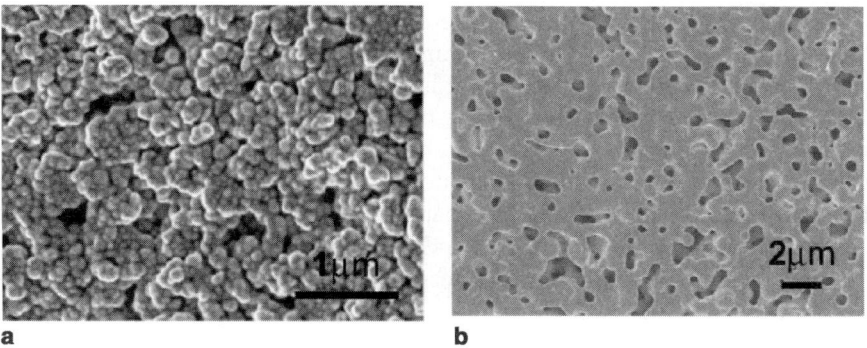

Fig. 12.10 SEM images of the nanoscale Ag (**a**) before and (**b**) after sintering on silicon substrate at 280°C [35] (courtesy of Bai et al.)

Table 12.3 Relative changes of thermal properties of sintered nanoscale Ag paste [35]

	Density (g/cm^3)	Specific heat [J/(g K)]	Thermal diffusivity (cm^2/s)	Thermal conductivity [W/(K m)]
Bulk Ag	10.5	0.235	1.74	429
Sintered Ag	8.6	0.233	1.19	240
Changes	−82%	−99%	−68%	−55%

at higher temperature [36]. The resistivity for lower temperatures is higher, and prolongation of curing time does not yield the effect (as it is shown in Fig. 12.1).

Similar changes of layer structure can be observed in the case of patterning by Cu-containing inks [1]. The sintering temperature of nonoxidized Cu particles dynamically reduces with decrease of their size. Cu nanoparticles were sintered by reductive atmosphere at more than 250°C to reach resistivity of 5×10^{-6} Ω cm at 300°C (after 15 min).

The sintering process may be accelerated by applying a high-efficiency heat source (e.g., a laser beam), which only directly affects the printed structures. Numerical modeling predicts that surface temperatures can reach more than 250°C, with the temperature just several tens of micrometers below only slightly higher than room temperature [37].

Tests with the laser beam were carried out on printed lines made with an ink consisting of 30 wt% Au particles (average diameter of the particles 2–5 nm), toluene, and a very small amount of surfactants to prevent agglomeration of nanoparticles [38–41]. A single-mode argon ion laser was used for evaporating the solvent and melting the nanoparticles and/or sintering them to form continuous, electrically conducting Au microlines on the substrate. The laser beam was focused at the printed line as close as possible to the point of droplet impact. As a result, in the middle region of the cured line the continuous film may degenerate to a fractal-like structure and finally to isolated and connected agglomerated Au particles in the order of hundred nanometers diameter size (Fig. 12.11) while the smooth layers near the edges of printed lines have good electrical conductivity.

It is also possible to achieve good electrical conduction at lower temperatures. Ag nanoparticles protected by dodecylamine, formed into paste, are successfully sintered at room temperature in air atmosphere. To remove the dodecylamine dispersant, Ag nanoparticles printed as lines on glass substrates were dipped in methanol for 10–7,200 s. Figure 12.12 shows the microstructural changes of Ag nanoparticles by the methanol dipping process. As shown in Fig. 12.12a, which is the initial state of the paste, Ag nanoparticles of average diameter of 7 nm are packed in a highly dense structure. After 3,600-s dipping, as shown in Fig. 12.12d, the Ag nanoparticles completely disappeared, and a high-density Ag structure was

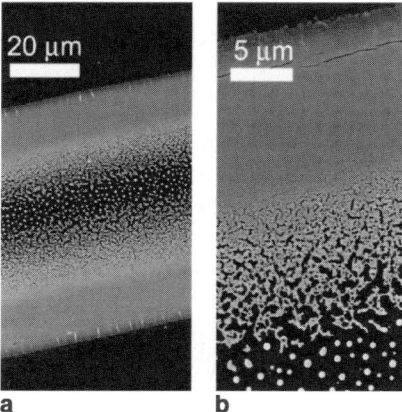

Fig. 12.11 SEM images of Au line cured with use of 300 mW power laser beam and translation substrate velocity of 1 mm/s: (**a**) whole structure and (**b**) close up view [39] (with permission)

12 Materials and Technology for Conductive Microstructures 255

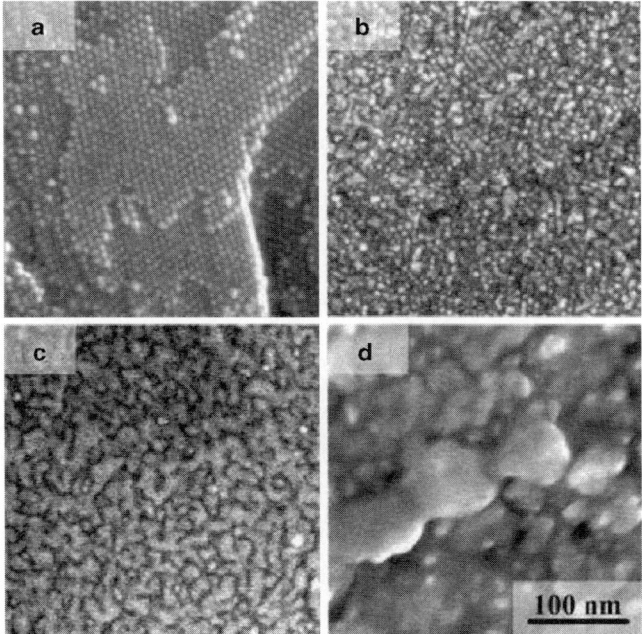

Fig. 12.12 The microstructural changes of Ag nanoparticle paste as observed by FE-SEM: (**a**) initial state, (**b**), (**c**), and (**d**) after dipping in methanol for 180, 600, and 3,600 s, respectively [42] (with permission)

achieved. As a result, the sintered wires possess excellent low resistivity, 7.3×10^{-5} Ω cm (after 7,200 s dipping) [42].

12.5 Conductive Microstructures and Contacts Using Nanosized Particles

Various types of printing macrostructures are expected to become key technologies for advanced electronics packaging. The main benefit is connected with fine pitch lines and space and easy scale-up of products. The ink-jet technology may be easily used for 3D stacking. It needs only repeating printing circuit layers over insulating layers by changing ink tanks. Vertical connection can be drawn by stacking ink drops without any contact between leads and substrates (via-holes can be filled with conductive ink drops).

In ink-jet technology, every singular ejected drop for each *shot* of printing action makes dots of highly uniform structure (Fig. 12.13 *left*). The series of printing shots makes the matrix of singular dots very regular. The shape of dots remains symmetric if the substrate moves only during the break between shots. Otherwise, if the

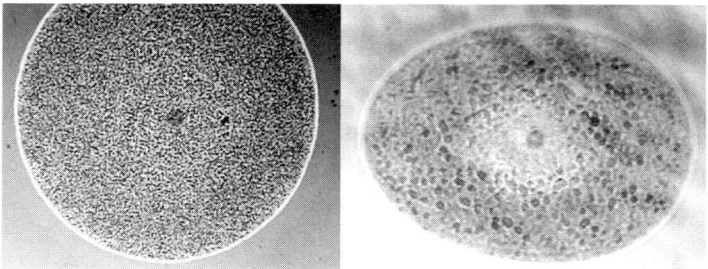

Fig. 12.13 One dot (0.23-mm diameter) of printed nano-Ag ink [3, 4]

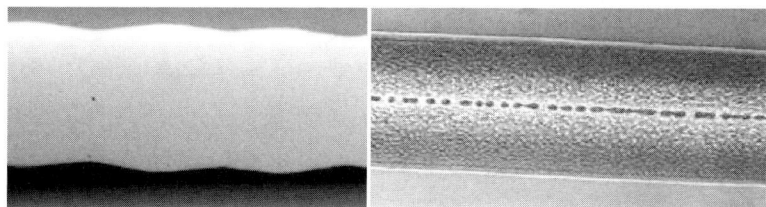

Fig. 12.14 Lines printed by nano-Ag ink [3, 4]

substrate is moving very quickly, the dot shape will be deformed, as seen in Fig. 12.13 (*right*).

The combination of shot frequency and substrate speed as well as number of passes may lead to forming printed lines with slightly deformed edges (Fig. 12.14 *left*), or – what is usually desirable – to lines with stable width and thickness (Fig. 12.14 *right*). The possibility of printing more complicated figures depends only on the control software of the *X-Y* table, which moves the substrates.

The quality of the microstructures printed on different substrate materials may differ fundamentally. Additionally, the finer the microstructure printing is the higher is the possibility of patterning errors such as bulge or bridge formation or fragmentation. Strong surface tension in the liquid is a powerful driving force to minimize surface energy by minimizing surface area, which results in these defects. There are two ways of eliminating such errors. The first one is a prepatterning of substrates by using surface treatment of hydrophobic and hydrophilic coatings. The other is a multi-pass patterning technique. In the former case, ejected droplets are self-aligned by strong surface energy on substrates, but such a solution makes direct patterning of the ink-jet process impossible. In the latter case, the objective image is formed in multiple passes by superimposing on each discrete dot's image, and its patterning resolution is limited by droplet size [43]. Such ways of avoiding defects in ink-jet technology make possible accurate and dimension-stable forms.

The ink-jet technologies using electrically conductive materials are believed to be suitable for the maskless production of various electronic devices. The *printable electronics* by ink-jet technologies makes big benefits such as the following:

12 Materials and Technology for Conductive Microstructures

- Direct printing of metal pattern on large substrates
- Manufacturing of a small quantity but many kinds of devices
- Real-time production by digital design
- Small loss of coated materials

Such benefits of reliable printing technologies are more and more attractive to flexible electronic industries. But the essential question is if such patterning made by nanometal particles can be competitive to *traditional* technologies such as conductive lithograph film Ag, thick-film Ag, thick-film Cu traces, and traditional etched Cu with regard to trace resolution, surface quality, microstructure, electrical conductivity, and – of course – to cost. The results of such comparison are presented in Table 12.4 [44].

Since there is no standard specification for acceptance of ink traces, criteria based on IPC-601311 [45] were set up to evaluate the trace resolution. An acceptable resolution should satisfy the following two requirements:

1. Deviation of the trace width from the nominal resolution is less than 20%. For example, for a nominal resolution of 75 μm, the trace width must be within the range of 60–90 μm.
2. Trace boundary is sharp and parallel, with no metal shifted out of the defined trace boundaries.

If either of the requirements is not satisfied, the printed circuit is deemed not acceptable. The acceptable resolution of examined samples is listed in Table 12.4.

Table 12.4 Pattern's features made by different technologies and materials [44]

Technology	Nano metal Ag	Lithograph film Ag	Thick-film Ag	Thick-film Cu	Etched Cu
Material	5–10-nm Ag particles with surfactant, ink	1–2-μm Ag particles with epoxy, ink	1–5-μm Ag particles with epoxy resin	0.2–1-μm Cu particles	Cu clad
Substrate	Polyimide	Polyester	Polyester/polyimide	Polyimide	Polyimide
Curing conditions	240°C, 1 h	125°C, 4 min	150°C, 3 min/320°C, 6 min	320°C, 6 min	
Nominal resolution (line/space width)	50–250 μm	50–250 μm	100–400 μm	100–400 μm	50–250 μm
Acceptable resolution (line/space width)	≥100	≥200	≥150	≥150	≥50
Resistivity	$4.5\text{–}4.9 \times 10^{-5}$ Ωcm	3.6×10^{-2} Ω cm	9.1×10^{-5} Ω cm/6.7×10^{-5} Ω cm	6.2×10^{-5} Ω cm	2.2×10^{-5} Ω cm

Those results as well as trace defects (according to [45]) and microstructures demonstrate that nano-Ag ink has shown technical advantages over the others. It has higher resolution, high conductivity, denser structure, and shining surface. Another advantage of nano-Ag over thick-film Cu is that it has higher resistance to humidity and corrosion. No cover layer is needed to protect it.

The industrial applications need special requirements with relation to Ag nanoparticle ink and technology [46]. For stable ejection of such ink, it is needed that aggregation and sedimentation of particles must be inhibited at any conditions and that the used ink does not increase the viscosity at the meniscus, which is the surface of liquid–air boundary near the nozzle of the print head. For high productivity of printable electronics, drop volume uniformity is important to using many nozzles or many heads. It is recognized that the drop velocity is proportional to drop volume, which determines the line width of printed pattern or the thickness of printed areas. The uniformity of drop volume on the level of ±1% and the drop placement accuracy of ±3 µm seem to be enough for industrial uses of ink-jet technologies.

Acceptable pitch resolution is in the range of a hundred micrometers. Such a value is reached by using conventional conductive pastes in screen printing technology. Ag nanoparticle pastes can easily form the fine pitch pattern with line and space widths less than 50 µm by screen printing. Figure 12.15 (*left*) shows an example of the fine line with line and space of 30 µm/30 µm formed on an alumina sheet after firing [7].

Finer patterns can only be achieved by ink-jet printing. Figure 12.15 (*right*) presents a set of wiring lines printed with nano-Ag ink on glass [21]. The width of the lines is about 3 µm on a 10-µm grid in the lattice area. The conductive ink contains Ag particles with an average size of 5 nm.

For such fine pitch patterns in electronic circuits, electromigration of Ag may be a problem negatively influencing long-term reliability. Migration-controllable film formation can be achieved by using Ag–Pd alloy as filler for the conductive formulation, e.g., NAGNPD8515, Table 12.1. In comparison with pure Ag, the sintering temperature is raised over 300°C by the alloy formation with Pd. After heating at this temperature for 30 min, the resistivity of printed lines on an alumina substrate is 12.1×10^{-6} Ω cm. The resistivity decreases to 8.4×10^{-6} Ω cm for 30 min sintering

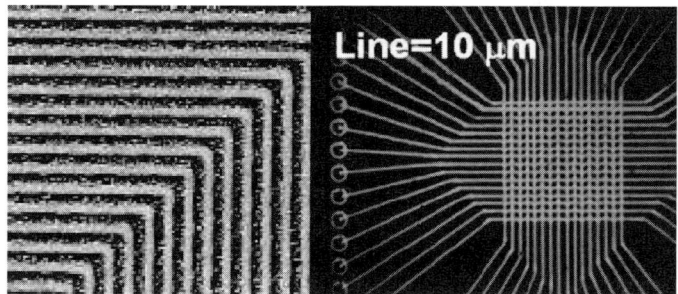

Fig. 12.15 Fine pitch pattern formation; with line and space 30 µm/30 µm formed by screen printing [7] (*left*) and by ink-jet printing (*right*) (with permission)

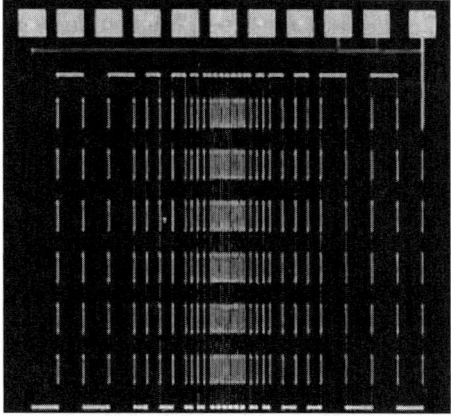

Fig. 12.16 Fragmental front view of a prototype 20-metal-layer circuit board [47] (with permission)

at 500°C. Reliability of the pattern made with use of the Ag–Pd alloy nanoparticles increased compared with the Ag nanometal, as confirmed by the dipping test [7].

Ink-jet technology can be applied to FPCs (Flexible Printed Circuits) or multi-layer circuit board production in general (e.g., direct circuit patterning based on the bit-map data transformed from CAD can be made [1]). For such production, many printing elements (nozzles) have to be used at the same time. In a printing machine [47], 2 lines of 180 nozzles each may produce multilayer circuit boards. During the first step of the production process, the metallic pattern is printed on the insulation layer, and then vertical wiring posts are formed, joining the layers of metal wiring. The second layer of insulation is made, avoiding the posts, and the second pattern layer is printed. By repeating the steps, it is possible to form multilayer circuit boards. Figure 12.16 presents the front view of a 20-metal-layer circuit board with a thickness of 200 μm. Metal wiring made with Au has a line width of 50 μm, thickness of 4 μm, and minimal line pitch of 110 μm.

So-called superfine inject technology [43] may be applied for making 3D microstructures. It is possible because of a fast drying effect; by using fine droplets of size 1 μm printed on silicon, the life-time of the wetting area is estimated at only about 0.1 s. A pillar with high aspect ratio (Fig. 12.17) can be formed by simply ejecting superfine droplets at the same spot. The height of the printed structure is linearly proportional to the ejecting time.

The 3D printing process is not limited to creating singular pillars or bumps. By moving the ink-jet head it is possible to construct more complicated microstructures such as microplugs or microsocket arrays [43], meander inductors and parallel capacitors [48], or three-dimensional MEMS systems [49].

SnPb solder was printed for making solder bumps at 220°C in both the continuous and drop-on-demand modes using a high-temperature ink-jet printing head.

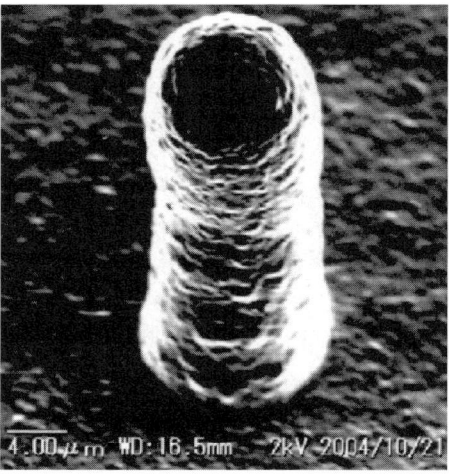

Fig. 12.17 SEM image of a gold pillar with diameter of about 6 μm fabricated by superfine inject system on silicon substrate [43] (with permission)

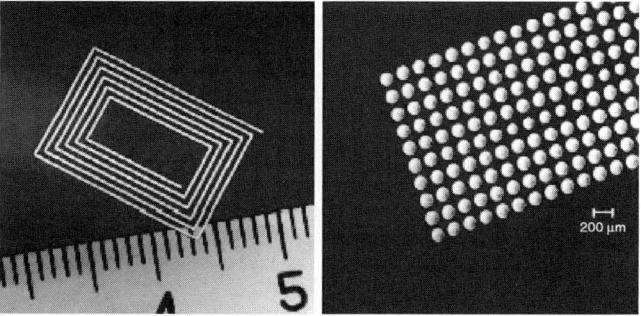

Fig. 12.18 Ink-jet printed antenna and matrix of dots with use of adhesive [17]

As a result, an array of 24-μm diameter bumps was formed with the deposition speed of over 400 bumps/s [50]. Microelectronic packaging is waiting for such technology to print small dots of adhesives.

Application of conductive adhesives via ink-jet makes specific demands on adhesive's formulations. The Ag particles must not exceed the maximum size determined by the possibility of clogging the printing nozzle; the viscosity of the formulation has to be less than 100 mPa s, and the adhesive should resist sedimentation for at least several hours at room temperature. Certainly, after printing and curing the microstructures must be electrically conductive. Tests with a specially prepared formulation, containing Ag particles with a diameter up to 5 μm, show that fulfilling these requirements is possible [17], and some patterns and matrices of dots have been printed (Fig. 12.18).

References

1. Saito H. Matsuba Y., *Liquid Wiring Technology by Ink-jet Printing Using NanoPaste*, 35th International Symposium on Microelectronics IMAPS, San Diego, 2006, TP65
2. www.harima.co.jp/products/electronics
3. Mościcki A. Felba J. Sobierajski T. Kudzia J. Arp A. Meyer W., *Electrically Conductive Formulations Filled Nano Size Silver Filler for Ink-Jet Technology*, Fifth International IEEE Conference on Polymers and Adhesive in Microelectronics and Photonics, Wroclaw, 2005, p. 40
4. Mościcki A. Felba J. Dudziński Wł., *Conductive Microstructures and Connections for Microelectronics Made by Ink-Jet Technology*, First Electronics System Integration Technology Conference, Dresden, 2006, p. 511
5. Tilaki R.M. Iraji Zad A. Mahdavi S.M., *Stability, Size and Optical Properties of Silver Nanoparticles Prepared by Laser Ablation in Different Carrier Media*, Applied Physics A, Vol. 84, 2006, pp. 215–219
6. Starowicz M. Stypuła B. Banas J., *Electrochemical Synthesis of Silver Nanoparticles*, Electrochemistry Communications, Vol. 8, 2006, pp. 227–230
7. Nakamoto M. Yamamoto M. Kashiwagi Y. Kakiuchi H. Tsujimoto T. Yoshida Y., *A Variety of Silver Nanoparticle Pastes for Fine Electronic Circuit Pattern Formation*, Polytronic 2007 – Sixth International Conference on Polymers and Adhesives in Microelectronics and Photonics, Tokyo, 2007, p. 105
8. Szczech J.B., Megaridis C.M., Gamota D.R., Zhang J., *Fine-Line Conductor Manufacturing Using Drop-on-Demand PZT Printing Technology*, IEEE Transactions on Electronics Packaging and Manufacturing, Vol. 25, No. 1, 2002, pp. 26–33
9. www.amepox-mc.com
10. www.cabot-corp.com
11. www.nanoscale.com/products/
12. www.anapro.com/english/product/product7.asp
13. Bai J.G. Calata J.N. Lei T.G. Lu G.Q. Creehan K.D., *Lead-Free Die-Attachment with High-temperature Capability by Low-temperature Nanosilver Paste Sintering*, 35th International Symposium on Microelectronics IMAPS, San Diego, 2006, TA43
14. Mei J. Lovell M.R. Mickle M.H., *Formulation and Processing of Novel Conductive Solution Inks in Continuous Inkjet Printing of 3-D Electric Circuits*, IEEE Transactions on Electronics Packaging and Manufacturing, Vol. 28, No. 3, 2005, p. 265
15. Nur H.M. Song J.H. Evans J.R.G. Edirisinghe M.J., *Ink-Jet Printing of Gold Conductive Tracks*, Journal of Materials Science Materials in Electronics, Vol. 13, 2002, p. 213
16. Kawazome M. Suganuma K. Hatamura M. Kim K.-S. Horie S. Hirasawa A. Tanaami H., *Low Temperature Printing Wiring with Ag Salt Pastes*, 35th International Symposium on Microelectronics IMAPS, San Diego, 2006, WP61
17. Kolbe J. Arp A. Calderone F. Meyer E.M. Meyer W. Schaefer H. Stuve M., *Inkjetable Conductive Adhesive for Use in Microelectronics and Microsystem Technology*, Fifth International IEEE Conference on Polymers and Adhesive in Microelectronics and Photonics, Wroclaw, 2005, p. 160
18. Yamaguchi K. Sakai K. Yamanaka T. Hirayama T., *Generation of Three-Dimensional Micro Structure Using Metal Jet*, Precision Engineering, Vol. 24, 2000, pp. 2–8
19. Ukita Y. Tateyama K. Segawa M. Tojo Y. Gotoh H. Oosako K., *Lead Free Die Mount Adhesive Using Silver Nanoparticles Applied to Power Discrete Package*, Advanced Microelectronics, 2005, pp. 8–11
20. Gaugel T. Bechtel S. Neuman-Rodekirch J., *Advanced Micro-Dispensing System for Conductive Adhesives*, First International IEEE Conference on Polymers and Adhesive in Microelectronics and Photonics, Potsdam, 2001, p. 40
21. Murata K. Matsumoto J. Tezuka A. Matsuba Y. Yokoyama H., *Super-Fine Ink-Jet Printing: Toward the Minimal Manufacturing System*, Microsystem Technologies, Vol. 12, 2005, pp. 2–7

22. Quinines H. Babiarz A. Fang L. Fiske E., *Jet Dispense for Electronic and Optoelectronic Packaging*, European IMAPS Symposium, Prague, 2004, p. 384
23. www.microdrop.de
24. Meyer W., *Micro Dispensing of Adhesives and Other Polymers*, First International IEEE Conference on Polymers and Adhesive in Microelectronics and Photonics, Potsdam, 2001, p. 35
25. Chen P.-H. Peng H.-Y. Liu H.-Y. Chang S.-L. Wu T.-I. Cheng Ch.-H., *Pressure Response and Droplet Ejection of a Piezoelectric Inkjet Printhead*, International Journal of Mechanical Science, Vol. 41, 1999, p. 235
26. Liou T.M. Shih K.C. Chau S.W. Chen S.C., *Three-Dimensional Simulations of the Droplet Formation During the Inkjet Printing Process*, International Communications in Heat and Mass Transfer, Vol. 29, No. 8, 2002, pp. 1109–1118
27. Le H.P., *Progress and Trends in Inkjet Printing Technology*, Journal of Imaging Science and Technology, Vol. 42, No. 1, 1998, pp. 49–62
28. www.optomec.com
29. Murata K., *Super-Fine Ink-Jet Printing for Nanotechnology*, International Conference on MEMS, NANO and Smart Systems (ICMENS'03), Banff, 2003, p. 346
30. Lee D.Y. Hwang E.S. Yu T.U. Kim Y.J. Hwang J., Structuring of micro line conductor using electro-hydrodynamic printing of a silver nanoparticle suspension, Applied Physics A, Vol. 82, 2006, pp. 671–674
31. Nagai T. Hoshino K. Matminoto K. Shimoyama I., *Direct Ink-Jet Printing of Electric Materials with Active Alignment Control*, 13th International Conference on Solid-State Sensors, Actuators and Microsystems, Seoul, 2005, p. 1461
32. Mościcki A. Felba J. Gwiaździński P. Puchalski M., *Conductivity Improvement of Microstructures Made by Nano-Size-Silver Filled Formulations*, Polytronic 2007 – Sixth International Conference on Polymers and Adhesives in Microelectronics and Photonics, 2007, p. 305
33. Measured by K. Nitsch and T. Piasecki from Faculty of Microsystem Electronics and Photonics, Wroclaw University of Technology
34. Buffat Ph. Borel J.P., *Size Effect on the Melting Temperature of Gold Particles*, Physical Review A, Vol. 13, No. 6, 1976, pp. 2287–2298
35. Bai J.G. Zhang Z.Z. Calata J.N. Lu G.-Q., *Low-Temperature Sintered Nanoscale Silver as a Novel Semiconductor Device-Metallized Substrate Interconnect Material*, IEEE Transactions on Components and Packaging Technologies, Vol. 29, No. 3, 2006, pp. 589–592
36. Kim K.-S. Hatamura M. Yamaguchi S. Suganuma K., *Curing Characteristics of Nano Paste for Finite Printed Circuits*, Third International IEEE Conference on Polymers and Adhesive in Microelectronics and Photonics, Montreux, 2003, p. 369
37. Marinov V.R., *Electrical Resistance of Laser Sintered Direct – Write Deposited Materials for Microelectronic Applications*, International Microelectronics & Packaging Society – JMEP, Vol. 1, No. 4, 2004, p. 261
38. Jaewon Chung J. Ko S. Bieri N.R. Grigoropoulos C.P. Poulikakosa D., *Conductor Microstructures by Laser Curing of Printed Gold Nanoparticle Ink*, Applied Physics Letters, Vol. 84, No. 5, 2004, p. 801
39. Bieri N.R. Chung J. Haferl S.E. Poulikakos D. Grigoropoulos C.P., *Microstructuring by Printing and Laser Curing of Nanoparticle Solutions*, Applied Physics Letters, Vol. 82, No. 20, 2003, p. 3529
40. Bieri N.R. Chung J. Poulikakos D. Grigoropoulos C.P., *Manufacturing of Nanoscale Thickness Gold Lines by Laser Curing of a Discretely Deposited Nanoparticle Suspension*, Superlattices and Microstructures, Vol. 35, 2004, pp. 437–444
41. Chung J. Bieri N.R. Ko S. Grigoropoulos C.P. Poulikakos D., In-tandem deposition and sintering of printed gold nanoparticle inks induced by continuous Gaussian laser irradiation, Applied Physics A, Vol. 79, 2004, pp. 1259–1261
42. Wakuda D. Hatamura M. Suganuma K., *Novel Room Temperature Wiring Process of Ag Nanoparticle Paste*, Polytronic 2007 – Sixth International Conference on Polymers and Adhesives in Microelectronics and Photonics, Tokyo, 2007, p. 110

43. Murata K. Shimizu K., *Micro Bump Formation by Using a Super Fine Inkjet System*, 35th International Symposium on Microelectronics IMAPS, San Diego, 2006, TP66
44. Peng W. Hurksainen V. Haskizume K. Dunford S. Quader S. Vatanparast R., *Flexible Circuit Creation with Nano Metal Particles*, 55th Electronic Component and Technology Conference, Lake Buena Vista, 2005, p. 77
45. IPC-6013 standard, *Qualification and Performance Specification for Flexible Printed Boards for Etched Cu Traces*
46. Nishi S.-i., *Direct Metal Pattering for Printable Electronics by Inkjet Technology*, 35th International Symposium on Microelectronics IMAPS, San Diego, 2006, TP63
47. Imai H. Mizuno S. Makabe A. Sakurada K. Wada K., *Application of Inkjet Printing Technology by Electro Packaging*, 35th International Symposium on Microelectronics IMAPS, San Diego, 2006, TP67
48. Kawamura Y. Sigezawa K. Tanaka T. Koiwai K. Mizugaki K. Sakarada K. Kobayashi T. Wada K., *LTCC Multilevel Interconnection Substrate with Ink-jet Printing and Thick Film Printing for High-Density Packaging*, 35th International Symposium on Microelectronics IMAPS, San Diego, 2006, TP64
49. Fuller S.B. Wilhelm E.J. Jacobson J.M., *Ink-Jet Printed Nanoparticle Microelectromechanical Systems*, Journal of Microelectromechanical Systems, Vol. 11, No. 1, p. 54
50. Hayes D.J. Grove M.E. Cox W.R., *Development and Application by Ink-Jet Printing of Advanced Packaging Materials*, International Symposium on Advanced Packaging Materials, Chateau Elan, 1999, p. 88

Chapter 13
A Study of Nanoparticles in SnAg-Based Lead-Free Solders

Masazumi Amagai

13.1 Introduction

Tin–lead (Sn-Pb) solder alloy has been widely used as an interconnection material in electronic packaging due to its low melting temperatures and good wetting behavior on several substrate platings such as Cu, Ag, Pd, and Au. Recently, because of environmental and health concerns, a variety of new lead-free solders have been developed. Lead-free solders lack the toxicity problems associated with lead-containing solders. However, unlike lead solders, the recently employed lead-free solders do not have a long history and manufacturing process, and also board level reliability has not been established well. Especially, drop test performance is a serious concern for mobile products such as cellular phones, cameras, video, and so on. Sn–Ag–Cu alloys are leading candidates for lead-free solders.

However, Sn–Ag–Cu alloys are not enough to meet severe board reliability requirements. Two lead-free solders were introduced in 2003 [1]. One was Sn–Ag–Ni–P. The other was Sn–Ag–Cu–P. Sn–Ag–Ni–P system has a balance of thermal cycling, bend, drop, and internal void test performance. On the other hand, Sn–Ag–Cu–P system has a significant advantage for drop test performance. The combination of Cu and P significantly reduced intermetallic compound (IMC) thickness.

Recent mobile electronic products (e.g., cellular phones) require a thermal aging process followed by drop and bend tests. Thermal aging affects IMC and Kirkendall voids. Kirkendall voids under IMC reduce the strength of solder joint and degrade drop test performance significantly. It was found that Kirkendall voids in lead-free solder joints could be greatly reduced by adding Ni and In to Sn–Ag–Cu. Sn–Ag–Cu–Ni–In, which was introduced in 2004 [2], improves drop test performance over Sn–Ag–Cu–P.

Previous lead-free solders we introduced were four- or five-element-based lead-free solders. Effects of additional elements on the growth of IMC were not identified with so many elements in the solders. In this study, we focused on three elements to study the growth of IMCs. Co, Ni, Pt, Al, P, Cu, Zn, Ge, Ag, In, Sb, and Au

---

M. Amagai
Tsukuba Technology Center, Texas Instruments, Ibaragi-ken, Japan

inclusions in Sn-3Ag-based lead-free solders were evaluated to study whether these nanoparticles increase IMC thickness and grain size after four solder reflows. Also, IMC element analyses were carried out to study if nanoparticles were dissolved in IMC after one solder reflow and four solder reflows. Then, Co, Ni, Pt, Al, P, Cu, Zn, Ge, Ag, In, Sb, or Au inclusions in Sn-3.0Ag-based lead-free solders were evaluated to study whether these nanoparticles can reduce the frequency of occurrence of IMC fracture in high-impact pull tests. In addition to IMC analyses, the thickness of these nanoparticles on solder ball hardness was studied, since large solder displacement under drop impact improves drop test performance. Solder hardness is relative to solder ball displacement, so low solder hardness (soft solder) improves drop test performance. Therefore, the hardness test was performed to study if nanoparticles increase solder hardness after one solder reflow and two solder reflows + 100°C thermal aging (0, 100, 200 h). Finally, Co, Ni, Pt, Al, P, Cu, Zn, Ge, Ag, In, Sb, or Au inclusions in Sn-1.0Ag-based lead-free solders were evaluated to study if these nanoparticles can improve drop test performance. Ni, Co, and Pt were very effective for drop test performance. Sn-1.0Ag was used to study drop test performance, since Sn-1.0Ag shows better drop test performance than Sn-3.0Ag [2].

In this study, it was found that adding Co, Ni, or Pt, located to the left of Cu in the periodic table, to SnAg-based solder alloys did not increase IMC thickness and grain size significantly after the solder reflow process and thermal aging. Hence, these nanoparticles resulted in good drop test performance compared with Cu, Ag, Au, Zn, Al, In, P, Ge, Sb [3].

13.2 Nanoparticle Effects on Solder IMC Grain Size and Thickness

Co, Ni, Pt, Al, P, Cu, Zn, Ge, Ag, In, Sb, or Au inclusions in Sn-3Ag-based lead-free solder balls were evaluated to study if these nanoparticles affect IMC thickness and grain size after four solder reflows. These solder ball samples were attached to OSP Cu solder pads through the reflow process (maximum temperature 245°C).

Figure 13.1 shows the sample preparation procedure to observe solder IMC grain size and thickness. Following the solder reflow process, solder balls were cut and polished with a sand paper. To expose IMCs, solder balls were chemically etched using Meltex HN-980M. The samples were then cleaned ultrasonically twice in water. IMC grain size and thickness were subsequently observed with scanning electronic microscopy (SEM).

Figure 13.2 shows IMC with Sn3.0Ag. (a) and (b) show a top view of IMCs and a cross section of IMCs after one solder reflow, respectively. (c) and (d) show a top view of IMCs and a cross section of IMCs after four solder reflows, respectively. (The same conditions will apply to a–d in subsequent figures.) Four solder reflows increased the grain size and thickness of IMCs compared with one solder reflow.

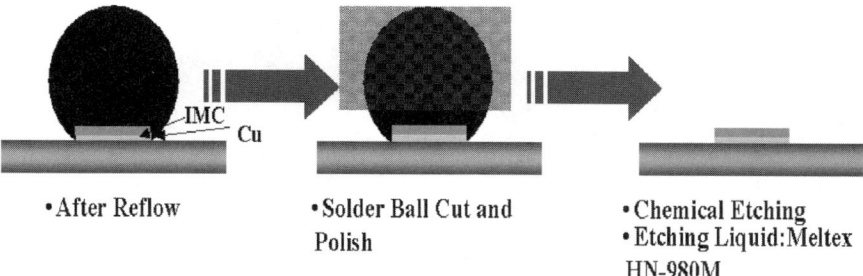

- After Reflow
- Solder Ball Cut and Polish
- Chemical Etching
- Etching Liquid:Meltex HN-980M

Fig. 13.1 Sample preparation procedure. (**a**) Top view of intermetallic compound (IMC) after one solder reflow, (**b**) cross section of IMC after one solder reflow, (**c**) top view of IMC after four solder reflows, (**d**) cross section of IMC after four solder reflows

Figure 13.3 shows IMC with Sn3.0Ag0.05Al. Four solder reflows increased the grain size and thickness of IMCs compared with one solder reflow. There is no significant difference between no aluminum and the inclusion of aluminum in Sn3.0Ag in the grain size and thickness of IMCs. Some voids were also observed for Sn3.0Ag0.05Al. This may be due to aluminum oxidation, since it is easy to oxidize aluminum. Based on the results, it was found that aluminum added to Sn3.0Ag could not reduce the grain size and thickness of IMCs compared with the no-aluminum case.

Figure 13.4 shows IMC with Sn3.0Ag0.03Ni. There is a significant difference between no nickel and nickel in Sn3.0Ag in increasing the grain size and thickness of IMCs. Ni (0.03 wt%) reduced the grain size and thickness of IMCs after four solder reflows, compared with the no-nickel case. Based on the results, it was found that the addition of nickel (in particular, 0.03 wt% Ni) to Sn3.0Ag was very effective for reducing the grain size and thickness of IMCs.

Figure 13.5 shows IMC with Sn3.0Ag0.5Cu. There is no significant difference between Sn3.0Ag and Sn3.0Ag0.5Cu in increasing the grain size and thickness of IMCs from one reflow to four solder reflows. Hence, it was definite that copper added to Sn3.0Ag did not affect the grain size and thickness of IMCs very much.

Figure 13.6 shows IMC with Sn3.0Ag0.03Co. As can be seen in the pictures, nanocobalt particles added to Sn3.0Ag were very effective in reducing the grain size and thickness of IMCs compared with the no-cobalt case after four solder reflows.

Figure 13.7 shows IMC with Sn3.0Ag0.3In. There is no significant difference between no indium and the inclusion of nanoindium particles in Sn3.0Ag in increasing the grain size and thickness of IMCs from one solder reflow to four solder reflows. Thus, it can be seen that nanoindium particles added to Sn3.0Ag did not affect the grain size and thickness of IMCs significantly.

Figure 13.8 shows IMC with Sn3.0Ag0.3Sb. There is no significant difference between no antimony and the inclusion of nanoantimony particles in Sn3.0Ag in increasing the grain size and thickness of IMCs from one solder reflow to four solder reflows. Thus, it was found that nanoantimony particles added to Sn3.0Ag did not affect the grain size and thickness of IMCs considerably.

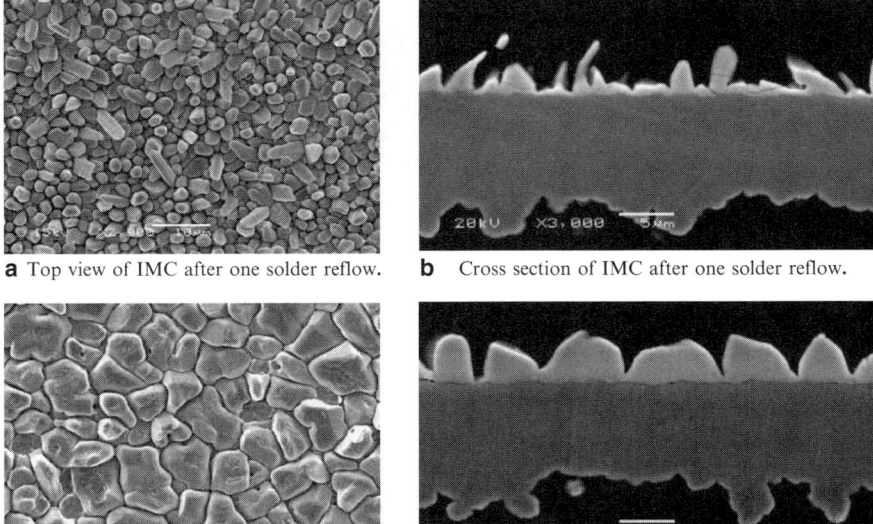

Fig. 13.2 IMC with Sn3.0Ag. (**a**) Top view of IMC after one solder reflow, (**b**) cross section of IMC after one solder reflow, (**c**) top view of IMC after four solder reflows, (**d**) cross section of IMC after four solder reflows

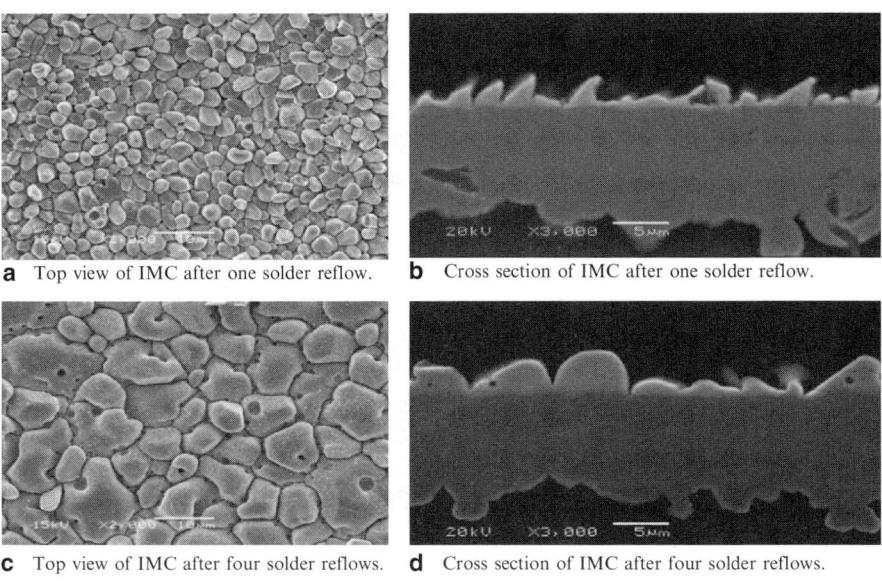

Fig. 13.3 IMC with Sn3.0Ag0.05Al.(**a**) Top view of IMC after one solder reflow, (**b**) cross section of IMC after one solder reflow, (**c**) top view of IMC after four solder reflows, (**d**) cross section of IMC after four solder reflows

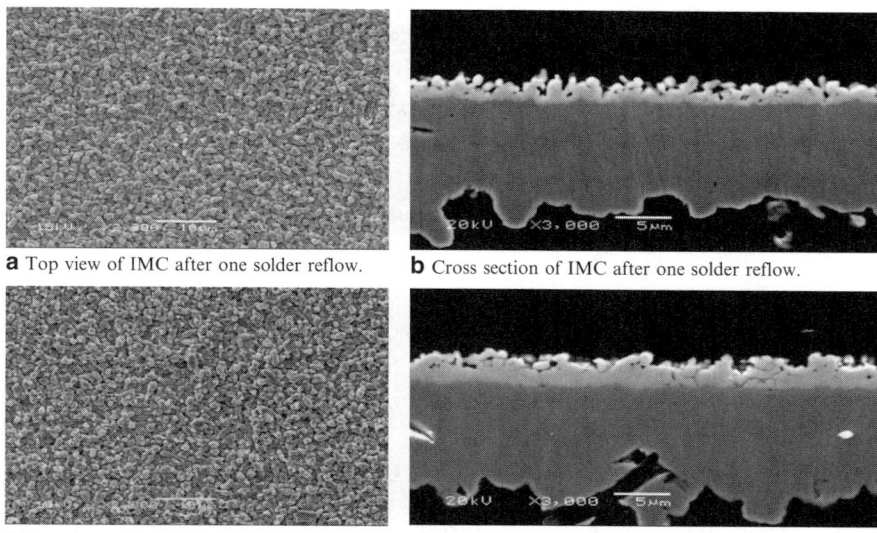

Fig. 13.4 IMC with Sn3.0Ag0.03Ni. (**a**) Top view of IMC after one solder reflow, (**b**) cross section of IMC after one solder reflow, (**c**) top view of IMC after four solder reflows, (**d**) cross section of IMC after four solder reflows

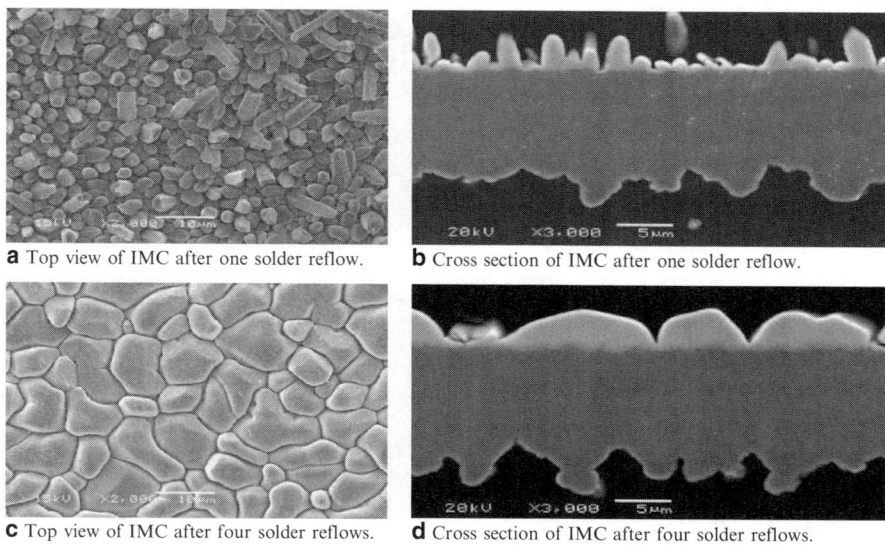

Fig. 13.5 IMC with Sn3.0Ag0.5Cu. (**a**) Top view of IMC after one solder reflow, (**b**) cross section of IMC after one solder reflow, (**c**) top view of IMC after four solder reflows, (**d**) cross section of IMC after four solder reflows

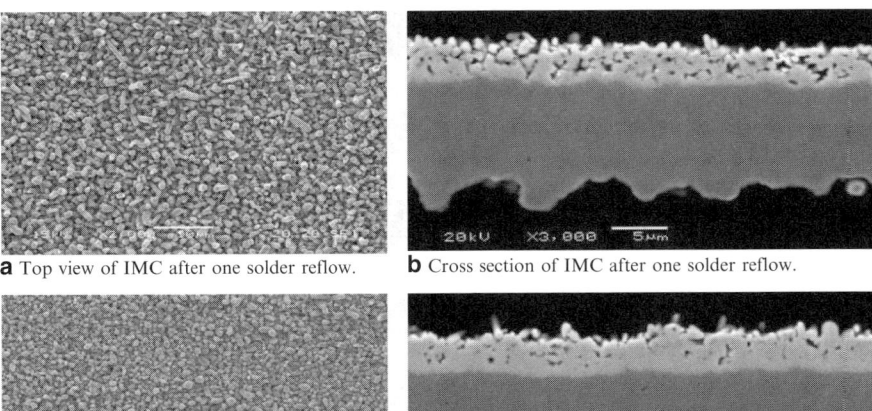

Fig. 13.6 IMC with Sn3.0Ag0.03Co. (**a**) Top view of IMC after one solder reflow, (**b**) cross section of IMC after one solder reflow, (**c**) top view of IMC after four solder reflows, (**d**) cross section of IMC after four solder reflows

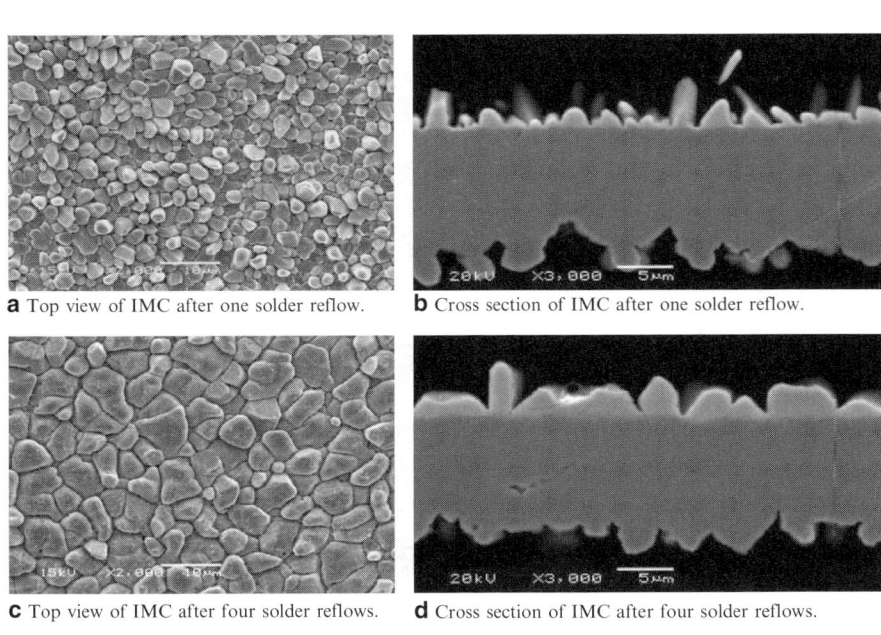

Fig. 13.7 IMC with Sn3.0Ag0.3In. (**a**) Top view of IMC after one solder reflow, (**b**) cross section of IMC after one solder reflow, (**c**) top view of IMC after four solder reflows, (**d**) cross section of IMC after four solder reflows

13 A Study of Nanoparticles in SnAg-Based Lead-Free Solders

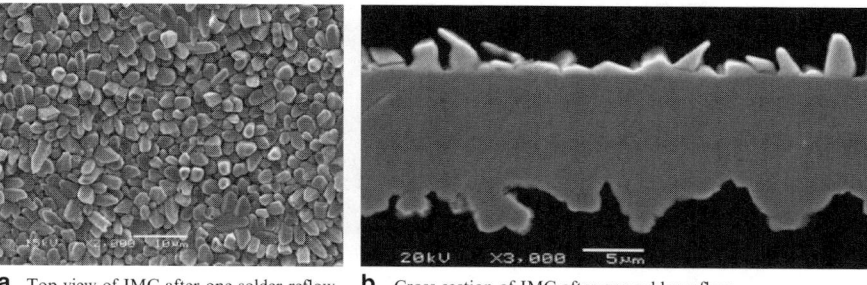

a Top view of IMC after one solder reflow. **b** Cross section of IMC after one solder reflow.

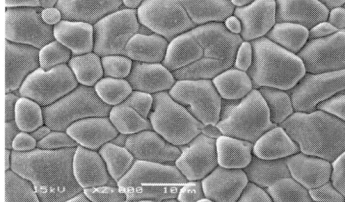 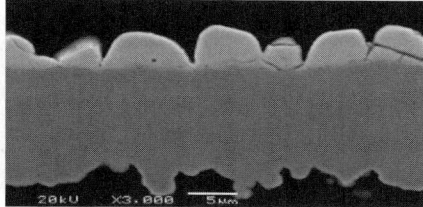

c Top view of IMC after four solder reflows. **d** Cross section of IMC after four solder reflows.

Fig. 13.8 IMC with Sn3.0Ag0.3Sb. (**a**) Top view of IMC after one solder reflow, (**b**) cross section of IMC after one solder reflow, (**c**) top view of IMC after four solder reflows, (**d**) cross section of IMC after four solder reflows

Figure 13.9 shows IMC with Sn3.0Ag0.1Zn. There is no significant difference between no zinc and the inclusion of zinc particles in Sn3.0Ag in increasing the grain size and thickness of IMCs from one solder reflow to four solder reflows. As with aluminum, some voids were also observed for Sn3.0Ag0.05Zn and Sn3.0Ag0.1Zn. This may be due to zinc oxidation, since it is easy to oxidize zinc. As can be seen in pictures, it was observed that nanozinc particles added to Sn3.0Ag did not affect the grain size and thickness of IMCs tremendously.

Figures 13.10 and 13.11 show IMC with Sn3.0Ag0.03P and Sn3.0Ag0.1Au, respectively. Gold and phosphorus could not prevent increasing the grain size and thickness of IMCs from one to four solder reflows. Thus, it is obvious that nanogold or phosphorus particle inclusions in Sn3.0Ag did not affect the grain size or thickness of IMCs significantly.

Figures 13.12 and 13.13 show the IMC with Sn3.0Ag0.05Pt and the IMC with Sn3.0Ag0.05Ge. As can be seen in the pictures, nanoplatinum particles added to Sn3.0Ag were very effective for reducing the grain size and thickness of IMCs compared with no-platinum case after four solder reflows. However, germanium could not prevent increasing the grain size and thickness of IMC from one to four solder reflows.

Table 13.1 summarizes the comparison of nanoparticles vs. no nanoparticles added to Sn3.0Ag for the grain size and thickness of IMCs, and also the change of the grain size and thickness of IMCs between one and four solder reflows. It is

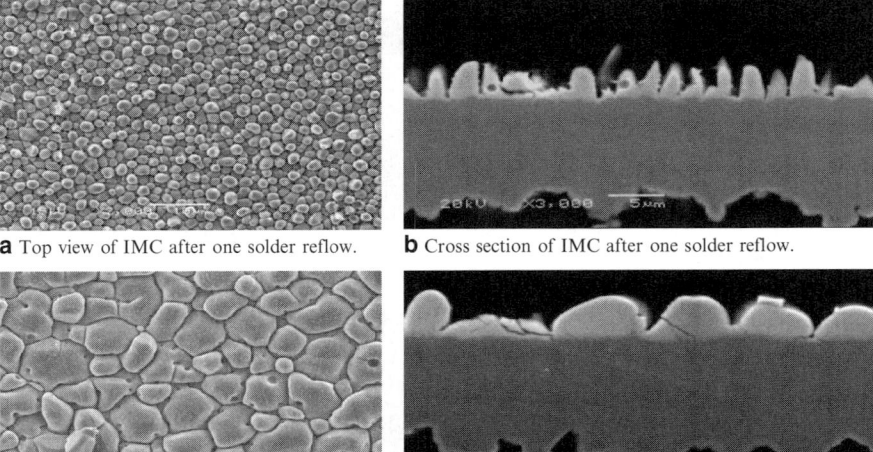

a Top view of IMC after one solder reflow. b Cross section of IMC after one solder reflow.

c Top view of IMC after four solder reflows. d Cross section of IMC after four solder reflows.

Fig. 13.9 IMC with Sn3.0Ag0.1Zn. (**a**) Top view of IMC after one solder reflow, (**b**) cross section of IMC after one solder reflow, (**c**) top view of IMC after four solder reflows, (**d**) cross section of IMC after four solder reflows

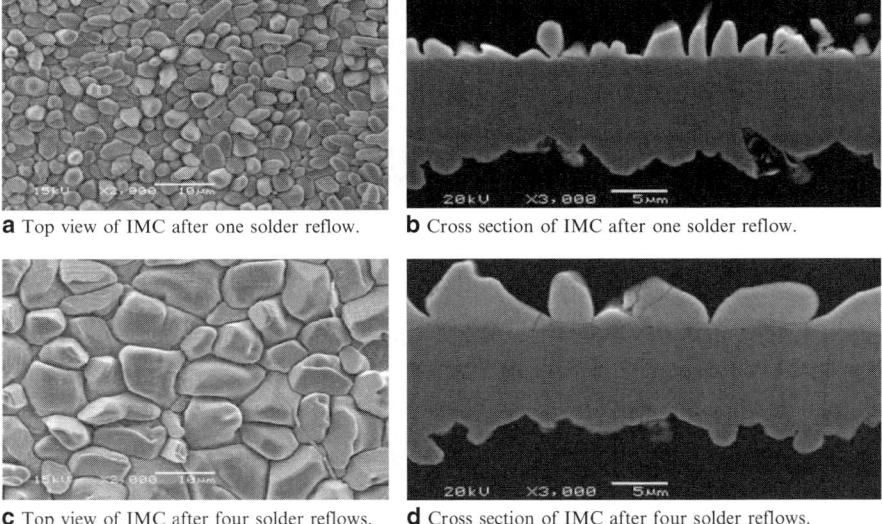

a Top view of IMC after one solder reflow. b Cross section of IMC after one solder reflow.

c Top view of IMC after four solder reflows. d Cross section of IMC after four solder reflows.

Fig. 13.10 IMC with Sn3.0Ag0.03P. (**a**) Top view of IMC after one solder reflow, (**b**) cross section of IMC after one solder reflow, (**c**) top view of IMC after four solder reflows, (**d**) cross section of IMC after four solder reflows

13 A Study of Nanoparticles in SnAg-Based Lead-Free Solders

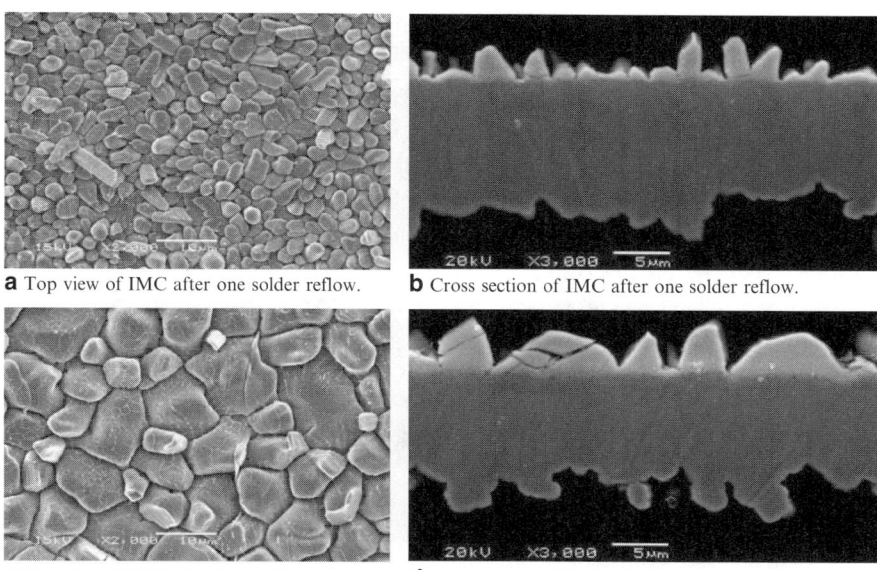

a Top view of IMC after one solder reflow. **b** Cross section of IMC after one solder reflow.

c Top view of IMC after four solder reflows. **d** Cross section of IMC after four solder reflows.

Fig. 13.11 IMC with Sn3.0Ag0.1Au. (**a**) Top view of IMC after one solder reflow, (**b**) cross section of IMC after one solder reflow, (**c**) top view of IMC after four solder reflows, (**d**) cross section of IMC after four solder reflows

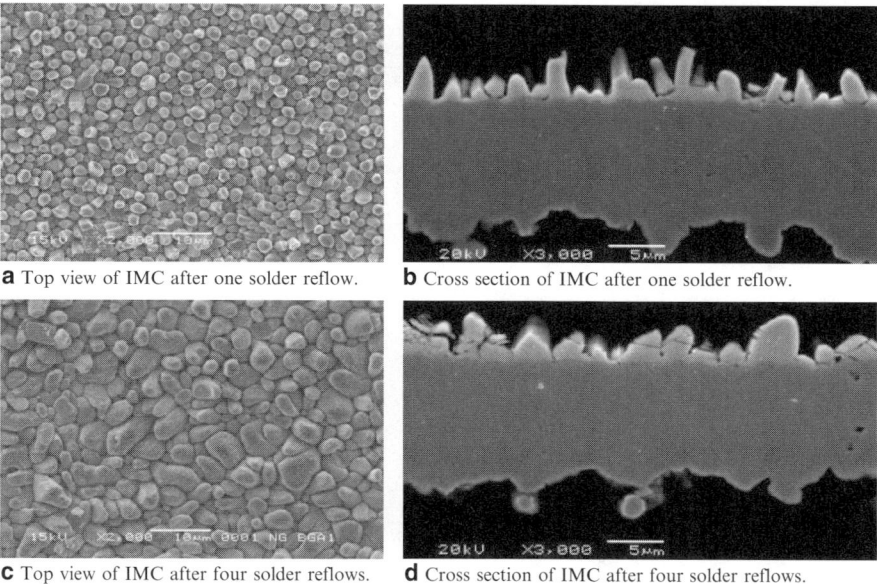

a Top view of IMC after one solder reflow. **b** Cross section of IMC after one solder reflow.

c Top view of IMC after four solder reflows. **d** Cross section of IMC after four solder reflows.

Fig. 13.12 IMC with Sn3.0Ag0.05Pt. (**a**) Top view of IMC after one solder reflow, (**b**) cross section of IMC after one solder reflow, (**c**) top view of IMC after four solder reflows, (**d**) cross section of IMC after four solder reflows

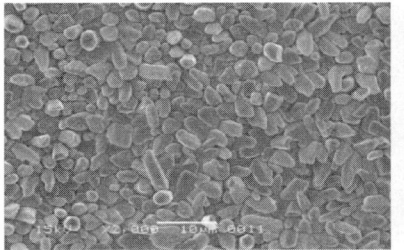
a Top view of IMC after one solder reflow.

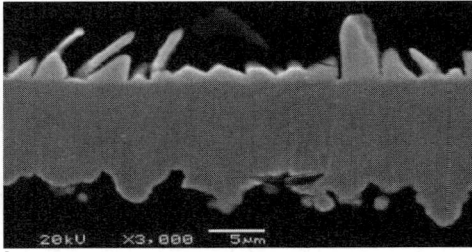
b Cross section of IMC after one solder reflow.

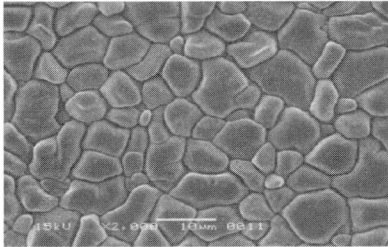
c Top view of IMC after four solder reflows.

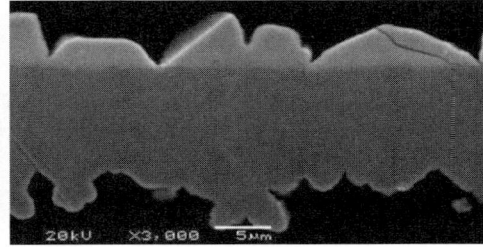
d Cross section of IMC after four solder reflows.

Fig. 13.13 IMC with Sn3.0Ag0.05Ge. (**a**) IMC element analysis after one solder reflow, (**b**) IMC element analysis after four solder reflows

Table 13.1 Summary of IMC effects of nanoparticle additions to Sn3.0Ag

Nanoparticle	Wt%	IMC comparison with Sn3.0Ag after 4 time reflow processes		IMC change from 1 time to 4 time reflow processes	
		Grain size	Thickness	Grain size	Thickness
Ni	0.01	Small	Low	Not large	Not large
	0.02	Small	Low	Small	Small
	0.05	Small	Low	Small	Small
Cu	0.1	Same	Same	Large	Large
	0.3	Same	Same	Large	Large
	0.5	Same	Same	Large	Large
	0.1	Same	Same	Large	Large
	1	Same	Same	Large	Large
Co	0.01	Small	Low	Not large	Not large
	0.03	Small	Low	Small	Small
In	0.1	Same	Same	Large	Large
	0.2	Same	Same	Large	Large
	0.3	Same	Same	Large	Large
Sb	0.1	Same	Same	Large	Large
	0.3	Same	Same	Large	Large
	0.5	Same	Same	Large	Large
Zn	0.05	Same	Same	Large	Large
	0.1	Same	Same	Large	Large
P	0.03	Same	Same	Large	Large
Au	0.1	Same	Same	Large	Large
Ge	0.05	Same	Same	Large	Large
Pt	0.05	Small	Low	Small	Small
Al	0.05	Same	Same	Large	Large

obvious that nickel, cobalt, and platinum play an important role for preventing the growth of IMC thickness and grain size.

13.3 Are the Nanoparticles Dissolved in the IMC?

IMC element analyses were performed to study if nanoparticles were dissolved in IMC after one solder reflow and four solder reflows. FE-SEM was utilized to observe elements in the IMCs.

Figure 13.14 shows IMC element analysis (wt%) for Sn3.0Ag0.03Ni. (a) and (b) show the element analysis after one solder reflow and four solder reflows, respectively. As can be seen in the pictures, it is obvious that nickel was dissolved in Cu6Sn5 and subsequently formed (CuNi)6Sn5. The ratio of Ni:Cu:Sn for wt% was 1.3–2.9:43–53:45–54 and 1.2–4.0:42–54:43–55 for one solder reflow and four solder reflows, respectively. It seems that four solder reflows increased Ni wt% in the IMC compared with one solder reflow.

Figure 13.15 shows the IMC element analysis (wt%) for Sn3.0Ag0.03Co. (a) and (b) imply the same as mentioned previously. As with nickel, cobalt was dissolved in Cu6Sn5 and subsequently formed (CuCo)6Sn5. The ratio of Co, Cu vs. Sn for wt% was 1.7–3.2:43–58:49–54 and 0.5–2.3:42–48:50–55 for one solder reflow and four solder reflows, respectively. Co wt% is smaller than Ni wt% in the IMC.

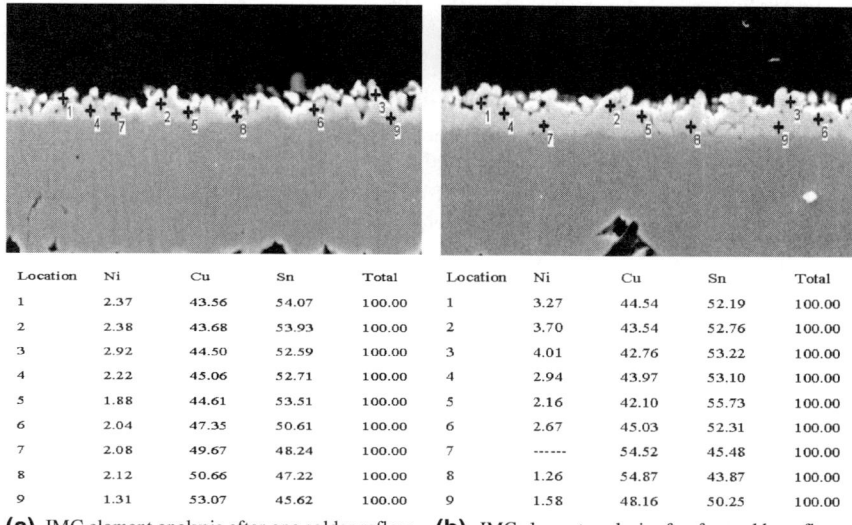

Location	Ni	Cu	Sn	Total	Location	Ni	Cu	Sn	Total
1	2.37	43.56	54.07	100.00	1	3.27	44.54	52.19	100.00
2	2.38	43.68	53.93	100.00	2	3.70	43.54	52.76	100.00
3	2.92	44.50	52.59	100.00	3	4.01	42.76	53.22	100.00
4	2.22	45.06	52.71	100.00	4	2.94	43.97	53.10	100.00
5	1.88	44.61	53.51	100.00	5	2.16	42.10	55.73	100.00
6	2.04	47.35	50.61	100.00	6	2.67	45.03	52.31	100.00
7	2.08	49.67	48.24	100.00	7	------	54.52	45.48	100.00
8	2.12	50.66	47.22	100.00	8	1.26	54.87	43.87	100.00
9	1.31	53.07	45.62	100.00	9	1.58	48.16	50.25	100.00

(a) IMC element analysis after one solder reflow. (b) IMC element analysis after four solder reflows.

Fig. 13.14 IMC element analysis (wt%) for Sn3.0Ag0.03Ni.(**a**) IMC element analysis after one solder reflow, (**b**) IMC element analysis after four solder reflows

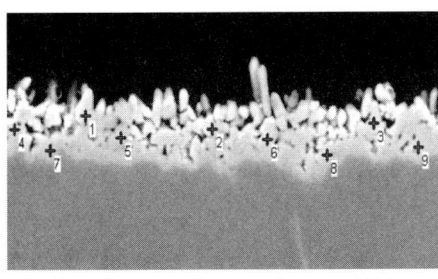

Location	Co	Cu	Sn	Total
1 (111,137)	------	45.08	54.92	100.00
2 (254,150)	3.29	44.21	52.50	100.00
3 (436,145)	1.87	43.74	54.39	100.00
4 (31,150)	2.14	44.93	52.92	100.00
5 (151,158)	2.53	45.20	52.28	100.00
6 (316,160)	2.43	46.91	50.67	100.00
7 (71,170)	2.08	48.22	49.71	100.00
8 (383,175)	2.06	48.21	49.74	100.00
9 (486,168)	1.77	45.51	52.71	100.00

(a) IMC element analysis after one solder reflow.

Location	Co	Cu	Sn	Total
1 (101,130)	2.11	42.75	55.14	100.00
2 (268,127)	1.99	44.53	53.48	100.00
3 (443,122)	2.38	43.58	54.03	100.00
4 (66,142)	2.65	45.01	52.34	100.00
5 (246,142)	1.42	45.88	52.71	100.00
6 (418,140)	2.02	44.82	53.16	100.00
7 (128,160)	0.88	48.13	50.99	100.00
8 (276,159)	0.50	47.99	51.51	100.00
9 (451,157)	1.66	45.36	52.99	100.00

(b) IMC element analysis after four solder reflows

Fig. 13.15 IMC element analysis (wt%) for Sn3.0Ag0.03Co. (**a**) IMC element analysis after one solder reflow, (**b**) IMC element analysis after four solder reflows

Figure 13.16 shows the IMC element analysis (wt%) for Sn3.0Ag0.05Pt. (a) and (b) imply the same as mentioned previously. As with nickel and cobalt, nanoplatinum particles were observed in the IMC. The ratio of Pt:Cu:Sn for wt% was 2.4–3.6:43–64:43–53 and 2.6–3.8:42–48:35–53 for one solder reflow and four solder reflows, respectively.

Figure 13.17 shows the IMC element analysis (wt%) for Sn3.0Ag0.5Sb. (a) and (b) imply the same as mentioned previously. Unlike nickel, cobalt, and platinum, nanoantimony particles were not observed in the IMC.

Based on the results of FE-SEM, Ni, Co, and Pt from the evaluated elements (Co, Ni, Pt, Al, P, Cu, Zn, Ge, Ag, In, Sb, Au) were dissolved into the IMC and then formed three-elements-based IMC.

13.4 Nanoparticle Effects on Solder Ball Hardness

Large solder displacement under drop impact improves drop test performance, since large solder displacement can reduce stress in IMCs. Solder hardness is relative to solder ball displacement, so low solder hardness (soft solder) improves drop test performance. Hardness testing was performed to study if nanoparticles increase solder hardness after one solder reflow, and two solder reflows followed by 100°C thermal aging (0, 100, 200 h).

13 A Study of Nanoparticles in SnAg-Based Lead-Free Solders

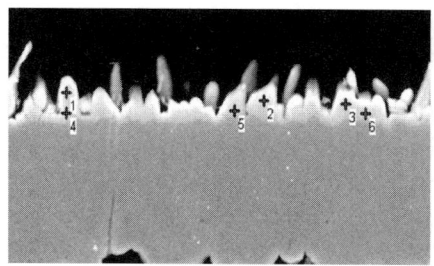
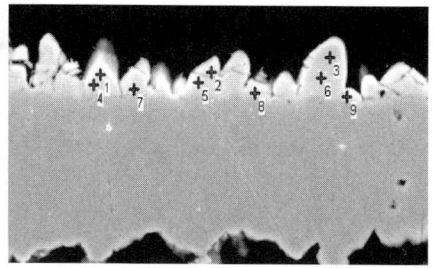

(a)

Location	Cu	Sn	Pt	Total
1 (89,120)	46.80	53.20	------	100.00
2 (318,127)	43.83	52.48	3.69	100.00
3 (413,130)	42.95	53.96	3.08	100.00
4 (89,140)	47.12	50.37	2.52	100.00
5 (284,137)	53.22	43.74	3.04	100.00
6 (436,139)	49.33	48.18	2.49	100.00

(a) IMC element analysis after one solder reflow.

(b)

Location	Cu	Sn	Pt	Total
1 (129,105)	43.94	52.18	3.88	100.00
2 (258,102)	43.05	53.45	3.50	100.00
3 (396,87)	43.43	53.09	3.48	100.00
4 (121,115)	46.77	50.49	2.74	100.00
5 (243,112)	46.26	51.06	2.68	100.00
6 (386,107)	44.83	52.52	2.64	100.00
7 (168,119)	61.25	38.75	------	100.00
8 (309,122)	52.98	47.02	------	100.00
9 (416,125)	64.28	35.72	------	100.00

(b) IMC element analysis after four solder reflows.

Fig. 13.16 IMC element analysis (wt%) for Sn3.0Ag0.05Pt. (**a**) IMC element analysis after one solder reflow, (**b**) IMC element analysis after four solder reflows

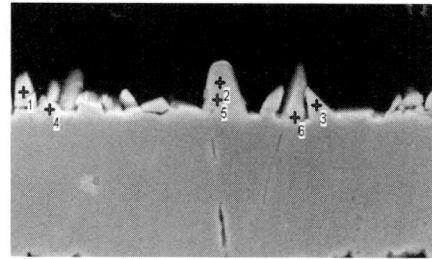
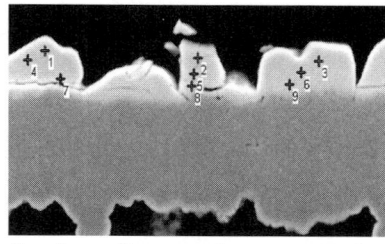

Location	Cu	Sn	Total
1 (37,163)	46.35	53.65	100.00
2 (264,153)	47.36	52.64	100.00
3 (376,175)	47.21	52.79	100.00
4 (67,180)	49.44	50.56	100.00
5 (261,170)	46.29	53.71	100.00
6 (351,187)	49.45	50.55	100.00

a IMC element analysis after one solder reflow.

Location	Cu	Sn	Total
1 (70,90)	46.24	53.76	100.00
2 (268,97)	45.97	54.03	100.00
3 (425,100)	46.31	53.69	100.00
4 (48,100)	45.58	54.42	100.00
5 (263,114)	46.72	53.28	100.00
6 (402,112)	47.34	52.66	100.00
7 (90,120)	47.96	52.04	100.00
8 (260,127)	48.13	51.87	100.00
9 (387,125)	46.63	53.37	100.00

b IMC element analysis after four solder reflows.

Fig. 13.17 IMC element analysis (wt%) for Sn3.0Ag0.5Sb

Figure 13.18 shows solder hardness (Vickers) after nanocopper particles (0.01–0.07 wt%) were added to Sn1.0Ag solders. Sn3.0Ag0.05Cu was used as a reference solder. Sn1.0Ag shows lower solder hardness than Sn3.0Ag. Since 0.01–0.07 wt% Cu added to Sn1.0Ag did not increase solder hardness compared with no Cu added to Sn1.0Ag, it is believed that 0.01–0.07 wt% Cu does not affect solder displacement under drop impact. It can be seen that two solder reflows followed by 100°C thermal aging decease solder hardness compared with one solder reflow.

Figure 13.19 shows solder hardness (Vickers) after nanoparticles (In, P, Sb, Co, Pt, P, Ni, Zn) were added to Sn1.0Ag solders. Except for Sb and Zn, solder hardness was not affected by nanoparticles. Sb, P, Ni, and Pt show that solder hardness was

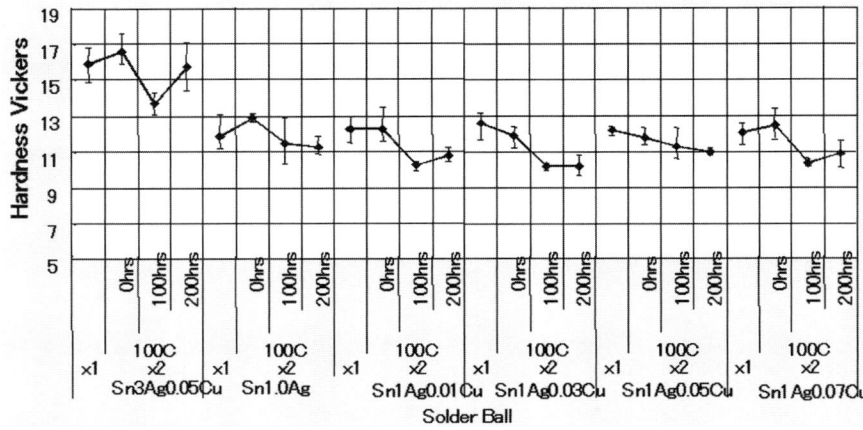

Fig. 13.18 Solder hardness (Vickers) after nanocopper particles (0.01–0.07 wt%) were added to An1.0Ag

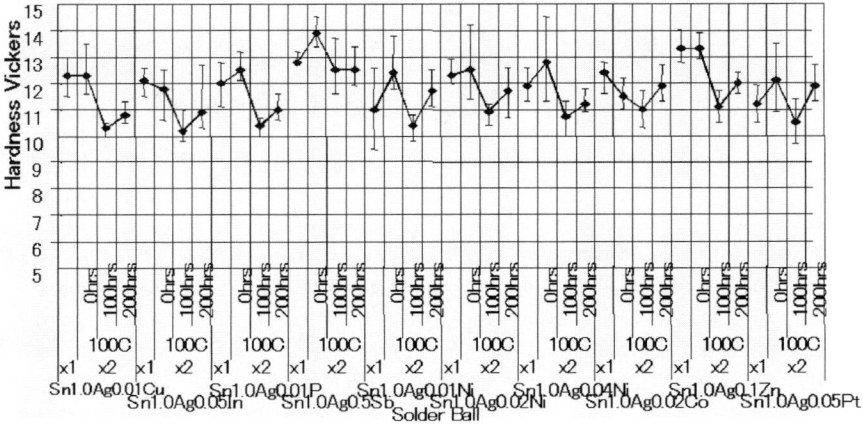

Fig. 13.19 Solder hardness (Vickers) after nanoparticles (In, P, Sb, Co, Pt, P, Ni, Zn) were added to Sn1.0Ag solders

increased after two solder reflows. It seems that solder hardness was increased after 100°C thermal aging for 100 h. Most of the nanoparticles show that solder hardness increased after 200 h, 100°C thermal aging.

13.5 Fracture in IMCs in High-Impact Pull Test

Co, Ni, Pt, Al, P, Cu, Zn, Ge, Ag, In, Sb, or Au inclusions in Sn-3Ag-based lead-free solders were evaluated to study if these nanoparticles can reduce the frequency of occurrence of IMC fracture in high-impact pull tests.

Figure 13.20 shows the apparatus for high-impact ball pull testing. A Dage 4000 bond tester was used in this test.

A pull jaw holds a solder ball and subsequently lifts up at 50 mm/s speed.

Figure 13.21 shows fracture modes after high-impact pull testing. Mode 1 is solder pad fracture. Mode 2 is fracture in solder. Mode 3 is no fracture, with pull jaw slip due to solder deformation. Mode 4 is fracture in the IMCs, which shows a failure (not good — NG) in this test.

Figure 13.22 shows high-impact pull strength and fracture mode before and after 100°C thermal aging (100 h) for Cu (0–1.0 wt%), Co (0.01–0.03 wt%), or Ni

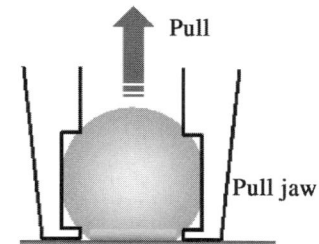

Fig. 13.20 High-impact pull test

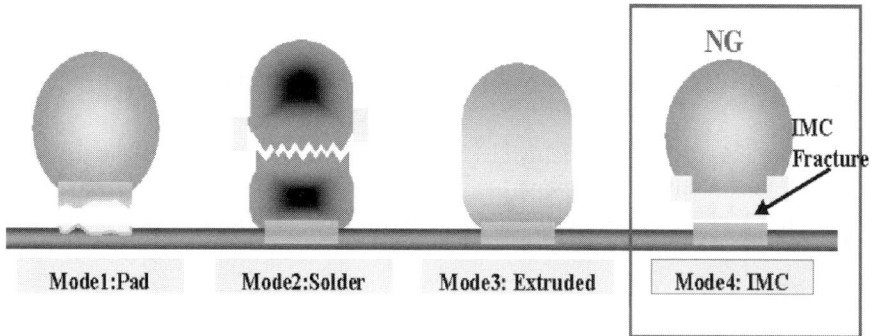

Fig. 13.21 Fracture mode

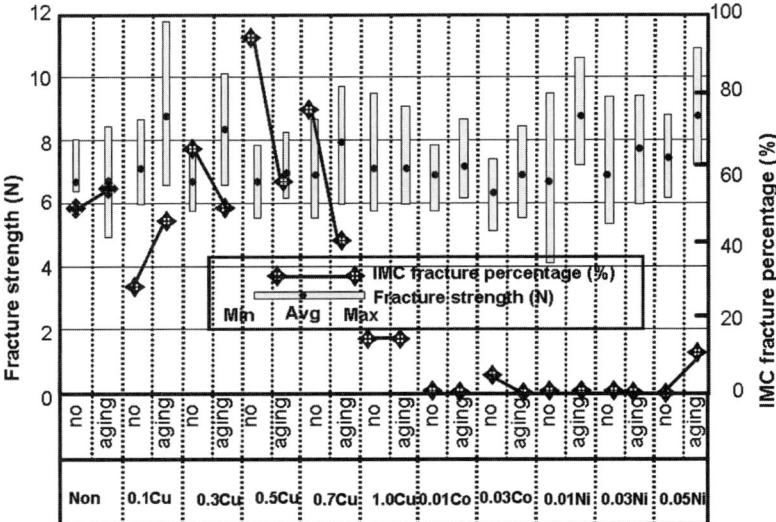

Fig. 13.22 High-impact pull strength and fracture mode before and after 100°C thermal aging (100 h) for Cu (0–1.0 Cu wt%), Co (0.01–0.03 wt%), or Ni (0.01–0.05 wt%) added to Sn1.0Ag

(0.01–0.05 wt%) added to Sn1.0Ag. One should pay attention to the fracture mode rather than fracture strength. As can be seen in the graph, nickel and cobalt show lower frequencies of occurrence of fracture in the IMCs than copper.

Figure 13.23 shows high-impact pull strength and fracture mode before and after 100°C thermal aging (100 h) for Sb (0.1–0.5 wt%), In (0.1–0.3 wt%), Bi (0.1–0.3 wt%), Ge (0.03 wt%), Al (0.005 wt%), Pt (0.05 wt%), or Zn (0.05–0.1 wt%) added to Sn1.0Ag. Platinum also reveals a lower frequency of occurrence of fracture in the IMCs than copper

Based on the result of fracture mode, it is obvious that nickel, cobalt, and platinum decrease fractures in IMCs compared with other nanoparticles.

13.6 Nanoparticle Effects on Drop Test Performance

Co, Ni, Pt, Cu, Zn, or Sb inclusions in Sn-1.0Ag-based lead-free solders were evaluated to study if these nanoparticles can improve drop test performance.

Figure 13.24 shows the drop test apparatus (Yoshida-seiki HDST-230). The package was a 12 mm× 12 mm BGA, a 30 mm × 120 mm × 0.8 mm PCB with a Cu + OSP (NSMD type) pad finish, and a solder paste (Senju M705-GRN360-K2V). The samples were left for 5 days before drop test was carried out. The daisy chain resistance was monitored. When the resistance exceeds 1.5 times the initial one, it was counted as a failure.

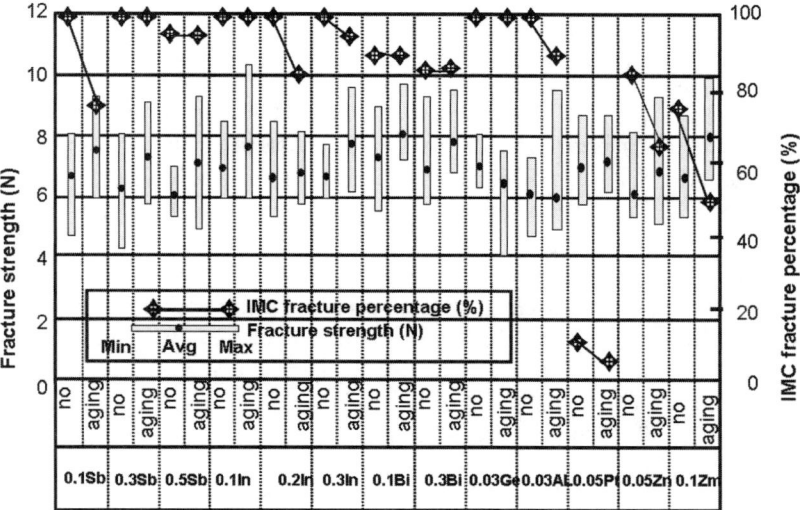

Fig. 13.23 High-impact pull strength and fracture mode before and after 100°C thermal aging (100 h) for Sb (0.1–0.5 wt%), In (0.1–0.3 wt%), Bi (0.1–0.3wt%), Ge (0.03 wt%), Al (0.005 wt%), Pt (0.05 wt%), or Zn (0.05–0.1 wt%) added to Sn1.0Ag

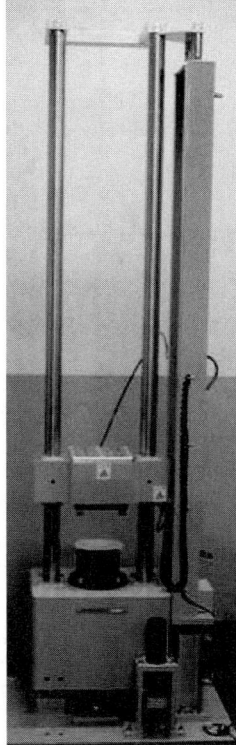

Fig.13.24 Drop test apparatus

Figure 13.25 shows the drop test table feature. The PCB was screwed down with a 15 cN m torque and the package was faced downward.

Figure 13.26 shows the drop test acceleration (1,500 G in this test).

Figure 13.27 shows a Weibull plot of drop test failure for 0.02Ni, 0.04Ni, 0.02Co, 0.5Sb, 0.1Zn, or 0.05Pt inclusions in Sn1.0Ag0.01Cu and a reference solder (Sn1.0Ag0.01Cu, SAC101) before 100°C thermal aging (100 h). As can be seen in the graph, Ni, Co, and Pt were very effective for drop test performance. In particular, 0.02Ni, 0.05Pt, and 0.02Co show better drop test performances than 0.04Ni, 0.5Sb, and 0.1Zn.

Figure 13.28 shows a Weibull plot of drop test failures for 0.02Ni, 0.04Ni, 0.02Co, 0.5Sb, 0.1Zn, or 0.05Pt inclusions in Sn1.0Ag0.01Cu and a reference solder (Sn1.0Ag0.01Cu, SAC101) after 100°C thermal aging (100 h). As with no thermal aging, Ni, Co, and Pt were also very effective for drop test performance. In particle, 0.02Ni, 0.05Pt, and 0.02Co show better drop test performance than 0.04Ni, 0.5Sb, and 0.1Zn.

A transition metal such as Co, Ni, Pt located to the left of Cu in the periodic table (Table 13.2) did not increase IMC thickness and grain size significantly after

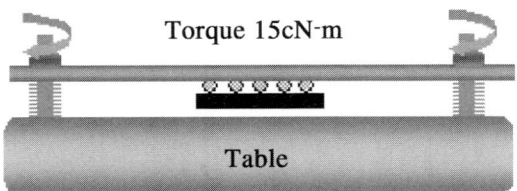

Fig. 13.25 Drop test table feature

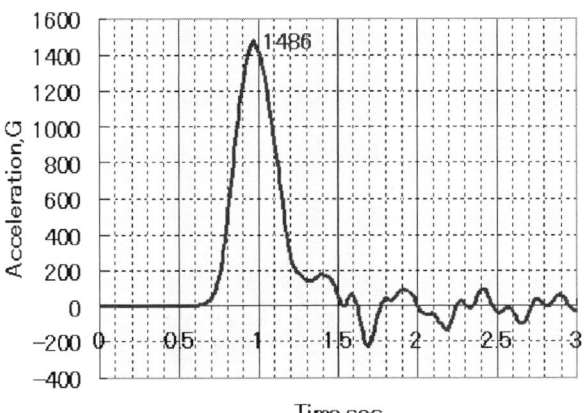

Fig. 13.26 Drop acceleration

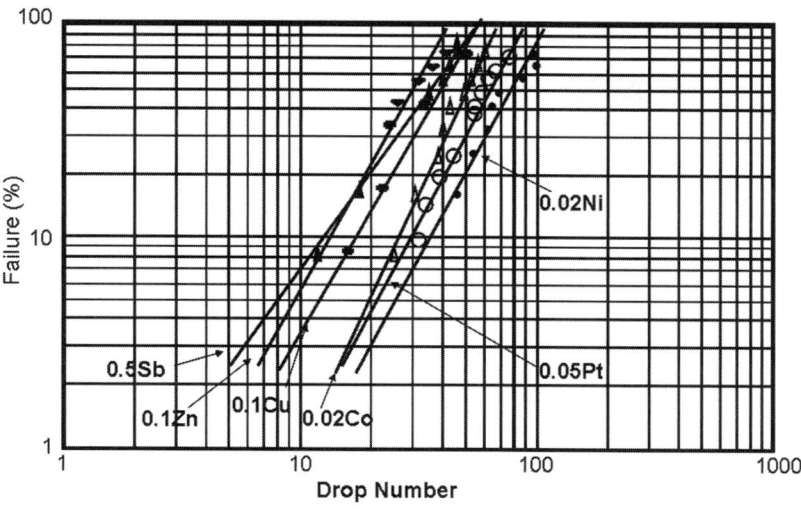

Fig. 13.27 A Weibull plot of drop test failures before 100°C thermal aging

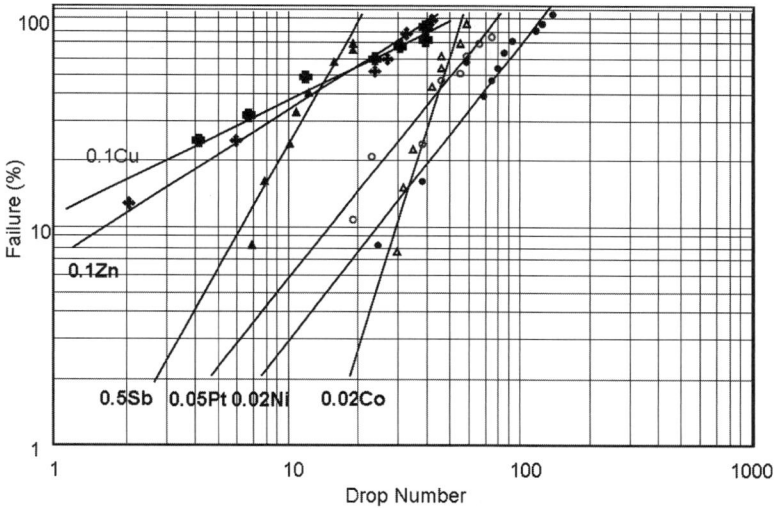

Fig. 13.28 A Weibull plot of drop test failures after 100°C thermal aging (100 h)

Table 13.2 A periodic table and evaluated nanoparticles

the solder reflow process and thermal aging when added to SnAg-based solder alloys. Hence, SnAgX solders show better drop test performance. (X means a transition metal such as Co, Ni, Pt.) On the other hand, a transition metal such as Cu, Ag, Au, Zn, a typical metal such as Al, In, or a metalloid such as P, Ge, Sb, located to the right of Cu in the periodic table, when added to SnAg-based solder alloys show increased IMC thickness and grain size after the solder reflow. Hence, SnAgY solders show poor drop test performance. (Y means Cu, Ag, Au, Zn, Al, In, P, Ge, Sb.)

13.7 Conclusion

Co, Ni, Pt, Al, P, Cu, Zn, Ge, Ag, In, Sb, or Au inclusions in Sn-3Ag-based lead-free solders were evaluated to study if these nanoparticles increase IMC thickness and grain size after four times solder reflow. Co, Ni, or Pt inclusions in Sn-3Ag-based lead-free solders did not increase IMC thickness and grain size significantly after four solder reflows. Al, P, Cu, Zn, Ge, Ag, In, Sb, or Au inclusions in Sn-3Ag-based lead-free solders increased IMC thickness and grain size after four solder reflows. IMC element analyses were carried out to study if nanoparticles were dissolved in the IMC after one solder reflow and four solder reflows. Co, Ni, and Pt were dissolved in IMC, which did not increase IMC grain size and thickness significantly after four solder reflows. Al, P, Ge, In, and Sb were not observed in the IMC, which increased IMC grain size and thickness after four solder reflows. Large solder displacement under drop impact improves drop test performance. Since solder hardness is related to solder ball displacement, low solder hardness (soft solder) improves drop test performance. Hence, hardness tests were performed to study if nanoparticles increase solder hardness after one solder reflow, and two solder reflows + 100°C thermal aging (0, 100, 200 h). Sn1.0Ag shows lower solder hardness than Sn3.0Ag. Cu, In, Ni, Co, and Pt did not increase solder hardness. Co, Ni, Pt, Al, P, Cu, Zn, Ge, Ag, In, Sb, or Au inclusions in Sn-3.0Ag-based lead-free solders were evaluated to study if these nanoparticles can reduce the frequency of occurrence of IMC fracture in high-impact pull tests. Co, Pt, or Ni inclusions in Sn3.0Ag-based lead-free solders show a low frequency of occurrence of IMC fracture compared with Cu, Sb, In, Bi, Ge, Al, and Zn in high-impact pull test. Co, Ni, Pt, Al, P, Cu, Zn, Ge, Ag, In, Sb, or Au inclusions in Sn-1.0Ag-based lead-free solders were evaluated to study if these nanoparticles can improve drop test performance. Ni, Co, and Pt were very effective for drop test performance.

The addition of transition metals such as Co, Ni, Pt (located to the left of Cu in the periodic table) to SnAg-based solder alloys did not increase IMC thickness and grain size significantly after the solder reflow process and thermal aging. Hence, these nanoparticles lead to better drop test performance than Cu, Ag, Au, Zn, Al, In, P, Ge, Sb.

References

1. Masazumi Amagai, Yoshitaka Toyoda, Takeshi Tashima, "High Solder Joint Reliability with Lead Free Solders," Proceedings of the IEEE 53rd Electronic Components and Technology Conference, New Orleans, LA, 2003, pp. 317–322
2. Masazumi Amagai, Yoshitaka Toyoda, Tsukasa Ohnishi, Satoru Akita, "High Drop Test Reliability: Lead-Free Solders," Proceedings of the IEEE 54th Electronic Components and Technology Conference, Las Vegas, NV, 2004, pp. 1304–1309.
3. Masazumi Amagai, "A Study of Nano Particles in SnAg Based Lead Free Solders," Proceedings of the IEEE 56th Electronic Components and Technology Conference, San Diego, CA, 2006, pp. 1170–1190

Chapter 14
Nano-Underfills for Fine-Pitch Electronics

Pradeep Lall(✉), Saiful Islam, Guoyun Tian, Jeff Suhling, and Darshan Shinde

14.1 Introduction

Packaging materials undergo dimensional changes under environmental exposure to temperature change. Thermomechanical cyclic loads induce stresses and damage interconnects. Underfills compensate for the mismatch in coefficient of thermal expansion (CTE) between silicon and the printed circuit board (PCB), and have been used as a supplemental restraint mechanism to enhance the reliability for flip-chip devices and chip-scale packages in a wide variety of applications including portable consumer electronics such as cellular phones, laptops, under-the-hood electronics, microwave applications, system in package (SIP), high-end workstations, and several other high-performance applications. Figure 14.1 shows an underfilled flip-chip assembly, with solder interconnects between the silicon chip and the PCB. It surrounds the solder balls. Underfill technology has evolved to meet the demand of decreasing feature size and increasing input/output (I/O) number in the integrated circuit (IC) chip.

14.2 Potential of Nano-Underfills

Capillary-flow underfills rely on the flow of the underfill into the gap between the bonded die and substrate. This process may be slow and often requires a batch cure after dispense. No-flow underfills are an attractive alternative to the use of capillary-flow underfills, and eliminate the need for postreflow batch cure operation and reduce production cycle time. Instead, no-flow underfills are cured during the reflow process. In addition, no-flow underfills typically contain a fluxing component, eliminating the need for flux dispensing and cleaning. No-flow underfill adhesives are mostly unfilled or filled with very low filler loading due to interference of fillers with solder joint yield.

P. Lall
Department of Mechanical Engineering, Auburn University, Auburn, AL, USA

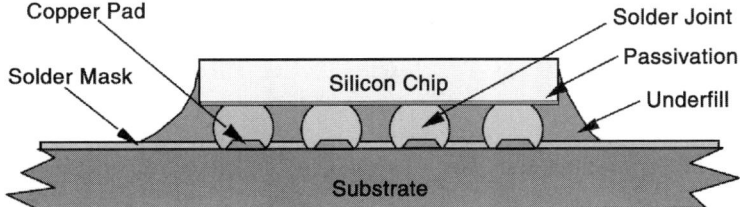

Fig. 14.1 Underfill between silicon and printed circuit board (PCB)

Epoxy is a common ingredient used in the underfill material, because of its desirable characteristics such as corrosion resistance and good adhesion, in addition to physical and electrical properties. However, epoxies by themselves possess a high CTE (above 80 ppm/°C), making them unable to meet the very first requirement of a good underfill material. For this reason, the epoxies are filled with filler particles that decrease the CTE of the adhesive. Common filler particles include silica and ceramics. Commercially available epoxies contain micron-sized filler particles. The no-flow underfills are largely unfilled or have very low filler loading because micron-sized filler particles interfere with the solder interconnection process [1, 2]. Low filler loading in no-flow underfills causes the CTE to be higher than capillary-flow underfills.

Nanosilica particles do not settle in an underfill formulation and do not interfere with the solder interconnection process, unlike micron-sized particles. Nanosilica imparts the same modulus enhancement and CTE reduction to adhesives as micro-sized silica particles. In addition, nanosilica particles can achieve much higher filler-particle loading than micron-sized particles without affecting the solder joint resistance, thus providing greater control over underfill properties. Underfill formulation including volume fraction, size, distribution, and material properties of the filler particles influences the elastic modulus, CTE, and mechanical deformation behavior and determine the thermomechanical reliability of flip-chip devices.

14.3 Nanoparticle Production

Production techniques are mainly classified into four categories including condensation from a vapor [3], chemical synthesis [4], solid state processes [5], and supercritical fluids (SCFs) [5]. Each of the processes is briefly described later.

14.3.1 Vapor Condensation

This method is used to make metallic and metal oxide ceramic nanoparticles. A solid metal is evaporated followed by rapid condensation to form nanosized clusters that settle in the form of a powder. Metal can be vaporized using various approaches including the exploding wire technique, vacuum evaporation on running liquids, and chemical vapor deposition (CVD).

In the exploding wire technique, an electrical arc is created at the surface of a metal wire with sufficient energy to explode or vaporize clusters of atoms. These clusters condense within an inert gas into nanoscale particles [6]. The method vacuum evaporation on running liquids uses a thin film of a relatively viscous material, an oil, or a polymer on a rotating drum. A vacuum is maintained in the apparatus and the desired metal is evaporated or sputtered into vacuum [7]. The CVD technique uses both liquid and gas forms of a substance, which are put into a reactor. Depending on several parameters like gas–liquid ratio, the order of gas, liquid addition, and heat application duration, different particle shapes and different sizes can be created [8].

Variation of the medium into which the vapor is released affects the nature and size of the particles. For example, inert gases are used to avoid oxidation when creating metal nanoparticles and reactive oxygen is used to produce metallic oxide ceramic nanoparticles. The main advantage of this method is low contamination. Particle size is controlled by variation of parameters such as temperature, gas environment, and evaporation rate.

14.3.2   Chemical Synthesis

The most widely used chemical synthesis technique consists of growing nanoparticles in a liquid medium composed of various reactants. Chemical techniques are better than vapor condensation techniques for controlling the final shape of the particles. The final size of the particles is controlled by stopping the process when the desired size is reached or by choosing chemicals that form particles that are stable and stop growing at a certain size [4].

14.3.3   Solid State Processes

Grinding and milling can be used to create nanoparticles [9]. The milling material, milling time, and atmospheric medium affect resultant nanoparticle properties. Bead mills are used to grind coarse particles into the nanometer range. Bead mills grind suspended solid particles by impact and shearing forces between moving grinding beads. Very fine particle size grinding media beads, in the range of 70–125 μm, are used. This process has a limitation; industrial equipment that can use these small beads on a continuous basis is not well known or well accepted due to the difficulty in handling the small beads, i.e., removing them for the suspension after dispersing the particles, or loading and discharging the small beads into the machine.

14.3.4 Supercritical Fluids

SCFs are used as a medium for metal nanoparticle growth [10]. SCF precipitation processes can produce a narrow particle size distribution. A gas becomes a SCF above a critical point, at a certain temperature (critical temperature, T_c) and pressure (critical pressure, P_c). CO_2 is used because of its relatively mild supercritical conditions ($T_c = 31°C$, $P_c = 73$ Bar). CO_2 is also inexpensive, nontoxic, noncorrosive, nonexplosive, and nonflammable.

14.4 Surface Modification of Underfills

To improve the rheological behavior of the nanosilica composite no-flow underfill, filler surfaces are treated using silane coupling agents [11]. The nanosize silica particles may have a high surface area covered by silanol groups. This hydrophilic surface does not process good compatibility with the polymer resin, and therefore the silica cannot be wetted very well by the resin. On the contrary, the silica particles with hydrophilic surface easily adhere to each other through hydrogen bonding and form irregular agglomerations. The agglomerations of the nanosilica can form a network through the whole polymer matrix and occlude liquid polymer in their interparticle voids, thereby affecting the rheology of the composite underfill and giving a significant rise to the viscosity as filler loading increases. The high viscosity of the no-flow underfill not only makes underfill dispensing difficult, but prevents the chip from collapsing and forming solder joints during the solder reflow process as well. The presence of filler agglomerations will decrease the maximum filler loading, resulting in an inferior thermal mechanical performance.

To decrease the viscosity of underfill and to increase the extent of filler loading, it is therefore necessary to reduce the degree of agglomeration. For nanosize filler, the mechanical mixing and dispersion methods such as high-speed shearing or milling are not effective to break down the agglomerations because the electrostatic forces holding the particles together are stronger than the shear force created by the velocity gradient. In such circumstances, chemical treatment of nanoparticle surface is necessary to achieve better compatibility and dispersion of the filler in epoxy resin.

Silane coupling agents are often used to treat the silica filler because of their unique bifunctional structure with one end capable of reacting with the silanol groups on silica surface and the other end compatible with the polymer. The modification process is described as a hydrolysis and condensation reaction between the silane coupling agents and the silica surface in a polar medium. The bonding between the silane and the silica surface removes the surface silanol groups and changes the hydrophilic surface into a hydrophobic surface. The ideal result of surface treatment is to reduce the filler–filler interaction and to achieve the homogenous distribution of the nanosize silica in the polymer. Figure 14.2a shows the nanosilica without silane treatment that formed large agglomerations, and Fig. 14.2b shows the nanosilica after silane treatment.

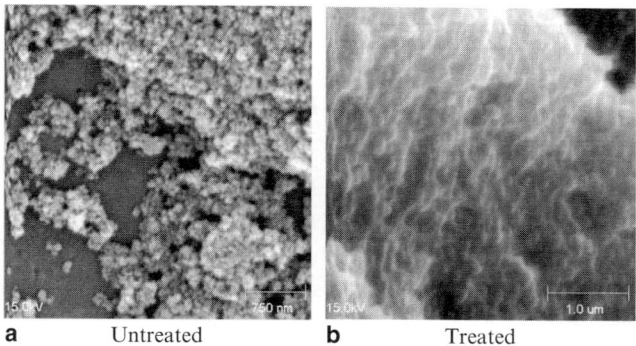

| a | Untreated | b | Treated |

Fig. 14.2 SEM photographs of nanosilica composite materials [12]

14.5 Computational Techniques for Property Design

Property prediction techniques for formulation of the underfills with desirable thermomechanical characteristics can significantly impact reliability of fine-pitch assemblies. In this section, reliability of nanosilica underfills and methodologies for property prediction have been discussed. In this section, models have been presented for prediction of underfill properties [13, 14]. The models are based on constituent component properties and enable the prediction of effective equivalent properties of statistically isotropic composites formed by random distribution of spherical filler particles. Drugan [15, 16] showed that the representative volume element (RVE) is an effective technique for prediction of elastic composites. Segurado and Llorca [17] also demonstrated the RVE with modified random sequential adsorption algorithm (RSA) as a reliable approach to estimate the equivalent properties. Lall et al. [13, 14] used an algorithm similar to the RSA to generate statistically isotropic cubic unit cells of underfill containing up to 38% nanofillers. Developed unit cells have been analyzed, and the elastic modulus and CTE of the underfill have been computed by using RVE implementation in implicit finite elements.

14.5.1 Unit Cell Generation

The modulus of elasticity and the CTE of nano-underfills can be predicted by the implicit finite element models of three-dimensional cubic unit cells. The unit cell is generated by randomly distributing spherical fillers in an epoxy matrix. The volume of the cube is L^3, N is the total number of particles, and r is the radius of the spherical particle. Volume fraction (γ) of the filler is determined as the ratio of total volume of the sphere to the volume of the cube.

$$\gamma = \frac{N\left(\frac{\pi}{4}r^3\right)}{L^3}. \tag{14.1}$$

Volume fraction of the cube is controlled by varying the total sphere number N as required by the value of L. The radius r of the sphere is kept the same for the analysis of all volume fractions. The fillers should be distributed in such a way that the unit cell should be isotropic, i.e., equivalent in all directions and it should be quite suitable for generating good finite element mesh. An algorithm based on modified RSA can be used to generate the random center coordinates of the nanosilica particles in the underfill [17]. According to this algorithm all the accepted random coordinates of the particles (for $\gamma = 0$–0.25) pass the following conditions: (a) If the particle surface touches the surface of the cube, or if they are very close, it may not be possible to mesh or the generated finite element mesh will be distorted or sometime meshing may not be possible at all. To avoid these, the particles are kept inside of the cube at some minimum distance d_2 from the surface of the cube.

$$d_2 = r + 0.1r. \tag{14.2}$$

To fulfill the above condition, center coordinates of the ith particle must pass the following check.

$$x_j^i \geq d_2; \quad j = 1, 2, 3.$$
$$\left| x_j^i - L \right| \geq d_2; \quad j = 1, 2, 3. \tag{14.3}$$

(b) If two adjacent particles overlap each other, they will violate the rigid sphere condition. In addition, two adjacent particles cannot touch each other. To fulfill this condition, center coordinates of the ith particle must pass the following check.

$$d_1 = 2.07r. \tag{14.4}$$

$$\left\| \left(\bar{x}^i - \bar{x}^k \right) \right\| \geq d_1; \quad k = 1, \ldots, (i-1). \tag{14.5}$$

The modified RSA method has been used to achieve volume fraction higher than $\gamma = 0.25$. In unit cells with volume fraction, γ, such that $0 < \gamma < 0.25$, all the particle centers are kept inside the cube, and no particle overlaps the outer surfaces of the cube. For unit cell volume fractions more than 0.25, the particles are allowed to overlap the surface boundary of the cube. If any particle overlaps the surface, the portion of the particle outside the cube is carefully cut into several sections and copied at a suitable position on the opposite surface of the cube. First, a cubic cell with volume fraction less than 0.25 is generated by fulfilling the earlier two conditions and then the cell is compressed to a smaller size while keeping the size of the spherical particle the same. The size of the sphere and total number of the sphere does not change during the operation, but the cell is compressed, giving a higher volume fraction of filler content. Length of the cubic cell is compressed first, by multiplying it by a user-defined shrinking factor c_f (<1). The new length of the cube is given by

$$L_n = c_f L. \tag{14.6}$$

The old position of the center of ith particle is given by $\vec{x}^i$, then the new position of the particle $\vec{x}_n^i$ will be given by

$$\vec{x}_n^i = \vec{x}^i c_f. \tag{14.7}$$

Once the particles have moved, they will be allowed to overlap the surface of the cube but no particle center will be outside of the cube. If the new coordinates of the ith particle $\vec{x}_n^i$ fall outside the cube then it is moved back into the cube at a random position between the surface and at an inward distance $(r - \alpha)$, where α is a user-defined constant and $\alpha > 0.1r$. If the particle center sits at a distance r from the surface of the cube then the particle surface will touch the cube surface and meshing will not be possible. If the new coordinates of the ith particle $\vec{x}_n^i$ fall inside the cube and at a distance r from the surface then it will be moved to a random position between the surface and at an inward distance $(r - \alpha)$, where α is a user-defined constant and $\alpha > 0.1r$. If the particle overlaps any other previously accepted position then it will be moved to a new random position in a random direction. The new position will be given by

$$\vec{x}_n^i = \vec{x}_n^i \pm \beta, \tag{14.8}$$

where β is a small random number. Smaller β ensures faster convergence. The new position of the particle will be accepted only if all the earlier conditions are fulfilled. Since the algorithm involves random movement and random positioning, each of the iterations may not produce an acceptable distribution. The algorithm also counts the number of iterations and stops the program if the number of iterations exceeds a certain number.

14.5.2 Isotropy of the Unit Cell

Once a valid distribution of the filler particles has been created, the algorithm also calculates the centroid and moment of inertia of the distributed particles. These quantities are calculated to check the isotropy of the distribution. Distributions with centroids at positions in the neighborhood of 0.5L are accepted. An isotropic distribution will have identical moment of inertia for the nanoparticles about three orthogonal axes. Figure 14.3 shows a plot of the coordinates of the centroids of the accepted distribution for different volume fractions. Table 14.1 shows the values of moment of inertia for the accepted distributions for several volume fractions. In almost all cases, moments of inertia about all three axes have very close values. This indicates that the generated unit cell is isotropic in all directions.

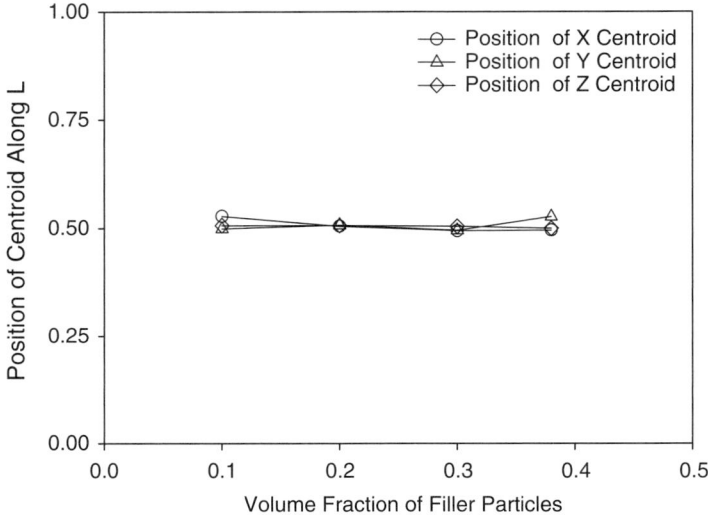

Fig. 14.3 Centroids for acceptable nanoparticle distribution

Table 14.1 Moment of inertia of acceptable nanoparticle distributions

Volume fraction	Moment of inertia about axis 1	Moment of inertia about axis 2	Moment of inertia about axis 3
0.10	100,105.25	85,311.77	86,064.94
0.20	208,298.14	195,363.03	210,952.10
0.25	278,355.74	259,266.68	254,355.15
0.38	168,768.24	179,810.48	186,470.79

14.5.3 Randomness of Filler Distribution

The radial distribution function describes on an average the presence of the filler particles around an arbitrary particle. To calculate the radial distribution function, a series of concentric spheres are drawn around an arbitrary particle at small fixed intervals Δr apart. Randomness of filler distribution can be used to ensure that no periodicity exists in the filler distribution. A radial distribution function of filler centroids, $g(R)$, is used for this purpose [18, 19]. The function is given as

$$g(r) = \frac{n(r)}{\rho 4\pi r^2 \Delta r}, \qquad (14.9)$$

where $g(r)$ is the radial distribution function, $n(r)$ is the mean number of particles in a shell of width Δr at a distance r, and ρ is the mean atom density. Figure 14.4 shows a plot of $g(R)$ for a distribution of filler volume fraction 0.20. The distributions are not periodic as very little presence of spikes of maxima and minima are seen.

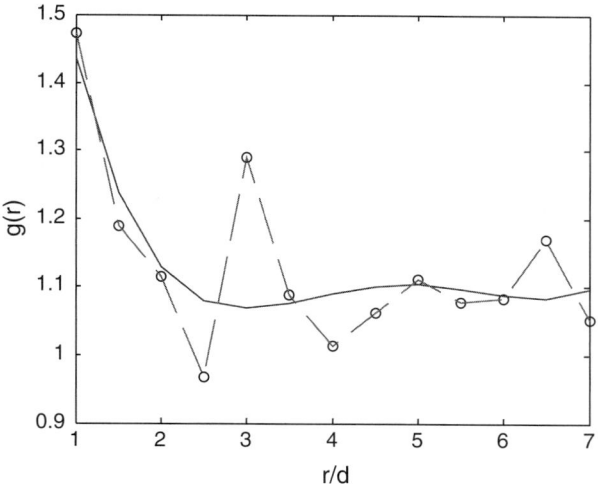

Fig. 14.4 Plot of radial distribution function

14.6 Finite Element Model of Unit Cell

Lall et al. [13, 14] created finite element models of the cubic unit cells to predict the CTE and elastic modulus of the nano-underfill. In these models, spherical silica fillers have been generated in an epoxy cube. The length of the epoxy cube is equal to the length of the unit cell considered in the unit cell algorithm. The filler particles have been generated by using the center coordinates and radius. For the models having volume fraction of filler more than 0.25, fillers overlapping the cube surface are carefully cut into pieces and copied at a suitable place onto the other side of the cube. The model geometry is meshed by eight-node tetrahedral brick elements. Figures 14.5–14.7 show representative models of the unit cell and filler distribution. Figure 14.8 shows an example of the mesh generated in the FEA models. Figure 14.9 shows the isotropic distribution of particles with volume fraction, $\gamma = 0.38$. Table 14.2 shows the material properties of the filler and epoxy matrix.

14.7 Prediction of CTE

Finite element models can be used to predict the CTE of the nano-underfill for different volume fractions of filler particles. Symmetric boundary conditions are used at the cube faces at x, y, and z equal to 0. The degrees of freedom are coupled at the faces at x, y, z equal to L. The temperature of the model is raised to a user-defined uniform temperature. This ensures that the cube faces will not be distorted after

Fig. 14.5 Isotropic view of filler distribution, $\gamma = 0.20$

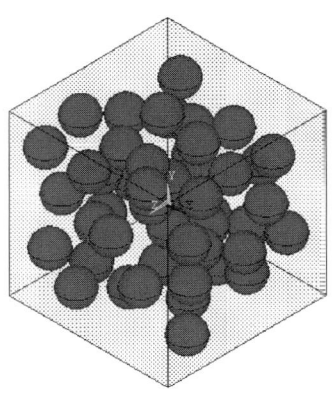

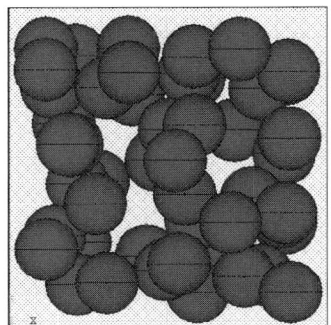

Fig. 14.6 Front view of filler distribution, $\gamma = 0.20$

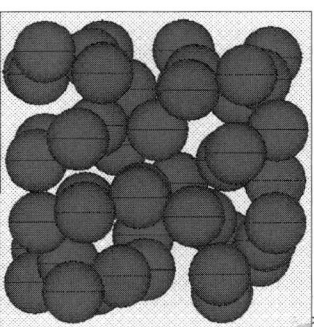

Fig. 14.7 Side view of filler distribution, $\gamma = 0.20$

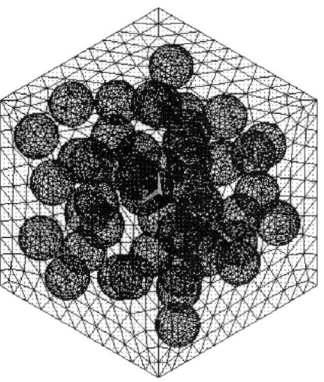

Fig. 14.8 FEA mesh, $\gamma = 0.20$

Fig. 14.9 Isotropic view of filler distribution, $\gamma = 0.38$

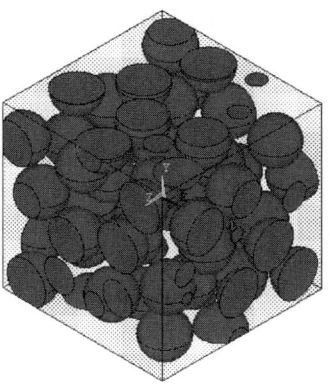

Table 14.2 Material properties for epoxy matrix and nanosilica particles

	Elastic modulus (GPa)	Poisson ratio	CTE (ppm/°C)
Filler	77.8	0.19	0.5
Epoxy	2.5	0.40	62.46

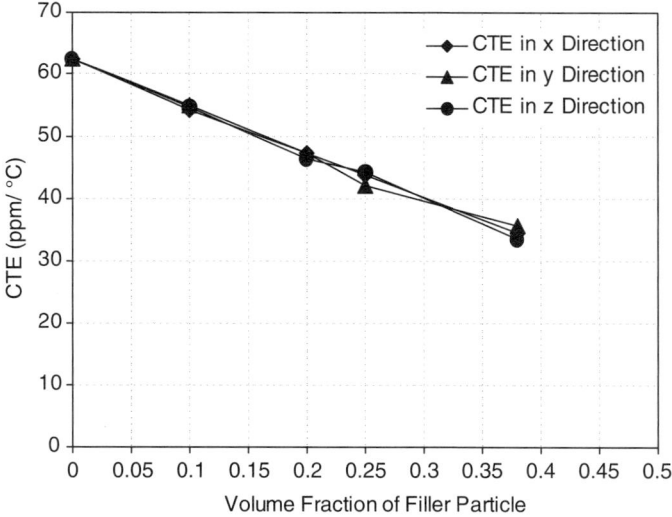

Fig. 14.10 Prediction of x, y, z CTE by finite-element analysis of unit cell

deformation. Calculated CTE values are presented in Fig. 14.10. The CTE of the underfill decreases linearly with the increase of volume fraction of the filler particles. Figure 14.10 shows predicted and experimental CTE values in x, y, z directions together on the same graph. This graph shows that CTE in all directions is almost identical, which once again proves that the generated unit cells are isotropic.

14.8 Prediction of Elastic Modulus

In this section, elastic modulus prediction of nano-underfill for different volume fractions has been presented. For this analysis, different boundary conditions than the CTE models are used. Displacement in the x-direction is 0 at the cube surface at x equal to 0. Fixed boundary condition is applied at one of the points at the mid position of the surface at x equal to 0. Tensile load is applied at the surface at x equal to L in the form of uniform pressure. A suitable small value of the pressure load will give better results because the modulus is defined as the initial slope of the linear part of stress–strain curves. Figure 14.11 shows the elastic modulus vs. volume fraction of filler particle. Elastic modulus increases exponentially with the increase of volume fraction of the filler particles.

14.9 Prediction of Bulk Modulus

In this section, nano-underfill bulk modulus calculations by finite element analysis of the earlier unit cell are discussed. A fixed boundary condition is applied on the node at the origin. The origin is located at one corner of the unit cell. The x, y, and z deflections are kept 0 on the surfaces located at $x = 0$, $y = 0$, and $z = 0$, respectively. A uniform tensile loading in the form of hydrostatic pressure is applied on the surfaces located at $x = L$, $y = L$, and $z = L$, respectively. The degrees of freedom are coupled in the corresponding directions at $x = L$, $y = L$, and $z = L$, respectively.

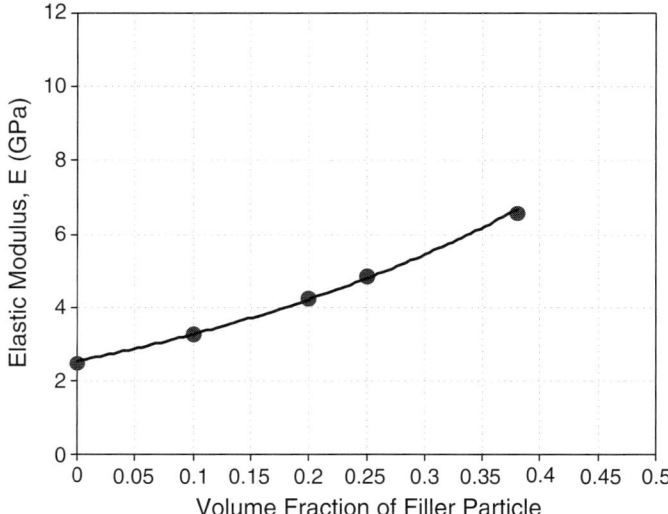

Fig. 14.11 Prediction of elastic modulus by FEA analysis of unit cell

14 Nano-Underfills for Fine-Pitch Electronics

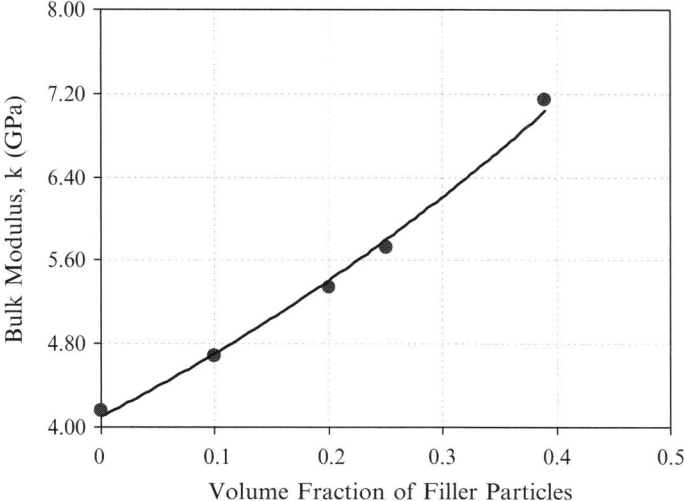

Fig. 14.12 Prediction of bulk modulus by FEA analysis of unit cell

The loading causes an increase in the volume of the cube without changing the shape at all. Since the unit cells are isotropic, it generated a uniform change in length in all directions. Bulk modulus values are calculated by using the following formula

$$k = \frac{\sigma}{\frac{\Delta V}{V}}, \qquad (14.10)$$

where k = bulk modulus, σ = value of applied hydrostatic stress, V = original volume, and ΔV = change in volume. The analysis has been performed for unit cells with $\gamma = 0, 0.1, 0.2, 0.25$, and 0.39. Figure 14.12 shows the predicted bulk modulus as a function of volume fraction of the filler particle. In this case, the bulk modulus of the epoxy without any filler is 4.16 GPa, and it increases mono-tonically with increases of volume fraction of the filler particles.

14.10 Prediction of Poisson's Ratio

The unit cell approach has been used to predict Poisson's ratio of the nano-underfill, and quantify the lateral deformation of the underfill under stress. Displacement in the x-direction is 0 at the cube surface at x equal to 0. Fixed boundary condition is applied at one of the points at the mid position of the surface at x equal to 0. Tensile load is applied on the surface at x equal to L in the form of uniform pressure. Nodal degrees of freedom on surfaces at $y = 0$ and L, $z = 0$ and L are coupled at

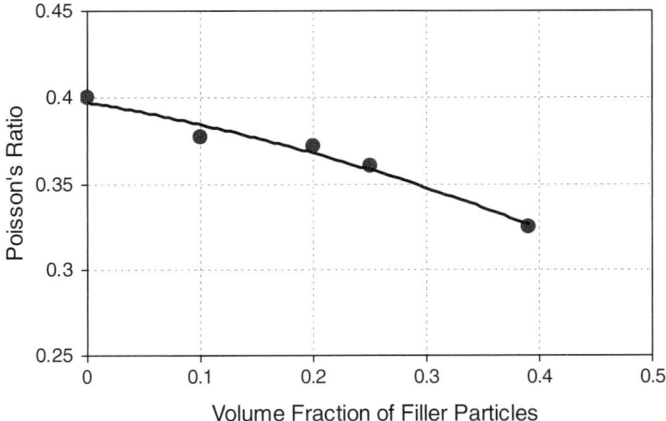

Fig. 14.13 Prediction of Poisson's ratio of nano-underfill by FEA unit cell analysis

corresponding directions. Loading causes extension in x direction and contraction in y and z directions. Poisson's ratio is calculated by using the extension and contraction values. Figure 14.13 shows that Poisson's ratio decreases nonlinearly with increase of filler volume fraction.

14.11 Viscoelastic Model for Nano-Underfills

Underfills exhibit time-dependent, strain-rate-dependent deformation response under thermomechanical loading. Nano-underfills are viscoelastic materials, which exhibit instantaneous elasticity, creep and recovery, stress relaxation, strain-rate dependence, and delayed recovery. Silica nanoparticles used in underfills exhibit linear elastic behavior. Viscoelastic property of the epoxy matrix is responsible for the time-dependent behavior of underfill.

Combinations of Maxwell and Kelvin Models have been used to represent viscoelastic material behavior in nano-underfills. The Maxwell model includes a linear spring element and a linear viscous dashpot element connected in series. Kelvin models include linear spring and linear dashpot connected in parallel. Generalized parallel Maxwell models and generalized series Kelvin Models have been used, since they are capable of representing instantaneous elasticity, delayed elasticity with various retardation times, and stress relaxation with various relaxation times in addition to viscous flow. Generalized parallel Maxwell model is better suited in cases where the strain history is prescribed since the same prescribed strain is applied to each individual element, and also the resulting stress is the sum of the individual contributions. The generalized series Kelvin model is more convenient for viscoelastic analysis in cases where the stress history is prescribed since the same prescribed stress is applied to each individual element, and the resulting strain is the sum of the individual strain in each element. A generalized series Maxwell Model formed by connecting several Maxwell elements in series is capable of exhibiting behavior equivalent to a single Maxwell element. Similarly, a generalized parallel Kelvin Model is capable of exhibiting behavior equivalent to a single-element Kelvin-Model.

14.12 Input Constants for Viscoelastic Material Model

ANSYS™ finite element analysis software has been used to compute the predicted time-dependent behavior of the developed unit-cell model. The nano-underfill material has been modeled using the Williams–Landel–Ferry (WLF) shift-function constants and Prony-Series constants of volumetric and shear response. WLF shift functions were calculated from elastic modulus relaxation data and time–temperature superposition method.

Figure 14.14 shows the stress-relaxation data used in calculation of the viscoelastic constants. Figure 14.15 shows the log–log plot of the temperature-dependent relaxation modulus vs. time data obtained from the data shown in Fig. 14.14. The curve at 25°C is taken as the reference curve, and other curves at 75, 100, and 125°C are shifted sideways parallel to the time axis to an appropriate distance and the single master stress-relaxation curve shown in Fig. 14.16 is formed. The magnitude of the total shift for each curve has been plotted against temperature in Fig. 14.17. WLF constants are calculated by substituting the temperature-dependent shift function a_T and temperature value into (14.11) and then performing

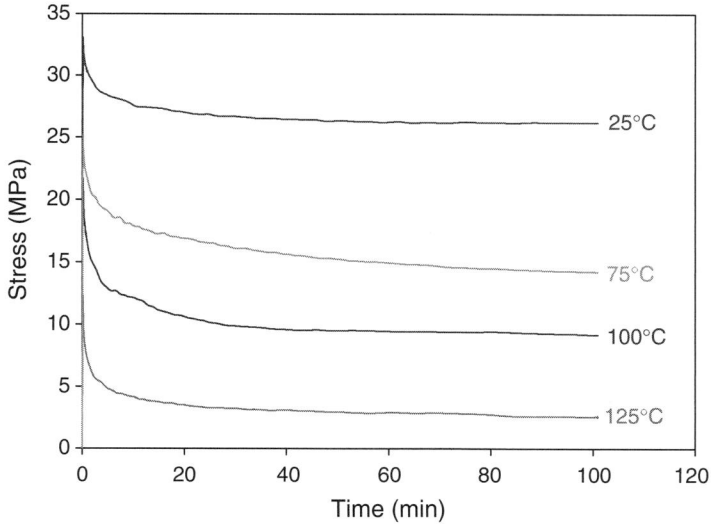

Fig. 14.14 Stress-relaxation data for calculating viscoelastic constants (1% strain)

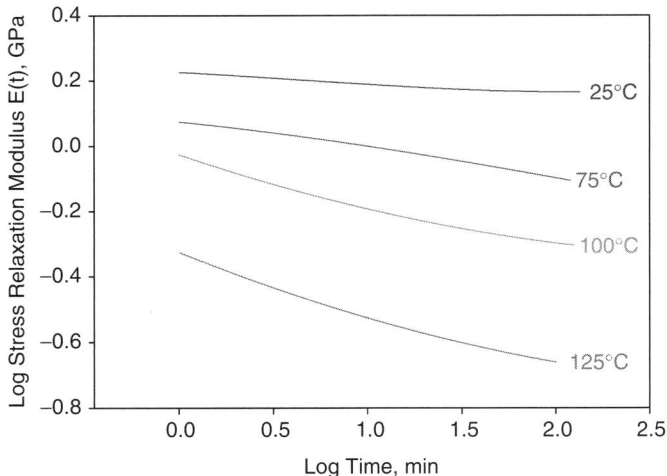

Fig. 14.15 Log–log plot of the relaxation modulus vs. time data

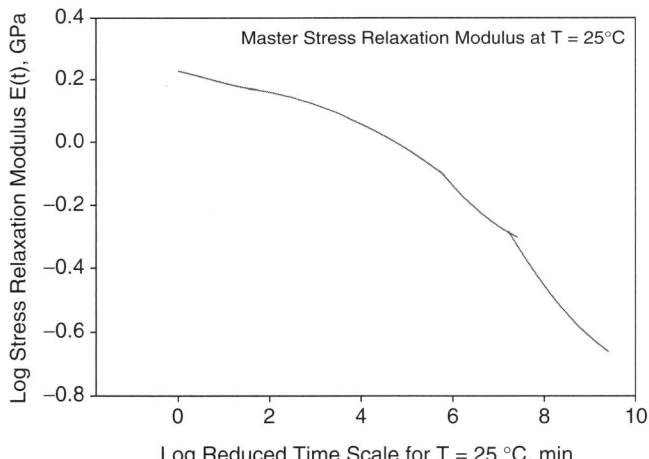

Fig. 14.16 Master relaxation modulus at 25°C

nonlinear regression of that equation. Equation 14.11 has been originally proposed by Williams, Landel, and Ferry and it gives the relation between the shift factor and the temperature. Calculated WLF constants for the nano-underfill studied are as follows: $C_1 = T_0 = 25°C$, $C_2 = -42.6$, $C_4 = 517°C$.

$$\mathrm{Log}(a_T) = \frac{-C_2(T-T_o)}{C_4 + T - T_o}. \quad (14.11)$$

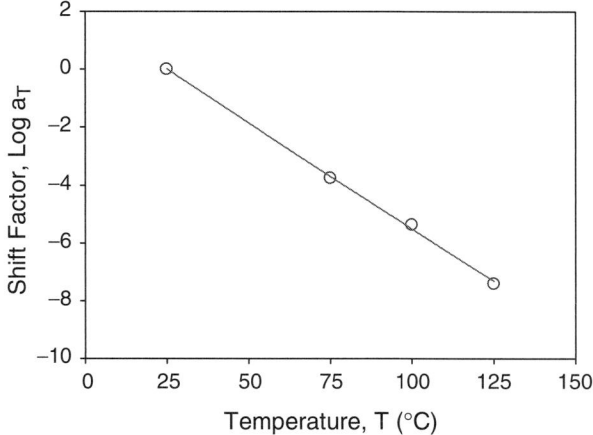

Fig. 14.17 Temperature-dependent shift factor

Viscoelastic constants to represent volumetric response and shear response are calculated by using Prony-Series. Prony-Series is derived from the solution of generalized Maxwell model in parallel. Stress–strain relation of spring and dashpot can be represented as

$$\sigma = k\varepsilon_2. \tag{14.12}$$

$$\sigma = \eta \dot{\varepsilon}_1. \tag{14.13}$$

Total strain,

$$\varepsilon = \varepsilon_1 + \varepsilon_2, \tag{14.14}$$

and strain rate

$$\dot{\varepsilon} = \dot{\varepsilon}_1 + \dot{\varepsilon}_2 \tag{14.15}$$

Inserting (14.13) and the time derivative of (14.12) in (14.15) will give

$$\dot{\varepsilon} = \frac{\dot{\sigma}}{k} + \frac{\sigma}{\eta}, \tag{14.16}$$

or

$$\frac{d\varepsilon}{dt} = \frac{1}{k}\frac{d\sigma}{dt} + \frac{\sigma}{\eta}. \tag{14.17}$$

Taking the Laplace transform of (14.17)

$$s\hat{\varepsilon}(s) - \varepsilon(0) = \frac{1}{k}\left[s\hat{\sigma}(s) - \sigma(0)\right] + \frac{\hat{\sigma}(s)}{\eta}.$$

Since $\varepsilon(0) = \sigma(0) = 0$,

$$s\hat{\varepsilon}(s) = \frac{1}{k}\left[s\hat{\sigma}(s)\right] + \frac{\hat{\sigma}(s)}{\eta}.$$

Rearranging gives

$$\hat{\sigma}(s) = \frac{ks}{s + \frac{k}{\eta}}\hat{\varepsilon}(s). \tag{14.18}$$

Assuming applied strain is a step function and that $\varepsilon(t) = \varepsilon_0 H(t)$, then

$$\hat{\varepsilon}(s) = \varepsilon_0 \frac{1}{s}.$$

Substituting in (14.18) gives

$$\hat{\sigma}(s) = \frac{ks}{s + \frac{k}{\eta}} \frac{\varepsilon_0}{s},$$

or

$$\hat{\sigma}(s) = \frac{k\varepsilon_0}{s + \frac{k}{\eta}}. \tag{14.19}$$

Taking the inverse Laplace transform of (14.19) gives

$$\sigma(t) = k\varepsilon_0 \exp\left(-\frac{k}{\eta}t\right).$$

Assuming $\tau = \frac{\eta}{k}$ gives

$$\sigma(t) = k\varepsilon_0 \exp\left(-\frac{t}{\tau}\right). \tag{14.20}$$

Fig. 14.18 Generalized parallel Maxwell Model with a free spring in parallel

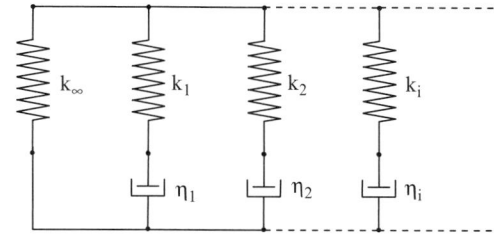

For the generalized Maxwell model, several Maxwell elements are connected in parallel as shown in Fig. 14.18. Since the Maxwell elements are in parallel, the stress-relaxation response will be obtained by summing the stress-relaxation response of the individual elements. For the combination

$$\sigma(t) = \varepsilon_0 \sum_{i=1}^{N} k_i \exp\left(-\frac{t}{\tau_i}\right). \tag{14.21}$$

Prony-Series representation of the elastic modulus can be obtained by dividing the stress response equation by applied constant strain. Derived Prony-Series will have the following form

$$E(t) = \sum_{i=1}^{N} k_i \exp\left(-\frac{t}{\tau_i}\right). \tag{14.22}$$

A generalized parallel Maxwell model with a free spring modulus, k_∞, connected in parallel (Fig. 14.18), will give the modulus relaxation response to a constant strain ε_0 as shown in (14.23),

$$E(t) = k_\infty + \sum_{i=1}^{N} k_i \exp\left(-\frac{t}{\tau_i}\right). \tag{14.23}$$

Denoting ks by Es will give

$$E(t) = E_\infty + \sum_{i=1}^{N} E_i \exp\left(-\frac{t}{\tau_i}\right). \tag{14.24}$$

ANSYS™ allows a maximum of ten Maxwell elements to approximate the relaxation function. Equations 14.25 and 14.26 are used to approximate the relaxation of shear modulus and bulk modulus.

$$G(t) = G_\infty + \sum_{i=1}^{n_G} G_i \exp\left(-\frac{t}{\tau_i^G}\right), \tag{14.25}$$

$$k(t) = k_\infty + \sum_{i=1}^{n_k} k_i \exp\left(-\frac{t}{\tau_i^k}\right), \tag{14.26}$$

where,

G_∞ = Final shear modulus (GPa)
k_∞ = Final bulk modulus (GPa)
n_G = Number of Maxwell elements to approximate G relaxation
n_k = Number of Maxwell elements to approximate k relaxation
τ_i^G and τ_i^k = Relaxation time for each Prony component (min).

For the present study, a total of five Maxwell elements are considered. Temperature-dependent Prony constants have been calculated by the nonlinear regression of (14.24) and (14.25). Calculated Prony series constants are given in Tables 14.3 and 14.4.

Table 14.3 Prony viscoelastic shear response

	25°C	75°C	100°C	125°C
G_1	0.0059	0.0208	0.1269	0.0215
τ_1^G	9.39	17.55	0.9385	2.92
G_2	0.0196	0.0278	0.0256	0.1146
τ_2^G	10.91	35.43	13.91	0.3227
G_3	0.0646	0.0887	0.0374	0.0396
τ_3^G	1.09	1.60	12.85	23.02
G_4	0.0195	0.0294	0.0376	0.0212
τ_4^G	12.12	32.07	12.85	2.94
G_5	0.018	0.0298	0.0093	0.0188
τ_5^G	32.23	34.81	66.02	2.73

Table 14.4 Prony viscoelastic volumetric response

	25°C	75°C	100°C	125°C
k_1	0.0972	0.0922	0.1148	0.07078
τ_1^k	16.09	31.15	13.24	2.84
K_2	0.0543	0.0804	0.4311	0.1345
τ_2^k	16.12	31.15	0.9412	23.06
k_3	0.2078	0.0922	0.1146	0.0709
τ_3^k	1.24	31.15	13.24	2.84
k_4	0.021	0.0919	0.1132	0.0679
τ_4^k	1.27	31.15	13.21	2.93
k_5	0.0429	0.3049	0.0288	0.3894
τ_5^k	16.03	1.71	58.52	0.3223

14.13 Material Property Measurement

A MT-200 tension/torsion thermomechanical test system from Wisdom Technology, Inc., has been used to test the samples in this study. The system provides an axial displacement resolution of 0.1 μm and a rotation resolution of 0.001°. Testing can be performed in tension, shear, torsion, bending, and in combinations of these loadings, on small specimens such as thin films, solder joints, gold wire, fibers, etc. Cyclic (fatigue) testing can also be performed at frequencies up to 5 Hz. In addition, a universal six-axis load cell is utilized to simultaneously monitor three forces and three moments/torques during sample mounting and testing. An environmental chamber provides a temperature range capability of ~−50 to 300°C. For uniaxial testing with the MT-200, forces and displacements are measured. The axial stress and axial strain are calculated from the applied force and measured cross-head displacement using

$$\sigma = \frac{F}{A}; \quad \varepsilon = \frac{\Delta L}{L} = \frac{\delta}{L}, \quad (14.27)$$

where σ is the uniaxial stress, ε is the uniaxial strain, F is the measured uniaxial force, A is the original cross-sectional area, δ is the measured cross-head displacement, and L is the specimen gage length (initial length between the grips).

14.14 Uniaxial Testing

In this section, uniaxial testing of the nano-underfill specimens is discussed. The specimen preparation procedure is presented in [20]. With the developed test specimen, tensile, creep, and relaxation tests have been performed in a wide temperature range. The nano-underfill material for detailed material testing is NUF1, and has a cure time of 30 min at 150°C. The NUF1 underfill has a volume fraction in the neighborhood of $\gamma = 0.22$. The glass transition temperature of this material is 156°C by the dynamic mechanical analysis (DMA) Method. The thickness of the cured uniaxial specimen is 75–125 μm (3–5 mil). The typical length and width of a specimen are 90 and 3 mm, respectively. In all uniaxial tests, the effective test length of the uniaxial specimen is 60 mm.

14.14.1 Stress–Strain Data

Figure 14.19 shows typical stress–strains curves for underfill NUF1 at 50°C, and a strain rate of $\dot{\varepsilon} = 0.001$ s^{-1}. The observed variation in the data between different tests is typical for cured polymeric materials. The elastic modulus E is the slope of the initial linear portion of the stress–strain curves. At 50°C and $\dot{\varepsilon} = 0.001$ s^{-1}, the

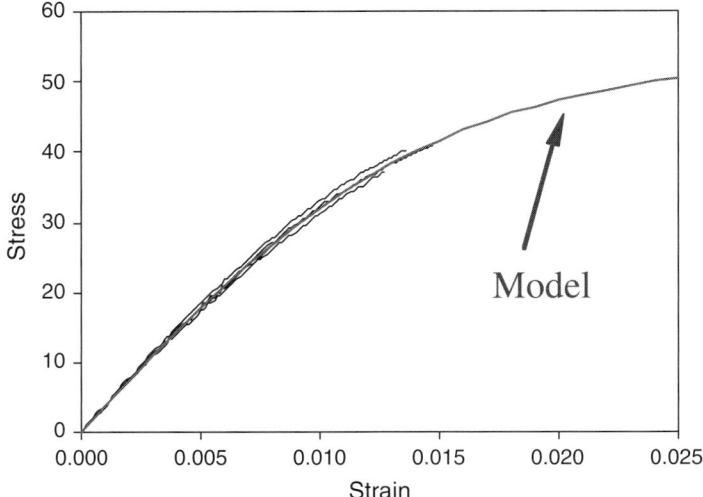

Fig. 14.19 Hyperbolic tangent model fit to typical underfill stress–strain data ($T = 50°C$, $\dot{\varepsilon} = 0.001$ s^{-1})

value for this underfill is measured to be $E = 3.74$ GPa. This value is found by averaging the results from five tests. An empirical three-parameter hyperbolic tangent model has been used to model the observed nonlinear underfill stress–strain data. Such a model has been used historically to model the stress–strain curves of cellulosic materials [21, 22]. The general representation of the hyperbolic tangent empirical relation is

$$\sigma(\varepsilon) = C_1 \tanh(C_2 \varepsilon) + C_3 \varepsilon, \tag{14.28}$$

where C_1, C_2, and C_3 are material constants. Differentiation of (14.28) gives an expression for the initial (zero strain) elastic modulus,

$$E = C_1 C_2 + C_3. \tag{14.29}$$

Likewise, constant C_3 represents the limiting slope of the stress–strain curve at high strains. For a given set of experimental data, constants C_1, C_2, and C_3 are determined by performing a nonlinear regression analysis of (14.28) through experimental data points. Based on the results from [22], the data from all of the stress–strain curves in a set should be fit simultaneously to obtain the best set of hyperbolic tangent model material constants.

The stress–strain data shown in Fig. 14.19 have been fit with the hyperbolic tangent model using a nonlinear regression analysis. Results from this calculation are $C_1 = 53.60$, $C_2 = 68.67$, and $C_3 = 10.68$ MPa. Excellent correlation is observed. Such results are typical for all of the temperatures at which testing has been performed.

Typical variation of the stress–strain curves of the tested nano-underfills NUF1 (20% volume fraction) and NUF2 (10% volume fraction) with respect to temperature is shown in Figs. 14.20 and 14.21. The strain rate for these tests is $\dot{\varepsilon} = 0.001\,\text{s}^{-1}$. The total test time (to failure) of a typical tensile test is less than 5 s for room-temperature experiments. Tests have been performed at $T = 25, 50, 75, 100, 125,$

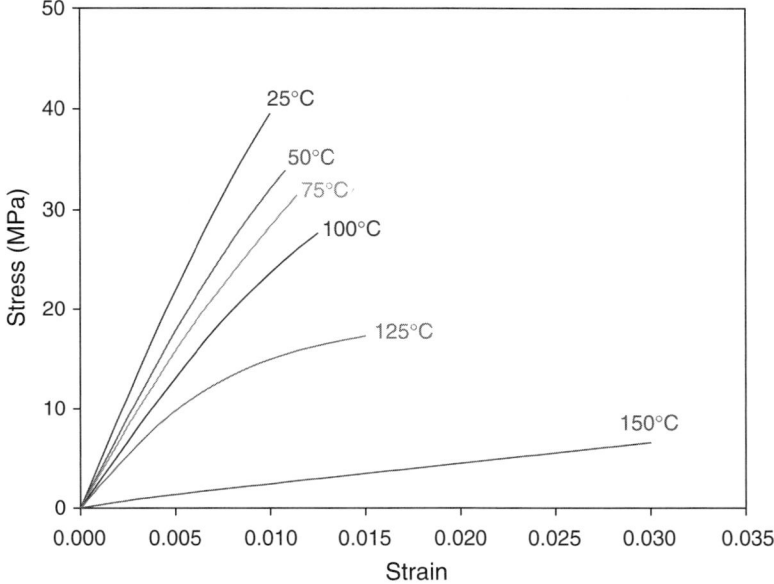

Fig. 14.20 Temperature-dependent average stress–strain curves of underfill NUF1 ($\dot{\varepsilon} = 0.001\,\text{s}^{-1}$)

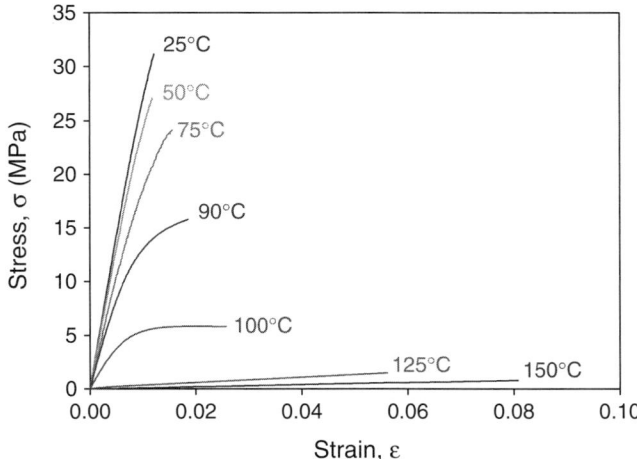

Fig. 14.21 Temperature-dependent stress–strain curves for nano-underfill NUF2 ($\dot{\varepsilon} = 0.001\,\text{s}^{-1}$)

and 150°C. The curves shown in Fig. 14.20 are the hyperbolic tangent empirical fits to the multiple curves measured at each temperature, and therefore represent an average measured stress–strain curve at each temperature. The stress–strain curves in Fig. 14.20 illustrate considerable softening and viscoplastic-type behavior as the temperature is increased.

Elastic modulus decreases approximately linearly with the increase of temperature from 25 to 125°C. Figure 14.22 shows temperature-dependent elastic modulus of underfill NUF1. The value of elastic modulus at 25°C is in the neighborhood of 4.65 GPa. This correlates well with the predicted value of 4.4 GPa from the unit cell model (Fig. 14.11). For temperature $T = 125°C$, elastic modulus decreases dramatically, and at $T = 150°C$ it is almost 0. This is typical as underfill approaches its glass transition temperature. Figure 14.22 shows a comparison of elastic modulus of the nano-underfill NUF1 vs. micron-filler underfill UF1, which has microsize particles as fillers. The volume fraction of filler particles in UF1 is almost double that in NUF1. Figure 14.22 shows that, due to higher filler concentration, UF1 has a higher elastic modulus than NUF1. Glass transition temperature of UF1 is 150°C, and its modulus drops greatly after 100°C but NUF1 has glass transition temperature as 156°C and its modulus drops greatly after 125°C.

Measurement of accurate mechanical properties at extreme low temperatures is very important for various applications, including the solar system exploration missions by NASA. In this present study, tests have been performed at very low temperatures down to −175°C. A newly developed environmental chamber is used with the MT-200 testing system for this purpose. Figure 14.23 shows stress–strain curves at −50, −100, and −175°C. The measured data have been plotted together with the room temperature and higher stress–strain curves shown earlier in Fig. 14.20. The test data shows that underfill NUF1 became more linear elastic as the temperature decreases to cryogenic temperature. Figure 14.24 shows the elastic

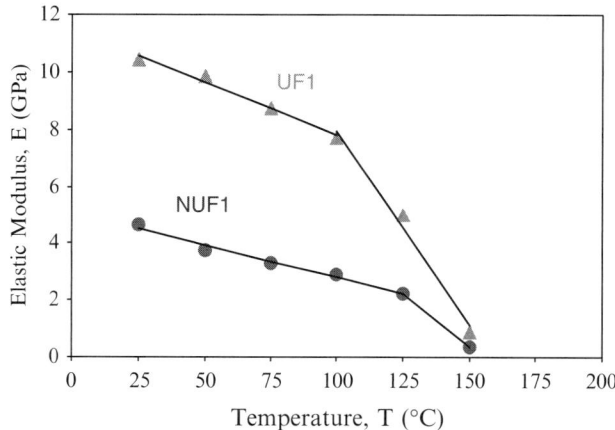

Fig. 14.22 Elastic modulus of a nanofiller underfill (NUF1, volume fraction = 0.22) and a microfiller underfill ($\dot{\varepsilon} = 0.001$ s^{-1})

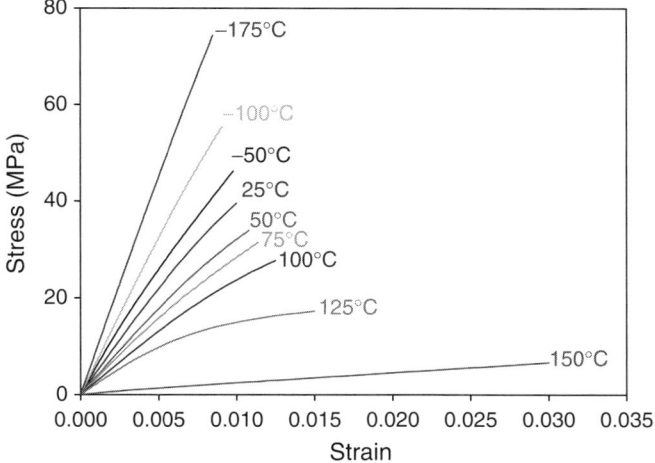

Fig. 14.23 Temperature-dependent stress–strain curves at extreme low temperatures ($\dot{\varepsilon} = 0.001$ s^{-1})

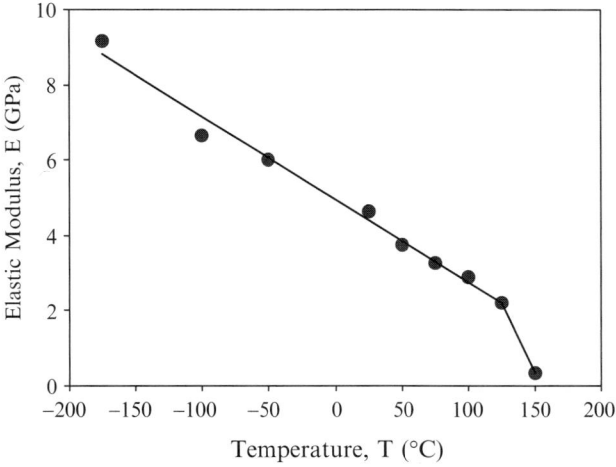

Fig. 14.24 Temperature-dependent elastic modulus in cryogenic temperatures ($\dot{\varepsilon} = 0.001$ s^{-1})

modulus of underfill NUF1 at extreme low temperatures. The elastic modulus also increases linearly with decrease in temperature from the room temperature to negative temperatures. At −175°C, elastic modulus of NUF1 became almost double the room temperature value.

14.14.2 Creep Data

Creep curves for the underfills NUF1 and NUF2 are shown in Figs. 14.25 and 14.26. All the tests have been performed at a constant stress level of 10 MPa. Both underfills show a strong influence of temperature upon the deformation under constant load.

At temperatures near the T_g, the creep compliance is greatly increased. The micro-underfill (UF1, 22% volume fraction) with comparable volume fraction to

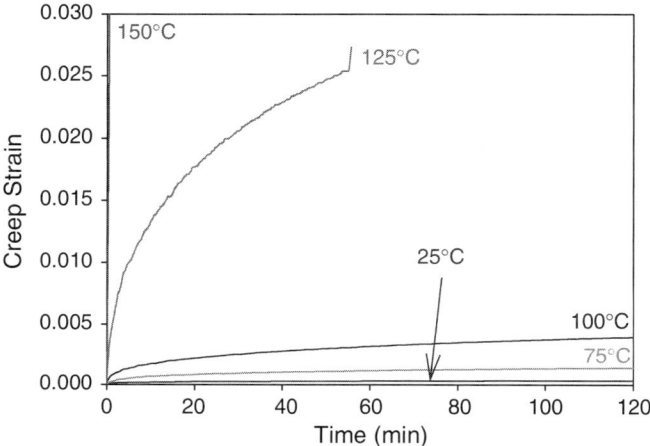

Fig. 14.25 Temperature-dependent creep data of NUF1

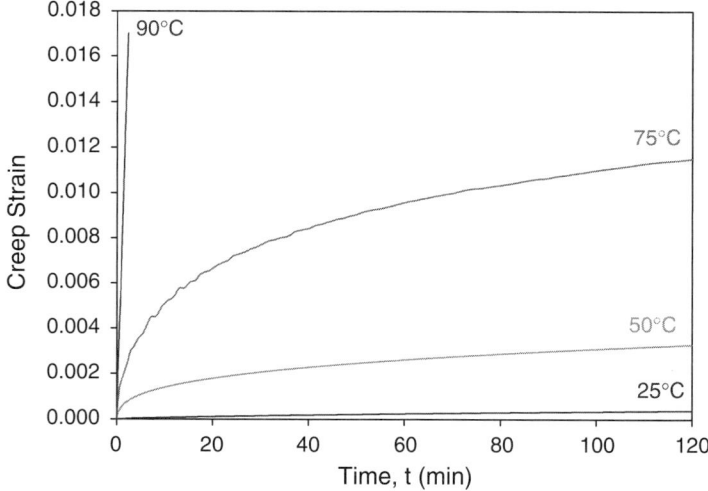

Fig. 14.26 Temperature-dependent creep data of NUF2

the nano-underfill (NUF1, 20% volume fraction) shows greater creep compliance (Fig. 14.27).

14.14.3 Preliminary Relaxation Data

Preliminary stress-relaxation test curves for the underfill NUF1 are shown in Fig. 14.28. All the tests are performed for a constant strain level, which is 1% in

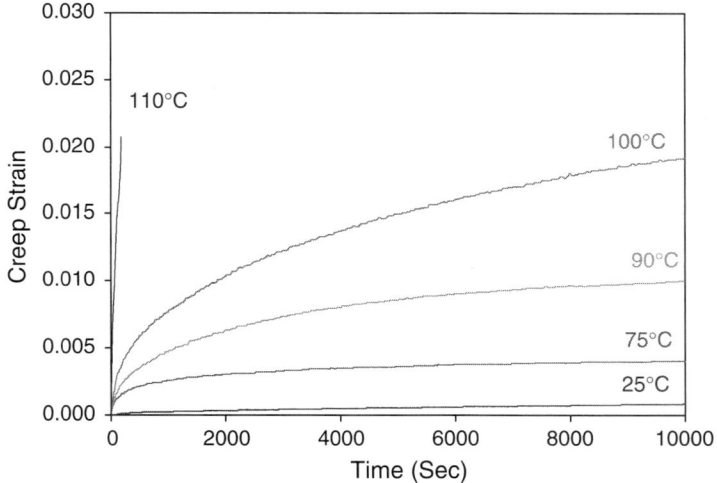

Fig. 14.27 Temperature-dependent creep data of UF1

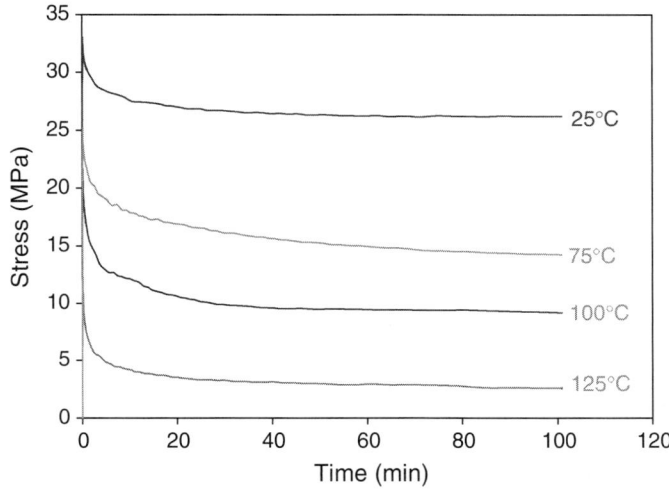

Fig. 14.28 Temperature-dependent stress-relaxation test data of underfill NUF1

this case. From test data, it is observed that the stress-relaxation rate increases with the increase of temperature. All the tests have been stopped after 100 min as it reaches a very small rate of change of stress after this time.

14.15 Correlation of Stress-Relaxation Behavior

The predicted relaxation behavior of 10% volume fraction (NUF2) and 20% volume fraction (NUF1) underfills has been correlated with experimental data. Fixed boundary conditions are applied at mid-side nodes of the $y = 0$ surface. The y-displacement has been set to 0 on the cube surface at $y = 0$. Tensile loading is applied at the surface at $y = L$ in the form of uniform nodal displacement. Step displacement loading is chosen instead of ramp loading to make sure instantaneous application of strain. The value of applied instantaneous strain is 1% and has been kept constant throughout the solution.

Nanofiller particles have been modeled as linear elastic. Viscoelastic material behavior of the generalized-parallel Maxwell model is applied for the epoxy matrix. To represent the viscoelastic properties of epoxy WLF constant data, Prony viscoelastic shear response data, and Prony viscoelastic volumetric response data of epoxy have been used. During the stress-relaxation test, the stress–strain value starts from 0 and ramps up to the applied strain level before undergoing stress relaxation. The instantaneous modulus continuously decreases with relaxation to a value significantly lower than the initial elastic modulus. In the calculation of Prony constants, the initial ramping of stress–strain is ignored, and instantaneous strain and stress is assumed to be applied at time equal to 0.

The elastic modulus value of the epoxy is 2.5 GPa, measured from the initial slope of the stress–strain curve of a tensile test. The aforementioned elastic modulus value may not be appropriate for viscoelastic analysis, since the applied strain loading is instantaneous. Use of the initial elastic modulus value may considerably overpredict the relaxation behavior of the underfill. Instantaneous value at the start of stress relaxation has been used for analysis. In this case, a value of 1.84 GPa has been used for analysis. Nonlinear solutions have been performed in this case with several time substeps. The nodal stress response of the surface at $y = L$ has been computed and plotted in Fig. 14.29. The predicted value for both the 10% and 20% volume fraction nano-underfill correlates well with the experimentally measured relaxation behavior.

14.16 CTE Measurement

CTE is the most critical thermomechanical property of underfill. Accurate measurement of CTE of the cured underfill is challenging, and in this section, a strain gage method has been applied to measure it. The strain gage technique is a very

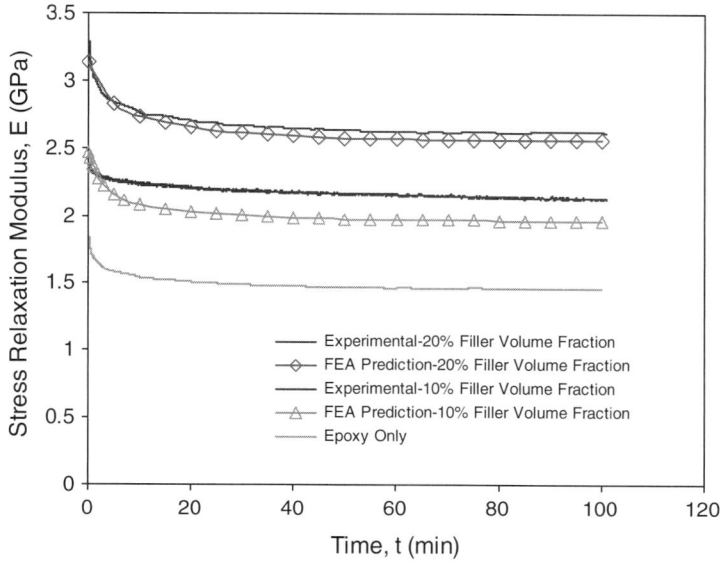

Fig. 14.29 Prediction of elastic modulus relaxation for NUF1 and NUF2 nano-underfills

accurate method for CTE measurement [23]. Hence, the specimen for this method should be very stiff; in this case, much thicker underfill specimens than the uniaxial test specimens are used. A 25 mm × 15 mm × 1 mm specimen has been cast and cured at 150°C for 30 min.

A strain gage is then placed on the cured specimen. A special type of CTE measurement strain gage is chosen for this purpose. A similar strain gage is placed on a reference material. The reference material is a TSB bar specially made for using in CTE measurement. The underfill specimen and the TSB bar are heated in an oven from room temperature to 120°C. Figures 14.30 and 14.31 show the strain gage readings at different temperatures for the gages placed on the reference material and underfill specimen, respectively. The CTE is calculated by the following equation [23],

$$\alpha_S - \alpha_R = \frac{\varepsilon_S - \varepsilon_R}{\Delta T}, \qquad (14.30)$$

where α_S and α_R are CTE of the test specimen and reference material, respectively. ε_S and ε_R are strain gage output of the gages on the test specimen and the reference material, respectively, and ΔT is the temperature difference. The calculated value of CTE of the underfill NUF1 is 39 ppm/°C. The value correlates reasonably with the CTE value of 45 ppm/°C predicted from the unit cell model (Fig. 14.10).

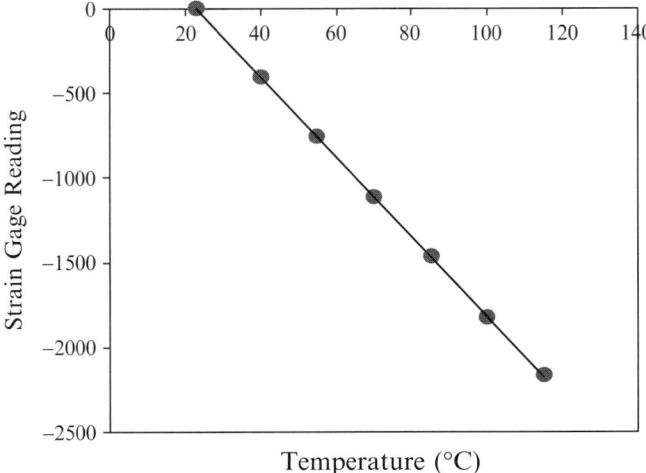

Fig. 14.30 Strain gage reading from the gage on the reference material

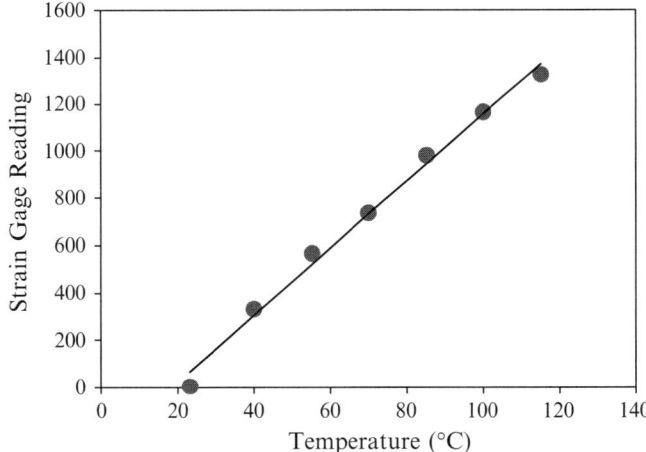

Fig. 14.31 Strain gage reading from the gage on the test specimen

14.17 Thermal Shock Reliability Testing

Thermal shock testing of PB8 board has been performed in a liquid-to-liquid thermal shock chamber. The chips were underfilled with NUF1. One complete cycle duration is 12 min. To complete a cycle, test boards are kept immersed for 5 min

14 Nano-Underfills for Fine-Pitch Electronics

Fig. 14.32 Cross section of flip-chip device assembled with nanosilica underfill. The top edge of the dark-gray area indicates the die-to-underfill edge

Fig. 14.33 Cross section of flip-chip device assembled with micron-filler underfill

alternately in cold and hot baths with extreme temperatures (−55 to 125°C, or −55 to 150°C) and 1-min holding time between. Figures 14.32 and 14.33 show the filler distributions in nanofiller and micron-filler underfills. Thermal shock data are discussed later.

14.17.1 Test Results, Failure Mechanism (Eutectic Solder)

The temperature range is from −55°C to 125°C. The die is a PB8 (5.08 mm × 5.08 mm) perimeter bumped flip chip with 88 bumps total. The pitch is 0.203 mm. The ball diameter is 0.127 mm. The solder alloy is 37Sn63Pb eutectic. The substrate is a high T_g laminate, with 10 chip sites on each substrate. Underfill delamination is observed to be the predominant mechanism causing electrical failure of flip chip on board assembly. Figure 14.34 shows the failed devices after 3,120 cycles of thermal shock and 100% failure of the population. Weibull distribution of failures is shown in Fig. 14.38. Underfill delamination from the chip interface provides the path for solder extrusion. Solder extrusion can be seen from x-ray inspection in Fig. 14.35 and SEM images of a flat-section sample in Fig. 14.36. Solder joint fatigue failures

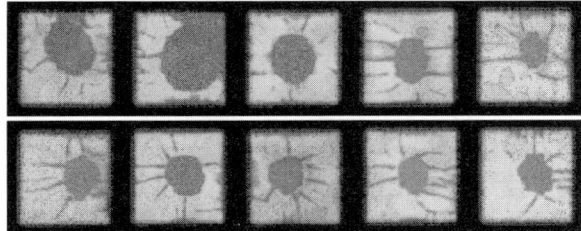

Fig. 14.34 Delamination at underfill and die interface

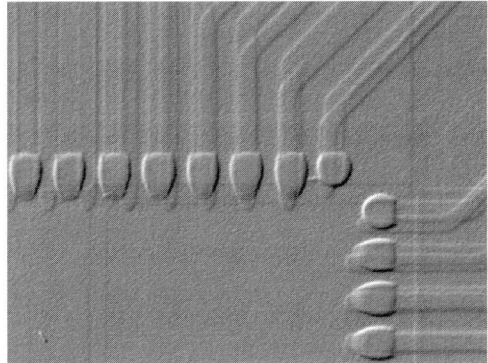

Fig. 14.35 Solder extrusion in underfill (X-ray)

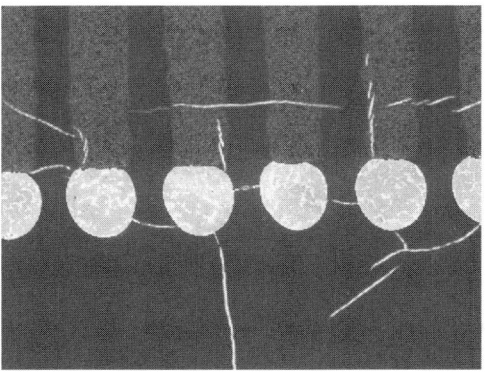

Fig. 14.36 Flat section showing solder extrusion in underfill (SEM image)

are caused by CTE mismatch between silicon and laminate substrate concurrent with underfill delamination. Fatigue cracks at the flip-chip interface are shown in Fig. 14.37.

14 Nano-Underfills for Fine-Pitch Electronics 319

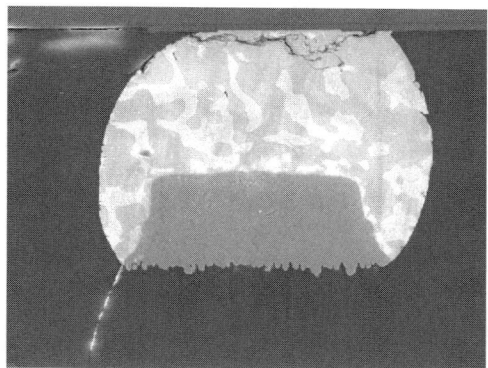

Fig. 14.37 Fatigue cracks at the flip-chip interface

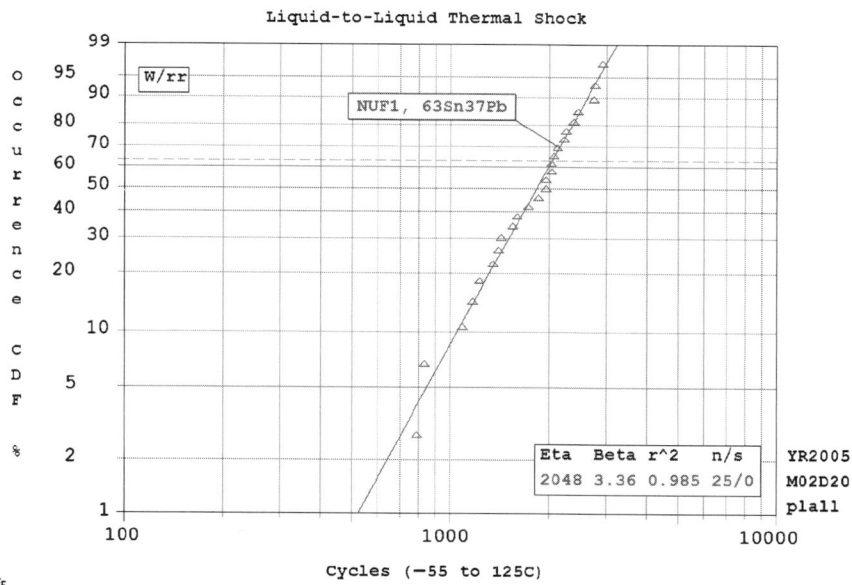

Fig. 14.38 Weibull distribution of failures for 63Sn37Pb solder alloy bumps

14.17.2 Test Results, Failure Mechanism (Lead-Free Solder)

The test die is a PB8 perimeter bumped flip chip with 88 bumps at 0.203-mm pitch. The ball diameter is 0.127 mm. The solder alloy is 95.5Sn3.5Ag1.0Cu. Test temperature extremes (−55 to 150°C) for lead-free solder bumped test vehicle are chosen to be higher than that for the eutectic test vehicle. Weibull distribution of failures is shown in Fig. 14.39. Similar to the failures in eutectic interconnects, cracks at the underfill-to-chip interface provide the path for solder extrusion. Figure 14.40 shows the delamination at the chip-to-underfill interface after failure. Solder extrusion can be seen from x-ray inspection in Fig. 14.41 and SEM images of a flat-section sample in Fig. 14.42. Solder fatigue failure caused by CTE mismatch between silicon and laminate substrate and by underfill delamination is shown in Fig. 14.43.

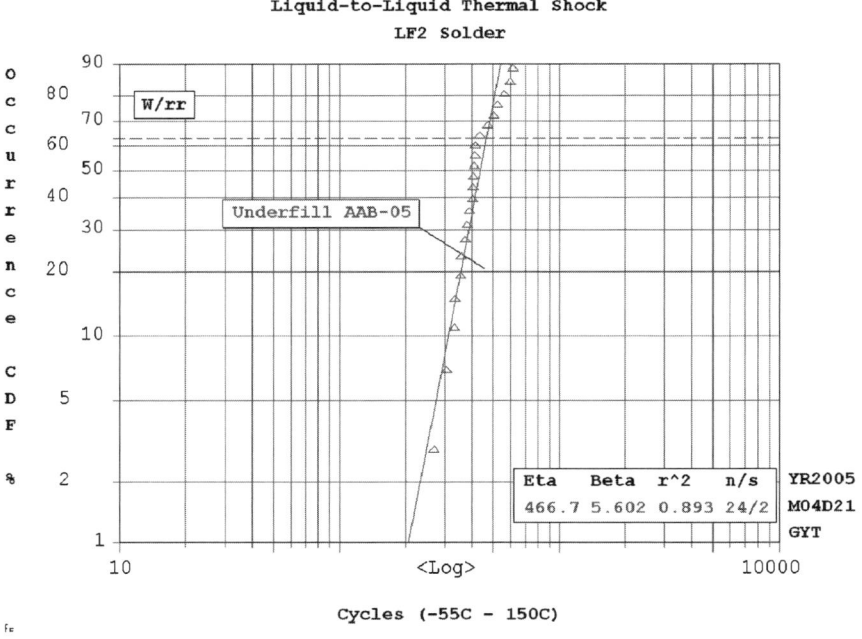

Fig. 14.39 Weibull distribution of failures for LF2 solder alloy bumps subjected to −55°C to 150°C thermal cycling

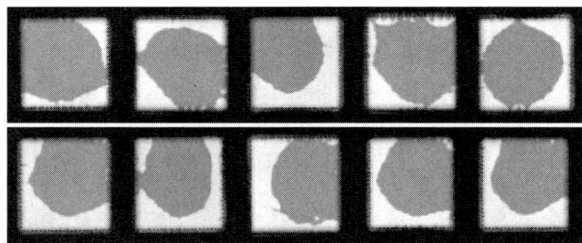

Fig. 14.40 Delamination at the chip-to-underfill interface after failure

14 Nano-Underfills for Fine-Pitch Electronics

Fig. 14.41 Solder extrusion in X-ray image for 95.5Sn3.5Ag1.0Cu solder bumps

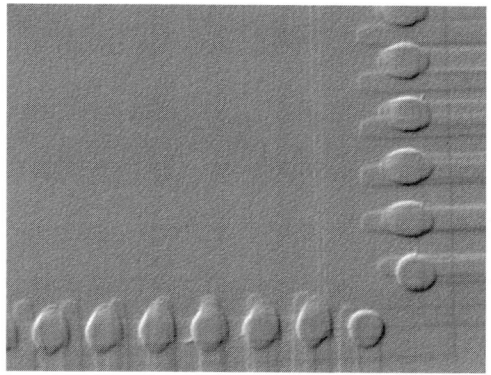

Fig. 14.42 Solder extrusion in underfill (SEM image) for 95.5Sn3.5Ag1.0Cu solder bumps.

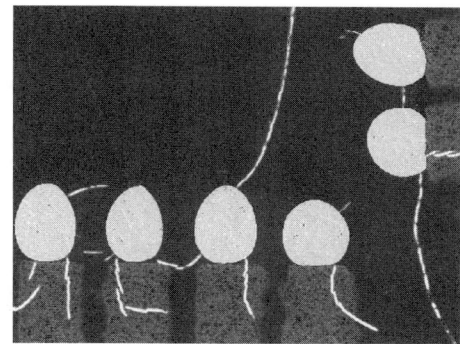

Fig. 14.43 Fatigue cracks at the flip-chip interface for 95.5Sn3.5Ag1.0Cu solder bumps.

14.18 Summary

In this chapter, a methodology based on a modified RSA has been discussed to generate statistically isotropic cubic unit cells of nano-underfills containing random-sized, randomly distributed nanofillers. Finite element models based on the random sequential addition algorithm have been presented for prediction of linear and nonlinear material properties of nano-underfill materials. Model validation of properties with experimental data has been discussed. CTE, elastic modulus, and viscoelastic properties including stress relaxation have been predicted and correlated with experimental data, as a function of the filler volume fraction. In addition, bulk modulus and Poisson's ratio have been predicted as functions of volume fraction. Temperature-dependent properties including, stress–strain and stress-relaxation data have been measured as a function of temperature between extremes of −175°C to +150°C. Liquid-to-liquid thermal shock testing with a temperature range −55°C to +150°C for lead-free and −55°C to 125°C for eutectic bumped flip-chip die is used to evaluate the thermomechanical reliability of the flip chip assemblies with nano-underfill.

References

1. Shi, S.H.; Wong, C.P., "Recent Advances in the Development of No-Flow Underfill Encapsulants – A Practical Approach Towards the Actual Manufacturing Application," IEEE Transactions on Electronics Packaging and Manufacturing, Vol. 22, pp. 331–339, 1999.
2. Liu, J.; Kraszewshi, R.; Lin, X.; Wong, L.; Goh, S.H.; Allen, J., "New Developments in Single Pass Reflow Encapsulant for Flip Chip Application," Proceedings of International Symposium on Advanced Packaging Materials, Atlanta, GA, pp. 74–79, March 2001.
3. Ha, J.-K.; Cho, K.-K.; Kim, K.-W.; Nam, T.-H.; Ahn, H.-J.; Cho, G.-B., "Consideration of Fe Nanoparticles and Nanowires Synthesized by Chemical Vapor Condensation Process," Materials Science Forum, Vol. 534–536, pp. 29–32, 2007.
4. Kim, D.-J.; Kim, K.-S.; Zhao, Q.-Q., "Production of Monodisperse Nanoparticles and Application of Discrete Monodisperse Model in Plasma Reactors," Journal of Nanoparticle Research, Vol. 5, pp. 3–4, August 2003.
5. Paul, H.; Weener, J.-W.; Roman, C.; Harper, T., "Nanoparticles," Technology White Papers, No. 3, pp. 1–11, October 2003, http//www.ceintifica.com.
6. Sen, P.; Joyee, G.; Alqudami, A.; Prashant, K.; Vandana, "Preparation of Cu, Ag, Fe and Al Nanoparticles by the Exploding Wire Technique," Proceedings of the Indian Academy of Sciences, Vol. 115, No. 5–6, pp. 499–508, 2003.
7. Yamada, S., "Structure of Germanium Nanoparticles Prepared by Evaporation Method," Journal of Applied Physics, Vol. 94, No. 10, pp. 6818–6821, November 15, 2003.
8. Adachi, M.; Shigeki, T.; Kikuo, O., "Nanoparticle Synthesis by Ionizing Source Gas in Chemical Vapor Deposition," Japan Journal of Applied Physics, Vol. 42, pp. 31–37, 2003.
9. Mende, S.; Stenger, F.; Peukert, W.; Schwedes, J., "Mechanical Production and Stabilization of Submicron Particles in Stirred Media Mills," Powder Technology, Vol. 132, pp. 64–73, 2003.
10. Inahara, J.; Ryuji, O.; Shin, C.K., Megumi, N.; Shinichi, H.; Takayuki, T., "Nano Particle Control on 300 mm-wafers in Super-Critical Fluid Technology," 212th ECS Meeting, Washington, DC, October 7–12, 2007.
11. Wong, C.P.; Sun, Y.; Zhang, Z., "Fundamental Research on Surface Modification of Nano-Size Silica for Underfill Applications," Electronic Components and Technology Conference, Las Vegas, NV, pp. 754–760, June 1–4, 2004.
12. Sun, Y.; Zhang, Z.; Wong, C.P., "Study and Characterization on the Nanocomposite Underfill for Flip Chip Applications," IEEE Transactions on Components and Packaging Technologies, Vol. 29, No. 1, pp. 190–197, March 2006.

13. Lall, P.; Islam, S.; Suhling, J.; Tian, G., "Nano-Underfills for High-Reliability Applications in Extreme Environments," Proceedings of the 55th IEEE Electronic Components and Technology Conference, Orlando, FL, pp. 212–222, June 1–3, 2005.
14. Lall, P.; Islam, S.; Suhling, J.; Tian, G., "Temperature and Time-Dependent Property Prediction and Validation for Nano-Underfills using RSA based RVE Algorithms," Proceedings of the ITherm 2006, Tenth Intersociety Conference on Thermal and Thermomechanical Phenomena, San Diego, CA, pp. 906–920, May 30 – June 2, 2006.
15. Drugan, W.J.; Wills, J.R., "A Micromechanics-Based Nonlocal Constitutive Equation and Estimates of the Representative Volume Element Size for Elastic Composites," Journal of the Mechanics and Physics of Solids, Vol. 44, pp. 497–524, 1996.
16. Drugan, W.J., "Micromechanics-Based Variational Estimates for a Higher-Order Nonlocal Constitutive Equation and Optimal Choice of Effective Moduli of Elastic Composites," Journal of the Mechanics and Physics of Solids, Vol. 48, pp. 1359–1387, 2000.
17. Segurado, J.; Llorca, J., "A Numerical Approximation to Elastic Properties of Sphere-Reinforced Composites," Journal of the Mechanics and Physics of Solids, Vol. 50, pp. 2107–2121, 2002.
18. Pyrz, R., "Correlation of Microstructure Variability and Local Stress Field in Two-Phase Materials," Material Science Engineering A Vol. 177, pp. 253–259, 1994.
19. Pyrz, R., "Quantitative Description of the Microstructure of Composites, Part I: Morphology of Unidirectional Composite Systems," Composites Science and Technology, Vol. 50, pp. 197–208, 1994.
20. Islam, M.S.; Suhling, J.C.; Lall, P., "Measurement of the Temperature Dependent Constitutive Behavior of Underfill Encapsulants," Intersociety Conference on Thermal and Thermomechanical Phenomena in Electronic Systems, ITHERM, Vol. 2, pp. 145–152, June 2004.
21. Andersson, O.; Berkyto, E., "Some Factors Affecting the Stress Strain Characteristics of Paper," Svensk Papperstidning, Vol. 54, No. 13, pp. 437–444, 1951.
22. Yeh, K.C.; Considine, J.M.; Suhling, J.C., "The Influence of Moisture Content on the Nonlinear Constitutive Behavior of Cellulosic Materials," Proceedings of the 1991 International Paper Physics Conference, TAPPI, Kona, Hawaii, pp. 695–711, September 22–26, 1991.
23. TN 513–1, Measurement Group, Technical Note, 2004.

Chapter 15
Carbon Nanotubes: Synthesis and Characterization

Yamini Yadav(✉) Vindhya Kunduru, and Shalini Prasad

15.1 Introduction

Carbon can form various types of structurally different frameworks due to the ability of the carbon atoms to form different species of valence bonds. The extremely organized coagulation process of carbon molecules resulting in the formation of the perfectly symmetric fullerene molecule despite the chaotic environment of the carbon arc is truly fascinating. Although many formation theories for the buckyball structure have been suggested, the "pentagon road model" is the most popular amongst many molecular physicists. The prominent features of this model are that carbon sheets have the tendency to accumulate isolated pentagonal carbon ring structures and grow into a carbon sheet with a large number of pentagons supporting its structure.

In the early years of the 1990s, Iijima, an electron microscopist from the NEC laboratories in Japan sifted through the soot formed inside the walls of the electric discharge chamber, which was almost completely amorphous containing fullerene molecules. Iijima finally found remarkable graphitic structures when he observed cylindrical deposits on the cathode of the arc discharge vessel [1]. A whole new era of intense research and experimental exploration began after the discovery of carbon nanotubes (CNTs) to understand their unique physical and electronic properties. Multiwalled carbon nanotubes (MWCNTs) and double-walled carbon nanotubes (DWCNTs) were produced during the initial arc discharge experiments by various research groups all over the world. Synthesis of single-walled carbon nanotubes (SWCNTs) [2] was reported by Iijima and Ichihashi of NEC, Japan [3] and Bethune's IBM group from California [4], independently.

This chapter describes the basic characteristics of various kinds of CNT synthesis techniques. Conventional techniques of CNT growth include vapor phase growth, corona discharge, catalyst-supported growth, pyrolysis of hydrocarbons, and laser ablation. More recent techniques such as plasma-enhanced chemical vapor deposition (PECVD) and chemical vapor deposition (CVD) not

---

Y. Yadav
Department of Electrical and Computer Engineering, Portland State University, Portland, OR, USA

only produce better quality CNTs but also can be patterned with the desired orientations on substrates [5]. This chapter attempts to explain the bottom–up approach of growing CNTs from primary growth mechanisms to the more sophisticated and modern techniques of controlled chemical synthesis of CNTs.

15.2 Synthesis of MWCNTs

Many growth mechanisms for MWCNTs have been studied, and subsequent experiments were also conducted for optimizing high-quality CNTs, which can also be bulk produced. The various techniques to synthesize MWCNTs are briefly described in this section. A mixture of benzene vapor and hydrogen was made to react in a reaction chamber in which a graphite rod was used as the substrate. A series of high-temperature treatments followed, which resulted in a material very similar to MWCNTs. This work by Oberlin et al. [6] created MWCNTs of good quality but relatively smaller yields when compared with the arc discharge method. Two graphite electrodes placed inside an arc discharge chamber were used in the arc discharge method. The arc struck between these two electrodes in an inert gas environment resulted in not only high-quality nanotubes but also produced better yields than any other methods. Further experimentation by Ebbesen and Ajayan in their 1992 paper "Large scale synthesis of carbon nanotubes" [7] showed that the yield of CNTs is extremely sensitive to the pressure of helium gas introduced into the arc discharge chamber. Other methods include electrochemical growth and catalytic growth of MWCNTs. Electrochemical growth was not very successful as it produced nanotubes with defective walls and also filled the innermost cylinder with chemical remnants. In the catalytic method, a pretreated substrate containing tiny catalytic particles on it is exposed to a temperature treatment chamber, which results in nucleation of the fiber-like growths on the substrate surface [8].

15.3 Synthesis of SWCNTs

Bethune and his team at IBM's research center in San Jose discovered SWCNTs [4]. The team's experiments began with exploring the electrical and magnetic properties of fullerene-related molecules, which enclosed metal particles. Blankets of soot, which clad the inner walls of their arc evaporation chamber, contained fullerene tubes of only one atomic layer. Subsequently, after more research on improving the technique of SWCNT growth by arc evaporation method, the same IBM team in collaboration devised a catalytic method of growth of SWCNTs.

15.4 Arc Discharge Method

The arc discharge method was first used to produce C_{60} and other fullerene molecules by changing the conditions of the arc discharge [2], which makes it one of the easiest and most common methods to produce CNTs in abundance. CNTs synthesized by a conventional arc discharge are mostly accompanied by a carbonaceous mix of carbon nanoparticles of undefined physical characteristics [2]. Graphite electrodes are evaporated in an inert gas environment by the application of homogenous electric voltage to produce a variety of fullerene molecules. The wide range of CNTs produced on the surface of the cathode have varying morphologies and suffer from a variety of defects such as amorphous carbon matter deposited on the inside and outside of the CNT walls. Scientifically useful specimens of CNTs can be retrieved from the crude end product of the arc discharge only after purification/distillation, which separates CNTs from carbon soot and other metallic residues.

A schematic representation of an arc discharge used by Ando et al. [9] from Japan is shown in Fig. 15.1.

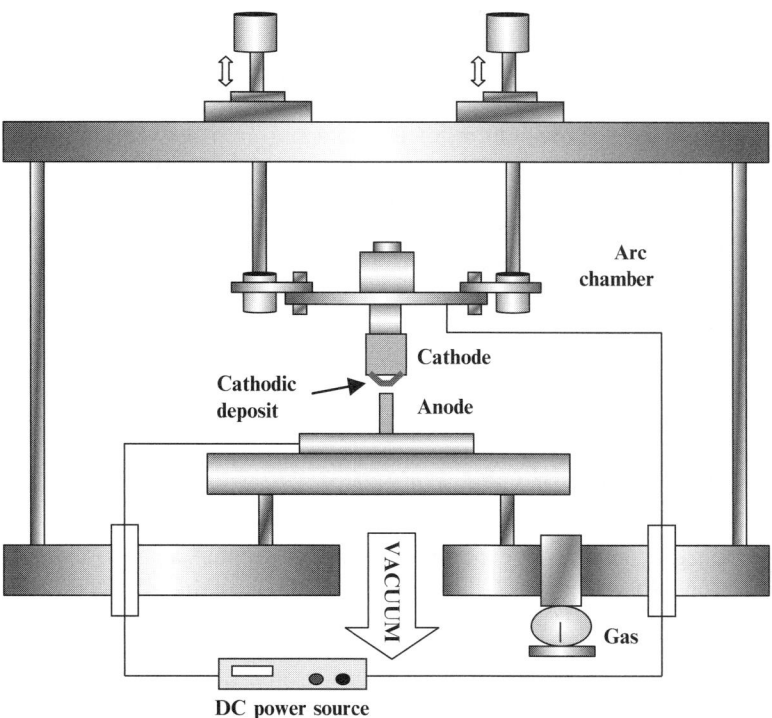

Fig. 15.1 Redrawn schematic of arc-discharge chamber used for producing carbon nanotubes (CNTs) [9]

15.4.1 Important Parameters for CNT Formation During Arc Discharge

The process of CNT creation near the cathode surface has been well explained by Gamaly and Ebbesen when they proposed a model for CNT formation based on the interaction of the bimodal carbon velocity distribution [10, 11]. According to Gamaly and Ebbesen, the carbon particle distribution in the raw rod-like carbonaceous deposit at the cathode is bimodal. While one mode comprises carbon nanoparticles of small size with varying shapes, the other mode represents CNTs of typical micrometer lengths with an outer diameter of 2–20 nm and inner diameters between 1 and 3 nm. This study of the mechanism of CNT formation revolved around answering key questions about interesting observations regarding the carbon deposits on the cathode.

Physical conditions of the arc discharge for efficient CNT production include parameters such as potential drop between electrodes, current density in the arc, the interelectrode spacing, plasma temperature, and pressure of the helium gas inside the chamber.

Electric parameters include the space-charge region near the cathode area where there is maximum potential drop due to the positive space charge around it. The gaseous mixture in the interelectrode region comprising neutral helium atoms, and neutral and ionized carbon species is important for determining the ionization potential in the arc. The condition of the potential drop between the electrodes being lesser than the first ionization potential is important to maintain the stability of the arc. The introduction of ions in the buffer gas might destabilize the ionic current, leading to the instability of the arc.

The study also confirms that the vapor layer near the surface of the electrode is most suitable for reactions involving carbon cluster formations. The effect of cooling of the system also has an impact on the quality and growth structure of the CNTs produced.

15.4.2 Mechanism of CNT Formation During Arc Discharge

After conducting numerous experimental and theoretical studies, Gamaly and Ebbesen proposed a sequence of events during the formation of CNTs in the arc discharge chamber [2, 11–13]. The vapor layer near the cathode surface is a result of evaporation of the solid graphite cathode consisting of saturated carbon vapors supporting maximum reactions to attach two groups of carbon particles having different velocity distributions. There are two competing sources of carbon – one group, which is a result of the evaporation from the cathode surface, has a Maxwellian velocity distribution, while the other group, composed of ions accelerated between the two potentials of the electrodes, has a single energy component oriented along the direction of the current. Thus, it can be noted that the absence of symmetry in the Maxwellian carbon groups will lead to the formation of random carbon clusters with unpredictable geometries, while the reacting particles from the directed current flux may form the elongated structures like CNTs.

The process of carbon deposition to form rod-like structures on the surface of the cathode is a result of many layers of carbon deposition. Following is the sequence of events responsible for the formation of CNTs during the arc discharge process.

Seed Formation: Seed structures are believed to be important for the growth of the carbon tubules. Initial heating and ionization of the electrodes and interelectrode gas play an important role in establishing a steady ion current [10]. As discussed earlier, the initial velocity distribution of the interacting particles, which is Maxwellian, is accompanied by the directed current fluxes that give rise to seed structures [11].

CNT Growth: The seed particle, which condenses on the cathode, interacts with the electric field in the cathode sheath to induce an electric dipole moment in the neighboring particles. The interaction between the various seed particles may align the electric field such that it results in the formation of linear carbonaceous structures [14]. Gamaly and Ebbesen also elaborated that the elongated carbon tubule formed due to the interaction of the directed ions with the solid surfaces is three orders stronger than that of the carbon cluster formation process. Hence, the arc discharge carbon formation process supports growth of carbon tubules along the axis of the symmetry more than the undirected formation of random carbon fullerene molecules. Simultaneously, the formation of Maxwellian velocity-distributed carbon particles continues with the tubule formation helping in the attachment of carbon particles to the growing tubules, thus forming MWCNTs.

CNT Kinetics: Although negatively charged CNT seeds are repelled in the cathode sheath, some seed particles can get deposited on the cathode surface due to the momentum provided by the ion flux and the CNT initial velocity at the edge of the cathode sheath [14]. Keidar and Waas from the University of Michigan established the parameters responsible for the formation of CNTs in different locations inside the arc discharge chamber [14]. It is summarized that CNTs with smaller aspect ratios can be influenced by the plasma jet and can get deposited on the chamber walls while those with longer aspect ratios get deposited on the cathode surface. The velocity that a CNT gains due to the interaction with the flowing plasma and this dependency on the aspect ratio of the CNTs determine the location of formation of the CNTS – on the chamber walls or cathode surface as explained in Fig. 15.2 later.

The resulting CNTs formed from the aforementioned methods are produced in abundance, but at the cost of quality, since the graphitization of the tube walls suffers from imperfections. Wang et al. from the Northwestern University [2] have improved the quality of the buckytubes as compared with those prepared by the arc discharge method by modifying certain physical and electrical components of the arc discharge chamber. A tungsten wire, which acts as an extension to a Tesla coil, was incorporated and made to point toward the arc region. The *stable glow discharge* was produced by a corona discharge triggered by the Tesla coil, thus overcoming the instabilities caused when the anode and cathode are made to strike each other

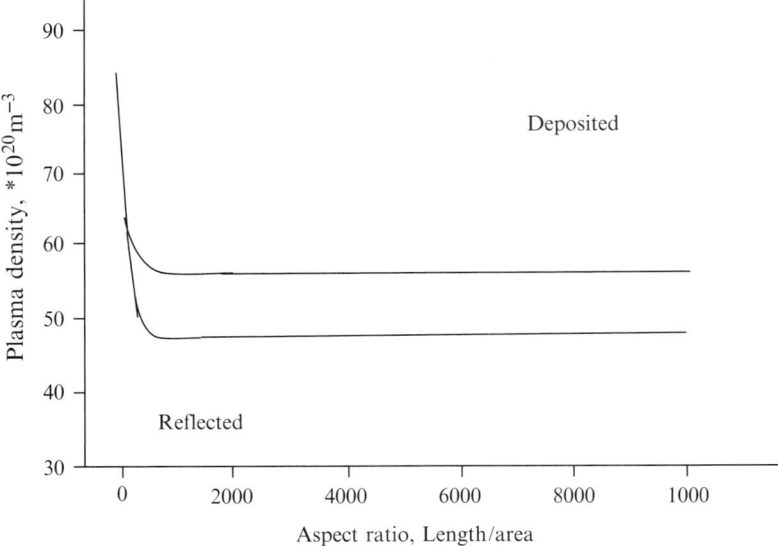

Fig. 15.2 Redrawn graph demonstrating that CNTs with large aspect ratios can overcome the potential barrier, therefore making a transition through the cathode sheath and deposited on the cathode [14].

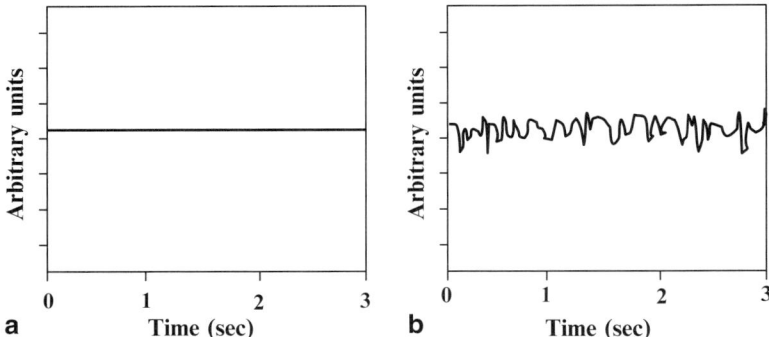

Fig. 15.3 Comparative redrawn graphical representation of current discharge as a function of time for (**a**) stable glow discharge and (**b**) arc discharge methods [2]

during a conventional arc discharge. Figure 15.3 demonstrates the differences in the time dependence of current across the interelectrode gap during the formation of CNTs between a conventional arc discharge and stable glow discharge. A Hewlett Packard 7090A Measurement Plotting System was used for recording the resultant glow discharge spectrum of both the arc discharge and stable glow discharge techniques. The glow discharge displays minimal current fluctuations with respect to time indicating that the process is homogenous and continuous, while on the other hand the continuous variations of current in the arc discharge method are indicative of the process being transient.

Many more research teams have studied and improved the method of arc discharge technique of producing nanotube in terms of experimental ease, quality of nanotubes, production quantity, and parameter control schemes. Journet et al. devised a method to produce SWCNTs by the electric arc technique [15] where the quality of CNTs produced was of the same as those generated by the laser ablation technique. Zhu et al. introduced interesting modifications to the arc discharge technique from China where CNTs were produced in a container with water, thus eliminating the need for vacuum or a water-cooled chamber. The water-arcing process [16] produced high-quality MWCNTs.

15.5 Laser Ablation Method

In the early 1990s, Smalley and coworkers from the Rice University were using energized power pulsed lasers to vaporize metallic targets to form various metal particles. The team was successful in creating MWCNTs when the metallic targets were substituted with graphite and a laser beam was impinged on the surface in an inert gas environment [17].

The experiment required used an Nd:YAG argon laser with a 50-cm long quartz tube inside a temperature-controlled furnace. The quartz tube is sealed after placing the graphite target, and the chamber is maintained at vacuum < 10 mTorr with temperatures at 12,000°C. The Gaussian laser beam energized with 250 mJ is pulsed at 10 Hz at 10-ns pulses. The laser beam is then made to scan across the surface of the target to deposit soot on top of a water-cooled conical copper-collecting rod. Figure 15.4 accounts for the experimental details provided.

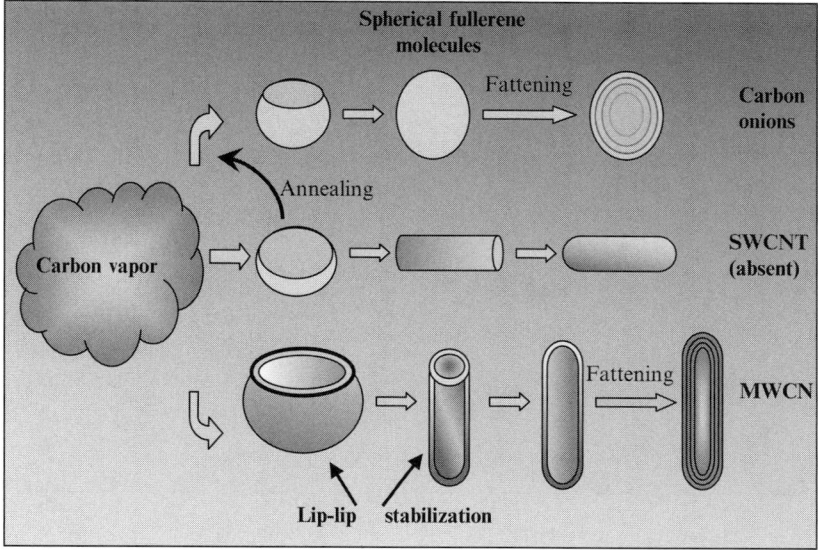

Fig. 15.4 Redrawn schematic of the experimental setup used by Smalley and coworkers [17] to prepare CNTS using the laser ablation technique

Spherical fullerene molecules are likely to form in temperature environments close to 10,000°C, and the formation of close-ended CNTs is more likely in such cases of high-temperature conditions. The team made a very interesting observation about the existence of long length (typical micrometer range) MWCNTs, which would normally be absent at temperatures ranging well over 10,000°C as observed from Fig. 15.4. Owing to the high annealing temperatures, the fullerene yield would rather comprise the more energetically stable spherical fullerene rather than long nanotubules, which were otherwise expected to get closed off at the ends with the nanotube precursor exposed to such high temperatures. The final fullerene yield consisted of MWCNTs and spherical carbon molecules called "carbon onions." The absence of SWCNTs provides a key explanation to the process of the formation of MWCNTs in the laser ablation technique. Figure 15.5 explains the theory behind the formation of CNTs despite the high temperatures of the furnace. On the basis of gas-phase mechanism, the team hypothesizes that the carbon atoms bridge between the adjacent edges of the growing graphene sheets, thus prolonging their open-ended structures.

In the same year, Guo et al. from the Rice Quantum Institute devised a procedure to catalytically grow SWCNTs by the laser ablation technique [18]. A mixture of transition metals such as cobalt and nickel catalyzed the carbon vaporization event during the direct laser vaporization process.

A few years later, Scott and his team of researchers from Houston theorized the growth mechanism for SWCNTs in the laser ablation process [19] using two Nd:

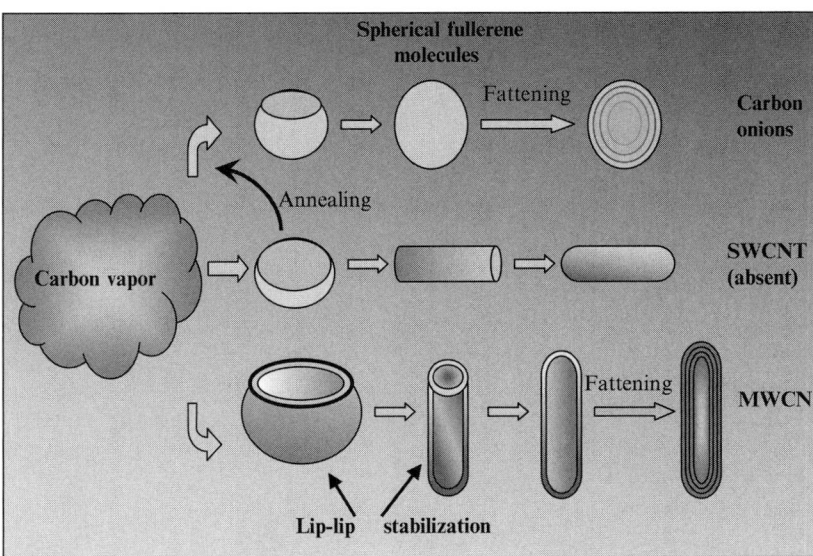

Fig. 15.5 Redrawn figure represents fullerene branching paths during the CNT formation in the laser ablation technique [17]

YAG pulsed lasers impinging on a graphite target containing a certain percentage of cobalt and nickel. The team's investigation suggested that the carbon, which is the source for nanotube formation, is found not only in the graphite target, but is also from the carbon particles present in the reaction zone. Fullerene molecules formed during the laser vaporization process may also participate as carbon feedstock, thus effectively reducing the contamination rates of SWCNTs during the tubule formation process. Contemporarily, Yudasaka et al. from NEC Corporation, Japan thoroughly delved into the process of SWCNT formation during the laser ablation process in the presence of nickel, cobalt, and nickel–cobalt catalysts [12]. The Japanese team of scientists along with Iijima revealed that the yield of SWCNTs depended on the target composition schemes of the catalysts, viz cobalt and nickel.

Figure 15.6 pictorially depicts the dependency of SWCNT yield on metal species [12]. Figure 15.6a: Cobalt (Co) does not melt very well in the carbon (C), thereby causing inhomogeneous distribution of Co and no growth of SWCNTs. Figure 15.6b: Nickel (Ni) or nickel–cobalt (NiCo) distribution was more homogenous in the molten carbon causing the formation of SWCNTs. Figure 15.6c: Chemical reactivity of Ni and C would be reduced at lower temperatures leading to reduced yield of SWCNTs. Figure 15.6d: Iron (Fe) dissolves in carbon, and the mixture remains as a solid solution at room temperature making it impossible for Fe clusters to segregate in the C–Fe droplets thereby eliminating the possibility of growth of any SWCNTs.

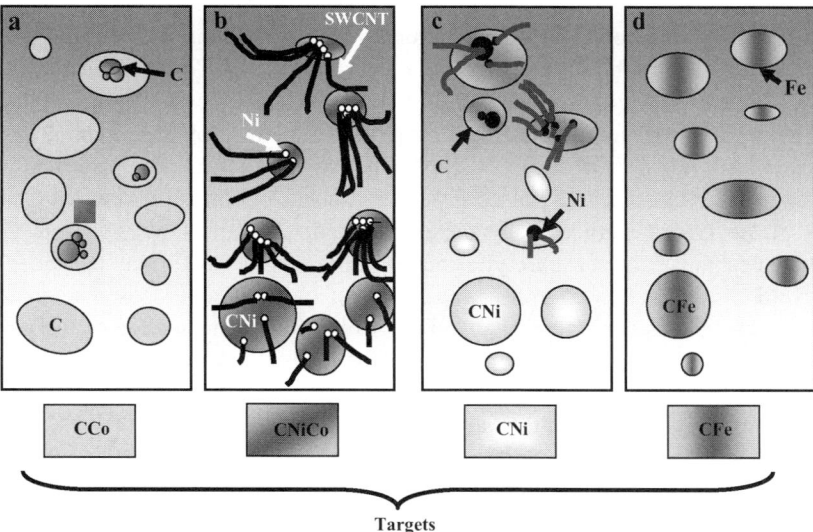

Fig. 15.6 Redrawn figure shows influence of metal species on single-walled carbon nanotubes (SWCNT) yields during laser ablation process [12]

15.6 Pulsed Corona Discharge Method

In 2004, Mishra et al. have used the pulsed corona discharge technique to prepare CNTs [20] by deposition of methane at atmospheric pressure. The process of successful production of hydrogen and CNTs has been attributed to the enhanced electric field near the surface of the inner electrode. The experimental technique paves the way for mass production of hydrogen as an energy source and CNTs for industrial and academic purposes.

A few years later, Sano and Nobuzawa from Japan made use of a needle electrode by atmospheric pressure corona discharge [21]. The carbon source was methane in a hydrogen stream, and CNT formation was observed at the tip of the needle-shaped cathode. MWCNTs were created at a localized surface on the tip of the needle-like electrode, and strong electric fields enhanced CNTs to form a freestanding MWCNT forest at the cathode tip.

Recently, Uhm et al. from Korea have devised a new technique to produce CNTs with a portable microwave plasma torch at atmospheric pressure [22]. Acetylene, which acted as a carbon source, was used in conjunction with iron pentacarbonyl acting as the metal catalyst. A high-temperature furnace was also incorporated to increase the yield of CNTs.

15.7 Other Methods

Catalytic Pyrolysis: In 1998, Cheng et al. produced bundles of SWCNTs by catalytic decomposition of hydrocarbons at temperatures around 12,000°C using the floating catalyst method [8]. Under different growth conditions, the addition of thiophene made it possible to enhance the production of SWCNTs and MWCNTs. Shortly thereafter, an Australian group of researchers synthesized large arrays of CNTs on glass and quartz substrates by pyrolysis of iron phthalocyanine in the presence of Ar/He at 800–11,000°C [23]. Aligned CNTs were also patterned into microarrays through a partially masked surface or contact printing process. De-Chang Li et al. devised a method to synthesize aligned CNT films by pyrolysis. The growth mechanism of the pyrolytic method of growing CNTs was theorized to involve the participation of iron nanoparticles of two different sizes. The smaller iron particles act as the active catalyst responsible for the nucleation of the nanotube while the larger iron particles behave as the feedstock of carbon for the formation of CNTs. The bamboo-like growths after pyrolysis are a result of surface diffusion of carbon atoms on the larger iron particle.

The arc discharge method, laser vaporization method, and the catalytically supported methods of synthesizing CNTs are capable of producing SWCNTs and MWCNTs in bulk and economically. The earlier described methods of fabricating CNTs are difficult to integrate, and factors such as the location and alignment of CNTs during their synthesis are under very little experimental control. The CVD method of synthesizing CNTs has shown promising results of producing high-quality CNTs in large quantities. Synthesizing CNTs using CVD schemes has resulted in producing organized CNT structures/arrays, which can be readily integrated into various electronic, sensing, mechanical, and chemical application-oriented devices.

15.8 CVD Method

The CVD technique is a most common catalyst-based technique used in research as well as commercial production of CNTs. This method has a capacity to economically grow large volumes of CNTs, with lengths up to 18 mm with controlled growth direction [24]. The parameters that affect the growth of CNTs are the catalyst materials, catalyst particle sizes, and the related support, temperature, pretreatment time, carbon source, and pressure of the carbon source.

15.8.1 Parameters Affecting the Growth of CNTs

15.8.1.1 Temperature

Temperature has an influence on the dimensions of the diameter as well as type of CNTs synthesized, i.e., whether they are SWCNTs or MWCNTs. It has been observed that there is an increase in the diameter of the CNTs with an increase in the temperature. Average CNT outer diameter has been found to increase from 20 to 150 nm and the growth rate from 1.6 to 28 μm/min with increasing temperature in a fixed-bed reactor [25]. In contrast, many researchers did not find a significant change in diameter due to temperature variation [26]. Nerushev et al. demonstrated that the diameters of CNTs depend on particle size, the growth temperature, and the carbon flux rate, rather than only on temperature. At a certain critical condition, CNTs with similar diameters were grown [27]. Therefore, further investigation is required for understanding the temperature dependence on the growth process. Figure 15.7 shows CNT growth rate and diameter dependence on temperature.

15.8.1.2 Pressure

The fluidized bed reactor (FDCVD) process is generally performed at atmospheric pressure. There has been no significant understanding that indicates that the FDCVD process is dependent upon pressure variations. It has been found that in the plasma-enhanced CVD technique, parameters such as diameter, quality, and growth rate can be changed under controlled pressure [28]. An increase in the diameter with improved quality and faster growth rate has been observed as the pressure

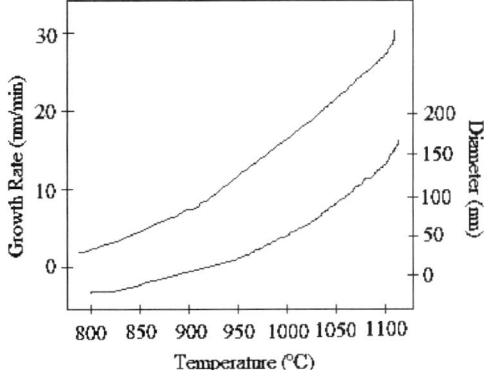

Fig. 15.7 Redrawn graphical representation of growth rate and diameter with respect to temperature [25]

decreases. Growth rates of 1–3 mm/min and CNT diameters of < 30 nm with well-graphitized features have been observed in the high-pressure system, as compared with lower growth rates, larger diameter sizes, 0.1 mm/min growth rate, 60–80-nm CNT diameters, and bamboo-like structures in the low-pressure system.

15.8.1.3 Carbon Source

Carbon monoxide, methane, ethylene, acetylene [23], benzene [29], camphor [30], methanol [31], and ethanol [31] are the different carbon sources used in the CVD processes. Different structural formations of the hydrocarbon chains as either the straight chain or the benzene ring have influence on the type of CNTs formed, as compared with the thermodynamic properties (e.g., enthalpy). It has been observed that methane and aromatic molecules favor the formation of SWCNTs. Nishii et al. [32] and Liu et al. [33] used carbon monoxide, methane, and ethylene as a carbon source for the growth of SWCNTs.

15.8.1.4 Metal Catalyst

Metal transition catalyst is used for the growth of MWCNTs and SWCNTs. Cobalt [34], nickel [35], and iron [36] metal nanoparticles and their alloys acts as a catalyst in CVD process. To achieve better selectivity, metals such as molybdenum, platinum, and copper have been used in combination with the metal catalysts. The type of metal catalyst is an important parameter in the synthesis of CNTs as it determines the rate of carbon decomposition, yield, selectivity, and quality of products. Alloys of different composites provide an added advantage than pure metal nanoparticles. The University of Cincinnati has demonstrated the synthesis of the 18 mm longest

MWCNTs using a novel catalyst composite. In addition, with the use of catalyst alloy such as FeZrN, low-temperature growth of CNTs is possible [37]. The metal catalysts–surface interaction helps in understanding the morphology of the catalyst on the base substrate, thereby the growth mechanism: base or tip. Different substrates, typically Al2O3, MgO, or SiO2, are used as catalyst supports to disperse the catalyst onto the surface for CNT growth in CVD techniques.

15.8.1.5 Particle Size

The catalyst particle size has a major role in determining the size as well as the type of CNTs: MWCNTs or SWCNTs [38]. The catalyst particle size that determines the nature of CNTs is still unclear and needs to be studied. It has been determined that the growth rate and the highest production yield can also be determined by the catalyst-metal nanoparticle size. In CVD, there is a contradiction in the correlation between the catalyst nanoparticle and the CNT diameter. For example, in the base growth mechanism with a catalyst particle diameter of 20 nm, CNTs with small diameter were synthesized, whereas the CNTs' diameter sizes were equal to those of the catalyst particle diameters in the tip growth mechanism. Furthermore, irrespective of particle size, the diameter of the CNTs can be changed by controlling the residing time of the reactive gas in the reactor chamber [39].

15.8.2 Basic Concepts

In CVD processes, CNTs are synthesized by using two gases that are injected into the reactor – the process gas and the hydrocarbon gas. Acetylene, ethylene, ethanol, and methane are the most common gases containing carbon whereas ammonia, nitrogen, and hydrogen are used as process gases. Prior to the inflow of the gas into the reactor, the substrate is prepared. Substrates are usually silicon, glass, or aluminum. Catalyst metal nanoparticles are deposited onto the substrate by solution, e-beam evaporation, or by sputtering, which serves as a catalyst for the growth of CNTs. Later the substrates are placed into the reactor tube at temperatures varying from 700 to 900°C and at atmospheric pressure. During the reaction process, injected hydrocarbons decompose near the catalyst nanoparticle and the carbon atoms are transported to the edge of the metal particle to form a novel structure. Figure 15.8 shows microscope images of SWCNTs using a scanning electron microscope (SEM).

15.8.3 Classification of CVD Methods

The synthesized process is classified based on catalyst type, the reactor alignment, the growth mode, and type of CNTs synthesized.

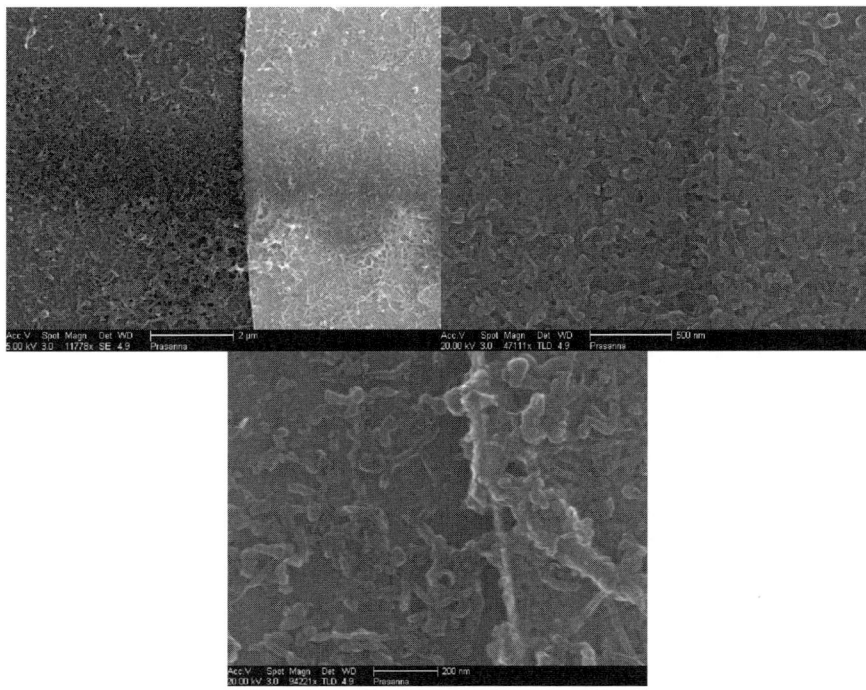

Fig. 15.8 SEM images of multi-walled carbon nanotubes (MWCNTs) synthesized using the CVD technique. The diameter of each MWCNT is ~12–15 nm

15.8.3.1 Classification Based on Catalyst

The size and shapes of CNTs are determined by the growth condition and the catalyst nanoparticles. The size and uniform distribution of the catalyst determine the radii of CNTs and preparation of pure CNTs with a uniform thickness, respectively. Sol–gel catalyst has assisted in synthesizing CNTs, which are aligned, isolated, and dense [40]. Mesoporous silica embedded with iron nanoparticles as the substrates are fabricated by the sol–gel method. After 48 h of growth, 2-mm long CNTs were formed [41]. Metal nanoparticle catalysts are embedded on the base substrate by physical evaporation or by sputtering in a simple and the most common method, whereas in the gas phase metal catalyst method both the catalyst and the hydrocarbon gas are injected into a tube reactor and the catalytic reaction takes place in the gas phase [42]. SWCNTs were produced by condensation of a laser-vaporized carbon–nickel–cobalt mixture at 1,200°C. The SWCNTs were formed when a graphite rod doped with Ni and Co was evaporated. SWCNTS have also been synthesized from a mixture of benzene and ferrocene ($C_{10}H_{10}Fe$) in a hydrogen gas flow. Similarly, the NEC Corporation, Japan [43] has synthesized CNTs by using reactant gases such as high-purity methane (CH_4) and hydrogen (H_2). Ferrocene was used as a catalyst at 750°C. The carbon decomposition was allowed to occur using methane gas in the presence of ferrocene gas. The reaction is as follows:

$$CH_4(g) + H_2(g) \xrightarrow[\text{Ferocene}]{600\text{-}1{,}200\ c} C(s) + 3H_2(g)$$

Of all the catalyst types discussed in the earlier section, large-scale synthesis has been achieved by gas phase metal catalyst, as the nanotubes are free from catalytic supports and the reaction can occur continuously.

15.8.3.2 Classification Based on Reactor

The reactor tube can be utilized either in the horizontal or in the vertical position for synthesis of CNTs. In a fixed-bed method [44] the reactor is aligned in the horizontal direction whereas in the fluidized bed method [45] it is in a vertical position. Figures 15.9 and 15.10 show schematics of the reactor setups for the fixed and fluidic bed methods, respectively.

Production volumes of CNTs in a fixed-bed reactor process are less than in the fluidized bed reactor. This is due to the fact that larger amounts of catalyst with respect to the surface area of the quartz boat would only increase the bed depth in a fluidized bed reactor. Thus, catalyst powders residing at the bottom of the quartz

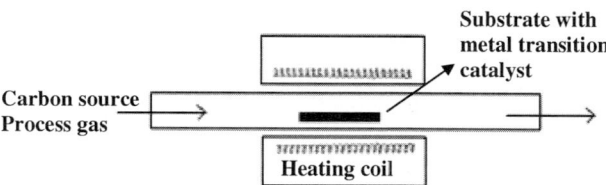

Fig. 15.9 Redrawn schematic of the chemical vapor deposition (CVD) reactor for the fixed-bed method. Both MWCNTs and SWCNTs of diameters ranging from 0.8 to 40 nm can be synthesized [45]

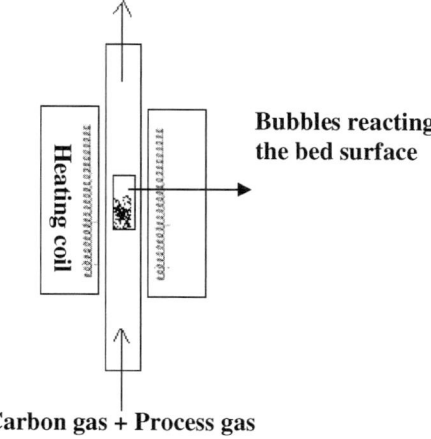

Fig. 15.10 Redrawn CVD reactor setup for the fluidized bed method. Using this technique, MWCNTs of diameter varying from 3 to 50 nm are synthesized [45]

boat face diffusion limitations that thereby lower the overall activity of the catalyst. The different parameters affecting the growth of CNTs have been summarized. It has been demonstrated by Kathyayini et al. [46] and Zeng et al. [47] that there is an increase in production of the CNTs, by using the same amount of catalyst on larger contact areas compared with a contact area reduced by half. This is due to the fact that reaction of gases occurs on a larger surface area in the prior case, whereas in a fluidized bed reactor due to continuous mixing there is an increase in reactive surface area and hence there is a large-scale production of CNTs. There is an elimination of diffusion limitation in the fluidized bed reactor method, and in addition, if the vertical reactor is used the flow rate of the process gas below the fluidized velocity can be achieved that in turn reduces the fluidic effects.

15.8.3.3 Classification Based on Growth Mechanism

The classification is based on the metal catalyst nanoparticle that is located either at the tip or at the base of CNTs. It depends upon the bulk/surface diffusion and also the interactive strength between the metal catalyst and the base substrate surface. The tip-growth mechanism is recognized due to weak catalyst–substrate interactions, i.e., low interfacial energy, while the base-growth mechanism is recognized due to strong interactions [48]. It has been demonstrated that there is a base growth mode between Fe metal nanoparticles and alumina substrates due to strong bonding between them. Figure 15.11 represents the growth model of CNTs. In addition to the earlier defined parameter, the growth mechanism is also dependent on the reaction temperature as previously discussed in Sect. 15.8.3.

15.8.3.4 Classification Based on Type of CNTs Synthesized

Both types of CNTs, i.e., SWCNT and MWCNT can be grown using the CVD method. Temperature has a strong influence on the formation of specific types of CNTs: SWCNTs and MWCNTs in this process. MWCNTs are generally observed

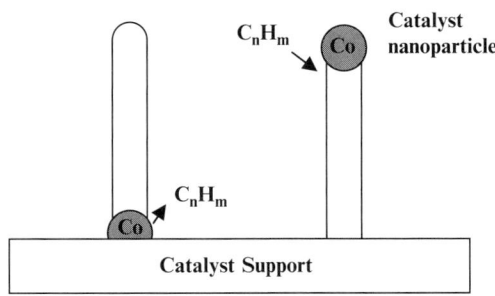

Fig. 15.11 Redrawn figure showing the growth mechanism, base growth, and tip growth [48]

15 Carbon Nanotubes: Synthesis and Characterization

Table 15.1 Comparison between arc discharge, laser ablation, and CVD techniques

Methods	Arc discharge	Laser ablation	CVD
Pioneer	Iijima (1991)	Guo et al. (1995)	Yacaman et al. (1993)
Methods	CNTs synthesized on the graphite rod by direct current arc evaporation in the presence of inert gas	Evaporation of graphite metal using laser ablation on the metal transition target	Deposition of hydrocarbon onto the metal transition target in the presence of process gas
Type of CNTs grown	Both MWCNTs and SWCNTs	Only SWCNTs	Both MWCNTs and SWCNTs
Percent yields	<75%	<75%	>75%
Optimal temperature	>3,000°C	>3,000°C	>1,200°C
Optimal pressure	50–7,600 Torr (generally under vacuum)	200–750 Torr	760–7,600 Torr (generally at atmospheric pressure)
Advantage	Simple, inexpensive; High-quality CNTs produced	Relatively high quality of SWCNTs at room temperature	Large-scale CNTs are produced using FDCVD technique
Disadvantage	Requires high temperature and cannot be scaled up	Expensive technique; not good for large-scale CNT production	Quality of CNTs is not as good

at moderate temperatures (those between 500 and 800°C), while SWCNTs tend to be observed at higher temperatures (>800°C). Also, size of the metal catalyst determines the nature of CNTs grown.

Table 15.1 shows the comparison between most of the common techniques used for the production of CNTs. The table describes the advantages and disadvantages between arc discharge, laser ablation, and CVD methods.

15.9 Vapor–Liquid–Solid (VLS) – CVD Method

In this technique, the nanoparticle catalyst is used as a vapor–liquid–solid (VLS) growth catalyst for the synthesis of CNTs. Uchino et al. demonstrated that the Germanium (Ge) nanoparticle can act as a seed for vapor–liquid growth of CNTs [13]. Using Raman measurement, SWCNTs were identified, and diameter ranges from ~1.6–2.1 nm. It has been noted that unlike metal catalyst nanoparticles there was no reduction in the melting temperature of Ge nanoparticles. Not much detailed study or research has been conducted to analyze the effect of different parameters on the growth of CNTs in this technique, and further investigation is required for better understanding of it.

References

1. Iijima S. (1991). Helical microtubules of graphitic carbon. *Nature, 354*(6348): 56–58.
2. Wang XK, Lin XW, Dravid VP, Ketterson JB, and Chang RPH. (1995). Stable glow discharge for synthesis of carbon nanotubes. *Appl. Phys. Lett., 66*(4): 427–429.
3. Iijima S and Ichihashi T. (1993). Single-shell carbon nanotubes of 1-nm diameter. *Nature, 363*(6430): 603–605.
4. Bethune DS, Klang CH, de Vries MS, Gorman G, Savoy R, Vazquez J, and Beyers R. (1993). Cobalt-catalysed growth of carbon nanotubes with single-atomic-layer walls. *Nature, 363*(6430): 605–607.
5. Dai H. (2002). Carbon nanotubes: synthesis, integration, and properties. *Acc. Chem. Res., 35*(12): 1035–1044.
6. Oberlin A, Endo M, and Koyama T. (1976). Filamentous growth of carbon through benzene decomposition. *J. Cryst. Growth, 32*(3): 335–349.
7. Ebbesen TW and Ajayan PM. (1992). Large-scale synthesis of carbon nanotubes. *Nature, 358*(6383): 220–222.
8. Cheng HM, Li F, Su G, Pan HY, He LL, Sun X, and Dresselhaus MS. (1998). Large-scale and low-cost synthesis of single-walled carbon nanotubes by the catalytic pyrolysis of hydrocarbons. *Appl. Phys. Lett., 72*(25): 3282–3284
9. Ando Y, Zhao X, Sugai T, and Kumar M. (2004). Growing carbon nanotubes. *Mater. Today, 7*(10): 22–29.
10. Gamaly EG and Ebbesen TW. (1995). Mechanism of carbon nanotube formation in the arc discharge. *Phys. Rev. B, 52*(3): 2083.
11. Harris PJF, Tsang SC, Claridge JB, and Green MLH. (1994). High-resolution electron microscopy studies of a microporous carbon produced by arc-evaporation. *J. Chem. Soc. Faraday Trans., 90*: 2799–2802.
12. Yudasaka M, Yamada R, Sensui N, Wilkins T, Ichihashi T, and Iijima S. (1999). Mechanism of the effect of NiCo, Ni and Co catalysts on the yield of single-wall carbon nanotubes formed by pulsed Nd:YAG laser ablation. *J. Phys. Chem. B, 103*(30): 6224–6229.
13. Uchino T, Bourdakos KN, de Groot CH, Ashburn P, Kiziroglou ME, Dilliway GD, and Smith DC. (2005). Metal catalyst-free low-temperature carbon nanotube growth on SiGe islands. *Appl. Phys. Lett., 86*(23): 233110.
14. Keidar M and Waas AM. (2004). On the conditions of carbon nanotube growth in the arc discharge. *Nanotechnology, 15*(11): 1571–1575.
15. Journet C, Maser WK, Bernier P, Loiseau A, Lamy de la Chapelle M, Lefrant S, Denlard P, Lee R, and Fischer JE. (1997). Large-scale production of single-walled carbon nanotubes by the electric-arc technique. *Lett. Nat, 388*: 756–757.
16. Zhu HW, Lia XS, Jianga B, Xua C, Zhua Y, and Chenb DWX. (2002). Formation of carbon nanotubes in water by the electric-arc technique. *Chem. Phys. Lett., 366*(5–6): 664–669.
17. Guo T, Nikolaev P, Rinzler AG, Colbert DT, Smalley RE, and Tomanek D. (1995). Self-assembly of tubular fullerenes. *J. Phys. Chem. B, 99*(27): 10694–10697.
18. Guo T, Nikolaev P, Thess A, Colbert DT, and Smalley RE. (1995). Catalytic growth of single-walled manotubes by laser vaporization. *Chem. Phys. Lett., 243*(1–2): 49–54.
19. Scott CD, Arepalli S, Nikolaev P, and Smalley RE. (2001). Growth mechanisms for single-wall carbon nanotubes in a laser-ablation process. *Appl. Phys. A 72*(5): 573–580.
20. Mishra LN, Shibata K, Ito H, Yugami N, and Nishida Y. (2004). Pulsed corona discharge as a source of hydrogen and carbon nanotube production. *IEEE Trans. Plasma Sci., 32*(4): 1727–1733.
21. Sano N and Nobuzawa M. (2007). Localized fabrication of carbon nanotubes forest at a needle electrode by atmospheric pressure corona discharge. *Diamond Relat. Mater., 16*(1): 144–148.
22. Uhm HS, Hong YC, and Shin DH. (2006). A microwave plasma torch and its applications. *Plasma Sources Sci. Technol., 15*(2): S26–S34.

23. Ren ZF, Huang ZP, Xu JW, Wang JH, Bush P, Siegal MP, and Provencio PN. (1998). Synthesis of large arrays of well-aligned carbon nanotubes on glass. *Science*, *282*(5391): 1105–1107.
24. Physorg (2007). Researchers shatter world records with length of latest carbon nanotube arrays, University of Cincinnati, Cincinnati, OH.
25. Lee YT, Park J, Choi YS, Ryu H, and Lee HJ. (2002). Temperature-dependent growth of vertically aligned carbon nanotubes in the range 800–1100°C. *J. Phys. Chem. B*, *106*(31): 7614–7618.
26. Lee YT, Kim NS, Park J, Han JB, Choi YS, Ryu H, and Lee HJ. (2003). Temperature-dependent growth of carbon nanotubes by pyrolysis of ferrocene and acetylene in the range between 700 and 1000°C. *Chem. Phys. Lett.*, *372*: 853–859.
27. Nerushev OA, Morjan RE, Ostrovskii DI, Sveningsson M, Jonsson M, Rohmund F, and Campbell EEB. (2002). The temperature dependence of Fe-catalysed growth of carbon nanotubes on silicon substrates. Paper presented at the Physica B: Condensed Matter, Tsukuba, Japan.
28. Hsun Lin C, Hsing Lee S, Ming Hsu C, and Tzu Kuo C. (2004). Comparisons on properties and growth mechanisms of carbon nanotubes fabricated by high-pressure and low-pressure plasma-enhanced chemical vapor deposition. *Diamond Relat. Mater.*, *13*(11–12): 2147–2151.
29. Tian Y, Hu Z, Yang Y, Wang X, Chen X, Xu H, Wu Q, Ji W, and Chen Y. (2004). In situ TA-MS study of the six-membered-ring-based growth of carbon nanotubes with benzene precursor. *J. Am. Chem. Soc.*, *126*(4): 1180–1183.
30. Andrews RJ, Smith CF, and Alexander AJ. (2006). Mechanism of carbon nanotube growth from camphor and camphor analogs by chemical vapor deposition. *Carbon*, *44*(2): 341–347.
31. Chhowalla M and Emrah UH. (2005). Investigation of single-walled carbon nanotube growth parameters using alcohol catalytic chemical vapour deposition. *Nanotechnology*, *16*: 2153–2163.
32. Nishii T, Murakami Y, Einarsson E, Masuyama N, and Maruyama S. (2005). *Synthesis of single-walled carbon nanotube film on quartz substrate from carbon monoxide*. Paper presented at the Conference on Experimental Heat Transfer, Fluid Mechanics, and Thermodynamics, Matsushima, Miyagi, Japan.
33. Liu C, Chang N, Chang Y, Hsu J, and Chang S. (2007). Preheated carbon source for carbon nanotube synthesis. In *Proceedings of the 35th International MATADOR Conference*, Taipei, Taiwan, pp. 3–6.
34. Yasuo K, Takeru N, Mizuhisa N, and Michio N. (2007). Infrared reflection absorption spectroscopy investigation of carbon nanotube growth on cobalt catalyst surfaces. *Appl. Phys. Lett.*, *90*(7): 073109.
35. Lee K-H, Baik K, Bang J-S, Lee S-W, and Sigmund W. (2004). Silicon enhanced carbon nanotube growth on nickel films by chemical vapor deposition. *Solid State Commun.*, *129*(9): 583–587.
36. Yunyu W, Zhiquan L, Bin L, Paul SH, Zhen Y, Li S, Eugene NB, and Robert JN. (2007). Comparison study of catalyst nanoparticle formation and carbon nanotube growth: support effect. *J. Appl. Phys.*, *101*(12): 124310.
37. Shiroishi T, Sawada T, Hosono A, Nakata S, Kanazawa Y, and Takai M. (2003). *Low temperature growth of carbon nanotube by thermal CVD with FeZrN catalyst*. Paper presented at the Vacuum Microelectronics Conference.
38. Li Y, Kim W, Zhang Y, Rolandi M, Wang D, and Dai H. (2001). Growth of single-walled carbon nanotubes from discrete catalytic nanoparticles of various sizes. *J. Phys. Chem. B*, *105*(46): 11424–11431.
39. Grill A, Neumayer D, and Singh D. (2003). US Patent No. 20050089467.
40. Liao XZ, Serquis A, Jia QX, Peterson DE, Zhu YT, and Xu HF. (2003). Effect of catalyst composition on carbon nanotube growth. *Appl. Phys. Lett.*, *82*(16): 2694–2696.
41. Xie SS, Chang BH, Li WZ, Pan ZW, Sun LF, Mao JM, Chen XH, Qian LX, and Zhou WY. (1999). Synthesis and characterization of aligned carbon nanotube arrays. *Adv. Mater.*, *11*(13): 1135–1138.

42. Kim SH and Zachariah MR. (2007). Gas-phase growth of diameter-controlled carbon nanotubes. *Mater. Lett.*, *61*(10): 2079–2083.
43. Qin LC. (1997). CVD synthesis of carbon nanotubes. *J. Mater. Sci. Lett.*, *16*: 457–459.
44. Khare R and Bose S. (2005). Carbon nanotube based composites – A review. *J. Minerals Mater. Char. Eng.*, *4*(1): 31–46.
45. See CH and Harris AT. (2007). A review of carbon nanotube synthesis via fluidized-bed chemical vapor deposition. *Ind. Eng. Chem. Res.*, *46*(4): 997–1012.
46. Kathyayini H, Nagaraju N, Fonseca A, and Nagy JB. (2004). Catalytic activity of Fe, Co and Fe/Co supported on Ca and Mg oxides, hydroxides and carbonates in the synthesis of carbon nanotubes. *J. Mol. Catal. A*, *223*(1–2): 129–136.
47. Zeng X, Sun X, Cheng G, Yan X, and Xu X. (2002). Production of multi-wall carbon nanotubes on a large scale. *Phys. B*, *323*: 330–332.
48. Baddour CE and Briens C. (2005). Carbon nanotube synthesis: A review. *Int. J. Chem. Reactor Eng.*, *3*: R3.

Chapter 16
Characteristics of Carbon Nanotubes for Nanoelectronic Device Applications

Vindhya Kunduru(✉) Yamini Yadav, and Shalini Prasad

16.1 Introduction

Carbon has the incredible ability to combine with itself in varied proportions to form molecules of distinctly disparate physical structures. The evolution of modern organic chemistry began with the growing interest amongst scientists to experiment with carbon clusters formed during synthesis of carbon compounds. These studies on carbon-related compounds began with the support of rigid understanding of a common form of carbon–graphite. Carbon molecule C_{60} which was discovered in trace amounts in the carbon clusters was a soccer-shaped fullerene molecule which had 60 carbon atoms arranged in a way that each atom was placed at a vertex of a truncated icosahedron [1]. The discovery of Kratschmer et al. to produce C_{60} in bulk served as a platform for extensive study on carbon-related molecules by scientists, chemists, and material science experts all over the world.

Significant findings were not recorded until Iijima [2], an electron microscopist from NEC Laboratories, Japan discovered tubular fullerenes in the cathode of an arc evaporation chamber. The procedure of arc evaporation resulted in mass production of high-quality carbon nanotubes (CNTs) [3]. A whole new era of CNT research began after Iijima published his first paper on CNTs – "Helical microtubules of graphitic carbon" in 1992. Pictorially, one can understand the structure of the CNT to be a more like an elongated fullerene molecule with half a bucky ball capped at either end to form a capsule-like carbon enclosure. Figure 16.1 shows a schematic 3D representation of a single-walled carbon nanotube. (Note that the image depicts the morphology of a single-walled carbon nanotube in a rudimentary fashion and does not incorporate the hexagonal building blocks of a nanotube.)

Scanning tunneling microscopy studies have provided strong evidence that CNTs occur in varied chiral configurations [4] and layers [2]. Atomic force microscopy studies have brought to light the amazing mechanical [5] and electrical properties [6]

V. Kunduru
Department of Electrical & Computer Engineering, Portland State University, FAB Suite 160, 1900 SW 4th Avenue, Portland, OR 97207-0751, USA

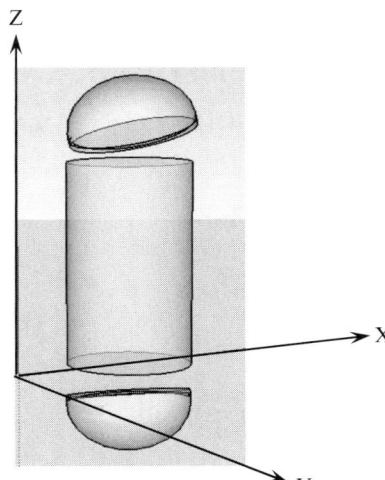

Fig. 16.1 3D representation of an SWCNT with half a fullerene molecule on each end

that CNTs exhibit. CNTs can be classified into two major types based on their layer properties – single-walled carbon nanotubes (SWCNTs) and multiwalled carbon nanotubes (MWCNTs). Though there are both catalytic and noncatalytic methods to synthesize CNTs, the noncatalytic method of arc evaporation is popular not only for its high quality and bulk production of CNTs, but also because the principle behind this mechanism is better understood than the other techniques.

SWCNTs promise a great future for the stream of "carbon nanotechnology" owing to their large surface area, high aspect ratios, and a good capacity for functionalization. SWCNTs have been used mostly by the electronic industry to replace interconnect networks with molecular quantum wires [7], which demonstrate high conductivity. Recent advancements in the field of transistors have proved that SWCNT-based field effect transistors (FETs) exhibit good device characteristics like high gain, good switching speeds and prove fully operational at room temperature [8].

MWCNTs, reinforced with more outer layers, are morphologically different from SWCNTs. Electron transport cannot be counted as one-dimensional conduction since there are many atomic layers of graphite surrounding the innermost cylinder as seen in Fig. 16.2. Most atomic force microscopy techniques involve using MWCNTs to conduct studies on their tensile and elastic properties [9, 10]. Such studies related to the mechanical properties of MWCNTs have resulted in several constructive applications in industrial commercialization of consumer products. MWCNTs are used as strengthening fiber materials and as field emission tips in scanning probe microscopes.

Finally, in summary, the properties of CNTs are determined by their synthesis methods and this in turn determines their applicability. Described below are the various ways in which CNTs can be classified based on their structural and electrical properties.

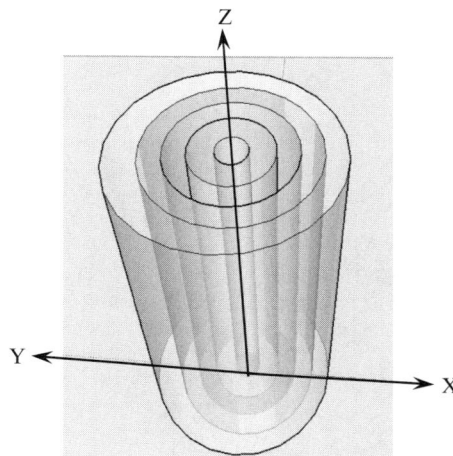

Fig. 16.2 3D representation of an MWCNT composing of five concentric cylinders, i.e., $N = 5$

16.2 Classification of Carbon Nanotubes

16.2.1 Classification Based on Layer Properties

CNTs can be classified mainly based on the number of layers which comprises each tubule.

Experimental observations on MWCNTs are more conclusive amongst researchers than that on SWCNTs. This is owing to the fact that MWCNTs were discovered earlier than SWCNTs. Studies and research data on SWCNTs are more recent and are relatively more conceptual than experimental. Figure 16.3 depicts schematically an ensemble of the various types of CNTs based on the number of layers each is composed of.

16.2.1.1 Multiwalled Carbon Nanotubes

MWCNTs are composed of concentric cylinders of varying diameters. They were first discovered occurring in bundles and with various other graphitic structures occurring in the cathodic soot inside the arc discharge chamber and as hard graphitic deposit on the electrode. The very first observation made by Iijima [2] showed that each layer was separated by a distance of 0.34 nm. This high-resolution TEM data showed that the outermost diameter of the cylinder ranged from being 2.5 nm to about 30 nm, with lengths of the nanotubes roughly occurring between a few nanometers and several micrometers.

Most MWCNTs did not retain their original number of layers all along their lengths. The reason behind this reduction in the number of concentric cylinders is that the innermost layer walls begin to fuse after a certain length along the length of the nanotube.

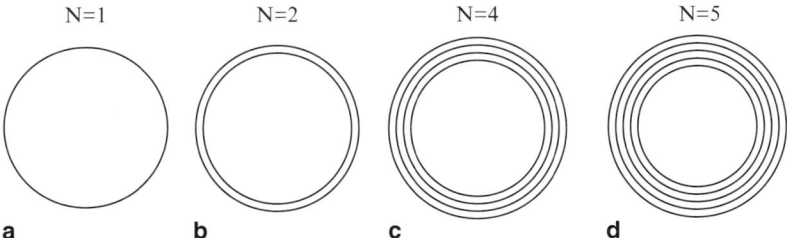

Fig. 16.3 Cross-sectional views of CNTs depicting layer properties, where N represents the number of layers of each CNT: (**a**) SWCNT, (**b**) double-walled CNT, (**c**) MWCNT with four layers, and (**d**) MWCNT with five layers

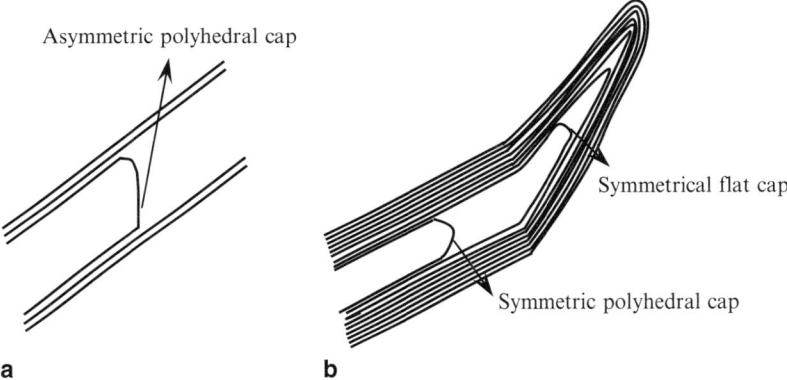

Fig. 16.4 Capping in longitudinal cross sections of MWCNTs: (**a**) MWCNT with $N = 3$ reducing to $N = 2$ after capping and (**b**) MWCNT with $N = 9$ having several caps reducing $N = 6$

This effectively capped the central cores of the nanotube resulting in an appearance of a reduced number of layers. Cap terminations occur in different shapes, which can be revealed by high-resolution transmission electron microscopy.

A schematic representation of MWCNT capping is shown in Fig. 16.4.

16.2.1.2 Double-Walled Carbon Nanotubes

Although double-walled carbon nanotubes (DWCNTs) can be classified as a subclass of MWCNTs, their behavior is mostly like that of SWCNTs. Their double-walled nature, however, gives them special electrical, chemical, and mechanical properties making them unique to certain applications. DWCNTs can be analogous to coaxial cables. The outer cladding of the coaxial cable provides insulation to the inner core. In the same way, the outer tube of the DWCNT protects the inner tube from environmental elements, thus maintaining its purity. In other words, the outer wall provides an interface with the outer electrical or gaseous experimental systems without affecting the inner core. The outer tube can also be used as a good host to functional groups, thereby making the outer shell a good functionalization surface while preserving the inner tube mainly for electron transportation purposes.

DWCNTs are being explored for possibilities of using them as molecular bearings and cylindrical molecular capacitors [11]. Many research groups have investigated further into the electronic and structural properties of DWCNTs and have discovered that "electron-libration" coupling in certain types of DWCNTs can be utilized as a potential possibility for achieving superconductivity in MWCNTs [12]. A cross-sectional view of a typical DWCNT has been shown in Fig. 16.3b.

16.2.1.3 Single-Walled Carbon Nanotubes

As indicated earlier, SWCNTs were discovered after MWCNTs and only after much study did certain research teams announce the synthesis of such structures. The very first SWCNTs were produced by two independent groups, which included Iijima from NEC and Donald Bethune from IBM, CA. The initial SWCNTs produced were either twisted or curled rather than straight and tube-like. With diameters of only 1 nm, they were also composed of several particles of carbon debris. Intense research on SWCNT synthesis has led to better quality yields presently. SWCNTs tend to frequently occur in arrays or tightly packed bundles. However, Bethune et al. [13] have significant findings of SWCNTs occurring independently. Most studies speculated the diameters of SWCNTs to be mostly circular. Ruoff and his team [14] observed that sometimes the cylindrical symmetry of the CNTs gets flattened out by the adjacent tubes due to van der Waals forces between the individual tube walls. High-resolution transmission electron microscopy images indicate that these deformations occur along the contact area over the walls between two SWCNTs. Another group from IBM, New York has supported its findings on the effect of van der Waals forces on SWCNTs which are lying on a substrate. Their studies signify that van der Waals forces on the walls of SWCNTs resting on a substrate cause radial and axial deformations, modifying the overall geometry of each nanotube. Such deformation of the SWCNTs may have further implications on the electrical properties of adsorbed nanotubes.

A cross-sectional view of a typical SWCNT is depicted schematically in Fig. 16.3a.

16.2.2 Classification Based on Chirality

CNTs are made of graphene sheets in which carbon atoms form a perfect ensemble very similar in structure to a chicken wire with carbon atoms occupying the vertices of the hexagonal unit cells. Considering the structure of an individual SWCNT as an ideal model, one can imagine its structure as a graphene sheet rolled up end to end to form a seamless tube, capped with hemispherical halves of a C_{60} fullerene molecule. While trying to understand the meaning of chirality, we neglect the caps of the nanotubes and assume that the tube length is much greater than the diameter. Chirality defines exactly how much twist is present in the graphene sheet when it is wrapped with respect to the axis of the tube. In other words, chirality depends on the amount of twist introduced with respect to the axis of the tube when rolling up the graphene sheet.

The "chiral vector" can calculate the chirality for each tube. A brief explanation is provided along with a few illustrations to understand the procedure to calculate the chiral vector of an ideal SWCNT. Here, it is assumed that a defect-free graphene sheet, which makes up the SWCNT under examination, is spread to represent a 2D lattice structure. The lattice comprises of hexagonal unit cells, which are fused such that the combination results in a sheet.

With reference to the book by Dresselhaus et al. [15], chirality vector defines the structure of any type of tube and is labeled C:

$$C = na_1 + ma_2, \qquad (16.1)$$

where a_1 and a_2 are the unit cell base vectors of the graphene sheet and n and m are the chiral indices, $n > m$. When the graphene sheet is wrapped, the ends of the chiral vector meet each other with the sheet forming the walls of the cylinder. Therefore, the chiral vector forms the circumference of a circular cross section of an SWCNT.

Different values of chiral indices can depict different types of twists in CNTs. The various types of chirality are discussed further and are furnished with illustrations. The type of chirality determines the electrical nature of CNTs. Further reading on the electrical properties of CNTs is provided in Sect. 16.2.3.

16.2.2.1 Armchair

The chiral indices n and m of an armchair SWCNT are always equal to each other, i.e., $n = m$ for all armchair structures. A schematic representation in Fig. 16.5 depicts a 2D graphene lattice which when wrapped with a definite chiral angle produces an armchair SWCNT. Figure 16.5, which is adapted from Dr. Harris' "Carbon nanotubes and related structures," depicts the armchair structure of a CNT. The rectangular box gives an idea about the orientation of the unit cells when the 2D graphene sheet is rolled along the longer side of the box as the axis of the cylinder.

16.2.2.2 Zigzag

The chiral index m of an armchair SWCNT is always equal to zero, i.e., $m = 0$ for all zigzag structures. A schematic representation in Fig. 16.6 depicts a 2D graphene lattice which when wrapped with a definite chiral angle produces a zigzag SWCNT.

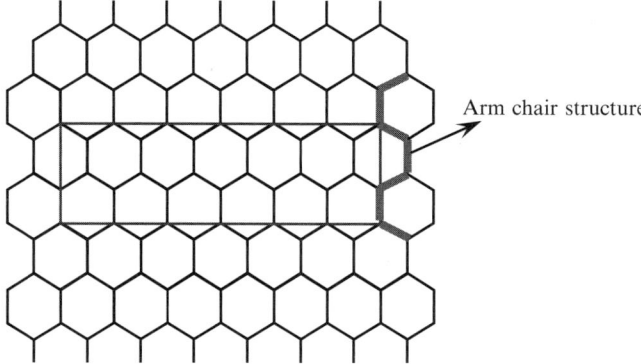

Fig. 16.5 Unit cells of an armchair SWCNT structure

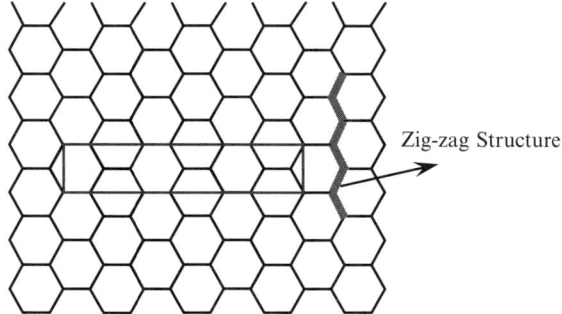

Fig. 16.6 Unit cells of a zigzag SWCNT structure

The figure is adapted from Dr. Peter Harris' "Carbon nanotubes and related structures" and depicts the zigzag structure of a CNT when rolled up along the rectangular box as the axis.

16.2.2.3 Chiral

All other SWCNTs that are neither armchair nor zigzag are invariably chiral in nature. Any lattice pattern, which does not appear to have a mirror image, can be called a *chiral CNT*. Here, the hexagonal unit cells of the graphene sheet are wrapped such that they twist helically around the axis of the nanotube. Such uniqueness in its twist angle gives chiral nanotube very different electrical properties from its counterparts.

16.2.3 Classification Based on Electrical Properties

16.2.3.1 Electrical Properties of MWCNT

Studies by Saito et al. [16] reveal interesting electrical behavior of DWCNTs. It was observed that two SWCNTs of zigzag chirality, which are inherently metallic in nature, result in a double-walled coaxial CNT, also metallic in nature when an interlayer coupling force is introduced between the two concentric cylinders. Similarly, the same hypothesis holds good for semiconducting nanotube as well. Saito's research team along with Fujita conducted further experiments with metal–semiconducting and semiconducting–metal DWCNTs, predicting that such structures also retain their original electrical character when introduced to intertubule coupling forces. Such robustness of DWCNTs can lead to a hypothesis that they can be used as a coaxial cable with an inner conducting core and an outer insulating cladding as mentioned in Sect. 16.2.1.

16.2.3.2 Electrical Properties of SWCNT

SWCNTs can be classified based on their mode of conduction. A summary is extracted from various studies involving SWCNT energy dispersion bands leading to theoretical conclusions about their electrical behavior and the same is provided in the next sections.

16.2.3.3 Metallic SWCNT

SWCNTs are monolayered in nature and hence do not suffer from electronic distortions occurring as a result of intertubule interactions, unlike MWCNTs. There exists a degenerate point between the valence and conduction bands at the point where the two bands cross in a normal energy dispersion spectrum of an SWCNT [17–19]. This existence of the degenerate point implies that the distance between the two bands is nearly zero, resulting in high conductivity. Hence, very low excitation energy is also sufficient to excite the electrons from the valence to the conduction bands. Hence, all armchair SWCNTs are theoretically anticipated to behave as metallic type nanotubes [16].

16.2.3.4 Semiconducting SWCNT

The semiconducting nature of an SWCNT is not as straightforward as that explained for their metallic behavior in Sect. 16.2.3. If the same calculation procedures are carried out for determining SWCNT's semiconducting properties, the results obtained are quite disparate. Studies have revealed that CNTs with different chiral properties exhibit different electrical behaviors [18, 20]. Consider two SWCNTs, both with zigzag chirality but having dissimilar chiral vectors (9, 0) and (10, 0). Observing the appearance of an energy gap in the energy dispersion band diagrams of a (10, 0) SWCNT, conclusions have been made that a (10, 0) SWCNT is semiconducting. On the contrary, a (9, 0) type SWCNT is predicted to exhibit metallic conducting properties.

16.3 Properties of Carbon Nanotubes

Their nanoscale size, unique structure, compositional elements, robustness, and immense surface area for functionalization are a few of the properties which give CNTs interesting prospects to be used in many varied applications. A brief idea about the different properties of CNTs is furnished further.

16.3.1 Electrical Properties

The electrical properties of CNTs are affected by their chirality and diameter as referred to in Sect. 16.2.3. CNTs can occur as metallic or semiconducting nanotubes. Theoretically, nanotubes are highly conductive and can have an electron density around thousand times higher than metals like copper. Although they can have stable high current densities of $J > 10^7$ A cm^{-2} [21], CNTs have a constant resistivity. Phaedon Avouris, a nanotube researcher at the IBM labs, also stated in one of his lectures that electron density can be pushed to a maximum of 10^{13} A cm^{-2}. Owing to its nanoscale dimensions, a nanotube can be thought of as a quantum wire, through which electron transport happens ballistically.

Sanvito and his research team [22] from the School of Physics and Chemistry, UK conducted experiments on MWCNTs using a scattering technique. Their findings provided significant proof which explained the reason behind observing unexpected integer and noninteger conductance values in MWCNTs. They elaborate on how the intertubule interactions block some of the quantum conductance channels in MWCNTs and redistribute the current nonuniformity over the individual tubes across the nanotube structure, thus giving rise to quantized conductance. Trygve et al. from Harvard University conducted experiments where "counting" atoms were possible using an SWCNT-based atom detector. Counting rates of 10^4 ions s^{-1} were achieved using a 5-μm long SWCNT. The capture of individual atoms occurred in quantized steps as seen in Fig. 16.7. The sharp steps resulted from the angular momentum quantization of the atoms, which are attracted toward the quantum wire. The steps observed in the "angular momentum quantum ladder" (see Fig. 16.7) demonstrate quantized conductance for a neural polarizable particle system.

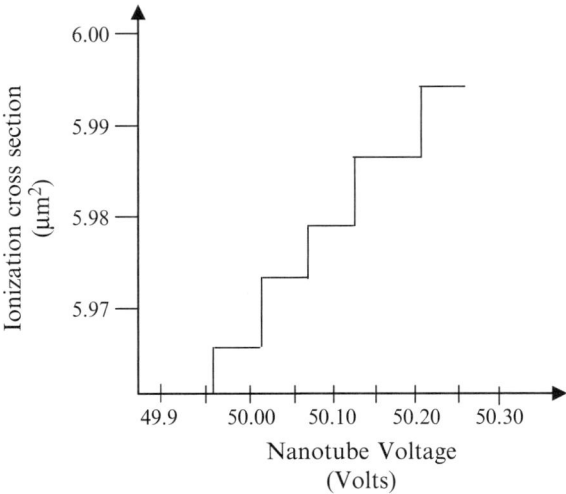

Fig. 16.7 Graph adapted from Ristroph et al. [23] depicting quantized conductance in CNTs

Table 16.1 Comparison of mechanical strengths of CNTs and common building materials (Applied Nanotech, Inc.)

Material	Young's modulus (GPa)	Tensile strength (GPa)
Single-wall nanotube	1,054	150
Multiwall nanotube	1,200	150
Steel	208	0.4
Epoxy	3.5	0.005
Wood	16	0.008

16.3.2 Mechanical Properties

Experiments and intense research on CNTs have proved that they are stiffer than steel and are extremely flexible. They are very robust and can resist damage from external factors of the surrounding environment. Scanning probe microscopy studies have proved that they are very elastic since they regain their original structure after the stress is removed. Several research teams conducted experiments to obtain the values of Young's modulus and tensile strength of CNTs. A collection of such data is provided in Table 16.1. The graph in Fig. 16.8 compares Young's modulus values of different nanomaterials.

Other groups like Wong et al. performed interesting atomic force microscopy techniques to determine the Young's modulus of MWCNTs. The factor responsible for the elastic modulus of MWCNTs and SWCNTs is still being studied because of

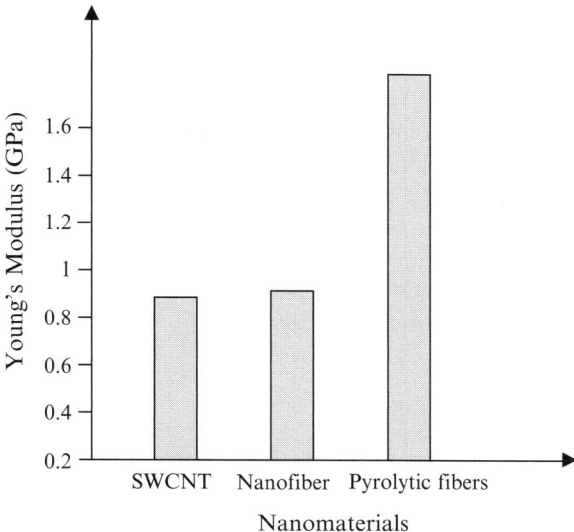

Fig. 16.8 Relative strengths of various nanomaterials in comparison with SWCNTs

controversies in theories proposed by different research teams. While some think it is the diameter and the shape of the MWCNT, which determines its elasticity, a few others claim that it is due to the disorder between the many layers of the MWCNT.

16.3.3 Thermal Properties

Common results exhibited by graphite when exposed to temperature conditions must be in a way similar to the kind of behavior CNTs will probably exhibit when subjected to the same conditions. However, since tube diameter in a CNT is orders of magnitude lesser than that of a unit crystal of graphite, the thermal behavior of CNT is much different. In addition, there is increased stress induced in the CNT walls due to the increased curvature of the tube as compared to that of planar graphite. Such stress factors tend to affect thermal properties of CNTs as well. Most researchers believe that thermal properties of CNTs depend upon the amount of current passing through it. Che and his team [24] theoretically deduced the thermal conductance of a zigzag (10, 10) CNT. Thermal conductance for such a CNT was found to be 2,980 W mK^{-1}. Berber and his team [25], however, provided experimental results supporting the thermal conductivity of a (10, 10) nanotube to approach 6,600 W mK^{-1} as the current applied to it is increased. A recent publication by Hone et al. [26] shows that the conductivity of CNTs is large even in bulk samples, whereas aligned bundles of SWCNTs show a thermal conductivity of over 200 W mK^{-1} at room temperature.

16.3.4 Chemical Properties

The open ends of a CNT are very susceptible to chemical agents and are mostly versatile to much number of functional groups. Functionalizing them with different chemical groups to suit certain unique applications can also chemically modify the walls of a CNT. Properties like a high surface area for functionalization give them important applicability in sensor devices and electrochemical device applications. Prof. Kwanwoo Shin from Sogang University in Korea and his team have been conducting experiments to study the effects of high intensity radiation on CNT-based electronic devices like transistor networks.

16.4 Applications

Researchers have proposed many plausible applications for CNTs ever since their discovery a decade ago. Although applications like hydrogen storage still remain debatable, most applications have revolutionized certain sectors of the IC in the industry. Their high current carrying capacity, high thermal conductivity, and mechanical stability make them good interconnect material in electrical circuits. CNTs are qualified to be used as carbon composites for strength enhancement in fabrics and building material. Common sports equipments like golf balls, baseball bats, and tennis racquets are made lighter and stronger by integrating them with CNTs. Scientists have also proposed the probable usage of CNTs while building space elevators. Rapid advancements have been achieved in manufacturing field emission displays with CNTs. Properties like sharp geometry, mechanical strength, and electrical conductivity have been exploited to make AFM and STM tips mounted with CNTs improving resolution drastically. One of the most fascinating applications lies in implementing CNTs in making sensors.

16.4.1 Sensors

Nanomaterial-based sensor systems of the future will radically improve in their sensitivity, selectivity, and rapid response criteria. These new age detector systems equipped with better control and cross-examination features will be not only capable of single molecule detection but also multiplexed detection of diverse signals necessary for diagnosing various analytes. CNTs are being used as chemical, gas, and biological sensors. The main advantage behind CNT-based sensors is the miniscule amounts of sensing material needed and an equally small amount needed for the purpose of sensing. Although most CNT sensor devices use the same sensing material, they differ in their sensing mechanisms and detection schemes. While some sensing mechanisms depend on mechanical properties of CNTs, others rely on their electronic transportation processes. Detection schemes, on the other hand, can vary from being optical or mechanical to observing variations in the overall frequency response of the sensor device. Although optical detection schemes are still evolving, electrical detection is more desirable due to its increased reliability. The electronic industry has incorporated CNTs while manufacturing semiconductor devices. Such CNT amalgams have been used as sensing material for various sensor applications. Most commonly, an interdisciplinary team of researchers is required to build sensors, since expertise in all sciences is required to develop a working sensor, which is commercially viable. The following sections provide a brief insight into a few state-of-the-art CNT-based sensor systems.

16.4.1.1 CNTs as Biological Sensors

Current flow in one-dimensional structures like CNTs is extremely sensitive to even the slightest changes occurring on its charge carrying outer surface. Hence, when biological analytes adsorb on the CNT walls, this binding event perturbs normal charge transport in the quantum wire. CNTs are thus an ideal sensor material for most applications. CNT-based biosensor is one of the most important applications of CNTs in real-life applications. These nanoscale sensing elements prove extremely sensitive and selective in many varied applications. Vast research is being conducted on CNTs to probe deeper into their applicability. CNTs have attracted attention for defense applications because of their ability to sense trace amounts of deadly biological agents and hence fight global terrorism. The cost of medical diagnostics and the time consumed for tiresome lab routines for diagnosis can be avoided by using nanomaterial-based sensors. Nanomonitors based on CNT sensing elements are being developed instead to hasten diagnosis.

Single molecule detectors using CNTs are being developed at the University of Illinois at Urbana-Champaign. Coated CNT-based sensors fluoresce when they come in contact with the target molecule and wavelength of such emissions is recorded to analyze the trapping of the target molecule. Sofia and Chaniotakis [27] have developed an amperometric biological sensor with aligned MWCNTs on a platinum substrate. CNTs display a dual role in this sensor by acting as a good medium by providing large surface area for immobilization of functional groups and as a medium for electron conduction.

DNA sensors have been developed and are under extensive study due to their vast applicability in the field of forensic science, genetic engineering, and gene therapy. Cai et al. [28] functionalized MWCNTs with a carboxylic group for the detection of a specific hybridization. This MWCNT-based DNA sensor demonstrated better charge transport characteristics and functionalization ability than an earlier model of a DNA sensor with carbon electrodes directly functionalized with oligonucleotides.

16.4.1.2 CNT FETs Gas Sensors

Section 16.4.1 briefly illustrates a few methods developed by research teams to devise gas sensors. These methods involve using CNTs as either nanocomposites or single CNT-based sensor devices. Interest in producing CNT FET-based gas sensors is also rising since CNTs can act as excellent transducers owing to their large charge sensitive surface area. This property of high surface area makes CNT FETs sensitive to ambient environment, especially to oxygen and oxygen-containing compounds [29, 30]. Research by Someya et al. [31] has shown that SWCNTs synthesized by chemical vapor deposition technique were used to fabricate FETs with channel lengths of 2.5 and 5 μm to detect alcoholic vapors. Detection is dependent on observing changes in the FET parameters like saturation current and threshold voltage, and this was achieved for a wide range of alcoholic vapors when the entire device is exposed to an alcoholic vapor environment. Most FETs suffer from a conductance limitation due to the Schottky barrier (SB) created in the CNT–metal interface. To overcome this current carrying limitation, Javey et al. [32] incorporated palladium on the sensor to interface with the SWCNTs. As a noble metal, palladium acts as a better wetting substance for the CNTs, thus reducing the charge barrier for electron conduction and improving current. CO_2 is a common gas, which is associated with most biological processes. Sensing CO_2 for domestic use is also important in many public access areas. Many researchers have undertaken research on CNT-based sensors already. However, a few research teams like Star et al. [33] constructed a CNT FET to detect CO_2. They have reported the use of chemically functionalized CNTs with a specific recognition layer for CO_2 sensing. They employed a polymer-coating method for functionalizing unlike other studies, to overcome modification of the physical properties of CNTs due to covalent bonding.

The following reading will provide information about more traditional CNT-based semiconductor devices. It gives us an idea of devices, which the IC industry is trying to incorporate in the current technology either to improve certain features of an existing technology or to devise an entirely new strain of technology, based on CNTs.

16.5 CNT with Single Junction

16.5.1 *Schottky Diode (CNT–Metal Junction)*

The contact formed at the metal–nanotube interface behaves as an electrical ohmic contact or as an active junction (Schottky diode). The nanotubes make several different types of contact to the metal; of these the configuration with side contact, where CNTs positioned on metal contact by weak van der Waals forces and end-bonded CNTs that are connected to the metal by covalent or metallic bonding, reduces dimensionality. They have a strong effect on active device formation. Fermi level pinning plays an important role in formation of these devices. Figure 16.9 shows the two different types of CNT and metal contact by two different bonding mechanisms.

16.5.1.1 Fermi Level Pinning

The barrier height at the planar interface due to Fermi level pinning is independent of metal work function given by

$$\phi_b = E_c - E_F, \quad (16.2)$$

where E_c and E_F are the energy conduction band and Fermi level, respectively.

In addition, the energy band diagram indicates the negligible band bending characteristics due to doping near the interface [34].

Now, considering the planar interface between metal and semiconductor, the barrier height that is dependent on the metal work function is given by

$$\phi_{b0} = X_m - X_s, \quad (16.3)$$

where X_m and X_s are the metal work function and semiconductor ionization potential, respectively.

However, there is a finite density of states at the metal–semiconductor interface inside the band gap of the semiconductor. These band gaps have a complex wave function that is called "metal-induced gap state" (MIGS). This wave decays exponentially into the semiconductor due to the boundary condition at the interface with the metal. Therefore, the induced charge at the interface due to MIGS creates a dipole ring or sheet at the interface. This charge changes the position of the Fermi level thereby raising or lowering the barrier where the charge near the semiconductor surface vanishes, and MIGS controls both the turn-on voltage and the band bending of the device. The barrier height is independent of the metal/work function and is dependent on the pinned Fermi level deep inside the semiconductor energy band gap.

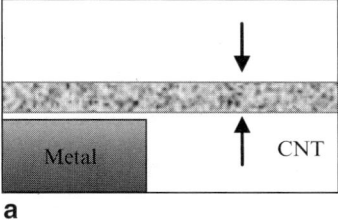
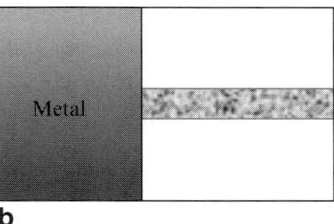

Fig. 16.9 Two types of nanotube/metal contacts: (**a**) CNT side contacted through the weak van der Waals adhesion and (**b**) CNT end bonded to the metal by covalent or metallic bonding (redrawn from [34])

16.5.1.2 CNT Devices with Planar Interface

For CNTs with the planar interface, the charge is modeled by a dipole and is given by

$$\sigma(z) = D_0(E_N - E_F)e^{-qz}, \qquad (16.4)$$

where D_0 is the density due to MIGS, and E_N is the neutrality level which depends upon the atomic structure level which varies with z due to electrostatic potential. E_F is the metal Fermi level and z is the distance from the metal–NT interface. q represents the average exponential decay value due to all the states in the band gap because of charge in the metal and due to MIGS at the interface. Léonard and Tersoff [34] demonstrated this in their work.

The barrier height is increased to

$$\phi_b = \phi_b^0 + E_F - E_N \qquad (16.5)$$

In the schematic diagram [34] for the planar Schottky barrier, charge neutrality level is at the center of the semiconductor band gap and the metal Fermi level is located at the midgap. Figure 16.10 shows a schematic diagram of the CNT planar interface devices.

16.5.1.3 CNT Device with Round-Ended Interface

Round end-bonded contacts between CNTs and metal are similar to the traditional planar contact where the semiconductor ends at the metal. For studying the Fermi level pinning in this type of contact, it has to be taken into consideration that the MIGS is a dipole ring instead of a sheet. Therefore, it has an overall effect of electrostatic potential and band bending. The electrostatic potential decays as the third power of the distance from a dipole ring [35]. There is rapid exponential decay of potential that vanishes after few nanometers. Thus, as the barriers are a few nanometers in size, the electrons tunnel through them, with the result that Fermi level pinning has a less significant effect on the end-bonded CNT/metal contacts. In summary, it can

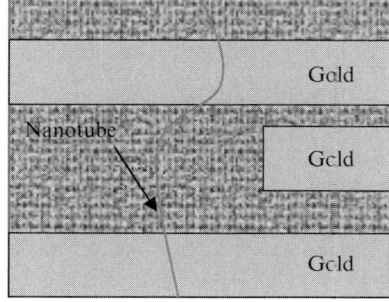

Fig. 16.10 CNT device that forms a Schottky barrier at planar interface between CNT and metal by weak van der Waals force. The figure is redrawn from [35]

be concluded that the type of contact produces different work functions that play an important role in device formation. The fabrication of the round end-bonded Schottky junction between CNTs and metal (Ti and Si) has been demonstrated [36].

16.5.1.4 Schottky vs. Ohmic Contact

The presence of Schottky barriers at these contacts gives a measure of temperature dependence of the current, where the current increases with the temperature. This is in contrast to ohmic contacts where the temperature dependence is opposite, i.e., the current decreases with an increase in temperature.

16.5.2 CNT-Based p–n Junction Diode

The p–n junction CNT diode shows rectification behavior, i.e., current–voltage characteristics with forward conduction and reverse blocking characteristics similar to the semiconductor diode.

For CNTs, the characteristics of the p–n junction depend on certain parameters: the dielectric constant of the material that it is embedded into, the doping fraction, $D(E, z)$ the CNT density at z position and $F(E)$, which is the Fermi function. This can be explained by (16.6), where the CNT charge density is given by

$$\sigma(z) = \frac{e}{\varepsilon}f - \frac{e}{\varepsilon}\int D(E,z)F(E)dE \tag{16.6}$$

and the density of states is given by

$$D(E,z) = \frac{a\sqrt{3}}{\pi^2 R V_0} \frac{|E|}{\sqrt{(E+eV(z))^2 - (E_g/2)^2}}, \tag{16.7}$$

where R is the CNT radius, E_g is the energy gap, and $V(z)$ is the electrostatic potential.

Based on these equations, it was determined that the same band bending characteristics [37] are observed in planar p–n junction devices for different doping fractions. As the screening of coulomb interaction is ineffective in one-dimensional CNTs, the charge distribution curve logarithmically decays away from the junction. There is a need for adding continuous charge to avoid dropping potential and to maintain the potential constant away from the junction. This phenomenon occurs due to the electrostatic dipole rings, in contrast to the dipole sheet as seen in the planar devices, whereas, in a normal planar device there is a constant charge near the junction, which vanishes outside the depletion region.

The same behavior is observed in a Schottky planar junction between metal and CNT. Figure 16.11 is redrawn showing the charge distribution at the interface of metal and CNT [8].

Figure 16.12 shows the depletion width of a CNT with respect to doping fraction. As the doping fraction reduces, the depletion barrier decreases, thus creating tunneling effect due to the reduced barrier and giving rise to negative differential resistance devices [38].

The calculated *I–V* curve demonstrates a negative differential resistance curve plotted using (16.7).

The negative differential resistance is calculated by

$$I = \frac{4e^2}{h} \int T(E)[F_L(E) - F_R(E)]dE, \tag{16.8}$$

where $T(E)$ is the transmission probability across the junction and $F_R(E)$ and $F_L(E)$ are Fermi functions in the right and left leads, respectively.

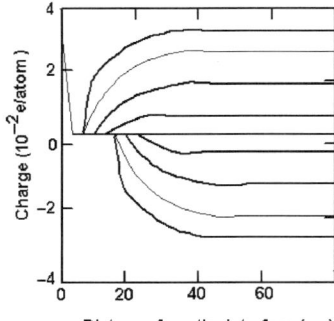

Fig. 16.11 Redrawn diagram represents the charge distribution from the electrode where there is a logarithmic decay away from the junction [8]

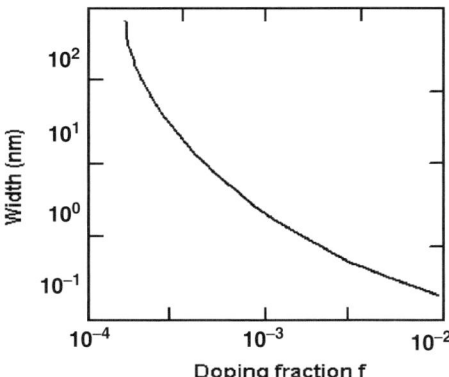

Fig. 16.12 Figure redrawn shows the depletion width for a CNT p–n junction of doping observed by Léonard and Tersoff [37] where the depletion width is directly related to the doping concentration

16.5.2.1 p–n Junction Depending on Type of Doping

The doping is controlled by the introduction of dopant atoms and charge transfer by metallic electrodes and hence determines the device characteristics. There are two type of doping:

1. Electrostatic doping
2. Chemical doping

p–n Junction: Electrostatic Doping

The electrostatic doped p–n junction shows an ideal device characteristic, whereas chemical doped shows electrical properties that result in the leaky behavior due to high doping concentration and abrupt junction formation. In this device, the gate voltage controls the doping strength. This is accomplished by fabricating two identical back gate electrodes [39] (Fig. 16.13).

p–n Junction: Chemically Doping

The rectifying p–n junction bipolar diodes are formed by modulated chemical doping of the CNTs [40]. This is performed by catalytically patterning CNTs onto SiO_2. The CNTs are formed as p-type due to an absorbed oxygen molecule from the atmosphere. The n-type CNT, on the other hand, is made by adding a poly(methyl methacrylate) (PMMA) layer on the first half p-type CNT, exposing the other half to potassium (K) dopant in vacuum. The potassium dopants are produced by electrical heating of the potassium source. The devices were fabricated by Dai et al. [41] in 2000. Figure 16.14 shows the layout diagram of this device.

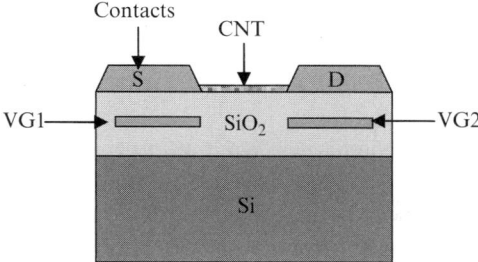

Fig. 16.13 Electrostatically doped p–n junction CNT device using two identical back gate electrodes (redrawn from [39])

Fig. 16.14 Chemically doped p–n junction device (redrawn from [41])

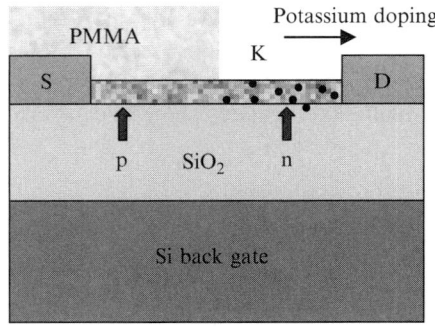

16.5.3 CNT Metal–Semiconducting Junction

Rectification diodes are formed by crossing metal and semiconductor CNTs [42] and by the formation of a kink-shaped intramolecular junction [43, 44]. The device characteristics of cross metal–semiconductor and intramolecular junction devices were demonstrated by McEuen et al. [42] and Yamada et al. [43], respectively.

16.5.3.1 CNT Crossed Metal–Semiconducting Junction

In the CNT crossed metal–semiconducting (MS) junction, rectification is shown as both the CNTs show the same graphene band structure thus having identical work functions. The metallic work function should align in between the band gap of the semiconducting CNT. Thus, in both the cases, the Fermi level is in the middle of the CNT band gap and is at the same energy as the Fermi level in a metallic tube. This leads to the presence of a Schottky barrier at the crossing point between the two CNTs. In addition, there is a barrier height approximately equal to half the band gap of the semiconductor CNT. This is due to barrier height associated with the depletion region. The nanoscale depletion region acts as a leaky tunneling barrier. The Schottky barrier in three-terminal MS junction devices offers a rectification to form a p–n junction Schottky metal–semiconductor diode. The device characteristic shows a nonlinear rectification behavior for metal–n-type semiconductor junction due to the Schottky barrier. Figure 16.15 shows the schematic representation of the device fabricated.

16.5.3.2 Intramolecular Metal–Semiconducting CNT Junction

The intramolecular semiconductor–metal junctions [45] are fabricated by introducing the pentagon and heptagon carbon ring defect pairs by mechanical deformation. The five to seven pair is placed onto the opposite side of the CNTs to have a large cap-like angle of curvature. A sharp kink of 40° was observed [46]. The device characteristic is shown in Fig. 16.16.

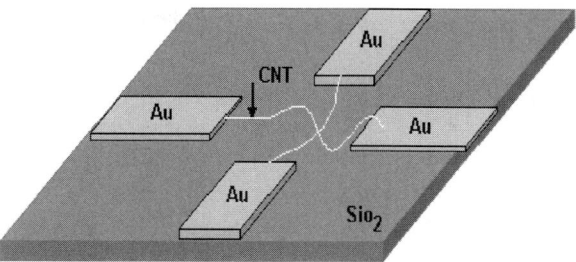

Fig. 16.15 Redrawn schematic diagram of metal–semiconducting CNT junction indicating formation of three-terminal device [42]

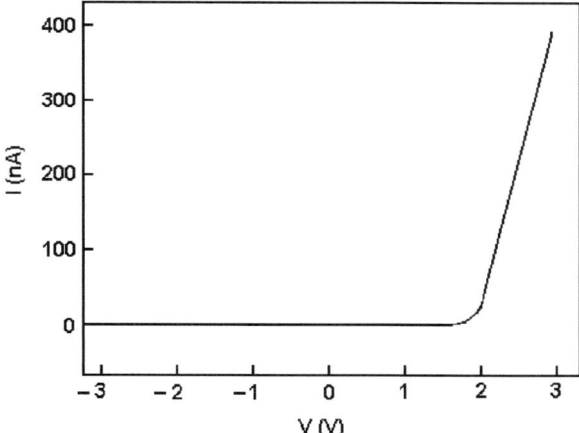

Fig. 16.16 Device characteristics of intramolecular MS CNT junction demonstrating ideal device characteristics similar to semiconductor diodes (redrawn from [46])

The junction formed causes the rectification behavior rather than the electrode contact. The *I–V* characteristics are similar to the p–n junction diode at room temperature.

16.6 Field Effect Transistors

FETs [47] are generally fabricated using semiconducting CNTs. The nanoscale CNT-based FET shows different characteristics than the traditional microscale semiconductor-based FET. The CNTs are bridged between two electrodes and the Si substrate is used as a back gate, which controls the switching action for the transistor. The drain current vs. drain voltage characteristics are studied for the transistor. When the gate voltage is varied from negative to positive, the CNT conductance characteristics are modified from high to low. The transistors are classified based on the type of contact and the doping profile.

16.6.1 Classification Depending on Types of Doping

16.6.1.1 Unipolar CNT Transistor (p-Type or n-Type)

The absorption of oxygen molecules onto CNTs results in the formation of p-type tubes. The transistors fabricated with these CNTs show unipolar p-type behavior. The transistor-doped holes show no electron transport at the high positive gate voltage, which results in a high Schottky barrier at the metal–semiconductor interface. This is due to the maximum Fermi level pinning near the valence band at the nanotube–metal interface. The n-type CNT is very essential in complementary logic devices and circuits. The p-type CNT can be converted into n-type by doping the CNT surface chemically by alkali metal or by the simple method of annealing in vacuum and in an inert gas. The annealing modifies the Schottky barrier height at the contact due to adsorption of oxygen. In doping, barrier thickness is changed and there is a shift in the threshold voltage [48].

The figure observed by Avouris [48] shows the transfer characteristics produced by annealing and by doping the CNT with potassium metal. First, the n-type unipolar device is fabricated using thermal annealing and then oxygen is introduced at each step. It was observed that as the oxygen is increased, the current at the positive bias decreased and increased at the negative bias.

The behavior of the n-type transistor produced by potassium doping is different from annealing process. At low doping, there is shift in the curve toward the more negative bias voltage. The device does not show any ambipolar behavior as there is no significant current during the intermediate doping. During the high doping, there is an increase in the current flow in the device.

16.6.1.2 Ambipolar CNT Transistor

During the addition of oxygen, at an intermediate step, the transistor behaves as an ambipolar transistor. There is conduction of holes and electrons depending upon the gate bias. An ambipolar device [30, 49] is hole conducting (p-channel) when the gate bias is sufficiently negative and electron conducting (n-channel) when gate bias is sufficiently positive.

16.6.2 Classification Depending upon Type of Contact

16.6.2.1 CNT Transistor with Ohmic Contact

The CNT transistors are studied by taking zero-bias conductance into consideration. The zero-bias conductance is given by

$$G = \frac{4e^2}{h} \int P(E) \left[-\frac{\partial f(E)}{\partial E} \right] dE. \tag{16.9}$$

The conductance characteristics and band bending curves are plotted by using the conductance equation by Léonard and Tersoff [50].

The plot shows three regions of interest. The traditional transistor behaves in only two regions, i.e., ON/OFF state, i.e., I and II region. At high negative bias voltage, the device exhibits high conductance. This corresponds to the ON state of the transistor. As the voltage is increased, the conductance drops to zero and the device turns into the OFF state. This is analogous to the conventional transistor. Region III is observed only in the CNT-based transistor and is a gate resonant tunneling effect. This is due to the tunneling through the localized states in the CNT with the coulomb blockade showing a negative differential resistance.

The band diagram explains the behavior of the transistor. At a low voltage, the metal work function is high and the Fermi level is below the valence band. Therefore, there is a hole transport in the channel without any barrier. Thus, the conductance is very high. As the gate voltage increases, the band is pulled down creating a barrier for hole transport, conductance drops to zero, and transistor is in OFF state. The third region shows interesting behavior. Gate voltage is increased; the conduction band is pulled down into the band gap creating an electrostatic quantum dot with a discrete localized state. The electron can tunnel through the quantum dot showing a sharp conductance peak at high voltage as explained earlier giving rise to strong negative differential resistance.

To summarize in both the conventional and the CNT FET, the current saturates due to an increase in the drain voltage. The saturation is due to "pinch-off" in normal transistors, whereas the current is limited by the number of carriers (holes or electrons) from the leads for CNT FETs. The presence of the resonating effect is observed only in the CNT transistor.

16.6.2.2 CNT Transistor with Schottky Contact

When the barrier is dominantly Schottky, the transistor operates as a SB transistor. The switching is due to the contact resistance rather than the channel conductance. There is also a significant change in band bending as compared to the ohmic contact transistor. The metal/nanotube contact and the applied voltage determine the performance of the Schottky CNT FET.

The band diagram shown by Avouris et al. [51] demonstrates sharp band bending near the contact due to the Schottky barrier instead of the observed CNT properties. Due to the sharp band bending, electrons can tunnel through it leading to a sharp increase in the current. In the ON state, the band is raised up and the hole carrier tunnels into the valence band due to the reduced tunneling distance. With the increase in gate voltage, the valence band is pulled down, creating a large tunnel barrier approximately equal to the channel length, making it impossible for the holes to reach the other end of the contact, thus allowing the transistor to operate in the OFF state. Here, it is also observed that the metal Fermi level is at the center of the valence and the conductance bands, thus creating a large SB. This transistor therefore operates as a SB FET. If the SB is small, the device works as the normal conventional FET. The figure below shows the sharp increase in the conductance as the gate voltage is varied, indicating the presence of a Schottky barrier at the contact of the device. The Schottky contact-based CNT FET [52, 53] operates differently than conventional transistor and ohmic contact CNT FET.

16.7 CNT-Based Single Electron Transistors

The molecular quantum CNT wire-based transistors are known as *single electron transistors* [54]. They consist of conducting islands connected by tunneling barriers to two metallic electrodes. If the temperature and bias voltage are less than the energy required to add an electron to the island, the device is in the blocked state. It is generally observed only in (n, n) armchair metallic CNTs. The quasiperiodic series of sharp peaks are observed in the conductance vs. gate voltage curve. The device characteristics show coulomb blockade and also indicates the coulomb charging of CNTs when the contact resistance is larger than the quantum resistance h/e^2 and the capacitance of the device is low enough to have significant charging energy due to the addition of a single electron.

The peak spacing is given by

$$\Delta V_g = (U + \Delta E)/e\alpha, \qquad (16.10)$$

where $U = e^2/C$ is the coulomb charging energy for every single electron addition in the dot, ΔE is the single particle level spacing, α ($=C_g/C$) is the rate of change of electrostatic potential of the dot due to the change in applied back gate voltage, and C and C_g are the total capacitance of the dot and the capacitance between dot and back gate electrode, respectively.

The peak amplitude of an isolated peak is approximately equal to e^2/h, where e and h are electron charge and Planck's constant, respectively. The rationale for the SET characteristics is as follows in Fig. 16.17.

The energy band diagram in Fig. 16.17 shows the relation of gate voltage V_g and drain voltage V. The dot filled with the N electrons and the energy separation of $U + \Delta E$ for adding the ($N + 1$)th electron to the excited single particle state. The above levels are evenly spaced with ΔE. The high conductance coulomb blockade peak is due to the alignment of the Fermi level of the metal contact with the lowest empty energy state, thus allowing single electron tunneling through the dot at $V = 0$. The low conductance valley indicates suppression of electron tunneling due to the single electron charging energy U. The voltage V dependence can be observed as the voltage V is increased, the right-hand contact is pulled below the energy level of the highest electron filled state allowing an electron tunneling giving rise to a conductance peak.

16 Characteristics of Carbon Nanotubes for Nanoelectronic Device Applications

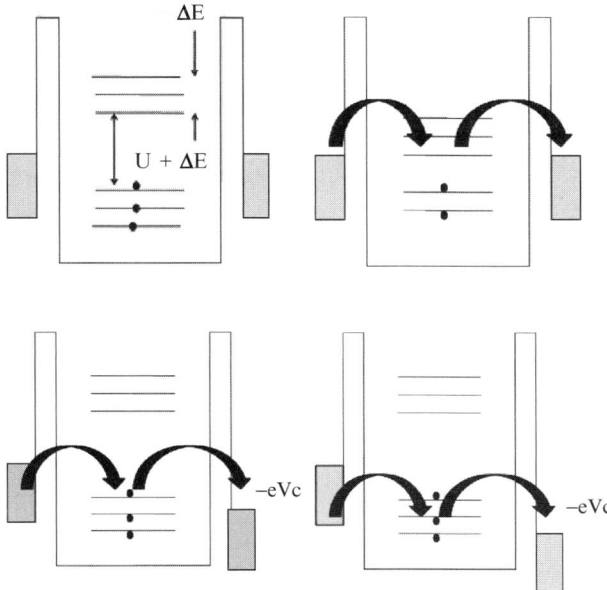

Fig. 16.17 Redrawn [55] schematic energy diagram with the Coulomb model. The electron transport indicated in the figure resulted into peaks in the conductance

With the additional increase in voltage V, the contact is further pulled down below the highest filled state allowing additional electron tunneling from the additional state, resulting in generation of additional resonating peak conductance. It is also observed [56] that the height of the conductance peak is dependent on the temperature. There is reduction in height and broadening of the peak as the temperature is increased. The conductance peak is held constant with increasing temperature if the density of the states is continuous as seen in the bulk semiconductor devices.

In summary, nanoelectronic devices using CNTs have more control and improve electrical characteristics over a normal operating condition. However, they show quick degradation outside the regular operating condition.

16.8 Integrated Device Fabrication

16.8.1 Nonvolatile Random Access Memory

Nonvolatile random access memory (NRAM) [57] is a high density electrostatically switchable wire array memory module. The molecular-scale device structure consists of a crossbar array formed by groups of CNTs on the substrate and groups of perpendicular CNTs suspended on a periodic support array. The bistable ON/OFF principle at the crosspoints is related to the two minima observed on the total energy vs. distance curve. The total energy equation is given by

$$E_T = E_{vdW} + E_{elas} + E_{elec}, \qquad (16.11)$$

where E_T is the total energy of the single crossbar device element, E_{vdW} is the van der Waals energy (vdW), E_{elas} is the elastic energy, and E_{elec} is the electrostatic energy. The ON state is determined by the minimum (vdW) force observed due to contact between two CNTs and the OFF state is determined by the minimum elastic energy observed when the distance of separation is finite. Figure 16.18 shows the schematic diagram of the NRAM and Fig. 16.19 shows the total energy diagram at each crossbar junction element indicating the ON and OFF state.

In addition, there is a significant variation in the resistance value due to reversible switching under normal ambient conditions at room temperature. The resistance values associated with the device ON/OFF state are observed in Fig. 16.20. The ON and OFF resistance is approximately 140 MΩ and 1.36 GΩ, respectively. The

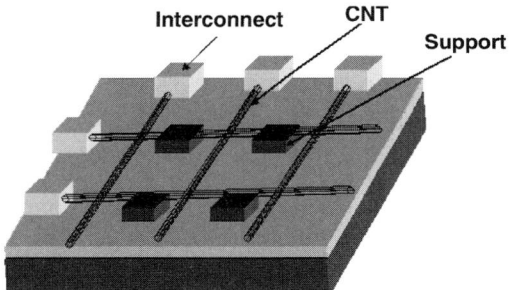

Fig. 16.18 Redrawn schematic diagram shows the crossbar junction formation by the two sets of perpendicular CNTs [57]

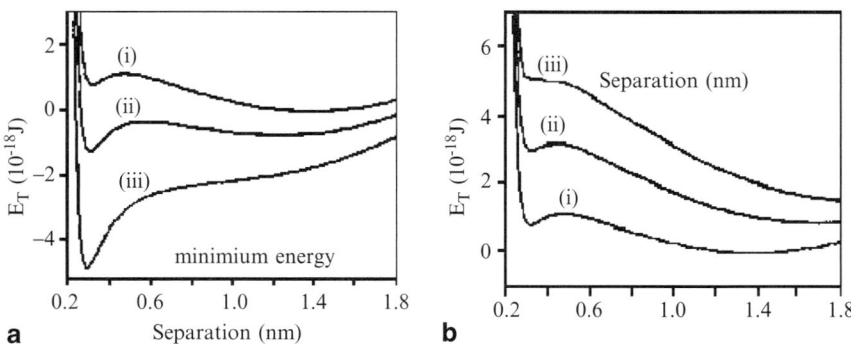

Fig. 16.19 Total energy vs. the distance of separation curves indicating the minimum energy for (**a**) ON state and (**b**) OFF state (redrawn from [57])

Fig. 16.20 ON/OFF resistance values due to reversible switching (redrawn from [57])

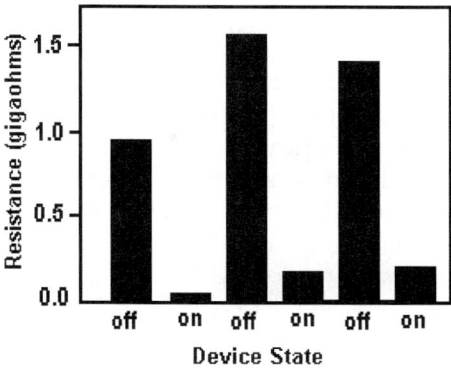

resistance value is in the higher range (MΩ GΩ) due to the large contact resistance which is generally observed between the CNTs and the metal contact [58].

This NRAM module shows promise in the next-generation semiconductor electronic world. Rueckes et al. have a patent number 6706402 for the NRAM technology. It was developed at Nantero, Inc., Woburn, MA.

16.9 Limitations to Carbon Nanotube Technology

Nanotechnologists all over the world are exploring various aspects of research ranging from the fundamental material research to their applications. Nanotechnology thrives due to the extensive interest shown by both academia and industry. In this chapter, we have focused on one type of nanomaterials, i.e., CNTs. This material has ubiquitous applications in the current sociopolitical environment: it promises improvement in current defense applications like fighting biological agents and making traditional weapons more effective. Clinical diagnostics have experienced a large-scale makeover in most of its detection schemes, which are faster, multiplexed, portable, and more reliable.

CNT technology has also been receiving extensive attention in making cutting edge sensor technology (see Sect. 16.5.1) and semiconductor-based devices, but despite its numerous applications it suffers from a few drawbacks. Like most other nanomaterial-based device technologies, it faces the problem of interfacing the nanoscale sensor materials and macroscale measurement systems. Although CNTs have shown promising results with respect to their electrical properties, integrating them with realistic electric circuits remains a challenge owing to the difficulty in alignment of these nanoscale structures in the desired positions. A method to mass-produce CNTs has not been established yet and the current means to do so, involve using expensive manufacturing equipment. Synthesizing CNTs of predetermined diameters and chirality remains the biggest challenge of all. Microscopy equipment is not only uneconomical but also suffers from technical limitations since imaging requires expertise and involves cumbersome techniques to achieve perfection. High intensity electron/ion beams used in different electron microscopes might burn out CNTs to form amorphous carbon while attempting to improve resolution. Transmission electron microscopy involves elaborate and difficult methods for sample preparation. It is capable of giving a good idea about the diameter of a CNT but can provide only limited data related to CNT chirality. Probing a single unidimensional wire like a CNT requires expertise and sometimes a good amount of fortune. Invasive techniques involving carbon-based nanoscale substances like CNTs for use in biological analytes have been believed to be hazardous to health of the living organism under examination.

Despite many disadvantages, fruitful research is being pursued with newer and better reasons to incorporate CNTs in most state-of-the-art nanoelectronic devices.

16.10 Nanopackaging

Producing reliable systems which maintain good signal integrity and are economy friendly are two important aspects for both the manufacturer and the buyer. The manufacturing industry thrives on making low voltage, high reliability, and high-speed systems, which can be integrated with future devices. When such is the goal, as discussed in the previous sections, traditional sensor devices based on CNT composites for gas detection and CNT arrays for biomolecular detection are highly sensitive to the agents under examination, even to the concentration levels of 1 ppb. Such sensitive systems behave quite differently under laboratory conditions. When being made commercially viable, however, care has to be taken that sensor systems are properly packaged to protect the sensor material from external contamination. Manufacturers need to take care of not only biological type of contaminants, but also electrical. Stray electric charges from other instruments or a static discharge from the person handling the device are also potential threats to unpackaged devices. The electronics linking the sensor element to the measurement systems must also be packaged well to avoid corruption in the sensitive signals that are exchanged between them. CNT-based semiconductor devices also require nanopackaging.

Nanopackaging for nanomaterial-based semiconductor devices is done in two levels: wafer packaging and system board level. Wafer level packaging uses nanomaterials to bring about improvements in the electrical, thermal, and mechanical properties in the chip to package interconnections. The advantages of wager level packaging are the small size and low cost of the product. These added benefits of wafer level packaging are because the package size is only a little more than the actual device itself and it costs less since all interconnections are done at this level in one parallel step. Such devices are also faster since there is reduced interconnect lengths. System level packaging, on the other hand, deals with packaging the entire device after wafer level packaging, to the various measurement systems. System level packaging requires matching of components and making compatibility possible between all kinds of measurement systems, thus making the device portable. Portability, however, brings with it the risk of contamination and damage to the device. Nanopackaging ensures both rigidity and resistance to contaminants.

References

1. Kroto HW, Heath JR, O'Brien SC, Curl RF, and Smalley RE (1985) C60: buckminsterfullerene. Nature, *318*(6042):162–163
2. Iijima S (1991) Helical microtubules of graphitic carbon. Nature, *354*:56–58
3. Ebbesen TW and Ajayan PM (1992) Large-scale synthesis of carbon nanotubes. Nature, *358*(6383):220–222
4. Maohui G and Klaus S (1994) Scanning tunneling microscopy of single-shell nanotubes of carbon. Appl. Phys. Lett., *65*(18):2284–2286
5. Falvo MR, Clary GJ, Taylor RM, II, Chi V, Brooks FP, Jr, Washburn S, and Superfine R (1997) Bending and buckling of carbon nanotubes under large strain. Nature, *389*(6651):582–584
6. Kim P and Lieber CM (1999) Nanotube nanotweezers. Science, *286*(5447):2148–2150
7. Dekker C (1999) Carbon nanotubes as molecular quantum wires. Phys. Today, *52*(5):22–28
8. Bachtold A, Hadley P, Nakanishi T, and Dekker C (2001) Logic circuits with carbon nanotube transistors. Science, *294*(5545):1317–1320
9. Kolmogorov AN and Crespi VH (2000) Smoothest bearings: interlayer sliding in multiwalled carbon nanotubes. Phys. Rev. Lett., *85*(22):4727–4730
10. Yu MF, Kowalewski T, and Ruoff RS (2001) Structural analysis of collapsed, and twisted and collapsed, multiwalled carbon nanotubes by atomic force microscopy. Phys. Rev. Lett., *86*(1):87–90
11. Chen G, Bandow S, Margine ER, Nisoli C, Kolmogorov AN, Crespi VH, Gupta R, Sumanasekera GU, Iijima S, and Eklund PC (2003) Chemically doped double-walled carbon nanotubes: cylindrical molecular capacitors. Phys. Rev. Lett., *90*(25):257403
12. Kwon Y-K and David T (1998) Electronic and structural properties of multiwall carbon nanotubes. Phys. Rev. B, *58*(24):R16001–R16004
13. Bethune DS, Klang CH, de Vries MS, Gorman G, Savoy R, Vazquez J, and Beyers R (1993) Cobalt-catalysed growth of carbon nanotubes with single-atomic-layer walls. Nature, *363*(6430):605–607
14. Ruoff RS, Tersoff J, Lorents DC, Subramoney S, and Chan B (1993) Radial deformation of carbon nanotubes by van der Waals forces. Nature, *364*(6437):514–516
15. Dresselhaus MS, Dresselhaus G, and Eklund PC (1996) Science of Fullerenes and Carbon Nanotubes. San Diego: Academic
16. Saito R, Fujita M, Dresselhaus G, and M Dresselhaus (1993) Electronic structure of double-layer graphene tubules. Appl. Phys. Lett., *60*:2204–2206
17. Saito R, Fujita M, Dresselhaus G, Dresselhaus MS, Hamada N, Sawada S-I, and Oshiyama A (1992) Electronic structure of chiral graphene tubules new one-dimensional conductors: graphitic microtubules. Appl. Phys. Lett., *60*:2204–2206
18. Saito R, Fujita M, Dresselhaus G, and Dresselhaus MS (1992) Electronic structure of graphene tubules based on C_{60}. Phys. Rev. B, *46*(3):1804
19. Dresselhaus MS and Dresselhaus AG (1981) Intercalation compounds of graphite. Adv. Phys., *30*:139–326
20. Mintmire JW, Dunlap BI, and White CT (1992) Are fullerene tubules metallic? Phys. Rev. Lett., *68*(5):631
21. Frank S, Poncharal P, Wang ZL and Heer WA (1998) Carbon nanotube quantum resistors. Science, *280*(5370):1744–1746
22. Sanvito S, Kwon Y-K, Tománek D, and Lambert CJ (2000) Fractional quantum conductance in carbon nanotubes. Phys. Rev. Lett., *84*(9):1974
23. Ristroph T, Goodsell A, Golovchenko JA, and Haul L (2005) Detection and quantized conductance of neutral atoms near a charged carbon nanotube. Phys. Rev. Lett., *94*(6):66102-1–66102-4
24. Che J, Cagin T, and Iii WAG (2000) Thermal conductivity of carbon nanotubes. Nanotechnology, *11*(2):65–69
25. Berber S, Kwon Y-K, and Tománek D (2000) Unusually high thermal conductivity of carbon nanotubes. Phys. Rev. Lett., *84*(20):4613

26. Hone J, Llaguno MC, Biercuk MJ, Johnson AT, Batlogg B, Benes Z, and Fischer JE (2002) Thermal properties of carbon nanotubes and nanotube-based materials. Appl. Phys. A, *74*(3):339–343
27. Sofia S and Chaniotakis NA (2003) Carbon nanotube array-based biosensor. Anal. Bioanal. Chem., *375*:103–105
28. Cai H, Cao X, Jiang Y, He P, and Fang Y (2003) Carbon nanotube-enhanced electrochemical DNA biosensor for DNA hybridization detection. Anal. Bioanal. Chem., *375*(2):287–293
29. Tans SJ, Verschueren ARM, and Dekker C (1998) Room-temperature transistor based on a single carbon nanotube. Nature, *393*:49
30. Martel R, Derycke V, Lavoie C, Appenzeller J, Chan KK, Tersoff J, and Avouris P (2001) Ambipolar electrical transport in semiconducting single-wall carbon nanotubes. Phys. Rev. Lett., *87*(25):256805
31. Someya T, Small J, Kim P, Nuckolls C, and Yardley JT (2003) Alcohol vapor sensors based on single-walled carbon nanotube field effect transistors. Nano Lett., *3*(7):877–881
32. Javey A, Guo J, Wang Q, Lundstrom M, and Dai H (2003) Ballistic carbon nanotube field-effect transistors. Nature, *424*:654–657
33. Star A, Han TR, Joshi V, Gabriel JC, and Grüner G (2004) Nanoelectronic carbon dioxide sensors. Adv. Mater., *16*(22):2049–2052
34. Léonard F and Tersoff J (2000) Role of Fermi-level pinning in nanotube Schottky diodes. Phys. Rev. Lett., *84*:4693–4696
35. Anantram MP and Léonard F (2006) Physics of carbon nanotube electronic devices. Rep. Prog. Phys., *69*:507–561
36. Zhang Y, Ichihashi T, Landree E, Nihey F, and Iijima S (1999) Heterostructures of single-walled carbon nanotubes and carbide nanorods. Science, *285*(5434):1719–1722
37. Léonard F and Tersoff J (1999) Novel length scales in nanotube devices. Phys. Rev. Lett., *83*(24):5174
38. Léonard F and Tersoff J (2000) Negative differential resistance in nanotube devices. Phys. Rev. Lett., *85*(22):4767
39. Lee JU, Gipp PP, and Heller CM (2004) Carbon nanotube p–n junction diodes. Appl. Phys. Lett., *85*(1):145–147
40. Zhou C, Kong J, Yenilmez E, Dai H, Léonard F, and Tersoff J (2000) Modulated chemical doping of individual carbon nanotubes negative differential resistance in nanotube devices. Science, *290*:1552–1555
41. Zhou C, Kong J, Yenilmez E, Dai H, Bockrath M, Hone J, Zettl A, McEuen PL, Rinzler AG, and Smalley RE (2000) Modulated chemical doping of individual carbon nanotubes chemical doping of individual semiconducting carbon-nanotube ropes. Science, *290*:1552–1555
42. Fuhrer MS, Nygård J, Shih L, Forero M, Yoon Y-G, Mazzoni MSC, Choi HJ, Ihm J, Louie SG, Zettl A, and McEuen PL (2000) Crossed nanotube junctions. Science, *288*:494–497
43. Yamada T (2002) Modeling of kink-shaped carbon-nanotube Schottky diode with gate bias modulation. Appl. Phys. Lett., *80*:4027–4029
44. Odintsov AA (2000) Schottky barriers in carbon nanotube heterojunctions. Phys. Rev. Lett., *85*(1):150
45. Chico L, Crespi VH, Benedict LX, Louie SG, and Cohen ML (1996) Pure carbon nanoscale devices: nanotube heterojunctions. Phys. Rev. Lett., *76*(6):971
46. Yao Z, Postma HWC, Balents L, and Dekker C (1999) Carbon nanotube intramolecular junctions. Nature, *402*:273–276
47. Martel R, Schmidt T, Shea HR, Hertel T, Avouris P, Tans SJ, Verschueren ARM, and Dekker C (1998) Single- and multi-wall carbon nanotube field-effect transistors room-temperature transistor based on a single carbon nanotube. Appl. Phys. Lett., *73*:2447–2449
48. Derycke V, Martel R, Appenzeller J, and Avouris P (2002) Controlling doping and carrier injection in carbon nanotube transistors. Appl. Phys. Lett., *80*:2773–2775
49. Babic B, Iqbal M, and Schönenberger C (2003) Ambipolar field-effect transistor on as-grown single-wall carbon nanotubes. Nanotechnology, *14*:327–331

50. Léonard F and Tersoff J (2002) Multiple functionality in nanotube transistors. Phys. Rev. Lett., *88*(25):258302
51. Appenzeller J, Knoch J, Derycke V, Martel R, Wind S, and Avouris P (2002) Field-modulated carrier transport in carbon nanotube transistors. Phys. Rev. Lett., *89*:126801
52. Heinze S, Tersoff J, Martel R, Derycke V, Appenzeller J, and Avouris P (2002) Carbon nanotubes as Schottky barrier transistors. Phys. Rev. Lett., *89*:106801
53. Appenzeller J, Lin YM, Knoch J, and Avouris P (2004) Band-to-band tunneling in carbon nanotube field-effect transistors. Phys. Rev. Lett., *93*:196805
54. Postma HWC, Teepen T, Yao Z, Grifoni M, and Dekker C (2001) Carbon nanotube single-electron transistors at room temperature. Science, *293*(5527):76–79
55. Bockrath M, Cobden DH, McEuen PL, Chopra NG, Zettl A, Thess A, and Smalley RE (1997) Single-electron transport in ropes of carbon nanotubes. Science, *275*(5308):1922–1925
56. Tans SJ, Devoret MH, Dai H, Thess A, Smalley RE, Geerligs LJ, and Dekker C (1997) Individual single-wall carbon nanotubes as quantum wires. Nature, *386*(6624):474–477
57. Rueckes T, Kim K, Joselevich E, Tseng GY, Cheung C-L, Lieber CM, Yang L, and Han J (2000) Carbon nanotube-based nonvolatile random access memory for molecular computing. Science, *289*:94–97
58. Yaish Y, Park JY, Rosenblatt S, Sazonova V, Brink M, and McEuen PL (2004) Electrical nano-probing of semiconducting carbon nanotubes using an atomic force microscope. Phys. Rev. Lett., *92*:46401

Chapter 17
Carbon Nanotubes for Thermal Management of Microsystems

Johan Liu(✉) and Teng Wang

17.1 Introduction

One important function of electronic packaging is to remove the heat generated by the integrated circuits (ICs). Efficient cooling requires both high heat conduction within the package and efficient heat removal from the package. Elevated temperature is damaging to the chip and its package. Material mismatch causes mechanical stress leading to fatigue, creep and finally failure; interconnects can melt and electromigration within the IC is speeded up. Efficient heat removal is not always the case. Plastic is a common packaging material as it is electrically insulating and cheap. The thermal conductivity is, however, low, about 0.2 W m^{-1} K^{-1} compared with that of metals (aluminium 220 W m^{-1} K^{-1} and copper 400 W m^{-1} K^{-1}). Other important factors are the heat spreading and thermal interface materials. The components are often mounted on a polymer board which is only cooled by air. The heat transfer coefficient is only 5–15 W m^{-2} K^{-1} for natural convection and 15–250 W m^{-2} K^{-1} for forced convection in gases [1].

As electronic circuits grow denser and the power consumption per unit area increases, new more efficient technologies for heat removal are necessary. Heat fluxes from the IC on the order of 100 W cm^{-2} (1,000,000 W m^{-2}) are not rare. A heat transfer coefficient (including a possible area enlarging factor) of 20,000 W m^{-2} K^{-1} is needed to accommodate a heat flux of 100 W cm^{-2} at a temperature difference of 50 K. The application of nanotechnologies is considered as a revolutionary approach to meet the tougher requirements for thermal management of microsystems (Fig. 17.1).

There are two possible approaches to improve the cooling of microelectronic packages. The first is to improve the thermal conductivity of the packaging material and package geometry, so that the thermal gradient within it becomes smaller.

J. Liu
Bionano Systems Laboratory, Department of Microtechnology and Nanoscience,
Chalmers University of Technology, Kemivägen 9 Room A517, Se 412 96, Gothenburg,
Sweden Shanghai University, Shanghai, China

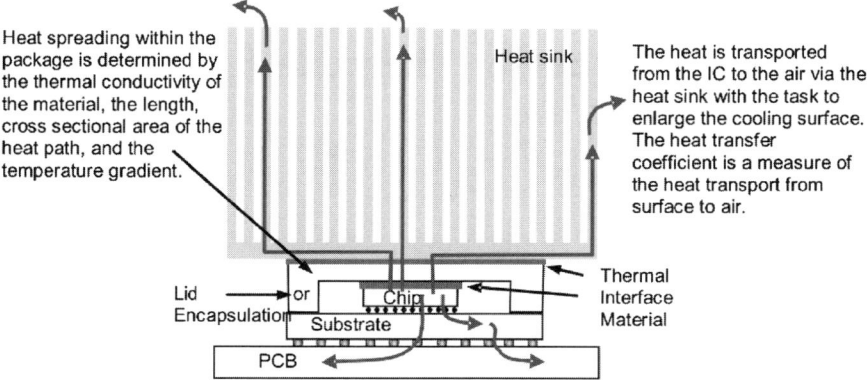

Fig. 17.1 Model of the heat path in a component mounted with a heat sink on a PCB with a heat sink on the back

The second is to improve the heat removal, i.e. increase the effect of the convective heat transfer, from the surface or from the inside of the package. Both approaches are discussed in this chapter. The first one is to make thermal interface materials (TIMs) by embedding carbon nanotubes (CNTs) or other nano-scale thermal conductive particles in nano-fibres prepared by electrospinning. The second approach is to build microchannel coolers based on CNTs. A brief introduction to the physical background of heat transfer is also included in this chapter.

17.2 Physical Background

There exist three basic heat transfer mechanisms: conduction, convection and radiation. In electronics, the first two mechanisms dominate, and consequently radiation is often neglected by engineers. Therefore, only conduction and convection are discussed in this section.

17.2.1 Conduction

Heat transfer by conduction Q, over a certain area A, depends on the thermal conductivity k, of the material and the temperature gradient ∇T. The law of heat conduction is also known as Fourier's law:

$$Q = k\nabla TA. \tag{17.1}$$

In one dimension with temperature difference ΔT over distance l,

$$Q = \frac{k\Delta T A}{l}. \qquad (17.2)$$

Heat transport in solids and fluids takes place via two effects: lattice vibration (phonons) and electron migration. Materials (metals) with high electrical conductivity are generally better thermal conductors as electrons are involved in the heat transfer. There is almost no transfer of electrons in insulators, so heat conduction must rely on lattice vibrations and is thus poorer.

17.2.1.1 Conduction in Composites

The denser and more uniform a material is the better its thermal conductivity. The conductivity of insulators is reduced by different types of phonon scattering processes. The scattering of phonons in composite materials is mainly due to acoustic mismatch between filler and matrix [2]. For a certain filler type and volume fraction, there are the following methods for increasing the thermal conductivity:

- Decreasing the number of thermally resistant junctions, e.g. by minimizing the number of polymer layers between fillers, for example by using larger filler particles
- Forming conducting networks by suitable packing
- Minimizing filler–matrix interfacial defects

Thermal conductivity can be estimated by the semi-empirical theory of Lewis and Nielsen [3]. According to this theory, conduction of the composite is dependent on volume fraction and heat conduction of the filler, heat conduction of the matrix and the filler shape. The basic estimate for composites can be done according to Lewis–Nielsen theory by using the following formulae:

$$k_C = k_M \frac{1 + AB\phi_F}{1 - B\psi\phi_F}, \qquad (17.3)$$

$$B = \frac{k_F / k_M - 1}{k_F / k_M + A}, \qquad (17.4)$$

$$\psi = 1 + \frac{1 - \phi_M}{\phi_M^2} \phi_F, \qquad (17.5)$$

where ϕ_M is the maximum filler load and A is a parameter depending on particle shape. Both are given in tables for different geometries (empirical and theoretical values). ϕ_F is the volume fraction filler and k_F, k_M and k_C are the thermal conductivities of filler, matrix and composite, respectively. In this theory, rods or fibres increase the thermal conductivity more than spheres of the same thermal conductivity and volume fraction. However, the theory does not explicitly take into account the size of particles or deal with more than two phases. Miloh and Benveniste [4] suggested a method for the estimation of effective thermal conductivity of three-phase composites with ellipsoidal inclusions.

Based on the above, it is obvious that elongated and well-dispersed particles forming networks are necessary to produce composites with high thermal conductivity. Filler–matrix interfacial contact can be improved by surface treatment of the particles and addition of coupling agents [5–7]. If untreated, the filler particles will form agglomerates in the epoxy matrix, so the particles are treated to render the surfaces hydrophobic. The coupling agent can also contain a functional group to form a chemical bond to the polymer.

17.2.1.2 Thermal Resistance

Thermal resistance R_{th} is a common way to define the cooling capabilities of a system. R_{th} is analogous to electrical resistance with the temperature drop ΔT as the driving force instead of electrical potential, and is defined as the temperature drop divided by the heat flow:

$$R_{th} = \frac{\Delta T}{q} = \frac{l}{kA}. \tag{17.6}$$

The cooling of electronic systems can be characterized by the thermal resistance between the heat source (the chip) and the ambient. The thermal resistance depends on the thickness, cross-sectional area and thermal conductivity of materials around the heat source and of the heat transfer coefficient h, to surrounding fluid (e.g. air or liquid coolant). In contrast to thermal conductivity, which is an inherent material quality, thermal resistance is a geometry-dependent property specific for each system.

17.2.2 Convection

Heat transfer by convection occurs between a solid surface and adjacent moving fluid and consists of two mechanisms: random molecular motion (diffusion) and macroscopic bulk motion in the fluid. Closest to the surface, the velocity is zero due to interaction with the solid and from there it increases to the bulk velocity v_∞. The same occurs for the temperature which changes from surface temperature T_s to the bulk temperature T_∞. These regions are called boundary velocity layer and thermal boundary layers and do not need to be same size.

Convective heat transfer can be by either natural convection or forced convection. The cause for natural convection is the density difference in fluids due to temperature variations. Forced convection is obtained by an external force that puts the fluid in motion. In electronics, this can, for example, be achieved with a fan for air or pump for fluid. Convective heat transfer can also be classified according to the flow regime: laminar or turbulent. Laminar flow is characterized by low pressure drop, neglecting mixing. Laminar flow occurs at low flow rates and small dimensions. When the flow velocity increases, the flow becomes unstable, vortices occur and at a certain point the fluid pattern changes to chaotic, i.e. turbulent. The pressure drop will be much higher but the heat transfer within the fluid will also mix much better. In electronic systems with their small dimensions and low flow rates, laminar flow is usually the case and most engineering rules of thumb are built on this assumption.

To determine the flow regime, the Reynolds number can be used. The Reynolds number is a dimensionless number that describes the ratio of the kinematic and viscous forces in the fluid. At a certain Reynolds number, the flow will undergo a transition from laminar to turbulent. The value varies for different geometries. From the definition, we see that low dimensions lead to low numbers, which is taken advantage of in microfluidics. The Reynolds number is defined as

$$Re \equiv \frac{vl}{\upsilon} = \frac{\rho vl}{\eta}. \quad (17.7)$$

Laminar flow can be described by the Navier–Stokes equations, which describe how the velocities v, u and w, in different direction, and pressure p are related.

The equations are the continuity equation

$$\frac{\partial u}{\partial x} + \frac{\partial v}{\partial y} + \frac{\partial w}{\partial z} = 0 \quad (17.8)$$

and the momentum equations in the x, y and z directions:

$$\left(\rho \frac{\partial u}{\partial t} + \rho u \frac{\partial u}{\partial x} + \rho v \frac{\partial u}{\partial y} + \rho w \frac{\partial u}{\partial z}\right) = \rho g_x - \frac{\partial p}{\partial x} + \mu \left(\frac{\partial^2 u}{\partial x^2} + \frac{\partial^2 u}{\partial y^2} + \frac{\partial^2 u}{\partial z^2}\right), \quad (17.9)$$

$$\left(\rho \frac{\partial v}{\partial t} + \rho u \frac{\partial v}{\partial x} + \rho v \frac{\partial v}{\partial y} + \rho w \frac{\partial v}{\partial z}\right) = \rho g_y - \frac{\partial p}{\partial y} + \mu \left(\frac{\partial^2 v}{\partial x^2} + \frac{\partial^2 v}{\partial y^2} + \frac{\partial^2 v}{\partial z^2}\right), \quad (17.10)$$

$$\left(\rho \frac{\partial w}{\partial t} + \rho u \frac{\partial w}{\partial x} + \rho v \frac{\partial w}{\partial y} + \rho w \frac{\partial w}{\partial z}\right) = \rho g_z - \frac{\partial p}{\partial z} + \mu \left(\frac{\partial^2 w}{\partial x^2} + \frac{\partial^2 w}{\partial y^2} + \frac{\partial^2 w}{\partial z^2}\right), \quad (17.11)$$

These equations are a set of coupled differential equations and in practice are extremely difficult to solve analytically. Instead, computational fluid dynamics (CFD) is used to solve flow problems. For an incompressible, Newtonian fluid with uniform viscosity and small temperature differences, the system can be described by the continuity and momentum equations. If the flow is compressible, or if heat fluxes occur, at least one extra equation is required.

17.2.2.1 Flow in Microchannels

The term 'micro' can be used for channels with hydraulic diameters of ten to several hundred micrometers. Fluid mechanics in microscopic scale is different from fluid mechanics at large scale. The Navier–Stokes equations and other models of fluid mechanics are based on the fact that the fluid can be treated as a continuum. However, as the scale shrinks, the number of molecules in the system becomes fewer and a point is reached when each molecule has larger chance to interact with surrounding walls that with another molecule of its own kind. At this situation, it could be awkward to model the fluid as a continuum and it should rather be modelled as individual molecules. According to Nguyen [8], this limit for water is reached at about 10 nm. At very small scale and high shearing, the Newtonian behaviour of the fluid breaks down and the macroscopic non-slip boundary condition (velocity is zero at interface) between fluid and surface cannot be used. According to Nguyen, this limit is reached well above the shear rates which were reached in the experimental setup. Recent experiments for circular and rectangular cross-section microchannels [9, 10] have shown that the transition from laminar to turbulent flows occurs at about same Reynolds number for flow at micro- and macroscales.

17.3 Nano-Thermal Interface Materials

Heat removal is crucial to the performance and reliability of microelectronic systems, as the heat generated by integrated circuits keeps increasing significantly. The objective of the present research work is to develop a new class of nano-thermal interface materials (nano-TIMs) using the electrospinning methodology by embedding nano-thermally conductive particles in nano-fibres to enhance heat removal between the chip and the heat sink/substrate [11].

Electrospinning is the process of subjecting a polymer solution to an electric field [12–16]. The applied voltage breaks down the surface tension at the capillary tip (called Taylor's cone) and as the charged solution moves in air, the solvents evaporate to produce nano-fibres on a collector. The electrospinning setup consists mainly of a high voltage source, a capillary tube, a needle and a collector (Fig. 17.2).

Nano-particles of silver, silicon carbide and multi-wall carbon nanotubes are added as thermal conductivity enhancement promoters. The developed nano-TIMs are soaked after spinning in some conducting liquids to enhance the thermal conductivity and wetting performance. To determine the maximum temperature, the resins can withstand, differential scanning calorimetry (DSC) and thermogravimetric analysis (TGA) have been utilized.

Figure 17.3 shows a SEM picture of a polymer resin nano-TIM. SEM pictures of nano-TIMs embedded with nano-silver particles and CNTs soaked in some conducting fluids are shown in Fig. 17.4.

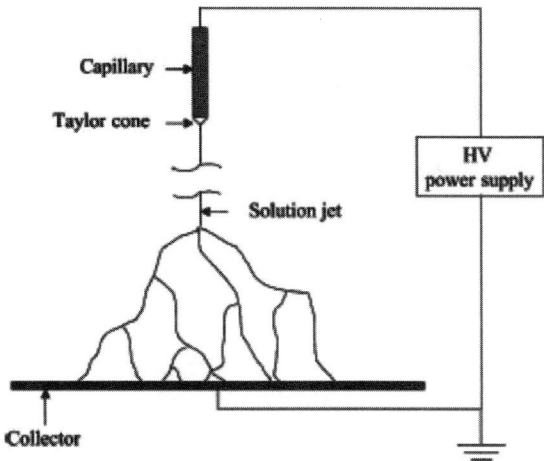

Fig. 17.2 Electrospinning setup

Fig. 17.3 SEM picture of nano-TIMs with a polymer resin

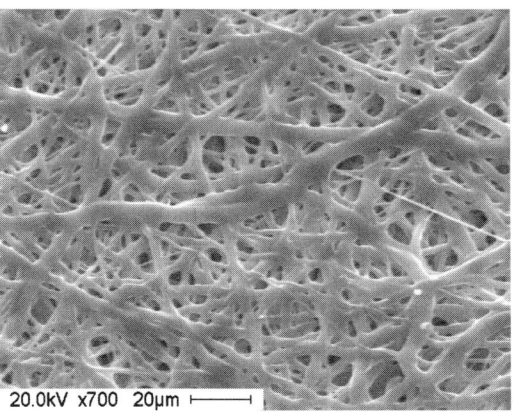

Fig. 17.4 SEM pictures of nano-TIMs embedded with nano-Ag (**a**) and CNTs (**b**)

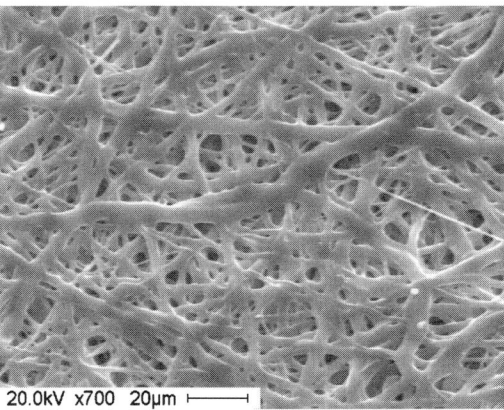

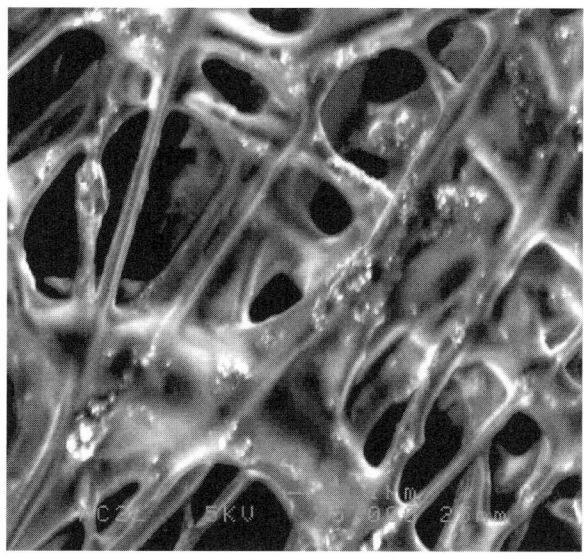

Fig. 17.5 Nano-TIM material with adhesive function in the material

Table 17.1 Properties of nano-TIMs in comparison with commercial materials

Property	Unit	Commercial sample A	Commercial sample B	Nano-TIM with CNT soaked in a conducting liquid	Nano-TIM with silver soaked in a conducting fluid
Thermal conductivity	W m^{-1} K^{-1}	4.0–5.2	4.4–6.5	0.72–4.33	1.02–2.73
Film thickness	μm	1,000	500	110	200
Thermal resistance	K W^{-1}	0.9–1.2	0.35–0.55	0.66–0.11	0.93–0.35
Operating temperature	°C	50–140	50–140	50–120	40–130
Degradation temperature	°C	400–500	400–500	400	300
Maximum stress during break	MPa	0.09	0.22	0.55	0.45
Maximum strain during break	%	54.1	30.	15	600
E-module	MPa	0.17	0.75	3.44	0.08
Colour		Grey	Grey	Black	Grey

To enhance the adhesion strength of the nano-TIM to the substrate, adhesives are also jetted in the same process as the formation of the nano-TIM. An example of the nano-TIM with adhesive function is shown in Fig. 17.5. By doing so, a complete nano-TIM tape is formed.

Table 17.1 summarizes the results for the various nano-TIMs in comparison with the two commercial samples, indicating that the nano-TIM with CNT and resin A can offer similar thermal conductivity, 3–9 times lower thermal resistivity, similar operation temperature range and degradation behaviour, 2–5 times better ultimate tensile strength and much higher Young's modulus. For the nano-TIM with nano-Ag particles and resin B, similar thermal resistivity, temperature operation range and degradation behaviour as compared with the commercial samples were obtained, but the mechanical properties are about 10–20 times better in terms of ultimate tensile strength, and elasticity and Young's modulus are about 50% to three times better. By adding more nano-silver particles into this TIM, it is believed that its thermal conductivity can be further improved.

In conclusion, nano-TIMs with various nano-particles have been manufactured using electrospinning. It is found that nano-particles can improve the heat transfer characteristics of the nano-TIMs. Such improvement depends on the quality and quantity of the nano-particles. Preliminary mechanical, thermal and degradation characterization of the nano-TIMs has shown that these materials can potentially offer better mechanical properties with equal or better thermal properties.

17.4 Microchannel Coolers Based on Carbon Nanotubes

The challenge of developing high-performance and low-complexity cooling solutions for microelectronic systems is becoming a key factor as the overall power consumption of integrated circuits continuously rises, despite the drop of the supply voltage. Many investigations about microchannels have been undertaken in the past several decades, showing that extremely high rates of heat transfer can be obtained by applying microchannel structures [17]. CNTs, a new form of carbon which can be described as rolled layers of graphite with 1–100 nm diameter [18], are a very ideal choice to make this kind of microchannel cooler due to many reasons. Firstly, well-structured CNTs are believed to have very high thermal conductivity based on theoretical predictions, although the experimental results vary dramatically in different references [19–21] for both single-walled and multi-walled nanotubes (SWNTs and MWNTs). If the unusually high thermal conductivity of CNTs can be achieved by a suitable synthesis process, the efficiency of the fins can be maximized, thus the total heat removal capability of the cooler can be promoted. Furthermore, CNTs can be grown directly on the surface of silicon and accurately according to pre-defined small-scale catalyst patterns normally transferred by standard photolithography processes. Therefore, micro-scale channels can be fabricated, making the cooler very compact and efficient. CNTs also provide a possibility of low-cost bulk production with a potential compatibility with standard CMOS technology [22].

This section briefly introduces the latest research work on using CNTs to build microcoolers [23–25], as well as supporting simulation work [26].

The overall manufacturing procedure of this CNT microcooler is illustrated in Fig. 17.6. The catalyst pattern is firstly transferred to the silicon substrate by photolithography and evaporation, and a lift-off process. The catalyst used in this application is iron and its thickness is roughly 1 nm. By using thermal chemical vapour deposition (TCVD), CNTs are then grown on the catalyst patterns, forming the fins of the microcooler. The growth temperature is about 750°C. The silicon substrate carrying CNT fins is then bonded with a lid to finish an entire cooler.

CNT microcoolers of different patterns and dimensions have been successfully manufactured. Figure 17.7 shows the SEM images of a CNT cooler with

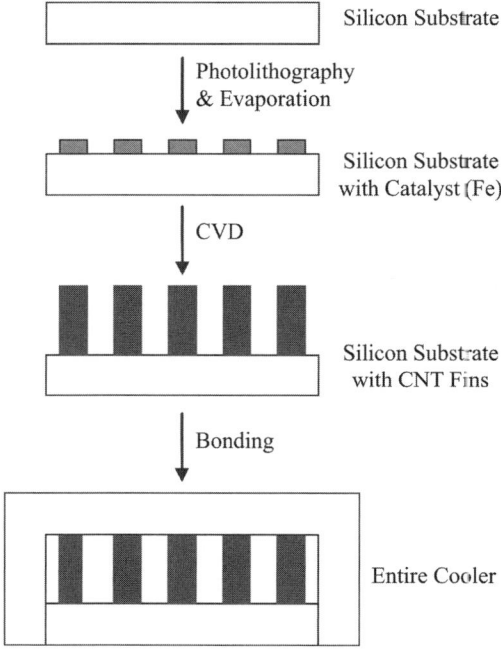

Fig. 17.6 Overall manufacturing process of carbon nanotube microcoolers

Fig. 17.7 SEM images of a carbon nanotube microcooler

17 Carbon Nanotubes for Thermal Management of Microsystems 387

50-μm wide channels and fins. The height of the fins is roughly 300 μm. Each fin contains numerous vertically aligned CNTs bunched together. The overall quality of the coolers is quite acceptable in terms of the uniformity and tube density of the fins.

The measurement was performed by using a power film resistor which can generate up to 30-W power as the heat source. The coolers are mounted on the exposed ceramic heat-dissipating surface of the resistor. No heat transfer paste is applied. The coolers are connected to plastic tubes on both ports and water proof tape seals the connections to prevent leakage of water. A micropump which can deliver up to 500 ml min^{-1} flow rate provides a continuous, stable and controllable water flow through the channels of the coolers. The temperature of the resistor is obtained by a thermocouple glued on the heat dissipation surface of the transistor.

A CNT microchannel cooler, a silicon microchannel cooler and a silicon cooler without fins, which are of similar dimensions, have been tested and compared under the same experimental conditions. The initial measurement results show that the cooling efficiencies of CNT and silicon microchannel coolers are roughly at the same level and much higher than that of the silicon cooler without fins.

Based on the initial experimental configuration, the power transferred from the heat source to the cooler cannot be measured or calculated precisely because an unknown portion of heat is dissipated by natural air convection. There are also some considerable thermal resistances existing between the resistor and the cooler, making accurate calculation more difficult. Therefore, some new coolers with heat resistors directly integrated onto the backside of the cooler have been designed and fabricated. The resistors are made of copper and formed in a serpentine pattern. The thermal resistance between the coolers and heat sources can thus be miniaturized. The new microcoolers are currently under test.

CFD simulation is a powerful tool to assist the study and design of microcoolers. Simulation work has also been done for CNT microchannel coolers to shorten the design cycle and lower the experimental cost.

Two types of fin arrays (1D and 2D nanotube fin arrays, respectively) have been simulated to study the relation between the cooling performance and the fin array structures, dimensions, fluid speed and thermal conductivity of the fins. The preconditioned conjugate residual (PRC) method in the FLOTRAN CFD program in ANSYS is utilized in this case. A numerical method to solve this conjugate heat transfer problem is to treat the solid and fluid as a unitary computational domain, and to solve the governing equations simultaneously. The FLUID141 element in ANSYS is selected to model the present fluid/thermal system with a fluid–solid coupling. Examples of the meshes generated in the CFD simulation of cooling assemblies for 1D and 2D fin arrays ($M = 5$, $N = 5$) are shown in Fig. 17.8.

Simulation results have shown that the thermal conductivity of the fins does have a significant effect on the cooling efficiency of the coolers. One example with five 1D fins shows that the maximum and average temperature of the fin array can be decreased dramatically by increasing the thermal conductivity of the fins (Fig. 17.9).

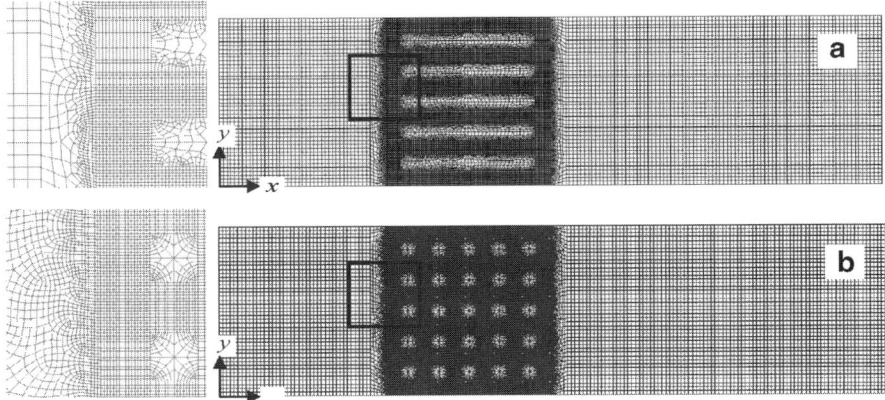

Fig. 17.8 Example of the models and meshes generated in the CFD simulation

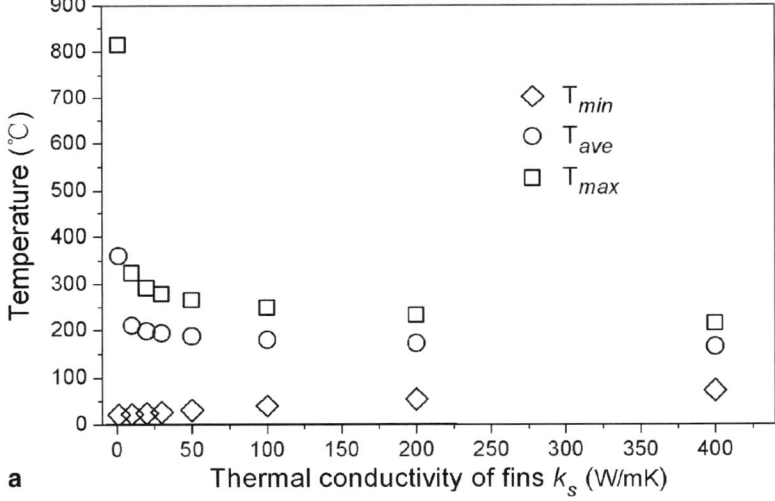

Fig. 17.9 Example of CFD simulation results for fins of different thermal conductivity

Another important factor affecting the performance of the microchannel coolers is the speed of the flow. As revealed in an example shown in Fig. 17.10, very high fluid speed must be applied to achieve good cooling performance. And the high fluid speed results in a very large pressure drop, which requires much higher pumping power. Applying high fluid speeds may also destroy the CNT fins as the nanotubes are basically bound together by weak van der Waals forces. This must be taken into account when designing the microcoolers. One possibility

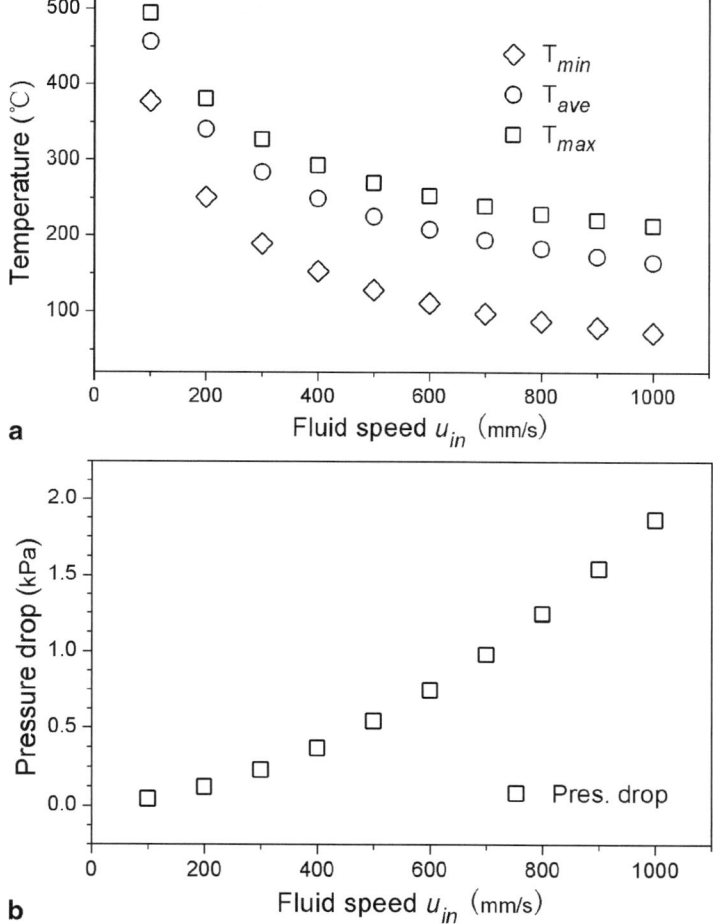

Fig. 17.10 Example of CFD simulation results revealing the effect of fluid speed on the temperature and pressure drop

is to coat the CNT fins, which increases the cost and complexity of manufacturing such coolers.

17.5 High Thermally Conductive Carbon Nanotube Bumps

Another important heat removal path from the chip is through the interconnects to the substrate. CNTs have also been proposed as bumps for flip chip interconnect due to their high thermal conductivity [27]. Besides the thermal conductivity,

CNTs also have high mechanical strength [28] and good electrical conductivity for metallic CNTs [29]. Moreover, CNTs can be aligned according to small-scale precisely pre-defined patterns by common technologies, making bulk production of fine pitch interconnects possible.

The simplified manufacturing process of patterned CNTs is illustrated in Fig. 17.11. Photolithography is used to transfer the patterns defined in the mask to the substrate. The catalyst is then deposited onto the substrate by an electron beam evaporator. A lift-off process is performed afterwards to remove the resist together with the unwanted catalyst. By using a DC-PECVD system, CNTs are grown on the catalyst patterns, forming the bumps of the chip.

In real applications, it is also necessary to grow CNTs on metal layers. The PECVD growth of carbon nanostructures can differ significantly on different metal layers and an insertion layer between the metal and catalyst particles can play a crucial role in the process [22]. A SEM picture showing the CNT bumps grown on both silicon and copper is shown in Fig. 17.12.

An important issue regarding using CNTs as flip chip interconnects is the high synthesis temperature of CNTs, typically higher than 650°C. Such high temperature is not compatible with certain temperature-sensitive processes and materials. Therefore, transfer of CNTs by solders [30] and conductive adhesives [31] has been proposed. Figure 17.13 shows some transferred CNT bundles by imprinted isotropic conductive adhesive (ICA). This technology developed at Chalmers University of Technology can successfully transfer CNT structures at a low temperature of 150°C, enabling the integration of CNTs into processes and materials that cannot withstand the high CNT synthesis temperatures [32].

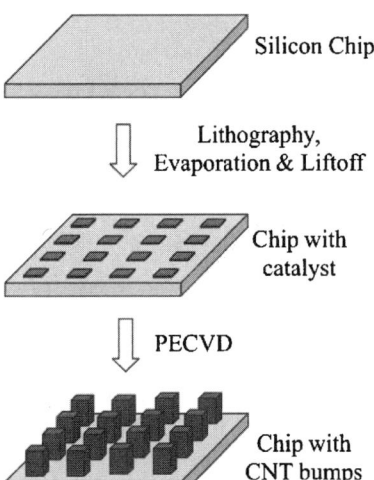

Fig. 17.11 Manufacturing process of CNT bumps

17 Carbon Nanotubes for Thermal Management of Microsystems

Fig. 17.12 SEM image of CNT bumps grown on silicon and copper

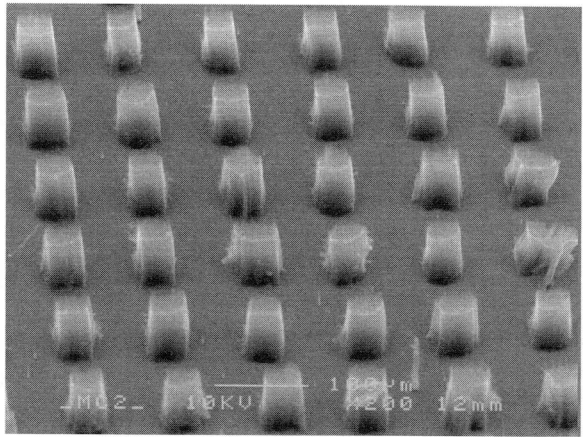

Fig. 17.13 SEM image of CNT bumps transferred by ICA at low temperature

17.6 Conclusions

Some new developments of applying nanotechnologies and nanomaterials in the field of thermal management of microelectronics packaging have been introduced in this chapter. These applications include nano-thermal interface materials and CNT-based microcoolers and flip chip bumps. The utilization of various nanotechnologies and nanomaterials may potentially revolutionize the electronics packaging and production industry to solve the needs of miniaturization, ultra-high density packaging and larger power dissipation.

Acknowledgements Financial support to the SMIT Center from its member companies including Avantec, France; Flextronics, Heller Industries, Intel, USA; Toyo-Kohan, Sumitomo Chemicals and Hitachi Chemicals, Japan; Mydata Automation and Foab Elektronik, Sweden is gratefully ac knowledged.Acknowledgements are also to the EU program EC-GEPRO with the contract no. 033349 SSA; the Shanghai Science and Technology Commission, PRC with the research contract no. 045107006 and 035007031; and Intel Higher Education Program with the contract No. CNDA# 4126007 for their financial support.

References

1. Welty JR, Wicks CE, Wilson RE, Rorrer GL (2001), Fundamentals of Momentum, Heat and Mass Transfer, 4th edition. Wiley, New York
2. Xu Y, Chung DDL, Mroz C (2001), Thermally conducting aluminium nitride polymer-matrix composites. Composites Part A: Applied Science and Manufacturing 32:1749–1757
3. Nielsen LE (1974), The thermal and electrical conductivity of two-phase systems. Industrial & Engineering Chemistry Fundamentals 13:17–20
4. Miloh T, Benveniste Y (1988), A generalized self-consistent method for the effective conductivity of composites with ellipsoidal inclusions and cracked bodies. Journal of Applied Physics 63:789–796
5. Liang S, Seung RC, Giannelis EP (1998), Barium titanate/epoxy composite dielectric materials for integrated thin film capacitors. In: Proceedings of the 48th Electronic Components and Technology Conference (ECTC), Seattle, WA, pp. 171–175
6. Hussain M, Nakahira A, Nishijima S, Niihara K (1996), Effects of coupling agents on the mechanical properties improvement of the TiO_2 reinforced epoxy system. Materials Letters 26:299–303
7. Kang S, Hong SI, Choe CR, Park M, Rim S, Kim J (2001), Preparation and characterization of epoxy composites filled with functionalized nanosilica particles obtained via sol-gel process. *Polymer* 42:879–887
8. Nguyen NT, Wereley ST (2002), Fundamentals and Applications of Microfluidics. Artech House, Norwood, MA
9. Sharp KV, Adrian RJ, Santiago JG, Molho JI (2001), Liquid Flows in Microchannels, CRC Handbook of MEMS. CRC, Boca Raton, FL
10. Lee S-Y et al. (2002), Microchannel flow measurement using micro particle image velocimetry. In: ASME IMECE Microfluidics Symposium, New Orleans
11. Liu J, Shangguan D (2006), New composite nano-fibrous materials and methods of manufacturing. Swedish Patent Pending
12. Chronakis IS, Frenot A (2003), Polymer nanofibers assembled by electrospinning. Current Opinion in Colloid & Interface Science 8:64–75
13. Formhals A (1934), US Patent 1,975,504
14. Hohman M, Shin M, Rutledge G, Brenner M (2001), Electrospinning and electrically forced jets. I. Stability theory. Physics of Fluids 13(8):2201–2220
15. Theron SA, Yarin AL, Zussmana E (2004), Experimental investigation of the governing parameter in electrospinning of polymer solutions. Polymer 45:2017–2030
16. Theron SA, Yarin AL, Zussman E, Kroll E (2005), Multiple jets in electrospinning: experiment and modeling. Polymer 45:2889–2899
17. Sobhan CB, Garimella SV (2001), A comparative analysis of studies on heat transfer and fluid flow in microchannels. Microscale Thermophysical Engineering 5:293–311
18. Iijima S (1991), Helical microtubules of graphitic carbon. Nature 354:56–58
19. Hone J (2004), Carbon nanotubes: thermal properties. In: Dekker Encyclopedia of Nanoscience and Nanotechnology. Dekker, New York, pp. 603–610

20. Berber S, Kwon YK, Tomanek D (2000), Unusually high thermal conductivity of carbon nanotubes. Physical Review Letters 84:4613–4616
21. Xie S, Li W, Pan Z, Chang B, Sun L (2000), Mechanical and physical properties on carbon nanotube. Journal of Physics and Chemistry of Solids 61:1153–1158
22. Kabir MS, Morjan RE, Nerushev OA, Lundgren P, Bengtsson S, Enokson P, Campbell EEB (2005), Plasma-enhanced chemical vapour deposition growth of carbon nanotubes on different metal underlayers. Nanotechnology 16:458–466
23. Wang T, Jönsson M, Nyström E, Mo Z, Campbell EEB, Liu J (2006), Development and characterization of microcoolers using carbon nanotubes. In: Proceedings of Electronics Systemintegration Technology Conference (ESTC), Dresden, Germany, pp. 881–885
24. Mo Z, Anderson J, Liu J (2004), Integrating carbon nanotubes with microchannel cooler. In: Proceedings of HDP'04, Shanghai, China, pp. 373–376
25. Mo Z, Morjan R, Anderson J, Campbell EEB, Liu J (2005), Integrated nanotube microcooler for microelectronics applications. In: Proceedings of the 55th Electronic Components and Technology Conference, Florida, USA, pp. 51–54
26. Zhong X, Zhang Y, Liu J, Wang T, Cheng Z (2006), Computational fluid dynamics simulation for on-chip cooling with carbon nanotube micro-fin architectures. In: Proceedings of the 8th International Conference on Electronic Materials and Packaging, Hong Kong, pp. 117–123
27. Iwai T, Shioya H, Kondo D, Hirose S, Kawabata A, Sato S, Nihei M, Kikkawa T, Joshin K, Awano Y, Yokohama N (2005), Thermal and source bumps utilizing carbon nanotubes for flip-chip high power amplifiers. In: International Electron Devices Meeting Technical Digest, Washington, DC, pp. 257–260
28. Xie S, Li W, Pan Z, Chang B, Sun L (2000), Mechanical and physical properties on carbon nanotube. Journal of Physics and Chemistry of Solids 61:1153–1158
29. Wei BQ, Vajtai R, Ajayan PM (2001), Reliability and current carrying capacity of carbon nanotubes. Applied Physics Letters 79:1172–1174
30. Kumar A, Pushparaj VL, Kar S, Nalamasu O, Ajayan PM, Baskaran R (2006), Contact transfer of aligned carbon nanotube arrays onto conducting substrates. Applied Physics Letters 89:163120-1–163120-3
31. Jiang H, Zhu L, Moon KS, Wong CP (2007), Low temperature carbon nanotube film transfer via conductive polymer composites. Nanotechnology 18:125203
32. Wang T, Jönsson M, Carlberg B, Campbell EEB, Liu J (2007) Development and characterization of carbon nanotube-based bumps for ultra fine pitch flip chip interconnection. In: Proceedings of the 16th European Microelectronics and Packaging Conference & Exhibition: EMPC2007, Oulu, Finland, June 17–20, 2007, pp. 723–727

Chapter 18
Electromagnetic Shielding of Transceiver Packaging Using Multiwall Carbon Nanotubes

Wood-Hi Cheng(✉), Chia-Ming Chang, and Jin-Chen Chiu

18.1 Introduction

The widespread deployment of low-cost optical access networks for fiber-to-the-home (FTTH) applications will necessitate a considerable reduction in the cost of key components such as optical transceiver modules. Optical transceiver module costs are primarily dependent on their packaging. Owing to its low-cost nature and ease of manufacture, plastic packaging technology has been considered as one of the major choices for reducing the costs of fabricating optical transceiver modules for use in FTTH applications [1–5]. However, plastics alone are inherently transparent to electromagnetic (EM) radiation, and hence provide no shielding against radiation emissions. To improve EM shielding for plastic packaging, electronic conductive properties have to be added into the plastic hosts. Currently available techniques for electromagnetic interference (EMI) shielding include conductive sprays, conductive fillers, zinc-arc spraying, electroplating or electrolysis plating on housing surfaces, modifications of electrical properties during the molding stage, and other metallization processes. Among these methods, the most popular one for EM shielding is to compound plastics with discontinuous electronic conductive fillers, such as metal particles, metal flakes, stainless fiber, graphitized carbon particles, graphitized carbon fibers, metal-coated glass, and carbon fibers [6, 7].

Experimental evidence has shown that carbon fibers with aspect ratios of 1,000 are good conductive fillers for providing high EM shielding [8, 9]. Optical transceiver modules employing nylon and liquid crystal polymer (LCP) reinforced with carbon fiber showed that the measured shielding effectiveness (SE) was over 20 dB with 25% of weight percentage of the carbon fiber mixed in the nylon and LCP composites [3, 4]. In addition, the SE performance will be higher for higher weight percentages of the carbon fiber and thicker material of the carbon fiber mixture. However, with such plastic composites, the dominant cost of the package is the

---

W.-H. Cheng
Institute of Electro-Optical Engineering, National Sun Yat-sen University, Kaohsiung, 80424, Taiwan, ROC

carbon fiber fillers. Therefore, developing a plastic composite housing with a low weight percentage of carbon fiber and with good shielding ability is necessary for fabricating a low-cost and high SE optical transceiver module. Recently, a low-cost, lightweight, and high EM shielding package for optical transceiver modules has been made by employing a woven continuous carbon fiber (WCCF) epoxy composite with compression-molding technology [4, 10]. Epoxy resins are one of the best matrix materials for carbon fiber composites because they adhere well to a wide variety of carbon fibers. By weaving the continuous carbon fiber in a balanced twill structure (BTS) with excellent conductive networks, we found that the SE of the package housing, while keeping a very low weight percentage of carbon fiber, can reach about 80 dB under the far-field source measurement and about 25 dB in the near-field source measurement [4, 10].

As the electronic and mechanical properties of carbon nanotubes (CNTs) are remarkable [11, 12], CNTs have been the focus of considerable research and development for use in nanoscale electronic and optoelectronic applications such as integrated circuit (IC) interconnections [13], optical emission devices [14], and electrical interconnect [15–17]. CNTs are also considered as one of the electronic conductive fillers for EM shielding of transceiver package applications because of their smaller diameter, higher aspect ratios, higher conductivity, and better mechanical properties [18, 19]. The aspect ratios of most CNTs could be higher than 1,000, which offers good prospects to form overlapping conductive CNT networks to provide high EM shielding. The electrical percolation behavior of the polymer-based CNT composites is also discussed in the literature [20–23]. The electrical conductivity changes dramatically, when the concentration is around a threshold value. Once the concentration is higher and away from the threshold, the conductivity increases slowly. The threshold value is usually small and less than 0.9 wt% of the CNT composite in thin films [22]. The low percolation threshold results from the homogeneous dispersion of CNTs in the matrix and the high aspect ratio of CNTs. For thick layers of the polymer-based CNT composites, the percolation threshold phenomenon usually becomes more complex, and the dispersion of CNTs in the composite may become more difficult.

In this chapter, a novel polymer-based multiwall carbon nanotube (MWCNT) with high SE and effective electromagnetic susceptibility (EMS) performance for use in packaging a high-speed plastic transceiver module is proposed. Further research may develop the lower cost, lighter weight, higher shielding, and better EMS performance of the optical transceiver module packages by employing MWCNT composites. The sections of this chapter are organized as follows. Section 18.2 describes the fabrication of MWCNTs. The measurement results of SE for plastic composites in the far-field source and the realistic package in the near-field source of the monopole type are presented in Sect. 18.3. Section 18.4 presents the shielding effectiveness and EMS performance of an optical transceiver module. Conclusion and discussion are given in Sect. 18.5.

18.2 Fabrication of MWCNT Composites

18.2.1 Material Properties of MWCNTs

CNTs have excellent mechanical and electrical characteristics, such as high yield strength, high current density, high electrical conductivity, and less heat dissipation [12]. Basically, a CNT is a hollow tube, which is constructed from carbon atoms. This kind of hollow structure of carbons is called *fullerene*. Usually, fullerenes consist of hexagons and pentagons that form spherical shapes. The most famous one is C_{60}, which is built by 60 carbon atoms [12]. A CNT is a fullerene, but it is in a long hollow tube structure and its ends are semispherical-shaped cups. The tube wall atoms are hexagonally bonded like graphite and the tube-tip atoms are mixed with hexagonal- and pentagonal-bonded-like C_{60} [12]. Basically, it explains the origin of the high strength and excellent electrical characteristics. The CNT has a high mechanical strength and a large aspect ratio. The latter aspect makes it suitable as a filler of plastic composite for conductive networks, which contribute to an increase in the conductivity of the composite for EM shielding purpose.

MWCNT means a tube with many CNTs inside. These CNTs include metallic and semiconducting properties, and the CNTs inside a MWCNT could be concentric or spirally spread. A MWCNT with both metallic and semiconducting CNTs inside is also metallic conductive [24], which can be used as a conductive filler to make plastic composites. In this study, the MWCNT is produced by the arc-discharge deposit method (ADM) [25]. The electrodes are graphite rods with a distance about 1–2 mm apart. The applied voltage is 15–30 V, the operated dc current is around 50–150 A, and the applied argon gas pressure is 500–760 mbars. The deposit at the cathode includes 25% MWCNTs, 10% carbon nanocapsules, and 65% amorphous carbon. The aspect ratio of these MWCNTs is about 200–500 under scanning electron microscope (SEM) observation. The MWCNTs produced by ADM are not like those produced by a chemical vaporization deposit (CVD) method where metal catalyst particles exist. The metal catalyst particles such as Fe and Co will influence the conductivity, which has been reported by Kim's group [26, 27]. The MWCNTs are needle-like, sticking out of many hollow carbon nanocapsules as shown in Fig. 18.1a. Figure 18.1b shows a transmission electron microscope (TEM) photograph of MWCNTs, which illustrates the multilayer CNTs.

18.2.2 Material Properties of MWCNT–LCP Composites

The polymeric materials applied in this study are LCPs [9]. The LCPs exhibit a highly ordered structure in both the melt and solid states and are often applied to replace materials such as ceramics, metals, composites, and other plastics, because of their outstanding strength at high temperature, chemical resistance, and their resistance to weathering, radiation, and flame. LCPs have been applied for many

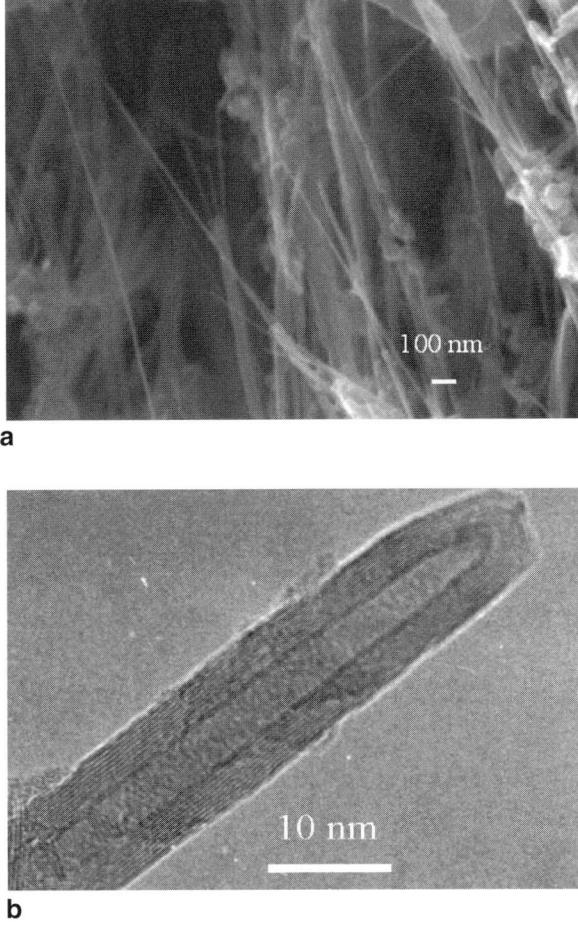

Fig. 18.1 (a) SEM photograph of MWCNTs and (b) TEM photograph of MWCNTs

injection- and compression-molded parts for their excellent properties [3–5, 9]. All materials applied in this study are powder for easy mixing and dispersion. A compression-molding machine with temperature controllers was employed to mold specimens. The processing temperatures were set at 300–350°C for LCP carbon-materials-filled composites [9]. A compression-molded circular specimen with a diameter of 133 mm and a thickness of 1 mm was made as an EMI specimen. An annulus 32 and 76 mm in inner and outer diameter, respectively, was cut for measuring the EM SE of the composites.

According to basic EM shielding theory, the higher conductivity has the higher SE. The more MWCNTs that are added, the more overlapping conductive MWCNT networks are formed, and hence the higher conductivity and the higher SE are obtained [4, 18, 19]. The conductivity of the MWCNT–LCP composite specimen was measured by a four-terminal technique. Comparing with the ADM MWCNT–LCP,

18 Electromagnetic Shielding of Transceiver Packaging

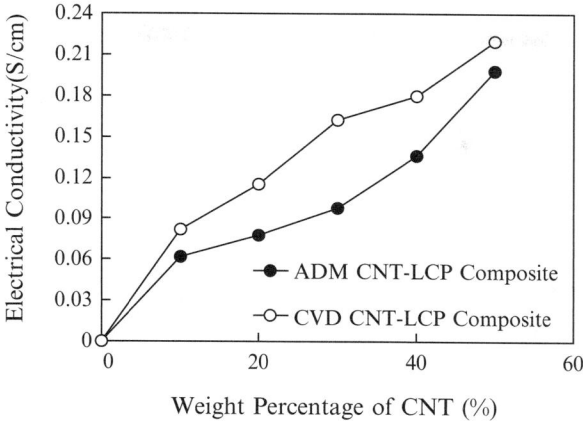

Fig. 18.2 Electrical conductivity as a function of mass fraction of MWCNT–LCP composite

the CVD MWCNT–LCP was made of purified MWCNTs in a CVD method from the Desunnano Ltd [25]. Figure 18.2 shows the relations between weight percentages of MWCNTs and electrical conductivity. It revealed that the higher weight percentage of MWCNTs exhibited a higher electrical conductivity. The higher conductivities of CVD MWCNT–LCP composite compared with ADM MWCNT–LCP composite are due to the higher aspect ratio and the remaining catalyst Fe enclosed in CVD MWCNTs. Using the SEM to examine the MWCNT–LCP composite, the morphology shows many trunk-like long fibers. However, in the MWCNT–LCP composite, it is hard to distinguish MWCNT from LCP, since the LCP itself has a highly ordered structure and envelops the MWCNTs, as shown in Fig. 18.3a. For the purpose of examining the real MWCNT dispersion, we use an ion technique to break the atomic bonds of LCP and MWCNT. Hence, the MWCNT morphology can be clearly observed as shown in Fig. 18.3b.

The mechanical strength of the MWCNT composite can be characterized by the tensile strength measurement. Figure 18.4 shows the relationship between the tensile strength and the weight percentage of MWCNT composite. The tensile strength increases as the MWCNT weight percentage increases. This is due to the reinforcement of the polymer-based MWCNT composites regardless of the MWCNT functionalization. Fillers such as MWCNTs, with a high aspect ratio, usually exhibit a high tensile strength and a good flexural strength. However, the MWCNT reinforcement material seldom carries over to the composite due to poor load transfer from matrix to MWCNT reinforcement, unless the MWCNTs are appropriately functionalized. The modulus of the composite usually improves with MWCNT reinforcement regardless of functionalization. The higher slope of CVD MWCNT–LCP composite as compared with that of the ADM MWCNT–LCP could be attributed to the high aspect ratio of the CVD MWCNT. However, the differences of conductivity and tensile strength of the ADM MWCNT–LCP and CVD MWCNT–LCP composites are small, as shown in Figs. 18.2 and 18.4. The ADM MWCNT–LCP composites still have a high aspect ratio advantage.

Fig. 18.3 (**a**) SEM photograph of MWCNTs embedded in LCPs and (**b**) SEM photograph of MWCNTs distribution in composite (locally)

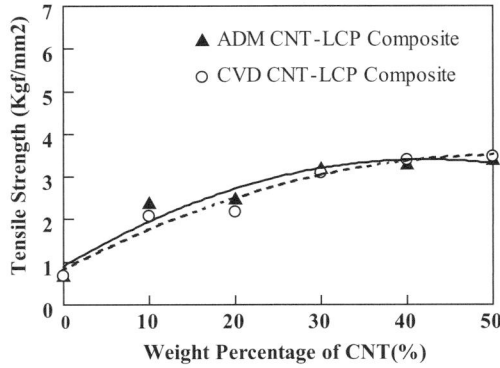

Fig. 18.4 Tensile strength as a function of MWCNT–LCP composite mass fraction

18.3 Electromagnetic Shielding Performance of MWCNT Composites

18.3.1 Shielding Measurements of MWCNT Composites in the Far-Field Source

Figure 18.5 shows a setup for the SE measurement and a cross section of the coaxial-type transmission-line holder. A flanged coaxial transmission-line holder was designed by the following ASTM D4935 method [28], which is used to measure the SE of a disk-shaped MWCNT–LCP specimen in far-field source. The testing frequency range is 1–3 GHz, since we focus on 2.5-Gb s^{-1} lightwave transmission applications. The diameter of the inner conductor was 33 mm, and the outer conductor had inner and outer diameters of 76 and 133 mm, respectively, according to the definition of ASTM D4935 [28]. The electromagnetic SE of the MWCNT–LCP composites was measured by an insertion of the disk-shaped specimen with a diameter of 133 mm and a thickness of 1 mm between the two identical flanges. The purpose of the SE test procedure is to quantitatively measure the insertion loss that results from introducing the test specimens. The result of the far-field measurement is shown in Fig. 18.6. The SE of the MWCNT–LCP composites was measured from 38 to 45 dB in the frequency range of 1–3 GHz, which is suitable for industrial use. The SE of the MWCNT–LCP composite is also comparable to the other shielding plastic composites such as carbon fiber (CF)–nylon, CF–LCP, and WCCF epoxy [4].

For a coaxial holder transmission-line circuit, there are significant parasitics between the holder and specimen. The equivalent circuit of a flanged coaxial holder with specimen can be modeled for theoretical calculation when the parasitic effects are considered [3, 8, 9, 19]. The calculated result of SE vs. frequency is also shown

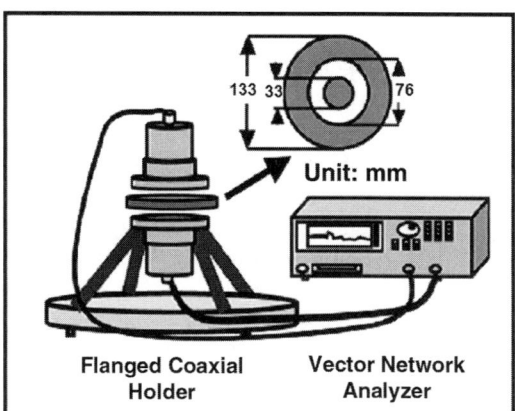

Fig. 18.5 Measurement setup for the shielding effectiveness of the plastic composites based on the ASTM D4935 method and a cross section of the coaxial-type transmission-line holder

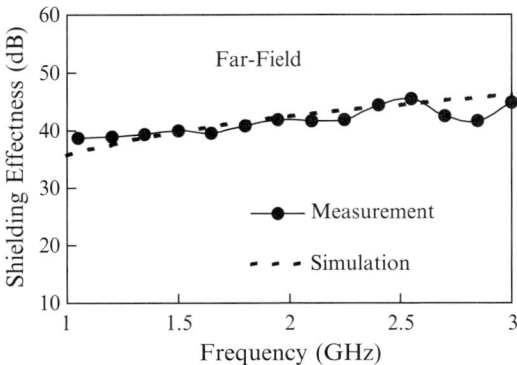

Fig. 18.6 Far-field measurement: 50% weight MWCNT–LCP composite

in Fig. 18.6. Both measured (solid line) and calculated (dashed lines) results of the far field are in good correlation.

18.3.2 Shielding Measurements of MWCNT Composites in the Near-Field Source

Near-field measurement is used to examine the SE of the specimen assembly, which is closer to the radiant source. It is more like the actual situation of a real application. A monopole-type antenna was used as the radiant source. Then, we put the monopole-type antenna into a module box built of the MWCNT–LCP composite to measure the difference of reference level and shielding level, which is the SE of the module box in the near-field situation. The SE measurement in a near-field source was carried out in a fully anechoic electromagnetic compatibility (EMC) chamber, as shown in Fig. 18.7 [3–5]. Because the hybrid absorbers combining the ferrite tiles and foam absorbers were aligned on the metal-shielded wall inside the chamber, good wave-absorbing performance could be achieved from 30 MHz to 18 GHz for the EMC chamber. The radiation source was put on a wooden table from which an antenna was seated at a distance of 3 m for receiving the radiated field. Because the interconnecting length on the optical transceiver module was generally about 1–2 cm, an electric monopole with a 2-cm length was used to emulate the radiation energy inside the molded housing.

Figure 18.8 shows the SE behavior vs. frequency of the realistic housing fabricated by the MWCNT–LCP composites for a near-field radiation source of the monopole type. The SE of the MWCNT–LCP composites was measured from 28 to 40 dB in the frequency range of 1–3 GHz, as shown in Fig. 18.8a. Figure 18.8b shows the received radiation with and without shielding boxes. The radiation with and without shielding boxes showed 58 and 92 dBµV m^{-1}, on average, respectively. Both values were far from the background white-noise level of around 30 dBµV m^{-1}.

18 Electromagnetic Shielding of Transceiver Packaging

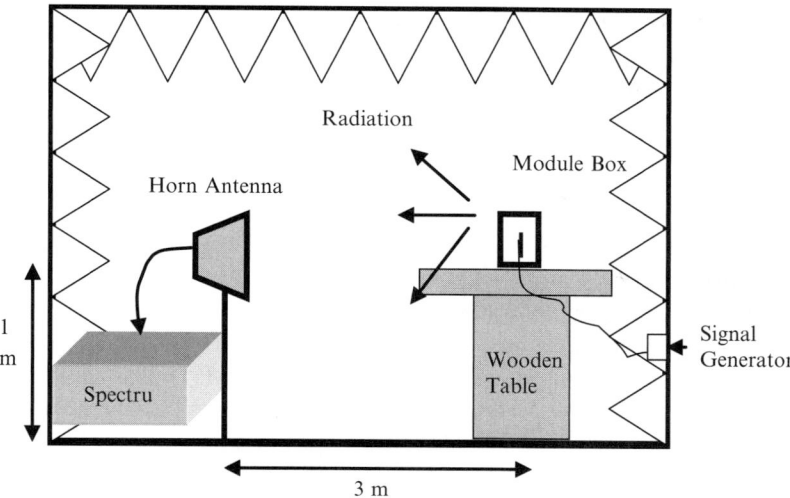

Fig. 18.7 SE measurement setup for near-field monopole-type source in a fully anechoic EMC chamber

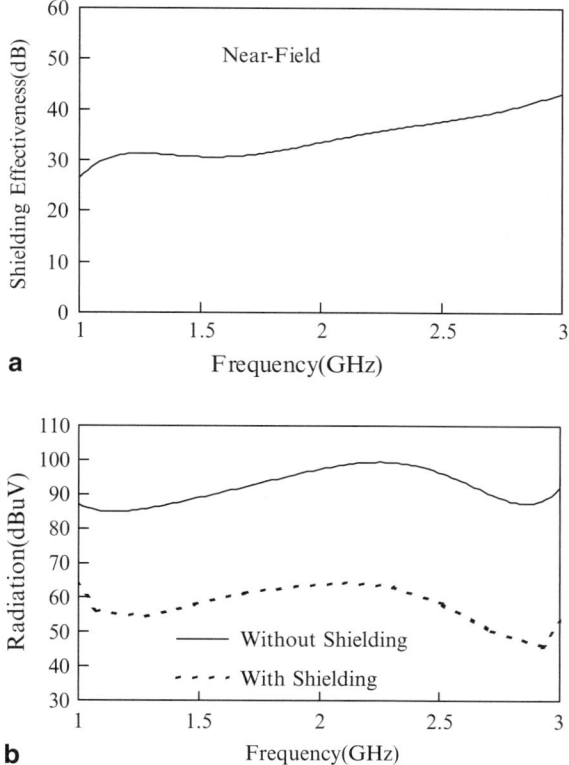

Fig. 18.8 (a) Near-field shielding effectiveness results and (b) received radiations of monotype antenna enclosed in a 50% weight percentage MWCNT–LCP box

18.4 Electromagnetic Shielding Performance of Transceiver Packages

18.4.1 Optical Transceiver Modules

An optical transceiver module consists of a 1.3-μm laser (TOSA), a photodiode (ROSA), two ICs, and a plastic housing [3, 4]. These high-speed electric signals with fast rising/falling edges would result in significant EMI problems as the transceiver modules are mounted onto the digital communication systems. Metallic housings are the general solution to reduce the EMI of the module for complying with the Federal Communications Commission (FCC) EMI regulation. But due to low-cost and lightweight considerations, plastic packaging or housing becomes one of the major trends for future optical transceiver module designs [1–5].

The optical transceiver modules were fabricated by the MWCNT–LCP composites. The shape of the molded package housing was a rectangular box with the dimensions of 70 × 30 × 20 mm. The weight percentage of MWCNT was 50%. In general, the SE increases as the mass fraction of MWCNT increases. However, when the mass fraction of MWCNT increases more than 50%, the MWCNT–LCP composite becomes easy to break when compressing the specimen. This might be due to less LCP to fill out all the space inside the composite specimen, which makes gaps or caves and reduces the strength of the specimen.

18.4.2 Electromagnetic Interference Measurement of Transceiver Module in the Near-Field Source

An optical transceiver module with transmission rate of 2.5 Gb s^{-1} is tested to evaluate the EM shielding against emitted radiation from the plastic packaging. The packaged transceiver communicated with a golden specimen module (GSM) with identical functions to the module under test (MUT) [3–5]. A pattern generator (Tektronix GTS1250) transmitted the differential pseudorandom bit sequence (PRBS) patterns (2.5 Gb s^{-1}) to the GSM. Through a fiber optic link, the data were received by the MUT working in the loopback mode inside the chamber.

An optical transceiver module usually consists of a transmitter and a receiver. In this work, due to the limited housing space in the MWCNT–LCP composite box, the SE measurement of the optical transceiver module is carried out by separate transmitter and receiver module measurements. The result of the near-field measurement of the 2.5-Gb s^{-1} transmitter module is shown in Fig. 18.9a. The SE was about 20 dB, on average, in the 2.5-GHz region. Figure 18.9b shows the received radiation at a frequency of 2.5 GHz for boxes with and without shielding to be 40 and 60 dBμV m^{-1}, on average, respectively. And, the radiation in the frequency range 1–3 GHz was 40 and 57 dBμV m^{-1}, respectively, for boxes with and without

18 Electromagnetic Shielding of Transceiver Packaging

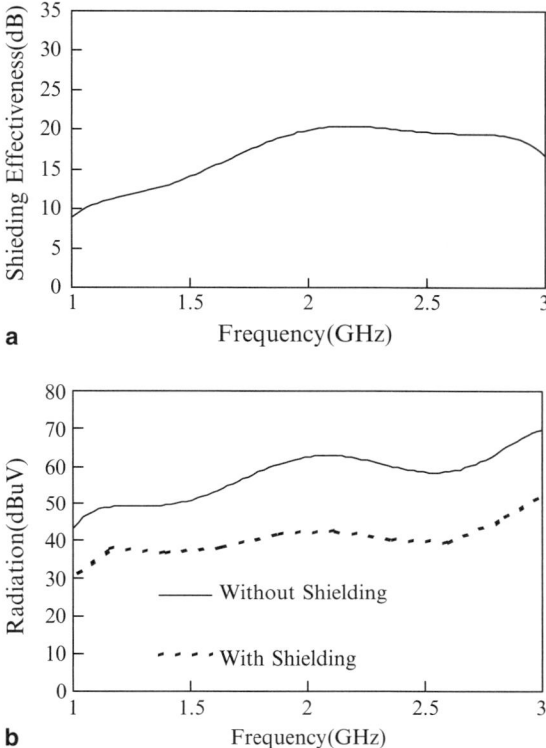

Fig. 18.9 (**a**) Near-field shielding effectiveness results and (**b**) received radiations of 2.5-Gb s^{-1} transmitter

shielding. The radiation of 40 dBμV m^{-1} is closer to the background level, especially at lower frequencies.

Figure 18.10a shows the results of the near-field measurement of the 2.5-Gb s^{-1} receiver module. The SE was about 14 dB, on average, in the 2.5-GHz region. Figure 18.10b shows the received radiation at a frequency of 2.5 GHz for boxes with and without shielding to be 40 and 54 dBμV m^{-1}, on average, respectively, and the radiation with and without shielding boxes was 40 and 54 dBμV m^{-1}, on average, respectively, in the frequency range 1–3 GHz. The radiation of 40 dBμV m^{-1} is closer to the background level, especially at lower frequencies.

The radiation received by the optical transceiver without a shielding box was measured at 59 dBμV m^{-1}, on average, in the frequency range 1–3 GHz, and 67 dBμV m^{-1}, on average, in the 2.5-GHz region. Based on the SE measurements of the transmitter and receiver modules from Figs. 18.9 and 18.10, the result of SE measurements for the optical transceiver module at a frequency of 2.5 GHz was expected to be 14 dB by using the worst-case SE value from the receiver module. However, the expected radiation of the optical transceiver module was 53 dBμV m^{-1}

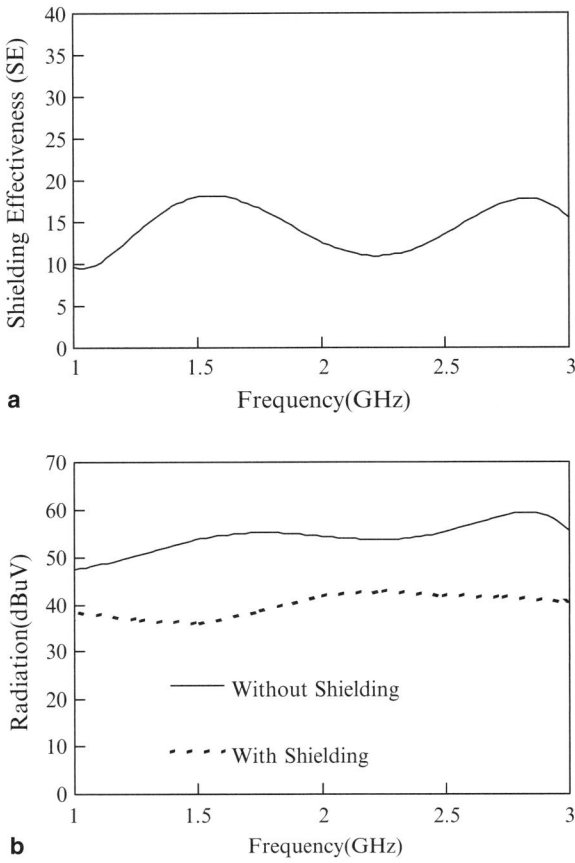

Fig. 18.10 (**a**) Near-field shielding effectiveness results and (**b**) received radiations of 2.5-Gb s^{-1} receiver

(67 dBμV m^{-1}–14 dB), which is lower than the FCC regulation of 54 dBμV m^{-1} [29]. It is also suitable for industrial use.

MWCNT–LCP composite used as a novel EM shielding material for packaging a plastic optical transceiver module was examined in both the near-field and the far-field sources. The results showed that the SE of MWCNT composites exhibited 38–45 dB in the far-field source and 28–40 dB in the near-field source at a frequency range of 1–3 GHz, and with an average of 14 dB for packaged optical transceiver modules at a frequency of 2.5 GHz. The SE behavior of the monopole antenna enclosed in the MWCNT–LCP composite box (Fig. 18.8) was higher than that of the optoelectronic devices (Fig. 18.9 for transmitter and Fig. 18.10 for receiver) in the 1–3 GHz frequency range. This is due to the radiation of optoelectronic devices, which is not as high as the monopole antenna (about 92 dBμV m^{-1}). The results of receivers enclosed in module boxes also showed a low SE (Fig. 18.10) when compared to that of the monopole antenna (Fig. 18.8). The adopted optoelectronic devices under test did not radiate as much as the monopole antenna. So, the results for the monopole antenna enclosed in the module box might reflect the real shielding ability of the module box in the near-field situation. In brief, the MWCNT–LCP module box composite has shown a good shielding ability to reduce the radiation from optoelectronic devices, such as transmitters, receivers, and transceivers under the testing frequency ranges of 1–3 GHz, and all the results meet FCC regulations [29].

18.4.3 Electromagnetic Susceptibility Measurement of Transceiver Module in the Near-Field Source

The EMS or electromagnetic immunity of the optical transceiver modules to EMI is one of the major concerns to maintain good signal quality over gigabits-per-second transmission rates [5, 30, 31]. In this work, the EMS performance of an optical receiver package fabricated by the MWCNT–LCP composite is experimentally evaluated by the eye diagram and bit-error-rate (BER) test for a 2.5-Gb s^{-1} lightwave transmission system. Figure 18.11 shows the measurement setup for the EMS performance of the proposed package. A pulse pattern generator (PPG, Anritsu MP1763C) is used to transmit 2.5-Gb s^{-1} signals to the transceiver module under test through an optical transmitter. The $2^{31} - 1$ PRBS pattern is given by the PPG. The received signal is electrically returned to a bit-error-rate tester (BERT, Anritsu MP1764C) and the eye patterns can also be measured by a sampling scope (Tektronix CSA 8000C). To perform the EMS measurement of the packaged receiver, the monopole radiator excited by the PPG is used to interfere with the packaged module at a distance of 3 cm.

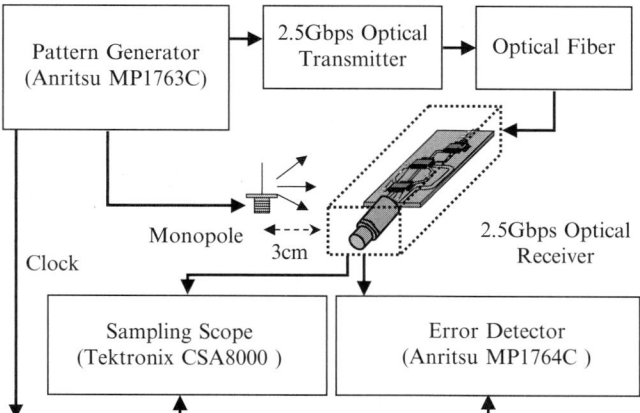

Fig. 18.11 EMS measurement setup

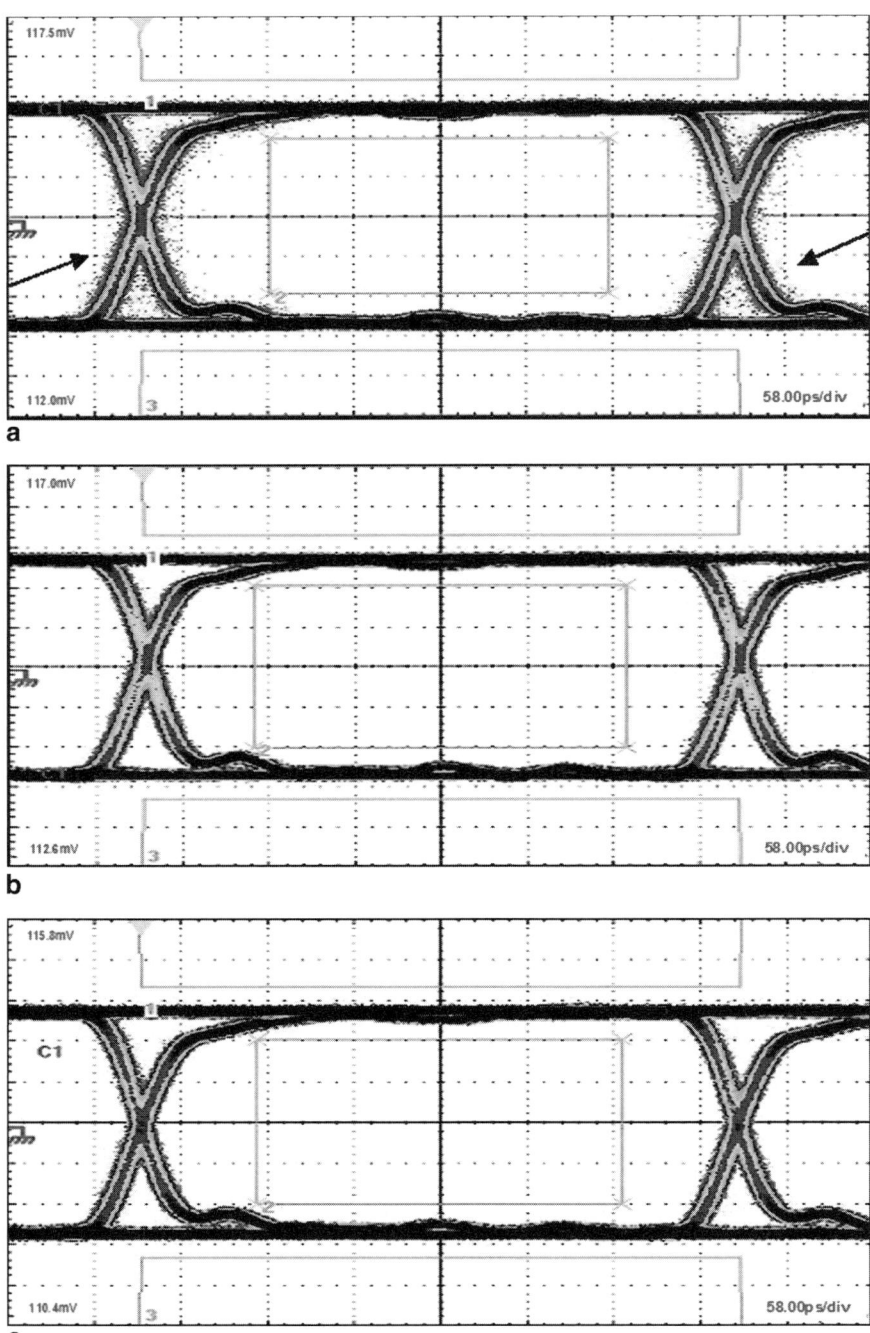

Fig. 18.12 Eye diagrams of the packaged module of different weight percentages of MWCNTs under radiation interference: (**a**) 20%, (**b**) 30%, and (**c**) 50%. The *arrows* indicate the worse jitters surrounding at X region

The display with eye mask gives qualitative values for noise, jitter, rise time, fall time, and pulse duration. Figure 18.12a, b, and c shows the eye diagrams of the packaged receiver boxes with 20, 30, and 50% weight percentages of MWCNTs in the MWCNT–LCP composites, respectively. In Fig. 18.12a, there was a lot of worse jitter surrounding the X region, as indicated by the arrows. The applied amplitude on the interference monopole-type antenna was 0.75 Vp-p. Figure 18.13 plots the eye mask margin as a function of the MWCNT weight percentage, based on Fig. 18.12. The mask margin is defined as the ratio of the maximum allowable mask to the standard mask. The mask margins were improved 46, 53, and 54% for 20, 30, and 50% MWCNT weight percentages, respectively. This shows that the mask margin increases as the MWCNT weight percentage increases. The result clearly indicates that the higher MWCNT weight percentage has the better EMS performance.

The superior EMS performance of the proposed package is also demonstrated through the measurement of the BER tests. Figure 18.14 shows the BER vs. the received optical power for three different cases (50% weight percentage of the MWCNTs): unpackaged module without radiated interference (case A), unpackaged module with radiated interference (case B), and packaged module with radiated interference (case C). As shown in Fig. 18.11, the radiated monopole is at 3-cm distance to the tested module with 1 Vp-p amplitude of the excitation. Comparing with case A, case B with significantly strong radiated noise needs larger optical power to keep the same BER. As shown in Fig. 18.13, the received optical power is about −11.1 and −8 dBm for cases A and B, respectively, to achieve the BER of 10^{-12}. However, case C with the package housing clearly shows that the BER performance is significantly improved. The optical power is about −10.6 dBm for the BER 10^{-12}. Comparing with case B, case C fabricated by the proposed CNT–LCP composite package significantly increases the electromagnetic

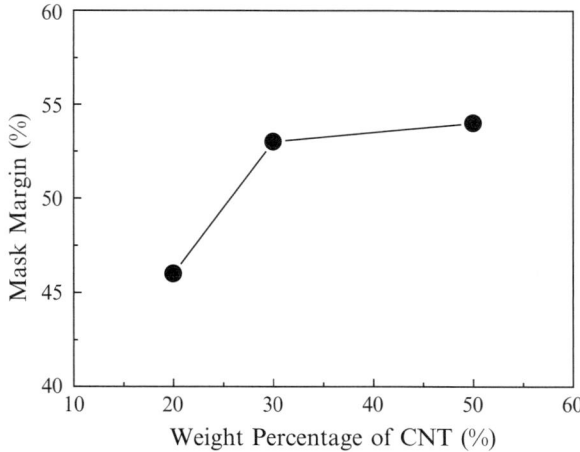

Fig. 18.13 Relationship between mask margin and MWCNT weight percentage at the amplitude of monopole-type antenna of 0.75 Vp-p

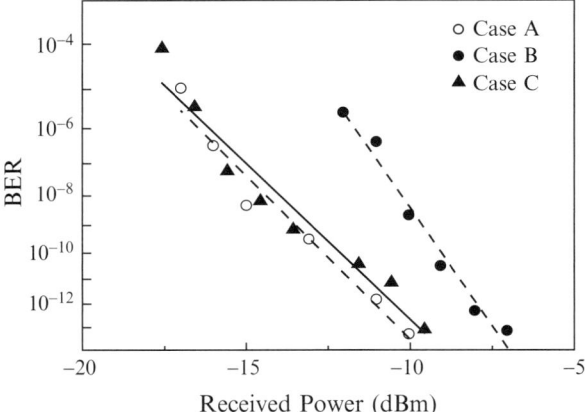

Fig. 18.14 BER vs. received optical power for three different cases: unpackaged module without radiated interference (*case A*), unpackaged module with radiated interference (*case B*), and packaged module with radiated interference (*case C*)

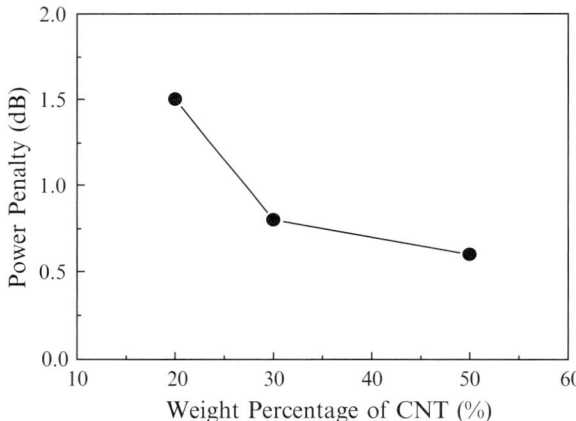

Fig. 18.15 Relationship between power penalty and MWCNT weight percentage at the amplitude of monopole-type antenna of 1 Vp-p

immunity to the radiated interference with about 2.6-dB optical power gains at BER 10^{-12}.

Figure 18.15 shows the relationship between power penalty and MWCNT weight percentage. The amplitude of interference monopole-type antenna was 1 Vp-p. The power penalty is defined as the received power difference between unpackaged boxes without radiated interference and packaged boxes with radiated interference in dB unit at BER equal to 10^{-12}. The result showed that the power penalty decreased as the MWCNT weight percentage increased. The power penalties are less than 1.5 dB for three different MWCNT weight percentages. The result indicates that the higher weight percentage of the MWCNTs, the better is their shielding ability and EMS performance.

8.5 Conclusion and Discussion

Using a novel polymer-based MWCNT with high SE and effective EMS performance in plastic packaging, a 2.5-Gb s^{-1} plastic transceiver module has been developed and fabricated. The SE of the MWCNT composite with 50% weight percentage MWCNTs exhibited 38–45 dB. Under strongly radiated interference, the package housing fabricated by MWCNT–LCP composites significantly improved the mask margin and power penalty for a 2.5-Gb s^{-1} lightwave transmission system. This demonstrated that the plastic transceiver modules with the higher weight percentage of the MWCNTs exhibited a higher SE, and hence showed effective EMS performance, a better mask margin, and a lower power penalty. This clearly indicates that the MWCNT–LCP composites with their high SE are suitable for packaging lightweight and high-performance EMS plastic transceiver modules used in FTTH lightwave transmission systems.

According to basic EM shielding theory, the higher conductivity has the higher SE. The more MWCNTs that are added, the more overlapping conductive MWCNT networks are formed, and hence the higher conductivity and higher SE are obtained [4, 18, 19]. However, it is well known and widely accepted that the electrical conductivity of polymer-based MWCNT composites has a percolation threshold at a very low concentration, where the electrical conductivity changes dramatically [20–23]. Once the concentration is higher and away from the threshold, the conductivity increases slowly. The threshold value is usually small and less than 0.9 wt% of the MWCNT composite in thin film. The low percolation threshold results from homogeneous dispersion of MWCNTs in the matrix and a high MWCNT aspect ratio. Published literatures on MWCNT–polymer composites mainly discuss thin films [20–23], which may be a two-dimensional system. In this work, the thickness of the MWCNT–polymer composite was more than 1 mm to offer a strong EMI protection of transceiver package. This transceiver package is a three-dimensional system. For thick layers of the polymer-based composites of MWCNT, the percolation threshold phenomenon usually becomes more complex, and the dispersion of CNT in the composite may become more difficult. Clearly, homogeneous dispersion in thick layers of the MWCNT–polymer composite is necessary and needs to be developed.

References

1. K. Tatsuno, K. Yoshida, T. Kato, T. Hirataka, T. Miura, K. Fukuda, T. Ishikawa, M. Shimaoka, and T. Ishii, High-performance and low cost plastic optical modules for access network system applications, J. Lightwave Technol., 17(7), 1211–1216 (1999)
2. M. Fukuda, F. Ichikawa, Y. Shuto, H. Sato, Y. Yamada, K. Kato, S. Tohno, H. Toba, T. Sugie, J. Yoshida, K. Suzuki, O. Suzuji, and S. Kondo, Plastic module of laser diode and photodiode mounted on planar lightwave circuit for access network, J. Lightwave Technol., 17(7), 1585–1590 (1999)
3. T.L. Wu, W.S. Jou, S.G. Dai, and W.H. Cheng, Effective electromagnetic shielding of plastic packaging in low-cost optical transceiver modules, J. Lightwave Technol., 21(6), 1536–1543 (2003)

4. W.H. Cheng, W.C. Hung, C.H. Lee, G.L. Hwang, W.S. Jou, and T.L. Wu, Low cost and low electromagnetic interference packaging of optical transceiver modules, J. Lightwave Technol., 22(9), 2177–2183 (2004)
5. T.L. Wu, M.C. Lin, C.W. Lin, T.T. Shih, and W.H. Cheng, High electromagnetic susceptibility performance plastic package for 10 Gibt/s optical transceiver modules, Electron. Lett., 41(8), 494–495 (2005)
6. P.B. Jana, A.K. Mallick, and K. De, Effects of sample thickness and fiber aspect ratio on EMI shielding effectiveness of carbon fiber filled polychloroprene composites in the 18-band frequency range, IEEE Trans. Electromagn. Compat., 34(11), 478–492 (1992)
7. P.F. Wilson, M.T. Ma, and J.W. Adams, Techniques for measuring the electromagnetic shielding effectiveness of materials. I. Far-field source simulation, IEEE Trans. Electromagn. Compat., 3(8), 239–247 (1988)
8. W.S. Jou, T.L. Wu, S.K. Chiu, and W.H. Cheng, Electromagnetic shielding of nylon-66 composites applied to laser modules, IEEE/TMS J. Electron. Mater., 30(10), 1287–1293 (2001)
9. W.S. Jou, T.L. Wu, S.K. Chiu, and W.H. Cheng, The influence of fiber orientation on electromagnetic shielding in liquid crystal polymers, IEEE/TMS J. Electron. Mater., 31(3), 178–184 (2002)
10. T.L. Wu, W.S. Jou, W.C. Hung, C.H. Lee, C.W. Lin, and W.H. Cheng, High electromagnetic shielding of plastic package for 2.5 Gbps optical transceiver modules, IEEE Trans. Adv. Packag., 28(1), 89–95 (2005)
11. E.D. Minot, Y. Yaish, V. Sazonova, J.Y. Park, M. Brink, and P.L. McEuen, Turning carbon nanotube band gaps with strain, Phys. Rev. Lett., 90(15), 154601–154604 (2003)
12. K.B.K. Teo, et al., Carbon nanotube technology for solid state and vacuum electronics, IEE Proc. Circuits Devices Syst., 151(5), 443–451 (2004)
13. J. Li, Q. Ye, A. Cassell, H.T. Ng, R. Stevens, J. Han, and M. Meyyappan, Bottom-up approach for carbon nanotube interconnects, Appl. Phys. Lett., 82(15), 2491–2493 (2003)
14. J.A. Misewich, R. Martel, P. Avouris, J.C. Tsang, S. Heinze, and J. Tersoff, Electrically induced optical emission from a carbon nanotube FET, Science, 300, 783–786 (2003)
15. R.T. Pike, R. Dellmo, J. Wade, S. Newland, G. Hyland, and C.M. Newton, Metallic fullerene and MWCNT composite solutions for microelectronics system electrical enhancement, In: Proceedings of the 54th ECTC, Las Vegas, NV, 2004, pp. 461–465
16. L. Zhu, Y. Sun, J. Xu, Z. Zhang, D. Hess, and C.P. Wong, Aligned carbon nanotube for electronical interconnect in thermal management, In: Proceedings of the 55th ECTC, Orlando, FL, 2005, pp. 44–50
17. L. Zhu, Y. Xiu, D. Hess, and C.P. Wong, In-situ opening aligned carbon nanotube films/arrays for multichannel ballistic transport in electrical interconnect, In: Proceedings of the 56th ECTC, San Diego, CA, 2006, pp. 461–465
18. C.M. Chang, J.C. Chiu, W.S. Jou, T.L. Wu, and W.H. Cheng, New package scheme of a 2.5 Gb/s plastic transceiver module employing multiwall nanotubes for low electromagnetic interference, J. Sel. Topics Quantum Electron., 12(5), 1025–1031 (2006)
19. C.M. Chang, M.C. Lin, J.C. Chiu, W.S. Jou, and W.H. Cheng, High-performance electromagnetic susceptibility of plastic transceiver modules using carbon nanotubes, J. Sel. Topics Quantum Electron., 12(6), 1091–1098 (2006)
20. D. Stauffer and A. Aharony, *Introduction to Percolation Theory*, 2nd ed. (Taylor & Francis, London, 1992)
21. R. Zallen, *The Physics of Amorphous Solids* (Wiley, New York, 1983)
22. G. Hu, C. Zhao, S. Zhang, M. Yang, and Z. Wang, Low percolation thresholds of electrical conductivity and rheology in poly (ethylene terephthalate) through the networks of multi-walled carbon nanotubes, Polymer, 47(1), 480–488 (2006)
23. E. Kymakis and G.A.J. Amaratunga, Electrical properties of single-wall carbon nanotube-polymer composite films, J. Appl. Phys., 99, 084302-1–084302-7 (2006)
24. P.G. Collins and Ph. Avouris, Multishell conduction in multiwalled carbon nanotubes, Appl. Phys. A, 74, 329–332 (2002)
25. W.S. Jou and C.F. Hsu, A novel carbon nano-tube polymer-based composite with high electromagnetic shielding, IEEE/TMS J. Electron. Mater., 35(3), 462–470 (2006)

26. H.M. Kim, K. Kim, C.Y. Lee, J. Jooa, S.J. Cho, H.S. Yoon, D.A. Pejakovic, J.W. Yoo, and A.J. Epstein, Electrical conductivity and electromagnetic interference shielding of multiwalled carbon nanotube composites containing Fe catalyst, Appl. Phys. Lett., 84(4), 589–591 (2004)
27. H.M. Kim, K. Kim, S.J. Lee, J. Joo, H.S. Yoon, S.J. Cho, S.C. Lyu, and C.J. Lee, Charge transport properties of composites of multiwalled carbon nanotube with metal catalyst and polymer: application to electromagnetic interference shielding, Curr. Appl. Phys., 4, 577–580 (2004)
28. Standard Testing Method for Measuring the Electromagnet Electromagnetic Shielding Effectiveness of Planner Materials, ASTM D4935-92 (ASTM, Philadelphia, PA)
29. C.R. Paul, *Introduction to Electromagnetic Compatibility*, pp. 42–77, ch. 2 (Wiley-Interscience, New York, 1992)
30. J.T. DiBene and J.L. Knighten, Effects of device variations on the EMI potential of high speed digital integrated circuits, In: *Proceedings of the IEEE 1997 International Symposium on Electromagnetic Compatibility*, August 18–22, 1997, pp. 208–212
31. D.M. Hockanson, X. Ye, J.L. Drewniak, T.H. Hubing, T.P.V. Doren, and R.E. DuBroff, FDTD and experimental investigation of EMI from stacked-card PCB configurations, IEEE Trans.

Chapter 19
Properties of 63Sn-37Pb and Sn-3.8Ag-0.7Cu Solders Reinforced With Single-Wall Carbon Nanotubes

K. Mohan Kumar(✉), V. Kripesh, and Andrew A.O. Tay

19.1 Introduction

As integrated circuit (IC) technology continues to advance, there will be increasing demands on I/O counts and power requirements, leading to decreasing solder pitch and increasing current density for solder balls in high-density wafer-level packages [1]. As the electronics industry continues to push for miniaturization, reliability becomes a vital issue. The demand for more and smaller solder bumps, while increasing the current, has also resulted in a significant increase in current density [2], which can cause the failure of solder interconnects due to electromigration [3].

Solders are extensively used in IC technology as mechanical and electrical interconnects because of their ease of processing and lower cost. However, because of their relatively low melting temperatures, creep is a major concern. When electronic devices are switched on and off, the electronic packages experience cyclic changes in temperature. Because of differences between the packages and the substrate, cyclic changes in thermomechanical stresses are induced in the package-to-board solder joints. Such cyclic stresses in the solder joints eventually lead to failure of the solder joints through thermomechanical fatigue [4, 5].

With the relentless trend toward very fine pitch IC packages, the cyclic stresses experienced by flip chip-to-board interconnects are increasing greatly, resulting in a drastic drop in fatigue life of solder joints. One way of overcoming this problem is to use new materials, which can provide enhanced mechanical, electrical, and thermal properties. Composite solders can offer improved properties [6]. Although a few researchers have investigated the influence of nanoparticles and nanotubes on the properties of solder [7–9], these investigators were mainly focused on the mechanical properties of the solders. In this study, the influence of nanotube addition on microstructural, mechanical, electrical, wetting, and thermal properties has been investigated. In addition to this, efforts have been made to evaluate the joint strength and creep strength of the composite solder joints.

K.M. Kumar
Nano/Microsystems Integration Laboratory, Department of Mechanical Engineering,
National University of Singapore, Singapore

Owing to their fascinating physical properties and unique structures, carbon nanotubes (CNTs) are receiving steadily increasing attention since their discovery [10]. Intense interest from researchers has been generated in utilizing these unique structures and outstanding properties, for example, in hydrogen storage, supercapacitors, biosensors, electromechanical actuators, and nanoprobes for high-resolution imaging and so on [11, 12]. In recent years, there has been a steadily increasing interest in the development of CNT-reinforced composites due to their remarkable mechanical, electrical, and thermal properties [13–16]. Depending on their length, diameter, chirality, and orientations, CNTs show almost five times the elastic modulus (1 TPa) and nearly 100 times the tensile strength (150 GPa) of high-strength steels [17]. The motive is to transfer the exceptional mechanical and physical properties of CNTs to the bulk engineering materials. Polymers, ceramics, and metals are favorable as matrix materials. CNT-reinforced polymer-based composites were widely synthesized by surfactant-assisted processing, repeated stirring, solution evaporation with high-energy sonication, and interfacial covalent functionalization [18–20]. Much of the research in nanotube-based composites has been on polymer or ceramic matrix materials and less on metal–matrix composites [21–23]. This is mainly due to the fact that uniform dispersion of CNTs in a metal matrix is quite difficult.

Nai et al. [9] demonstrated that the dispersion and homogenous mixing between MWCNTs and a lead-free solder matrix could be obtained by mixing nanosized matrix powders with CNTs. They showed that the powder metallurgy process was a very promising technique for full densification of CNT/lead-free solder nanocomposites, which showed remarkable enhancement of yield strength compared with that of unreinforced lead-free solders.

The current work provides an insight into the usage of SWCNTs as a reinforcing material for the enhancement of the properties of the solder material to be used in wafer-level chip-scale packages (WLCSP). The aim of this work is to fabricate and characterize CNT-reinforced nanocomposite solders and show their improved physical, thermal, electrical, mechanical, and wetting properties compared with the original Sn–Pb and Sn–Ag–Cu solders.

19.2 Experimental Aspects

19.2.1 Materials

The starting materials used in this study were Sn–Pb and Sn–Ag–Cu solder powders of type 7 (2–11 μm). The SWCNTs employed in this study were prepared using the chemical vapor deposition (CVD) technique and typically have an average diameter of 1.2 nm and lengths between 5 and 10 μm.

19.2.2 Preparation of Composite Solders

The solder powder and SWCNTs were weighed to the approximate weight percent ratio. Different compositions were prepared with varying SWCNT content ranging from 0.01 to 1 wt%. The preweighed SWCNTs and solder powders were blended homogeneously using a V-cone blender operated at a speed of 50 rpm. The homogenously blended composite solder powders were consolidated by uniaxial cold pressing with a pressure of 110 bar in the case of Sn–Pb composite solders while Sn–Ag–Cu composite solders were compacted at a pressure of 120 bar. The consolidated *green* composite solder compacts of diameter 35 mm were sintered at 150°C for Sn–Pb composite solders and at 180°C for Sn–Ag–Cu composite solders. Sn–Ag–Cu composite solders were sintered at 180°C to approach a reasonable rate of solid-state sintering. The sintered compacts were finally extruded at room temperature with an extrusion ratio of 20:1.

19.2.3 Scanning Electron Microscopy

Samples were cut from the extruded solder bars with a diamond saw, and mechanically polished with diamond pastes after cutting, and finishing with 0.02-μm grade. Microstructural observations were performed by scanning electron microscopy (SEM) using a Hitachi FE-SEM 4100 operated at 10 kV. The elemental analysis of the phases was carried out using energy-dispersive X-ray spectroscopy (EDX) equipped with FE-SEM.

19.2.4 Thermomechanical Analysis (TMA)

The linear thermal expansion coefficient of composite solders was measured using a Perkin-Elmer TMA-7 thermal mechanical analyzer operated in expansion mode. Cylindrical samples of diameter 8 mm were employed. TMA data were obtained in the heating range of 25–125°C in the case of Sn–Pb composite solders, while a 25–150°C heating range was employed for the Sn–Ag–Cu composite solders at a rate of 5°C/min. All TMA experiments were performed with a small loading force of 5 g to avoid deformation of the samples during testing. The CTEs of the composite solder specimens were obtained from the slope of the curve over a linear temperature range.

19.2.5 Differential Scanning Calorimetry (DSC)

The melting behaviors of the composite solder specimens were examined by a Perkin-Elmer DSC-7 system. DSC experiments were carried out at a heating rate of 10°C/min from 25 to 250°C. The heat flow as a function of temperature was recorded and analyzed. The entire scanning was carried out under an inert nitrogen atmosphere.

19.2.6 Electrical Properties

Electrical conductivity was measured on strips having dimensions of 50 mm × 10 mm cut from rolled composite solder preforms with a thickness of ~0.13 mm using a four-point probe technique.

19.2.7 Wettability

Solder alloys were cold rolled to preforms of thickness 1 and 0.13 mm for the joint tensile testing, wetting, and creep-rupture analyses. The solder preforms were remelted four times to get a uniform structure and composition. Approximately 0.2 g of the remelted solder preforms were weighed using an electronic balance. The weighed solder preforms were cleaned with acetone in an ultrasonic bath. The substrate used was a thin copper plate of 99.9% purity and dimensions of 25 mm × 25 mm × 0.1 mm. These small substrates were polished sequentially with silicon carbide sandpaper of up to 800 abrasive number, and then cleaned ultrasonically in acetone for 10 min to achieve an ultraclean substrate for wetting experiments.

The measurement of contact angle was performed using the following technique. First wetting was carried out on a hot plate. Rosin mildly activated (RMA) flux was applied on a copper substrate. Some flux was then applied on the surface of the preweighed solder preform before placing it on the copper substrate. In preparation for the reflow, the substrate containing the solder and the flux was first preheated to 100°C, and then to the reflow temperature of 240°C. After the time of reflow, the specimen was quickly removed, allowed to solidify, and later quenched to room temperature. The solder after reflow on the copper substrate was cleaned with alcohol for 10 min to remove the flux residues. After each test, the solder drop was cut perpendicular to the interface, mounted in resin, and polished to examine the morphology and contact angle of solder on copper substrate. Then the photograph of the specimen was taken and analyzed with the help of commercially available software for measuring spreading area.

19.2.8 Microhardness Testing

The sintered samples were polished to a mirror finish prior to the microhardness indentation tests. Microhardness of the composite solder specimens was measured using a Digital Micro-Hardness Tester with a Vicker's indenter. The samples were indented with a load of 10 g, and an average of seven indentations was made at different locations of the composite solder specimens for further analysis.

19.2.9 Tensile Testing

The samples for tensile testing were machined from the extruded bars. Dog-bone-shaped specimens of gauge length 25 mm and diameter 5 mm were prepared. Tensile experiments were carried out at room temperature on the specimens using an Instron 5569 tensile tester at a constant cross-head displacement of 1 mm/min. Five samples of each composite solder were tested. All samples were tested to failure.

19.2.10 Tensile Strength of Solder Joint

Cu samples of length 45 mm were cut from a 99.9% pure, half-hardened Cu bar with rectangular cross section (10 mm × 1 mm). These were etched in 50% sulfuric acid to get rid of the surface oxide layer. The mating surfaces were fluxed immediately with commercial RMA flux and the rest of the surfaces were coated with solder resist to prevent them from being wetted by molten solder. Solder alloys were rolled into thin sheets of thickness 1 mm and sliced into pieces that approximately covered the mating surface area of the Cu samples. Then, the sliced solder pieces were placed between the mating surfaces of two Cu samples in an aluminum mold and were heated in a furnace to a temperature 50°C above the liquidus of the solder. After holding in the molten state for 2 min, the samples were gently soldered with the help of a screw-driven mold to maintain a joint thickness of 500 μm by adjusting the screws placed at each end of the mold to obtain a good joint, and were cooled in the furnace. It was found that the tensile-testing specimens thus prepared resulted in joints with solder of thickness between 300 and 400 μm.

19.2.11. Creep Rupture Analysis

The creep rupture tests were conducted using composite solder lap joints between two dog-bone-shaped copper pieces, which were fabricated as follows. Two 99.9% pure, 0.1-mm thin copper sheets were first wire-cut into the shape of a dog bone. The composite solder alloys were cold rolled to obtain performs of thickness 0.13 mm and cut into square specimens of dimensions 1 mm × 1 mm. The dog-bone-shaped copper substrates were cleaned with dilute sulfuric acid and rinsed with acetone. The narrow ends of the copper substrates were coated with solder resist to obtain a cross-sectional area of 1 mm^2. Then, RMA flux was applied to each narrow end of the substrate and the composite solder preform was sandwiched between the two copper substrates. Reflow soldering was performed in a programmable oven. The creep-rupture life tests were performed at room temperature with a dead load stress of ~10.4 MPa.

19.3 Results and Discussion

19.3.1. Microstructural Observation

The SEM and TEM microstructures of the as-received SWCNTs used in the present study are shown in Fig. 19.1. The FE-SEM microstructure of the original Sn–Pb solder is shown in Fig. 19.2a showing white contrast for tin grains and dark contrast for lead grains. The average grain size of the as-cast Sn–Pb solder was 5.12 μm.

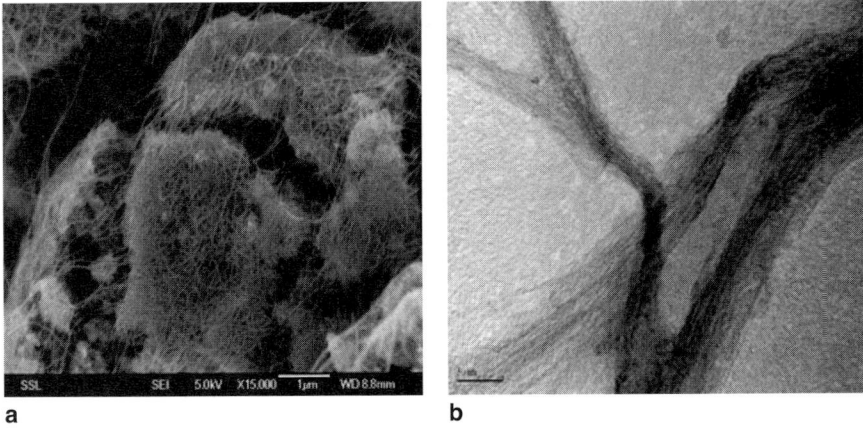

Fig. 19.1 Images of SWCNTs: (**a**) scanning electron microscopy (SEM) image of SWCNT, (**b**) TEM micrograph of SWCNT produced by a chemical vapor deposition (CVD) process

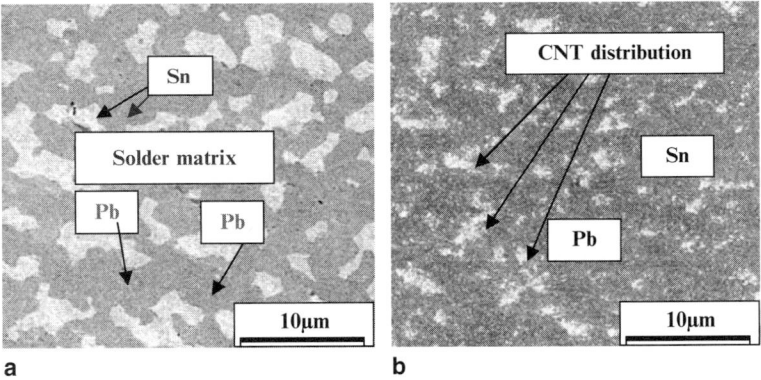

Fig. 19.2 FE-SEM micrographs of 63Sn-37Pb solder with (**a**) 0 wt% SWCNT, (**b**) 0.3 wt% SWCNT

Figure 19.2b shows the highly refined microstructure of 0.3 wt% SWCNT-doped Sn–Pb composite solder, which is a consequence of homogenous dispersion of the nanotubes. The average grain size of the composite is measured to be 1.08 μm by employing image analysis software. An obvious difference between the microstructures of the solder alloys with and without addition of nanotubes can be observed.

There is some porosity observed in the solder matrix. This is mainly attributed to the sintering process. During the sintering process, the matter of the solder matrix flows and the SWCNTs act as solid impurities [24]. The van der Waal forces cause the SWCNTs to get entangled with one another. Because of this phenomenon, it is very difficult to achieve a higher degree of homogeneous dispersion of the SWCNTs throughout the solder matrix. In this manner, the entangled SWCNTs may have resulted in the formation of pores in the solder matrix, which is being observed in the micrographs.

Figure 19.3 compares the microstructures of the Sn–Ag–Cu + SWCNT composite solders and pure Sn–Ag–Cu solder. The higher magnification micrographs in Fig. 19.3a reveal that the microstructure of Sn–Ag–Cu solder is composed of a dark-gray phase (Cu_6Sn_5) and brighter light-gray grains (Ag_3Sn) dispersed evenly in the β-Sn solder matrix. For the Sn–Ag–Cu pure solder sample, the average grain size of the secondary phase varied between 3.75 and 4.25 μm. The average grain size of the secondary phase was found to be 0.5–0.8 μm with 1 wt% addition of nanotubes to the Sn–Ag–Cu solder as shown in Fig. 19.3b. In the SWCNT-reinforced solder samples, the SWCNTs are distributed at the boundaries of the Ag_3Sn equiaxed grains. They can be identified by the difference in contrast, which is mainly associated with the different atomic numbers of the individual phases under consideration. Brighter regions correspond to the higher atomic numbers while darker phases correspond to the lower atomic numbers. The elemental analysis obtained by EDX is shown in Fig. 19.4. The intense "C" peak represents the presence

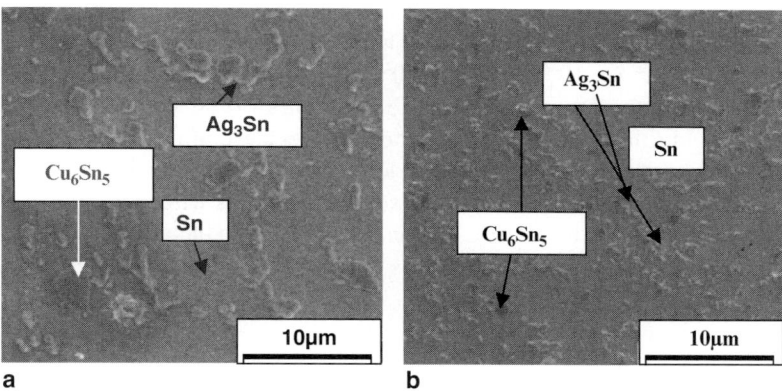

Fig. 19.3 FE-SEM micrographs of Sn-3.8Ag-0.7Cu with (**a**) 0 wt% SWCNT, (**b**) 1 wt% SWCNT

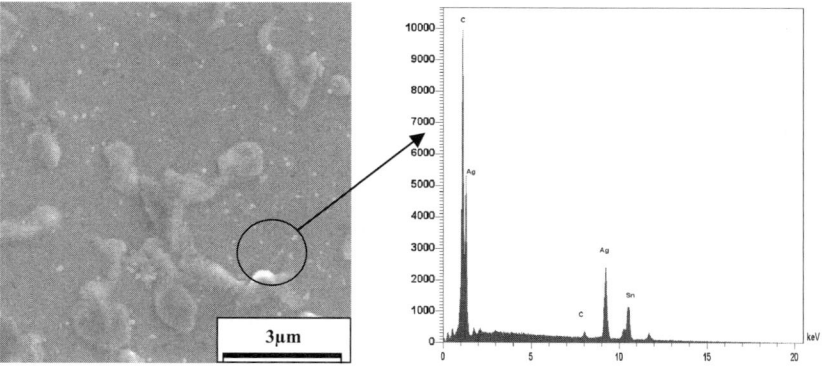

Fig. 19.4 Phase identification of SWCNT at the grain boundary of Ag_3Sn in sintered Sn–Ag–Cu/CNT composite: EDS of the white region showing the presence of carbon

of SWCNTs at the boundaries of the Ag_3Sn grains. This shows that the SWCNTs remained inside the solder matrix after sintering, but were concentrated at the boundaries of the Ag_3Sn grains.

The possible reason for the size refinement is as follows. SWCNT is a ceramic material. While processing the composite solder specimens, the surface diffusion of the Ag_3Sn can be suppressed by the extremely quick translations of ceramic materials through the temperatures that exist during the sintering process [25]. The reinforcement of the microstructure, as shown in Fig. 19.2 with the varying content of SWCNT, demonstrates a strong dependence of the sintered microstructure of the composite solders on the initial composition and morphology of the starting materials.

19.3.2 Coefficient of Thermal Expansion (CTE)

The CTE was measured using TMA and was obtained from the initial linear slope of the thermal strain–temperature plot. The CTEs of pure Sn–Pb and Sn–Ag–Cu were found to be 25.8×10^{-6} and $18.7 \times 10^{-6}/°C$, respectively, which are comparable with those in the literature [26, 27]. The variation of CTE with weight percent of SWCNT addition for the Sn–Pb and Sn–Ag–Cu composites is shown in Fig. 19.5. The composite solders exhibit lower CTE values than the parent alloys. It was observed that the CTE of both the solders decreases with increasing content of SWCNT. In general, the lower CTE can be attributed to the rigidity of the nanotubes and the fine dispersion of nanotubes in the solder matrix, which can obstruct

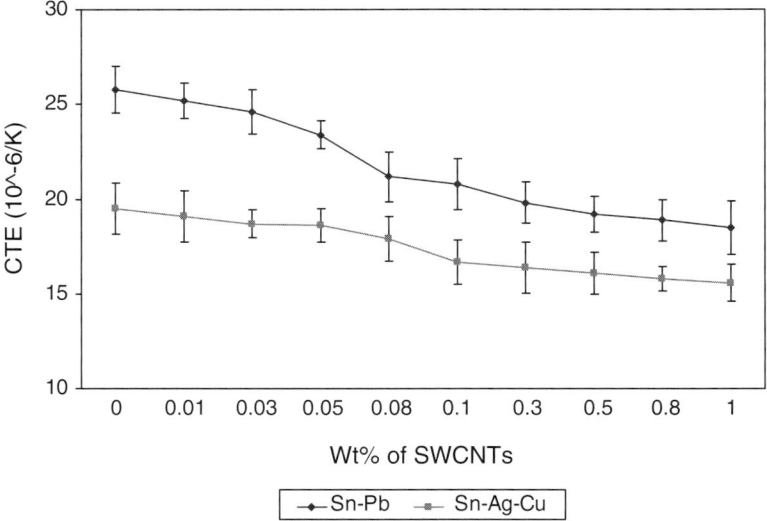

Fig. 19.5 Variation of CTE of both Sn–Pb and Sn–Ag–Cu composites with weight percent of SWCNTs

the expansion of the solder matrix at elevated temperatures. However, factors such as the adhesion of nanotube-matrix interfaces at testing temperatures, the apparent lack of orientation of the nanotubes, and the inevitable agglomeration at higher nanotube loads might affect the CTE values of nanocomposite solders and need to be confirmed by further studies and analysis.

19.3.3 DSC Analysis

DSC measurements were carried out to determine the thermal properties such as melting point and onset temperature of both Sn–Pb and Sn–Ag–Cu composite solders containing varying amounts of SWCNTs. The results are given in Table 19.1. Typical DSC thermograms of the Sn–Pb and Sn–Ag–Cu solders and their composites with SWCNTs are shown in Figs. 19.6 and 19.7. The shapes of the thermograms closely resemble one another. They are characterized by a sharp endothermic peak associated with the onset temperature and a peak temperature that exactly corresponds to the melting temperature of the solder or composite solder. It can be seen from Table 19.1 that the melting point of the composite solders as well as the onset temperature decreases with increasing content of SWCNTs. A similar decreasing trend in melting point was recently reported for the addition of nanoalumina, and nano-SiO_2 to polyether ether ketone (PEEK) [28].

The possible reasons for the reduction in melting point of solders could be due to the increase in the surface instability with the higher surface free energy rendered by the addition of SWCNTs. Also, the size effect of CNTs can significantly alter the grain boundary/interfacial characteristics of solders, resulting in such a change in physical properties [29–31].

It can be seen from Table 19.1 that both nanocomposite solders show decreasing melting points and onset temperatures with increasing nanotube content. However, the Sn–Ag–Cu/SWCNT system shows a much lower melting temperature than the Sn–Pb/SWCNT system. This was mainly attributed to the good adhesion between the nanotube and Ag_3Sn of the lead-free solder matrix.

Table 19.1 Onset and melting temperatures of Sn–Pb, Sn–Ag–Cu composite solders

Wt% SWCNT	63Sn–37Pb		Sn–3.8Ag–0.7Cu	
	Onset temp. (°C)	Melting temp. (°C)	Onset temp. (°C)	Melting temp. (°C)
0	181.1	183.3	217.7	221.0
0.03	181.1	182.8	217.1	220.1
0.08	180.2	182.2	216.4	219.8
0.1	179.5	182.0	216	219.3
0.3	179	181.9	215.6	218.9
1	176.3	181.1	213.4	217.9

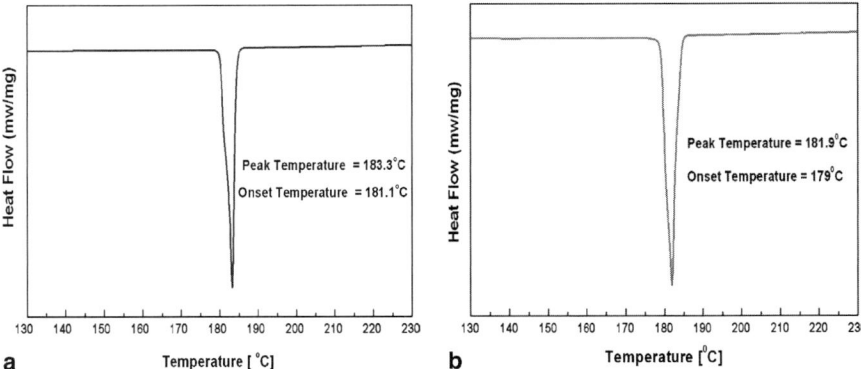

Fig. 19.6 DSC thermographs for a heating rate of 10°C/min for (**a**) 63Sn-37Pb solder, (**b**) 63Sn-37Pb + 0.3 wt% SWCNT

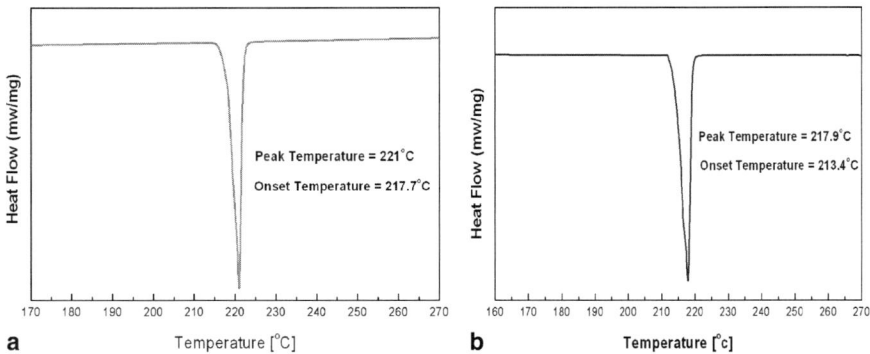

Fig. 19.7 DSC thermographs for a heating rate of 10°C/min for (**a**) Sn-3.8Ag-0.7Cu, (**b**) Sn-3.8Ag-0.7Cu + 1 wt% SWCNT

Addition of SWCNTs has resulted in lowering the melting point of the Sn–Ag–Cu and Sn–Pb composite solders by only 3.4 and 1.5°C, respectively. This lowering is not large and the resultant nanocomposite solders can readily be adopted with the current recommended reflow conditions.

19.3.4 Electrical Conductivity

Figure 19.8 shows the variation of the electrical conductivity of the Sn–Pb and Sn–Ag–Cu composite solders with SWCNT weight percent. The electrical conductivities of Sn–Pb and Sn–Ag–Cu solder are 10.58% IACS (International Annealed

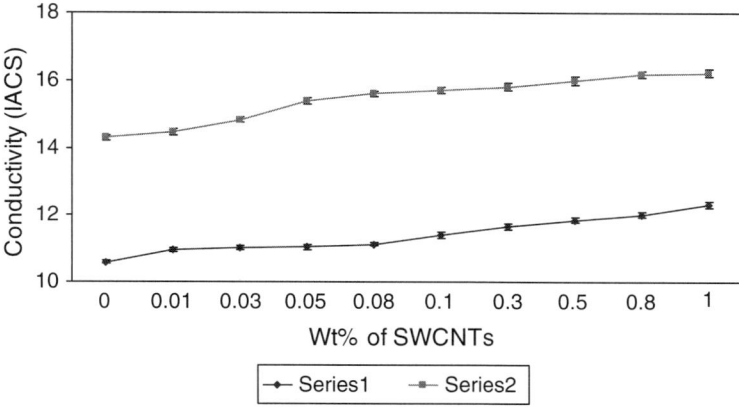

Fig. 19.8 Electrical conductivity of Sn–Pb and Sn–Ag–Cu composite solders vs. SWCNT content

Copper Standard) and 14.29% IACS, respectively. It is evident from the graph that increasing the SWCNT content increases the electrical conductivity of the composite solders. The electrical conductivity of the Sn–Pb-based composite solders increased from 10.58% IACS to 11.04% IACS with the addition of 0.05 wt% of SWCNTs. Further addition of SWCNTs increased the conductivity value to 11.86% IACS. This increment is ~12% higher than the parent Sn–Pb alloy. A similar behavior is noted for the Sn–Ag–Cu-based composite solders. The room temperature conductivities of the bare Sn–Pb and Sn–Ag–Cu are comparable with those in the literature [27]. Similar increments in conductivity have been observed with epoxy nanotube composites [32].

The trend of increasing electrical conductivity of the nanocomposite solders with increasing amounts of SWCNT addition can be explained by the fact that at the percolation threshold, there is a network structure of nanotubes surrounded by the immobilized solder matrix. Even if the nanotubes do not touch each other, conductivity of the nanocomposites is increased as long as the distances between the tubes are lower than the hopping distance of the conducting electrons [33].

19.3.5 Contact Angle

Figure 19.9 shows the contact angles measured for both Sn–Ag–Cu and Sn–Pb solders as a function of SWCNT content. As can be seen, the contact angle for both composite solders first decreases with SWCNT content up to about 0.1 wt% before increasing. The minimum value for Sn–Pb solder composite was 15.8° at 0.08 wt% SWCNT, while the minimum for Sn–Ag–Cu solder composite was 27° at 0.1 wt% SWCNT. The contact angle measured for Sn–Ag–Cu was 34.2°, which is very similar to the value reported by other researchers in the literature [27].

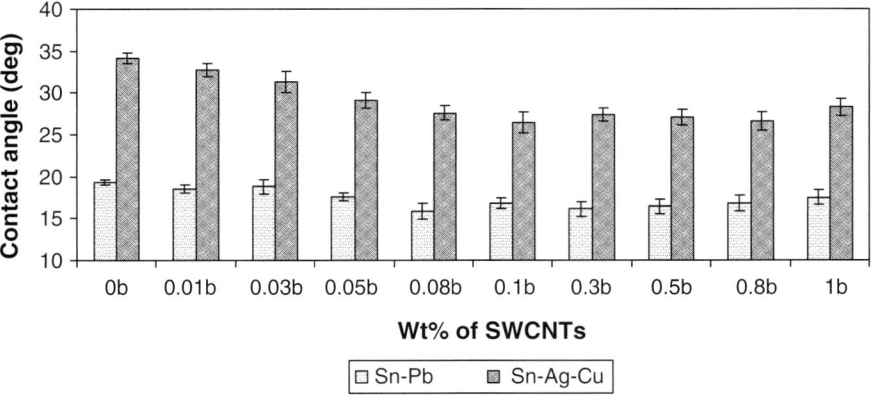

Fig. 19.9 Contact angles of composite solders on copper substrates with varying contents of SWCNT

The effect of nanotube addition on the wettability can be explained as follows. An increase in the flux-copper surface energy or decrease in flux-solder surface energy will decrease the contact angle, thus increasing the wettability. SWCNT addition that promotes these changes will result in improved wettability. It is expected that increasing the SWCNT content to a certain extent could greatly increase the flux-solder surface tension because of chemical reactions. Similar findings have been reported by Loomans [34] for lead-free solder systems.

19.3.6 Wettability

The spreading area and wetting area of a fixed mass of solder were used to evaluate wettability [35]. Figure 19.10 shows the spreading area measured for both composite solders. This figure reveals a similar trend as the contact angle results. For both composite solders, the spreading area increases with SWCNT content up to about 0.1 wt% before decreasing. The maximum value for Sn–Pb solder composite was 159.5 mm^2 at 0.08 wt% SWCNT, while the maximum for Sn–Ag–Cu solder composite was 128.5 mm^2 at 0.1 wt% SWCNT.

It is believed that the addition of nanotubes enforced the orbital interaction between the tin atoms and the copper atoms and greatly improved the spreading area. However, further addition of nanotubes beyond a critical concentration deteriorated the wetting properties and reduced the spreading area of the composite solders. If there is a high level of nanotubes in the composite solder, tin atoms can no longer play the important role of base metal, since the orbital reaction between nanotubes and tin atoms is not strong. Therefore, the spreading area and wettability are worse with increasing levels of nanotubes beyond a certain critical value.

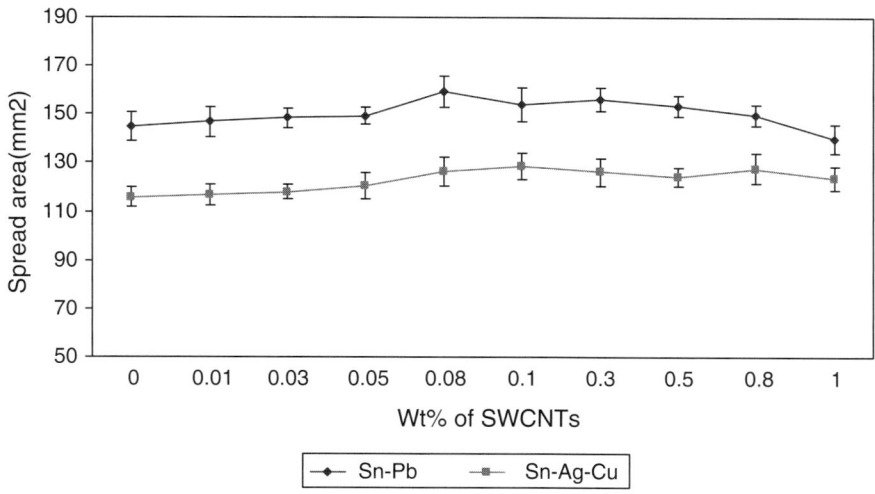

Fig. 19.10 Variations of the spreading area for composite solders with different weight percent of SWCNT

19.3.7 Microhardness

Figure 19.11 gives the microhardness values of the composite solder alloys as a function of SWCNT loading. Each value presented was obtained from an average of seven readings at different locations within each composite solder sample. A trend of increasing microhardness with SWCNT content was observed for both composite solder alloys. Microhardness tests show indeed that the SWCNT-reinforced Sn–Ag–Cu composite solders have slightly higher hardness than the SWCNT-reinforced Sn–Pb solder. The average microhardness of the Sn–Pb + 0.5 wt% nanotube composite solder is ~16.5% higher than that of the Sn–Pb solder alloy, whereas 1 wt% addition of SWCNT to Sn–Ag–Cu solder resulted in nearly 18% improvement in the microhardness value as compared with the original Sn–Ag–Cu solder alloy.

19.3.8 Tensile Properties

19.3.8.1 Yield Strength

The influence of nanotube addition on the yield strength of both composite solders has been investigated and plotted in Fig. 19.12. It can be seen that the yield strength of both composites increases with the nanotube content. It is interesting to note that in the case of Sn–Ag–Cu composite solders the yield strength increases continuously

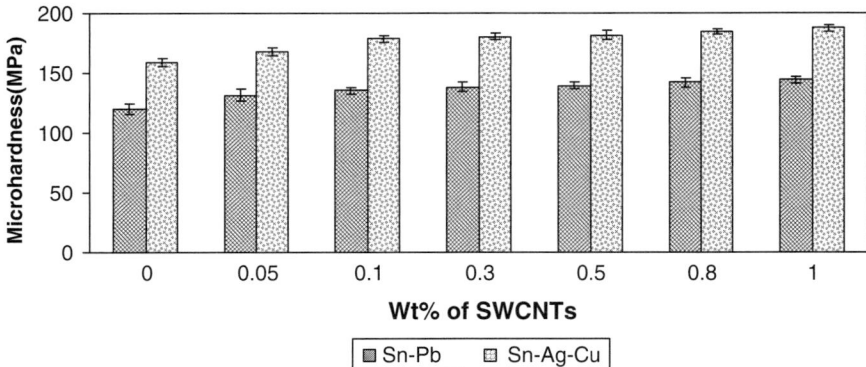

Fig. 19.11 Microhardness variation of the composite solders with different weight percent of SWCNT

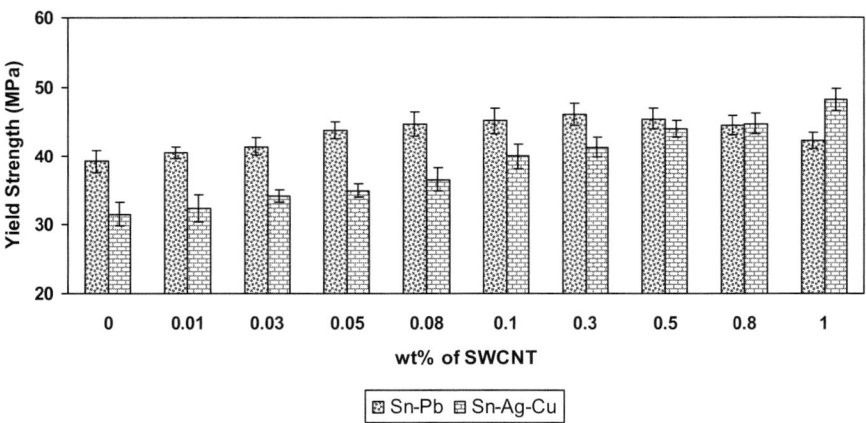

Fig. 19.12 Variations of yield strength of composite solders with weight percent of SWCNT

with nanotube content, while for Sn–Pb solders, the yield strength reaches a maximum value at 0.3 wt% of nanotube addition before decreasing. In both cases, it was found that it was impossible for the solders to absorb more than 1 wt% of SWCNT. The maximum increase in yield strength for Sn–Ag–Cu solder composite at 1 wt% of nanotube reinforcement was 52.9% higher than that of its pure counterpart, whereas the maximum increase in yield strength achievable for Sn–Pb solder was ~18% at 0.3 wt% of nanotube addition.

19.3.8.2 Ultimate Tensile Strength

Figure 19.13 shows typical variations of the ultimate tensile strength (UTS) of Sn–Pb-based composite solders in comparison with Sn–Ag–Cu-based composite solders

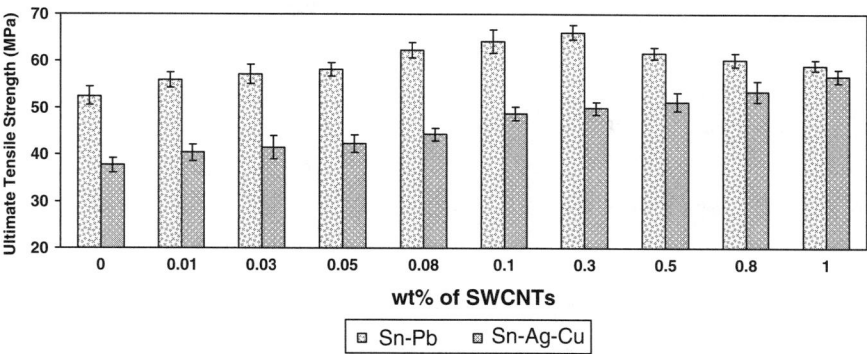

Fig. 19.13 Effect of SWCNT loading on UTS of Sn–Pb and Sn–Ag–Cu composite solders

as a function of different concentrations of SWCNTs. All the measured tensile strengths of nanocomposite solders exhibited very small deviations. The trends observed are similar to the effect of nanotube addition on the yield strength. The UTS of Sn–Ag–Cu solder specimens increase monotonically with increasing SWCNT while the UTS of Sn–Pb reached a maximum at 0.03 wt% of SWCNT before declining. The maximum UTS of Sn–Pb-based composite solders achieved with the addition of 0.3 wt% of SWCNTs was ~26% higher than that of undoped Sn–Pb solder. The effect of SWCNT on the UTS of solder is similar to its effect on polymers [32]. A maximum increase of about ~51% was observed for the Sn–Ag–Cu solder. The improvement in the tensile strength may be caused by the strong interactions between the solder matrix and the SWCNTs, which leads to good dispersion of SWCNTs in the nanocomposites. These well-dispersed SWCNTs may be the reason for the increase in the tensile strength. However, when the content of SWCNT is too high, the SWCNTs cannot be properly dispersed in the solder matrix and agglomerate into clusters because of the huge surface energy of SWCNTs [36]. This probably caused the decrease of tensile strength as observed in the case of Sn–Pb solder doped with more than 0.3 wt% nanotube.

19.3.8.3 Tensile Modulus

The variation of tensile modulus of both composite solders (Sn–Pb and Sn–Ag–Cu) with weight percent reinforcement of SWCNT is shown in Fig. 19.14. As with the yield strength and UTS, the tensile modulus for Sn–Ag–Cu composite solder increases monotonically with weight percent reinforcement of SWCNT while that for Sn–Pb has a maximum at 0.3 wt%. The general increase of tensile modulus with weight percent of SWCNT is probably due the reinforcing effect imparted by the nanotubes that allowed a greater degree of stress transfer at the grain boundaries. A possible explanation can be given for the behavior for Sn–Pb composite solders by assuming a similar state of high-quality dispersion for all nanocomposites after

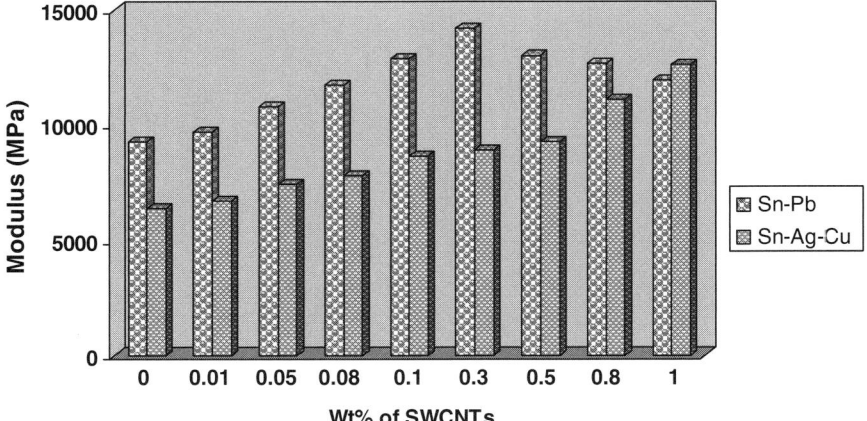

Fig. 19.14 Variation of the tensile modulus of composite solders with weight percent of SWCNT

the sintering process. An increasing amount of agglomerates in the sintered composite was observed for the SWCNT/Sn–Pb solder composites with nanotube content above 0.3 wt%. We propose these agglomerates to be a result of reagglomeration, which reduces the Young's modulus more significantly. The largest modulus of Sn–Ag–Cu composite solder was 12,642 MPa at 1 wt% SWCNT, which is almost 98% higher than the value of 6,385 MPa for the original solder. The largest modulus for Sn–Pb composite solder was 14,216 MPa at 0.3 wt% CNT, which represents an increase of 53% over the value of 9,276 MPa for the original solder.

19.3.8.4 Ductility

Ductility was quantified by measuring the plastic strain to failure. Appreciable ductility was measured for both Sn–Pb and Sn–Ag–Cu composite solders. A plot of ductility (percent elongation) as a function of SWCNT loading is shown in Fig. 19.15. The tests demonstrated a downward trend of percent elongation with increase in the SWCNT content for both composite solders. Sn–Ag–Cu solder shows a 33.3% elongation at break. Almost 26.6% elongation at break was found for 0.1 wt% of SWCNT added. The elongation to failure was observed to be 23.8% at 1 wt% of SWCNT addition, which is ~27% lower than that of the virgin Sn–Ag–Cu solder matrix. This indicates that adding the nanotubes to Sn–Ag–Cu solder material increases brittleness, which is consistent with the previous studies of composite solders reported by Chen et al. [37]. As shown in Fig. 19.15, the percent elongation of 0.03 wt% reinforced Sn–Pb solders was obviously lower than that of pure Sn–Pb solders; the elongation decreases with increasing SWCNT content from 0.03 to 0.5 wt%. At 0.5 wt% SWCNT, the ductility is ~24% lower than that of the Sn–Pb solder. From this, it is evident that both composite solders showed an

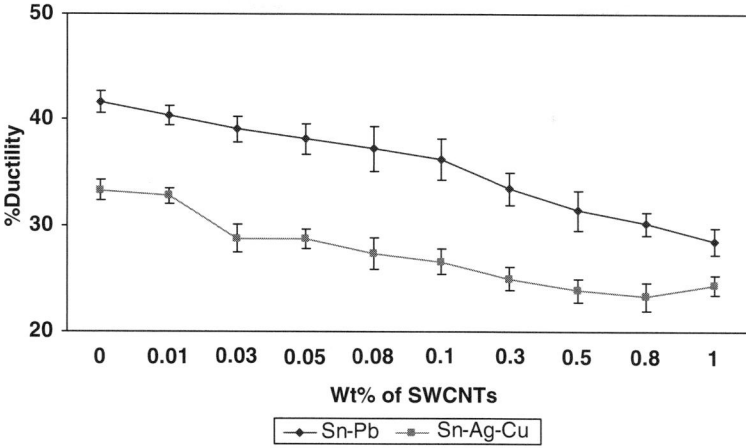

Fig. 19.15 Ductility of nanocomposite solders as a function of SWCNT content

increase in brittle characteristics due to the rigidity of the composite solder matrix as SWCNT content increased. In all the cases, the elongation to failure decreases. The major reason is that SWCNTs included into the solder matrix behave like physical constraints and restrict the deformation of the solder matrix.

19.3.8.5 Work of Fracture

Figure 19.16 shows a plot of the work of fracture vs. various SWCNT loadings for the Sn–Pb and Sn–Ag–Cu composite solders. However, the work of fracture of Sn–Ag–Cu composite solders did not vary linearly with increase in SWCNT content. Maximum and minimum values were observed in this case. The minimum work of fracture was observed at 0.5 wt% addition of SWCNT while the maximum was observed for the undoped Sn–Ag–Cu solder. For Sn–Pb solder, the minimum work of fracture occurred at 0.8 wt% CNT while the maximum was observed for the undoped Sn–Pb solder.

19.3.9 Strengthening Mechanisms

19.3.9.1 Grain Size Refinement

Table 19.2 gives the grain size values measured using image analysis for both composite solders with and without reinforcement of nanotubes. As is common in composite materials, the grain size decreases as the weight fraction of reinforcement addition increases. Since nanotubes may act as nucleation sites for recrystallized

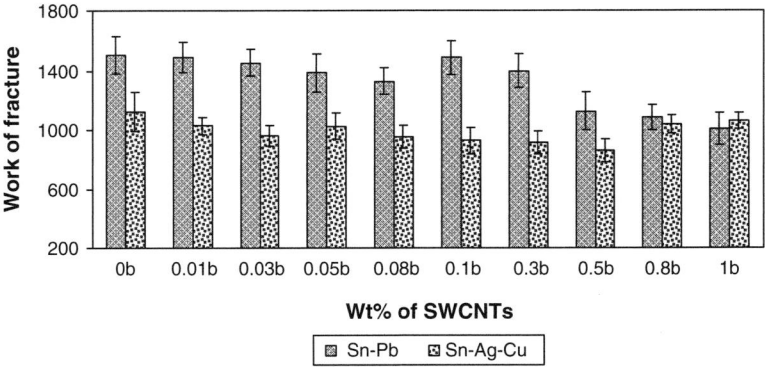

Fig. 19.16 Influence of SWCNT content on the work of fracture of Sn–Pb and Sn–Ag–Cu composite solders

Table 19.2 Grain sizes of the Sn–Pb, Sn–Ag–Cu composite solders

Solder alloy	Grain size (μm)
63Sn–37Pb	5.12
63Sn–37Pb + 0.3 wt% CNT	1.08
Sn–3.8Ag–0.7Cu	3.75–4.25
Sn–3.8Ag–0.7Cu + 1 wt% SWCNT	0.5–0.8

grains during sintering, the volume fraction of recrystallization grains increases when the reinforcement volume fraction increases; it can be observed that as the weight fraction of the SWCNT rises, the grain size of the composite solder diminishes, causing strengthening by the Hall–Petch mechanism. SWCNTs control the grain size of the composite solders, since they prevent grain growth. This grain size refinement can be clearly seen in the microstructure of the composite solders shown in Fig. 19.3. The Hall–Petch relation can be formulated as follows:

$$\sigma_H = \sigma_o + \frac{k}{\sqrt{D}}$$

where σ_H = yield stress, σ_o = friction stress, k = constant, D = grain size.

The yield stress increases as the grain size of the composite diminishes.

19.3.9.2 CTE Mismatch

63Sn-37Pb solder has a CTE of $25.8 \times 10^{-6}/°C$, while SWCNTs exhibit a much lower CTE of $-1.5 \times 10^{-6}/°C$ [36]. Hence, in the SWCNT-doped Sn–Pb solders there exists a significant CTE mismatch between the SWCNT reinforcement and the solder matrix. This CTE mismatch can result in the prismatic punch of the dislocations at the interface, which in turn can lead to the work hardening of the solder matrix. The dislocation density that is generated due to the CTE mismatch between the reinforcement and the solder matrix is directly proportional to the surface area of the reinforcement. The diameter of a SWCNT is very small leading to a lower density of Griffith flaws. Because of the lower Griffith flaws, the number of dislocations generated is likely to be higher, which in turn could result in the increased strengthening effect.

The dislocation density can be formulated as

$$\rho^c = \frac{10A\varepsilon f_{SWCNT}}{(1-f_{SWCNT})bd_{SWCNT}},$$

where f_{SWCNT} is the weight fraction of the SWCNTs, ε is the misfit strain due to the difference in the CTE values of SWCNT and solder matrix, b is the Burgers vector, and d_{SWCNT} is the diameter of the SWCNT.

The increment in stress can be indicated by

$$\Delta\sigma_c = \sqrt{3}\alpha\mu b\sqrt{\rho^c},$$

where μ = modulus of rigidity of the solder, b = Burgers vector, α = constant.

19.3.9.3 Orowan Mechanism

The interaction between the dislocations and the SWCNTs can inhibit the motion of the dislocations, leading to bending of the dislocations between the nanotubes. Bending of dislocations produces a back stress, which could prevent further dislocation migration and result in an increase in yield stress. The Orowan mechanism is less significant in the metal matrix composites where the reinforcements are generally coarser in shape and the interparticle spacing is large, but it is more effective in the SWCNT-reinforced composites as the nanotubes effectively represent the fine particles having very narrow diameters of the order of a few nanometers. In this manner, SWCNT can effectively strengthen the solder matrix by interacting with the dislocations.

Thus, the increment in the shear strength of the composite solders can be written as follows:

$$\Delta\tau = K\mu A^{1/2}b/r\ln(2r/r_o),$$

where K = a constant characterizing the transparency of the dislocation forest for basal–basal dislocation interaction, μ = modulus of rigidity of the solder matrix, r = volume equivalent radius of SWCNT = 7.087 nm, b = Burgers vector, A = constant = 0.093 for edge dislocations, and 0.14 for screw dislocations.

19.3.9.4 Residual Stresses

The CTE mismatch between the nanotubes and solder matrix resulted in the presence of residual stresses in the composite solder. The solder matrix remains in tension and the reinforced nanotubes in compression. A similar situation was observed when the metal matrix composites are reinforced with ceramic reinforcements [34]. Presence of residual stresses can also lead to the increment in the yield stress of the composite solders.

It is possible to conclude from the earlier discussion that the increment in yield strength and UTS of the composite solders has three main contributions: the grain size refinement, increase in dislocation density due to the CTE mismatch between the solder matrix and nanotubes, and the Orowan looping mechanism.

The decrease in ductility of the composite solders can be explained by the following mechanism. The reduction of ductility with the higher reinforcement addition is a very common phenomenon observed in metal-matrix-based composites [31]. The main reason may be the limited ductility exhibited by SWCNTs [32]. In addition, the SWCNTs may restrict the movement of dislocations either by inducing the large difference in the elastic behavior between SWCNTs and the matrix or by creating the stress fields around the dislocations.

19.3.10 Fracture Studies

Fracture surfaces of Sn–Pb composites are shown in Fig. 19.17a, b. The lower magnification topographies, which are represented, indicate extreme ductile fracture modes, characterized by dimples on the surface. Closer observations at higher magnifications shown in Fig. 19.18a, b demonstrated the breakage of nanotubes.

Figure 19.17a shows the fractograph of the Sn–Pb solder specimens. The fracture surface shows evidence of high ductility with the dimples. However, the

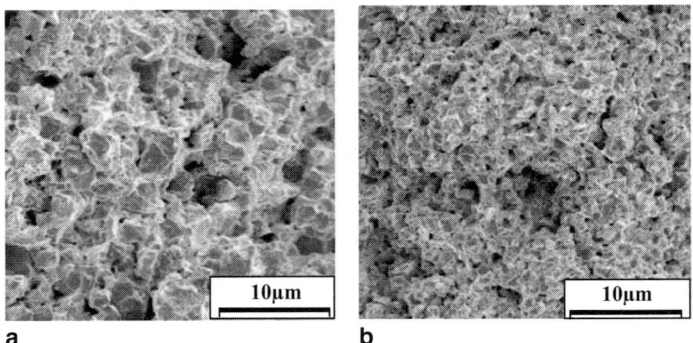

Fig. 19.17 Low magnification FE-SEM micrographs of the fracture surfaces of the Sn–Pb solder composite specimens with (**a**) 0.01 wt%, (**b**) 0.5 wt% SWCNT

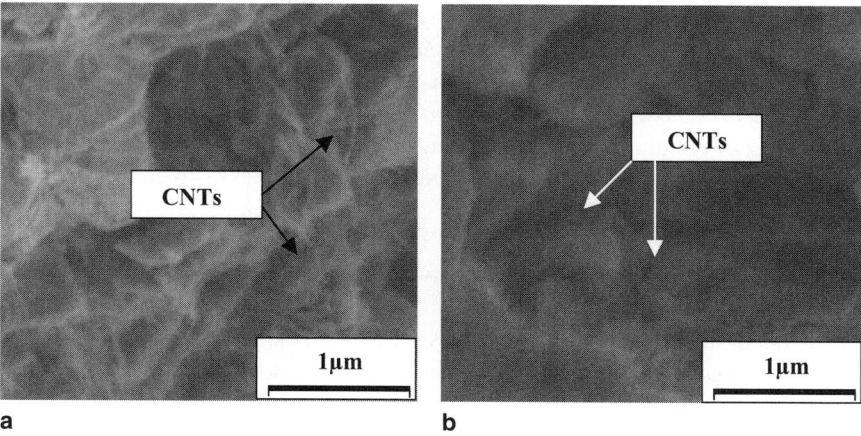

Fig. 19.18 High magnification FE-SEM fractrographs of Sn–Pb composite solders with (**a**) 0.01 wt%, (**b**) 0.5 wt% SWCNT

composite solders reinforced with nanotubes exhibited a limited ductility; the fracture occurred normal to the loading axis as is shown in Figs. 19.17 and 19.18. Figure 19.18 indicates that after tensile deformation, the solder matrix adheres well to the nanotubes due to the highly ductile nature of the matrix alloy. Dimples were observed on the composite solder surfaces in the region between the nanotubes and solder matrix. The nanotubes appeared to have been cut at the surfaces. From this observation, it can be inferred that the fracture took place in the solder matrix by void generation, propagation and finally resulting in the shearing of the nanotubes adhered to the matrix. This indicates that strong interface bonding has been developed in the nanotube-reinforced solder composites.

19.3.10.1 Fracture Mechanism for Sn–Pb Composite Solders

It is evident from the fractographs shown in Fig. 19.18 that the direction of the internal cracks in the composites is normal to the tensile loading axis. Here no nanotube pullouts were observed. According to these findings, it is noted that the crack was initiated in the solder matrix and then propagated and sheared through the nanotube reinforcements. These findings are consistent with the strong interfacial bonding between the solder matrix and the nanotubes.

According to Lloyd [33, 34], there are three possible ways that fracture behavior can be observed in composite materials. (1) If the interface between the reinforcement and the matrix is weak, the crack can initiate and can propagate through the interface. (2) If the interface and the matrix are both strong, the reinforcement can be loaded up to the fracture stresses and then be cracked. (3) If the matrix is weaker than the interfacial and reinforcement strengths, the fracture may occur in the matrix by void coalescence and growth mechanism. In the present SWCNT-based composites, the fracture mechanism observed can be described as follows. The fracture probably initiated in the relatively weaker solder alloy matrix rather than at the solder–nanotube interface or in the nanotube. The fracture can be initiated by void nucleation and propagation. When the crack reaches the nanotube–solder matrix interface, the nanotube–solder matrix interface does not separate due to the high interfacial bonding. Consequently, high stresses will be developed at the nanotubes causing them to be sheared off when their failure stress is reached.

19.3.10.2 Fracture Mechanism for Sn–Ag–Cu Composite Solders

The detailed fracture behavior of Sn–Ag–Cu composite solder specimens was revealed by extensive fractographic observations. SEM was performed at high magnification to probe the fractured specimens that were deformed during tensile loading. Typical FE-SEM micrographs of fracture surfaces at lower magnification with the various additions of nanotubes are shown in Fig. 19.19. As can be seen in Fig. 19.20, the fractured surfaces of the composite specimens mainly consist of matrix dimples and

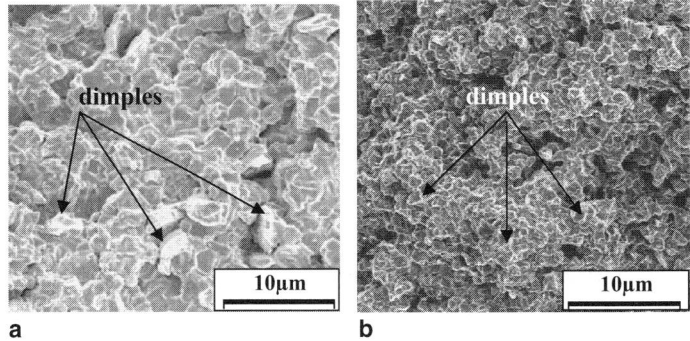

Fig. 19.19 Low magnification FE-SEM micrographs of the fracture surfaces of the Sn–Ag–Cu composite solder specimens with (**a**) 0.01 wt%, (**b**) 1 wt% SWCNT

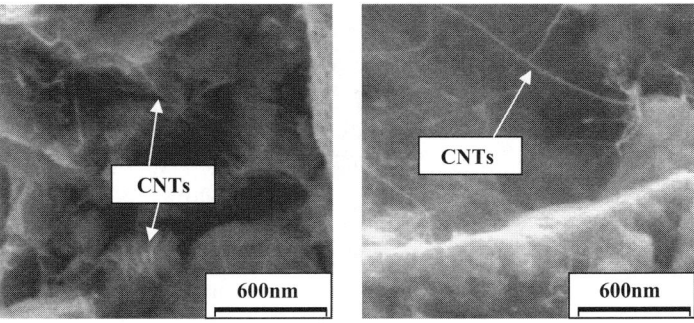

Fig. 19.20 High magnification FE-SEM micrographs of the fracture surfaces of the Sn–Ag–Cu composite solder specimens with (**a**) 0.01 wt%, (**b**) 1 wt% SWCNT

fractured nanotubes. These observations are consistent with those reported in the literature where MWCNTs were employed [31]. From the fractured surface, it is observed that SWCNTs are vertically aligned to the fracture surface during the tensile deformation of the composite solder specimens and this alignment might be one of the reasons for the increase in the strength of the composite solders. A similar effect has been observed in SiO_2-based CNT composites [31].

Figure 19.20 shows that cracks run through the SWCNTs that remain in the matrix of the solders. It is evident that fracture occurred mostly by the failure of the matrix and not by the debonding of the interface between SWCNT and solder matrix. There is some evidence of partial debonding at the interface between Ag_3Sn and the lead-free solder matrix, but none at the interface between SWCNT and the solder matrix.

Because of the high aspect ratio of SWCNTs, microcavities may form at the ends and this is one of the ways by which microcavities are formed inside the matrix. From Fig. 19.3, it is clear that Ag_3Sn has the equiaxed grain shape, compared with SWCNTs. As the deformation increases, the microcavities that already exist in the matrix grow parallel to the tensile loading axis, and the deformation becomes localized into intense shear deformation zones in which the Ag_3Sn grains can be completely debonded to form voids. Subsequently, these voids in the shear deformation zones combine with the microcavities at the ends of the SWCNTs to cause the failure of the solder matrix. Final failure occurs through the breakage of the SWCNTs.

19.3.11 Solder Joint Strength with Copper Substrate

Figure 19.21 illustrates the influence of the nanotube addition on Cu–solder–Cu joint strength under tensile loading conditions. It can be seen that the strength of the solder joints increased after the incorporation of nanotubes into the solder alloys. For Sn–Ag–Cu solder, the strength of the joint increased monotonically with the content of SWCNTs. The joint strength at 1 wt% was 59.1 MPa, ~32% higher

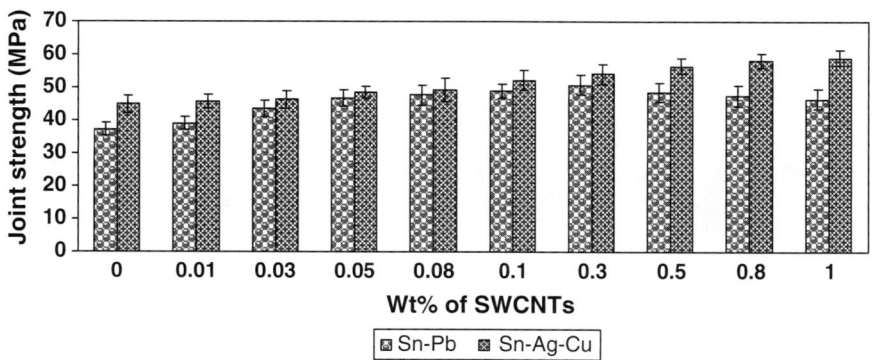

Fig. 19.21 Tensile joint strength values of Sn–Pb and An–Ag–Cu composite solders as functions of weight percent of SWCNT

than the value for the undoped Sn–Ag–Cu solder. For Sn–Pb composite solders, the joint strength first increased with SWCNT content, reached a maximum, and then decreased. The maximum joint strength reached was 50.8 MPa with 0.3 wt% SWCNT, which was ~37% higher than that for the undoped Sn–Pb solder.

19.3.12 Creep-Rupture Analysis

The creep-rupture life of both the Sn–Pb and Sn–Ag–Cu composite solders as a function of SWCNT content is presented in Fig. 19.22. Creep rupture life of both composites increased with increasing SWCNT content. In the case of Sn–Ag–Cu composite solders, the increase is monotonic, reaching a value of 14,043 min with a SWCNT content of 1 wt%, which is about six times the creep rupture time for the undoped Sn–Ag–Cu solder. In the case of Sn–Pb composite solders, however, the creep rupture time first increases, reaches a maximum at 0.1 wt% CNT, and then decreases with further increase of SWCNT content. The maximum creep-rupture life attained for Sn–Pb composite solder was 3,324 min, which is 8.29 times higher than the value for the undoped Sn–Pb solder.

19.4 Conclusions

The goal of this work was to produce and characterize novel SWCNT-reinforced composite solders for fine-pitch wafer-level packaging applications. Composites of Sn–Pb and Sn–Ag–Cu solders and SWCNTs were prepared by a sintering process.

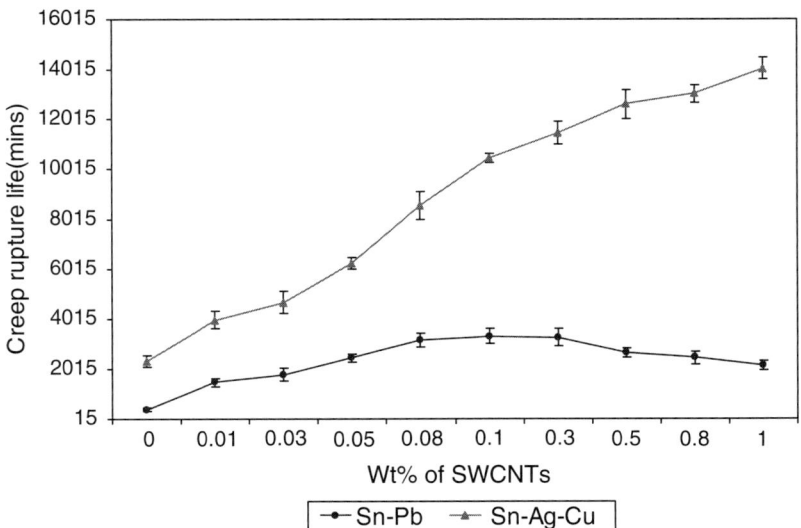

Fig. 19.22 Creep-rupture times of the Sn–Pb and Sn–Ag–Cu composite solders as a function of weight percent of SWCNT

Microstructural studies of the composite solders confirmed the uniform and homogenous distribution of nanotubes in the solder matrix. Nanotube addition also resulted in grain refining. CTEs of the composite solders were found to decrease with increasing weight content of nanotubes. It was found that the melting point of composite solders was lowered with increasing content of SWCNT but the decrease is not substantial and can readily be integrated with existing manufacturing conditions. The contact angles and wettability of composite solders on copper substrates were studied. Among the solders tested, Sn–Pb + 0.08 wt% SWCNT and Sn–Ag–Cu + 0.1 wt% SWCNT exhibited the lowest contact angle and highest spreading area, indicating excellent wettability. Microhardness values improved with the amount of nanotube addition for both the composite solders due to homogenous dispersion of nanotubes throughout the solder matrix. Mechanical properties such as modulus, yield strength, and UTS showed improvement with the nanotube addition. It was found that for Sn–Ag–Cu solder, the improvement in mechanical properties increased monotonically with SWCNT content while for Sn–Pb solder, the improvement first increased and reached a maximum before decreasing. SEM fractographs of the composite solder specimens revealed the ductile fracture mode of the composites, which is characterized by dimples. The addition of nanotubes significantly improved the creep-rupture life of both Sn–Pb and Sn–Ag–Cu composite solders.

References

1. International Technology Roadmap for Semiconductors – Assembly and Packaging, 2003, http://public.itrs.net/Files/2003 ITRS/Home 2003.html.
2. Tummala R.R., Fundamentals of Microsystems Packaging, New York: McGraw Hill, 2001.
3. Yeth C.C., Choi W.J., Tu K.N., Appl. Phys. Lett. 2002; 80:580.
4. Lau J.H., Flip Chip Technologies, New York: McGraw Hill, 1997.
5. Stam F.A., Davitt E., Microelectron. Reliab. 2001; 41:1815.
6. Mavoori H., Jin S., J. Electron. Mater. 1998; 27:1216.
7. Reno R.C., Panunto M.J., J. Electron. Mater. 1997; 26:11.
8. Chen K.-I., Lin K.-L., J. Electron. Mater. 2002; 31:861.
9. Nai S.M.L., Wei J., Gupta M., Mater. Sci. Eng. 2006; A423(1–2):166.
10. Ijama S., Nature 1991; 354:56.
11. Lu J.P., Phys. Rev. Lett. 1997; 79:1297.
12. Saether E., Frankland S.J., Pipes R.B., Compos. Sci. Technol. 2003; 63:1543.
13. Despres J.F., Daguerre E., Lafdi K., Carbon 1995; 33:87.
14. Iijima S., Brabec Ch., Maiti A., Bernholc J., J. Phys. Chem. 1996; 104:2089.
15. Dai H., Wong E.W., Lieber C.M., Science 1996; 272:523.
16. Ebbesen T.W., Lezec H.J., Hiura H., Benett J.W., Ghaemi H.F., Thio T., Nature 1996; 382:54.
17. Yakobson B.I., Brabec C.J., Bernholc J., Phys. Rev. Lett. 1996; 76:2511.
18. Salvetat-Delmotte J., Rubio A., Carbon 2002; 40:1729.
19. Zhong R., Cong H., Hou P., Carbon 2003; 41:848.
20. Dong S., Zhang X., Trans. Nonferrous Meter. Soc. China 1999; 9:457.
21. George R., Kashyap K.T., Rahul R., Yamdagni S., Scripta Mater. 2005; 53:1159.
22. Peigney E., Laurent Ch., Flahaut A., Rousset A., Ceram. Int. 2000; 26:677.
23. Xu C.L., Wei B.Q., Ma R.Z., Liang J., Ma X.K., Wu D.H., Carbon 1999; 37:855.

24. Couchman P.R., Ryan C.K., Philos. Mag. A 1978; 37:327.
25. Upadhya K., Proceedings of Symposium Sponsored by the Structural Materials Division of TMS Annual Meeting, Denver, CO, February 1993, p. 22.
26. Wu W.F., Lin Y.Y., Young H.T., Proceedings of Seventh Electronics Packaging Technology and Conference (EPTC2005), Singapore, 2005, p. 625.
27. NIST solder database, http://www.boulder.nist.gov/div853/lead%20free/solders.html.
28. Kuo M.C., Tsai C.M., Huang J.C., Chen M., Mater Chem Phys 2005; 90:185.
29. Lindemann F.A., Z. Phys. 1910; 11:609.
30. Couchman P.R., Ryan C.K., Philos. Mag. A 1978; 37:327.
31. Ziman J.M., Principles of the Theory of Solids, London: Cambridge University Press, 1972.
32. Sandler J.K.W., Kirk J.E., Kinloch I.A., Shaffer M.S.P., WIndle A.H., Polymer 2003; 16(3):248.
33. Potschke P., Abdel-Goad M., Alig I., Dudkin S., Lellinger D., Polymer 2004; 45:8863.
34. Loomans M.E., Vaynman S., Ghosh G., Fine M.E., J. Electron. Mater. 1994; 23:741.
35. Werner E., A Guide to Lead-Free Solders Physical Metallurgy and Reliability, London: Springer, 2006.
36. R.S. Ruoff, D.C. Lorents, Carbon 1995; 33(7): 925–930
37. Zhigang Chen, Yaowu S, Xia Z, Yan Y, Journal of Electronic Materials 2003;32:235.
38. Lloyd DJ, Int Mater Rev 1994;39:1–23.
39. Lloyd DJ. Processing of Particle-Reinforced Metal Matrix Composites. In: Mallick PK, editor. Composite Engineering Handbook New York NY; Marcel Dekker: 1997, 631–670.
40. Lloyd DJ, Lagace HP, McLeod AD. Interfacial Phenomena in Metal Matrix Composities. Proceedings of ICCI-III.: Cleveland, Ohio: USA; Elsevier; 1990:359–376
41. J. Ning, J. Zhang, Y. Pan, J. Guo, Mater. Sci. Eng. A 2003; 357:392.
42. Cadek M, Coleman JN, Barron V, Hedicke K, Blau WJ, Appl Phys Lett 2002; 81(27): 5123–5.

Chapter 20
Nanowires in Electronics Packaging

Stefan Fielder(✉), Michael Zwanzig, Ralf Schmidt, and Wolfgang Scheel

20.1 Introduction

In the light of continuous miniaturization of traditional microelectronic components, the demand for decreasing wire diameters becomes immediately evident. The observation of metallic conductor properties for certain configurations of carbon nanotubes (CNT) and their current carrying capability [1] set the minimal diameter of a *true* wire to about 3 nm (compare Chap. 15). Investigations are in progress even below that diameter on nanocontacts, formed by single metal atoms, i.e., quantum wires. Quantum wires can be produced by mechanical wire breaking [2], its combination with etching and deposition [3], or other techniques. The properties of quantum wires are only about to be understood theoretically [4]. Doubtless, they are worth considering for packaging solutions in molecular electronics to come [5]. In this chapter, we focus on metal wires and rods in the size range above 10 nm up to submicron diameters, evaluated already to be attractive for microelectronic packaging purposes. Techniques to generate, to characterize, and to handle them, as well as their interaction with electromagnetic fields will be useful for packaging applications in the age of nanotechnology. With the wealth of information available, this review focuses on general trends and starting points for deeper study. Although the cited references are representative, they cannot be complete, since numerous activities are ongoing to produce and to characterize new kinds of wire-like geometries from different materials.

Packaging-specific applications of nanowires (NWs) lie mainly in the fields of interconnect formation, sensor development, and photonics. Given the common understanding of a wire, one would expect NWs to be usually cylindrical conductive strands with diameters below 100 nm, ideally of infinite length, but at least elongated. Whereas common wires are drawn from metal rods, NWs cannot be produced by wiredrawing and do not necessarily consist of a metal or one single material. But approaches to *draw* electrically conductive polymer NWs in electronic circuitry by initiating chain polymerization with a STM cantilever do exist [6].

S. Fielder
Department of Module Integration and Board Interconnection Technologies, Fraunhofer Institute for Reliability and Microintegration, Berlin, Germany

Although rod-like colloidal structures are often mentioned as NWs in the literature, they should rather be regarded as rods or crystal needles. If they consist of metals, we include at least typical publications. Nevertheless, we exclusively focus on metallic wires and wire-like structures, even if they are fixed to a solid substrate as pillars or come as brushes or lawn-like structures. Supramolecular wire-like geometries are also depicted as *molecular NWs*, such as in the case of tropomyosin fibers, whose length and diameter can be directed by Na^+ or Mg^{2+} concentration [7]. Deoxyribonucleic acid (DNA) can form molecular NWs [8], which in turn can be used as templates to produce true metal NWs (see later). Especially alien to traditional electronic engineering are charge-transfer complexes of wire-like geometries. Such supramolecular NWs, e.g., porphyrin NWs generated by ionic self-assembly, perhaps can be used in microelectronic devices – thanks to their photocatalytic activity [9] and hence switchability. Those NWs fall beyond the scope of the present work. If molecular wires are largely short structures, seldom extending the micrometer scale, CNTs can reach even millimeter length scales and are therefore just as interesting for microelectronic packaging. They have been proposed, e.g., as transistor elements in logic circuits, field emitting structures, or vias [10–15]. NWs and nanotubes can be produced from semiconductor materials such as silicon, gallium nitride, or others. Because of the familiarity of microelectronics with those materials they could become even more important for sophisticated future microelectronic applications [16, 17]. Their synthesis and integration into classical planar technology, e.g., by the superlattice NW pattern transfer (SNAP) [18] and resulting application perspectives have been reviewed recently [19–22]. Excluded from this compilation are all sorts of oxide and multicomponent oxide NWs, e.g., ZnO. We consider them to be more important for sensors due to well-measurable conductivity changes with analyte adsorption [23].

Our own results in production, characterization, and application of gold submicron wires in the shape of nanolawn have been included to share the excitement of NW packaging research, connecting usually separated fields like low-temperature joining and interfacing electronics with biological cells.

20.2 Nanowires and Packaging Research

Reliability issues arising from contemporary microelectronic applications have widened the scope of packaging over the last decades remarkably. Modern packaging research for the development of sustainable technologies covers photonics, optical waveguide and fiber integration, (bio)microfluidics, joining, thermal management, wire-, wafer-, and flip-chip bonding, soldering and encapsulation, foil batteries and energy harvesting, and includes also solid mathematical modeling and simulation. This is shown in too many publications and annual reports to be reviewed here.

Metal NWs can be attractive for packaging in nearly every field mentioned [24] due to unique properties in comparison with mesoscaled and bulk materials. Their functional role as interfaces has been envisioned for future microelectronic applications toward 3D nanostructure integration [25]. Characteristics, production methods, and proposed applications of metallic NWs have been reviewed in depth previously [26–35]. Recent international research activities, evaluated and ranked by citation, indicate US leadership until 2005 [36]. Main international research institutes, engaged in microelectronic packaging in alphabetical order of the country, are as follows:

- Inter-University Micro Electronics Centre (IMEC), Belgium
- VTT Technical Research Centre of Finland/VTT Electronics, Finland, and Oulu University Electronics Materials, Packaging and Reliability Techniques, Finland
- Laboratoire d'Electronique de Technologie de l'Information (LETI), France
- Fraunhofer Institute Reliability and Microintegration (IZM), Germany
- Central Electronics Engineering Research Institute (CEERI), India
- Tyndall National Institute, Ireland
- Korea Advanced Institute of Science and Technology/Center for Electronic Packaging Materials (CEPM) and Samsung Advanced Institute of Technology (SAIT), Korea
- Philips Research Laboratories, Eindhoven, The Netherlands
- Institute of Microelectronics (IME), Singapore
- Industrial Technology Research Institute (ITRI), Taiwan
- Packaging Research Centre at Georgia Institute of Technology, USA

20.3 Nanowires: Fabrication

In the plethora of production principles and approaches, nevertheless typical ones can be distinguished and will be presented later. The reproducible generation of metal NWs with identical diameters can be dated back until 1970, when Possin described metal deposition inside etched tracks of high-energy charged particles in mica and proposed to use this method to form NWs in track-etched polymers as well [37]. The technological importance of such tracks had been foreseen even earlier [38]. Many more applications for swift ions in nanoscale microelectronics have been designed independently [39]. Unilaterally etched pores in flex substrates have been proposed for improved copper adhesion [40]. For wire diameters above some tens of nanometers, the use of exotemplates is still the most important production technique so far [41]. Depending on the application, such templates can serve as the scaffold remaining after metal filling by the formation of composites, e.g., dipole storage devices [42]. But the exotemplate can also be dissolved yielding a lost form approach to produce suspended single wires or more complex metal nanostructures. Exotemplates are also suited to produce wires consisting of conductive polymers [43, 44]. The most important (hard) exotemplates are anodic aluminum oxide (AAO) and track-etched polymer membranes (TEM).

AAO [45] offers electrochemically tunable nanopores in a rigid matrix and therefore finds wide application for single wire and wire array production [46–49] even at a very large scale [50]. Dispersions of high aspect ratio wires can be produced [51], or layers of anisotropically conductive or magnetically polarizable materials in dielectric matrices can be prepared. Beside the standard aqueous metallization baths used, AAO is especially suited for plating from aprotic media, due to its high stability in organic solvents [52], opening a way toward electrochemical deposition of NWs consisting of metals with low redox potential (below hydrogen), like Al or Ti [53]. Because of its high stability, AAO has been also used for NW production by high-pressure filling with molten metals; for a compilation, see [35].

Etched ion track polymer membranes (TEM) are other practically important exotemplates. Polyethylene terephthalate, polycarbonate [54, 55], or even polyimide [56] are typically used for their reproducible etchability [57]. Isodiametric and nearly monodisperse shape distributions can be generated following standard etching protocols [58]. Pore diameters in those materials reach about 0.002–1 μm for AAO and 0.010–20 μm for TEM. The density of the stochastically distributed pores in TEM can be chosen from a single pore [59, 60] up to ~109 cm^{-2} depending on the desired pore diameter [61]. The percolation-based electrochemical pore etching in aluminum allows pore densities of AAO templates up to ~1,011 cm^{-2} (e.g., commercially available ANOPORE™ and ANODISK™ inorganic aluminum oxide membrane filters). Whereas the distribution of pores in TEM follows statistics fulfilling Poissonian distribution criteria [62, 63], pores in AAO are always densely arranged. They can even be hexagonally packed over small domains, and if combined with imprinting [64] even over the whole area of a wafer [65]. The typical distances between single pores (i.e., insulating material around neighboring wires) reach the same dimension as the pore diameter for wires generated in AAO. The typical distances of pores in TEM without special precautions (e.g., mask or shutter) will always vary due to the inherent statistics of high-energy particles used. Therefore the distances between metal wires in ensembles generated with TEM, like the nanolawn introduced by us [66], are varying too. With commercially available TEM (Nucleopore™, SPI-pore™, Cyclopore™ – to name a few brands only) pore-to-pore distances vary between 0 and 2 μm at a pore density of 106 cm^{-2} on track-etched foils.

Supramolecular assemblies can work as exotemplates as well: Self-assembling calix [4] hydroquinones form in aqueous photochemical solutions chessboard-like arrays of very narrow rectangular pores. Such pores have been used as silver-ion reducing templates in a process resembling photochemical development. Stable NW arrays consisting of 0.4-nm wires grown up to micrometer length have been prepared [67].

Molecular endotemplates [68], characterized by inner (bio)molecular scaffolds (especially proteins, lipids, and DNA) in combination with a *toning approach* [69–76], or bioparticles, e.g., tobacco mosaic virus, are suitable for metal wire production [77] and should be mentioned as alternatives. Such a nanobiotechnological approach in combination with microelectronic technologies offers another additional advantage: the localized maneuverability of inorganic (conducting) structures, as shown for gold wires, driven by highly specific biochemical molecular machines (e.g., actin–myosin interaction) and a molecular fuel [78]. The metallization of (bio)polymer endotemplates offers additional advantages for complex 3D arrangements of wires and wire networks at the microscale, because of inherent self-assembly principles and specifically directed labeling (addressing). Complex nanotubular networks as generated *naturally* by living cells on artificial substrates [79, 80] or produced artificially by manipulation of liposomes [81–83] can be transformed into hard-wired circuitry by gentle metallization. This approach has been demonstrated for DNA [69] and lipid tubules [84, 85]. NWs can be grown also epitaxially at high temperatures by diffusion in grain boundaries [86]. The similar whisker growth is a notorious failure mechanism causing microelectronic reliability issues, worth mentioning in this context.

The colloid-chemical approach, effectively producing homogeneous – regarding their composition – one-dimensional nanomaterials from salt solutions is usually diminished to the lower nanoscale and low aspect ratios. However, even several micrometer-long gold wires measuring only 15 nm in diameter have been produced by chemical reduction [87]. Seed-mediated growth has been described for gold rods when the seeding nanoparticles have been attached to a solid substrate [88]. The introduction of rod-like micellar templates, e.g., the cationic detergent CTAB, allows the production of suspended cylindrical gold NWs [89]. Comparably long silver wires have been produced in a diameter controlled (20–500 nm) manner by a modified polyol process [90] using different growth-controlling modifiers. A somewhat similar molecular shielding (templating) strategy, i.e., soft exo-templates, comprises the use of temporarily arranged supramolecular ensembles in block-copolymer solutions [91]. For a review of nanostructure-producing techniques with block copolymers, see [92].

Classical photolithography and successors like deep UV lithography [93, 94], colloid mask/nanosphere lithography [95], and competing technologies, with nanoimprint (cold) lithography being the most ripened one among them [96], have been used to generate metal NWs directly on planar substrates. Typically, the NWs and NW grid arrays produced are oriented parallel to the substrate plane. Hence, they can be useful for applications as subwavelength metal gratings or directly as plasmonic waveguides and photonic crystals [97, 98] depending on a guiding medium in close proximity. In a top–down approach, photolithographically generated trenches in a resist layer can be filled forming stretched wires on a planar substrate [99]. A maskless alternative based on substrate steps to fabricate wires by deposition has been introduced as step-edge lithography (SEL) [100], the principles of which have been used to produce molybdenum NWs (15 nm–1 μm diameter and length up to 500 μm). NW composites have been formed starting with electrodeposited molybdenum oxide wires and their reformation in hydrogen and subsequent liftoff in polystyrene layers [101]. Palladium wires embedded in a cyanoacrylate film sensitive toward hydrogen have been prepared as well [102]. A similar growth of metal wires occurs along material cracks [103]. Among maskless NW production techniques the direct writing approach, either by direct atomic metal deposition [104], e-beam induced CVD [105], or by indirect structuring of metal substrates with small molecules via dip-pen technique [106] should be mentioned. With a sweeping AFM-cantilever, copper NWs have been assembled from deposited nanoparticles on a polymer substrate and cut afterward at will [107]. Direct writing techniques have been under intense investigation. To overcome seriality drawbacks, e-beam arrays [108] and cantilever arrays [109] have been proposed. Near-field laser nanofabrication has been scaled down to 80-nm resolution [110]. Two-dimensional photonic crystal structures in polymer have been produced by a combination of interference and e-beam lithography techniques [111]. Meanwhile, similar structures are becoming attractive for wire fabrication via nanoimprint techniques.

Other techniques to generate metal wires include assembly from suspended metal particles by dielectrophoresis [112, 113]. Plasma-enhanced growth techniques, e.g., by vapor–liquid–solid growth or vapor–solid transitions, cannot be covered here but could become more important for packaging, if applied at reduced temperatures. Single crystalline Ni NWs with diameters of 40 nm confined inside multiwall CNTs have been grown by a CVD process with lengths of some tens of micrometers [114].

20.4 Metal Nanowires: Materials

Nearly every electrochemically reducible cation has been deposited inside the pores of different exotemplates already as a wire.

The galvanic deposition of metals inside nanopores has been thoroughly investigated from aqueous electrolytes [115] and from nonaqueous ionic liquids [116] extending the range of available wire materials. Therefore, beside the colloid-chemical approach suitable for mass fabrication of nanoscale metal rods and wires [117], exotemplate methods can be used as well to prepare single wire contacts [118] or special polymer composites, requiring gram amounts of monodisperse wires [119].

Envisioned applications of periodic arrays of magnetizable wires [120] embedded in a dielectric matrix (e.g., AAO) are information storage via perpendicular data recording, characterized, e.g., for arrays of ferromagnetic Ni and Co NWs by high remanence and coercivities [121, 122]. The preparation of similar e-beam written pillar arrays proposed for high-density data storage [123] is much more time consuming and hence expensive. Based on their giant magnetoresistance properties, magnetic multistack layers have been considered as high-density storage elements [124]. Magnetic polymers are another application of polymer composites, e.g., with Ni wires [125]. The integration of oriented wires into polymer films can be used for anisotropically conductive interposer fabrication useful for chip interconnection [126].

20.5 Segmented Metal Nanowires

Such stacks, representing multilayered wires, can be prepared with exotemplates. The common practice of sequential layer plating of different metals or crystal morphologies yielding functional multilayers works well for exotemplate wires, too. Segmented NWs may possess different spectral characteristics, depending on the orientation of polarized light in relation to the axis of the wires [127, 128]. They therefore can serve as embedded identification tags (barcodes) or labels, not disturbing usual fluorescence detection of biomolecules for lab-on-chip application [129, 130]. With the stable material composition of each single wire in a large batch, alloy preformation for solder pastes, which is otherwise difficult to accomplish seems easy. Usually observed component demixing upon storage with NWs can be prevented. A wet chemical, electroless plating approach also can yield multilayered, rather stacked wires [131] with a similar application potential. By a combination of template techniques and etching, segmented wires have been produced with *on-wire lithography* [132], bearing potential for applications in plasmonics, since etched gaps can be selectively filled with different dielectrics. The catalytic activity of NWs and their use as catalysts [133] and their stability in microreactors have been shown [134].

20.6 Metal Nanowires: Structure and Configuration

NWs grown in free solution by anisotropic crystallization will always be single crystals, possessing smooth crystal planes. On the contrary, the structure of NWs cast inside a hard exotemplate naturally will reflect the inner pore structure and arrangement typical for a lost form approach as shown for nanolawn, i.e., nanowires on a flat substrate (Fig. 20.1).

As for metal plating with commercial baths, grain size can be different, depending on bath composition, temperature, current density, and regime. Extreme differences can be illustrated by smooth, amorphous pore filling with a commercial platinum bath and coarse grain pore filling with a gold plating bath (Fig. 20.2). Obviously, conditions can be found to generate wire-like crystal needles with diameters of single grains according in dimension to the template pore diameter. With overplating, large single crystals emerge. That has been described before for larger wires [135, 136].

20.7 Metal Nanowires: Mechanical Properties

The mentioned twinning is a good illustration of altered mechanical properties of an entire (nano)wire, caused by its single-grain spanned diameter. The electrical conductivity of NWs will be crucially influenced by the electrons' mean free path

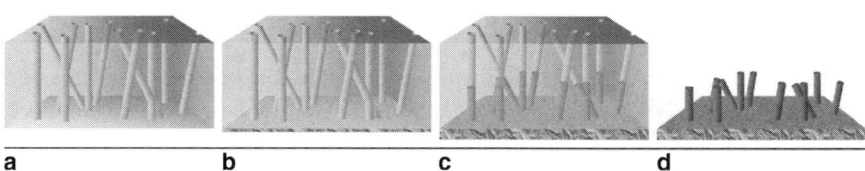

Fig. 20.1 Generation of a metal nanolawn on a track-etched polymer template. (**a**) Isoporous membrane, (**b**) sputter coating, (**c**) plating, (**d**) stripping

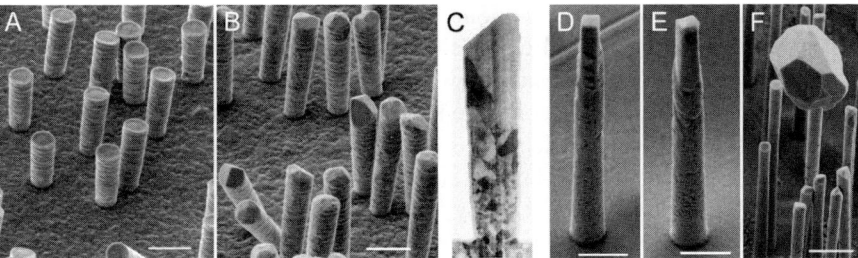

Fig. 20.2 Crystallinity of metal wires plated in track-etched polymer pores. (**a**) Fine crystalline (Pt), (**b**) stacked grains of different sizes (Au), (**c**) FIB dissection of a typical gold wire, as in (**b**), (**d**), (**e**): single crystals grown inside pores (Au), (**f**) single Au crystal grown by overplating. Scale bars (**a–e**): 1 µm, (**f**): 2 µm

and hence crystal defects [137]. The influence of crystal growth direction on twin formation during plating has been studied in this context for single crystalline Cu, Au, and Ag NWs with 30–300-nm diameters [138]. In our cases, probably, plating bath additives are segregating at the grain boundary during annealing and facilitate crystal plane slip. Properties of single metal grains in wires and layers are currently under intense study [139], since necessary high-resolution analysis techniques, e.g., nanoindentation and EBSD detection in scanning electron microscopy are becoming broadly available only now.

Contemporary techniques to characterize mechanical properties of one-dimensional nanostructures such as nanotubes and NWs have been recently reviewed [140]. Since NWs have much in common with single grain layers, theoretical considerations [141] are similar. Their deep understanding is essential for micro-nanoreliability issues in future packaging [142–144]. Theoretical approaches for planar single grain layers [145] and gold NWs [146] should be identical. More investigations, as requested by Uchic et al. in 2004, are still necessary to separate the critical influence of sample size from the observed material properties [147–149]. As already mentioned, the wire preparation conditions heavily influence the grain structure and crystallinity. So high-pressure injection cast wires are rather single-crystalline [150].

20.8 Metal Nanowires and Temperature

So far, melting point depression, known from nanoparticle behavior [151] has not been observed for NWs yet. However, Rayleigh instability, i.e., fragmentation of spatially confined NWs into loose pearl-chain-like structures [152] – at least for multimetal NWs a dramatic increase of intermetallic diffusion effects – has to be considered. Also for homogeneous NWs, heating can cause typical recrystallization rearrangements. We observed dramatic grain growth at the tip of multicrystalline

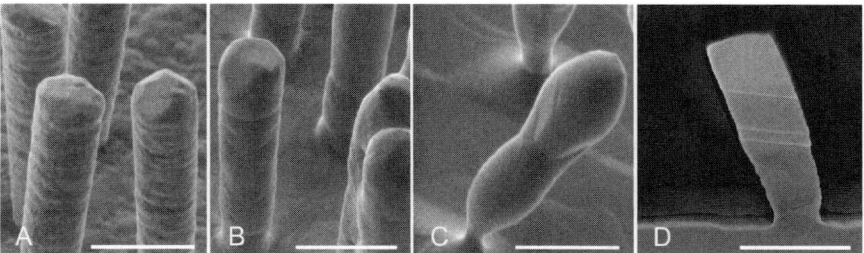

Fig. 20.3 Recrystallization of single wires of a typical lawn structure (Au). (**a**) Wires after template removal (at 65°C), diameter corresponding to pore size, (**b**) sample passing 400°C exposure, (**c**) sample recrystallization after 600°C, growth of large crystals fed by small basal grains, (**d**) twins found after recrystallization at 482°C. Duration of all incubation steps: 10 min. All scale bars: 1 μm

wires (Fig. 20.3) and crystal plane slip, i.e., twinning without additional mechanical load. Similar crystal plane slip has been attributed to mechanical stress during sample preparation [136] and has been surprisingly predicted for NWs of much smaller size [153].

20.9 Electrical Properties

Concerning the tip geometry of any NW, the advantages of miniaturized electrodes (and electrode arrays) for charge injection are obvious. Sharp wire tips are common in electrochemical scanning probe microscopy and with highest resolution in STM. Arrays of oriented CNT found their way into field emitting devices. Their metal counterparts have been proposed for field emission too [154–158]. Beside the tip geometry, surface enhancement by NW-decorated electrodes has been investigated as well [159].

Concerning their current carrying capacity, metal NWs will be rather limited in comparison with CNTs. If moving top–down along the size scale, common wire handling for (wedge or ball) bonding with diameters well above 5 μm (usually 12.5–17.5 μm) is still challenging (compare Chap. 23). However, the reliable current carrying capacity of much smaller metal microwire bundles, configured like brushes, has been demonstrated earlier with via feed throughs for board level interconnects by copper deposition in etched ion tracks [160]. Precise conductivity measurements of individual metal NWs became accessible by the single pore etching and plating, established at GSI Darmstadt, Germany [161]. Theoretical modeling in good agreement with experimental data has been done, e.g., for Bi NWs [162]. A nanoindent contact has been applied on electrochemically filled and polished AAO templates and used in systematic single wire measurements [163], where wire length had been reduced down to 100 nm by polishing. In this way, significant charges up to 109 A/cm^2 could be injected into single Co/Cu NWs to study magnetotransport properties. Other measurement techniques often suffer from contact resistance problems. The majority of measurements on NW conductivity has been undertaken on stochastically arranged wires, thrown on a substrate and contacted by connector depositions [164] or on planar electrode arrays, making contact with appropriate laying ones by a subsequent lithography-based metallization. FIB deposition has been used for that purpose as well, as for instance on 200-nm wires with preferably <111> orientation [165]. A comparison of resistance at room temperature for platinum NWs obtained by different authors [166] shows significant differences, reaching from 61 to 5,000 µΩ cm resistivity. The need of standardized measurement procedures is evident. A stage design with fixed contacts has been demonstrated recently to enhance reliability and throughput with preformed resist trenches. Wires longer than 60 nm have been arranged in appropriate positions [167]. A NW manipulation technique for four-point measurements with individually moved tips has been presented before [168]. Obtained data are detrimental in understanding and design of future devices, and are practically important already to develop sensors, based on analyte-depending properties of semiconductor materials [23, 24]. Conductivity changes of neat gold NWs have been used for ionic mercury detection [169]. To our knowledge, a systematic study of individual NWs, especially single crystalline wires, i.e., grains along different crystal axes is still lacking. However, it should be important for reliability issues in nanoelectronics (Chaps. 2–4) and future sensor developments.

20.10 Manipulation of Nanowires

For applications of NWs as admixtures in a composite, e.g., to manage electrostatic or magnetic properties or the conductivity of a polymer, they certainly do not have to be manipulated individually. However, if NWs have to be placed one by one and specifically to ensure the envisioned function, appropriate touchless handling techniques are necessary. The contact-free manipulation of sensitive objects such as metal NWs requires the combination of different physical principles, allowing directed transport, orientation, and fixation. Similar approaches have been developed successfully for single cell manipulation and measurement earlier [170, 171]. Main physical principles for contact-free handling of delicate individual components are magnetophoresis, electrophoresis, dielectrophoresis, capillary forces, and interaction with focused laser light (laser tweezers), often in combination with hydrodynamic streaming. For a review, see [172]. Dielectrophoretic field cages inside fluidic channel structures are especially suited to sort and align different types of micro- and nano- objects individually [173] as has been well proven for living animal cells; for review, see [174]. Positive dielectrophoresis has been used to assemble gold wires from 70 to 350 nm in diameter in electrode gaps [175–177]. Magnetic fields have been applied to arrange and to assemble large quantities of Ni NWs in parallel or, with a bit more luck, even individually [178]. Combinations with independent solution structuring, e.g., by a liquid crystal [179] can improve the alignment and ordering of large numbers of particles, important for collective interaction phenomena of NWs with external electromagnetic fields. Surfactants can be used to build assemblies of NWs [180]. Since gold can easily be modified with thiolates, gold-capped NWs can be labeled for self-assembly [181]. But placement alone does not yet yield the trick of contacting NWs.

20.11 Nanowires: Bonding and Joining

It seems as if NWs will require specialized bonding or joining techniques to be contacted reliably. Surprisingly, ultrasonic bonding in an almost conventional manner has been demonstrated to produce low contact resistance between 1-μm long CNT and metal electrodes, reaching 8–24 kΩ [182]. Nevertheless, it is obvious that practical applications of NWs in electronic packaging would require some paradigm shift, new approaches, and technical solutions. But metal NWs arrived already at the level of practical applications: interconnect formation has been demonstrated with 30-nm nickel NWs, generated with AAO and assembled magnetically between neighboring contact pads. The resistance of the formed contacts after annealing under reducing conditions decreased from >10 MΩ to ~800 Ω [183]. The use of vertically aligned NWs in an adhesive interposer foil has been envisioned [184] and realized [185] independently. We could show a new technical principle of decorating flip-chip bonding pads directly with NWs, omitting external force fields for their arrangement, to be successful for joining of gold nanolawn model structures [66] (Fig. 20.4).

Deleterious effects of recrystallization phenomena (see later), crucial for single wire contacts, can obviously be overcome and can even be used if applied with intercalated NW ensembles. The observed contact formation is based on Ostwald ripening,

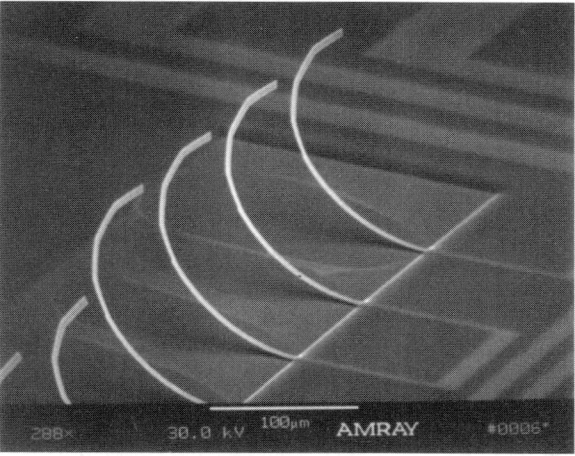

Fig. 20.4 FIB dissections of intercalated wire decorated surfaces (all gold), using flip-chip bonding principle. Bonding at ambient temperature with pressure as indicated. (**a**) Close-up for 100-MPa sample, (**b**) intergrain diffusion and contact formation after annealing. All scale bars: 1 μm

as indicated by FIB/SEM dissections (Figs. 20.3 and 20.4). Solid phase diffusion bonding, well established in microelectronics, hence can supplement the contact formation techniques described earlier, even without large pressures to be applied.

However, intermetallic diffusion and corrosion have to be considered as to be crucial for interconnects between individual NWs. Four different bonding techniques for individual single 100-nm gold NWs have been compared regarding their practical feasibility and efficiency [186], still leaving the need for further research.

Concerning the wire-to-wire bonding, gold- or Au–Ni–Au wires, have been joined together and soldered under reflow. Resistance measurement (lithographically patterned contacts) indicated solder contact formation for 200-nm diameter wires. The resistance of individual contacts was reduced to about 13 Ω, starting from 300 to 106 Ω before reflow [187].

If the NW template can be generated directly on surfaces to be decorated with NWs, the previous discussion on manipulation and joining to connecting electrodes becomes obsolete. A technique to generate nanoporous AAO spots directly on wafer substrates has been presented recently. Copper NWs have been generated in those AAO pads and proposed for contact formation with solder pads in a flip-chip bonding approach [188, 189]. Doubtless, microinterconnects by multiple metallic NWs will be useful for fine-pitch applications.

20.12 Nanowire Interaction with Electromagnetic Fields

Because of the high aspect ratio of NWs resonance frequency shifts can be observed for structures with dimensions below one-fourth of the wavelength of the external field. Nonlinear interaction of NWs with light- or high-frequency electromagnetic fields, useful for generating active and passive photonic structures [190] or HF structures [191] has been described previously, e.g., for enhancing the sensitivity of Raman scattering over several magnitudes in chemical analysis (SERS) [192, 193]. The possibility to use NWs to produce stop-band filters or antenna structures [194–196] should be mentioned as well. NWs could be used in new integrated optical devices because of their negative magnetic permeability and dielectric permittivity in the visible and the near-infrared if arranged into parallel pairs even randomly in a matrix [197]. Resonance phenomena of NWs in the infrared have been discussed in depth and investigated experimentally on copper and gold NWs of 100-nm diameter [198]. Significant antenna-like plasmon resonances and skin effects have to be considered for practical applications in light confinement or estimation of effective refractive indices in the antennas surrounding. For a review of general near-field optical properties of nanostructures, see [199].

The optical parameters of nanorod dispersions are changing according to the degree of orientation in the dispersion. The degree of orientation can be externally influenced by electric fields [200]. An observed change of phase shift as well as selective absorption could be useful in preparation of shielding materials to prevent electromagnetic interference (EMI). Polymer composites of conductive CNTs [201] and of copper wires in flex have been proposed [202]. The application of carbon nanotube NWs and nanotube antenna arrays, consisting of individual wires of different lengths and hence different resonance frequencies has been proposed for in/out signal demultiplexing and wireless interconnection of integrated nanosystems to the microelectronic periphery. The authors point out that this approach could translate interconnection challenges from the spatial to the frequency domain, consequently eliminating a technological bottleneck [203].

Size-dependent magnetoresistance measurements of single Ni-NWs (30 and 200-nm diameter) bridging Ni electrodes have been studied at 10–300 K [204]. With decreasing diameter, transverse and longitudinal magnetoresistance differ from values obtained for thin layers. Perpendicular giant magnetic resistance (GMR) studies on layered Co/Cu nanowire arrays have been undertaken, e.g., by [205]. The results obtained indicate the attractive possibilities of NW arrays for new memory device architectures.

20.13 Future Prospects

Because of the known field inhomogeneities at electrode edges and spherical vs. parallel diffusion at ultra-microelectrodes, ensembles of NWs, depicted as nanoelectrode ensembles, have been evaluated for applications in analytical chemistry [206], and for applications in wet chemistry and gas analysis.

Biophysics and cell biology are other exciting application fields for NW-decorated surfaces, since nanotopographies are known to trigger cell differentiation or even phagocytosis [207]. Traction forces generated by moving animal cells have been measured with microneedles [208], and could be accessible even more accurately on NW arrays.

In cooperation with cell biologists, we started to study the behavior of mammalian cells on gold nanolawn [209], especially the influence on cell substrate adhesion. For

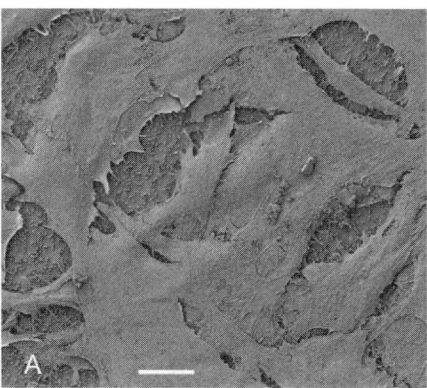 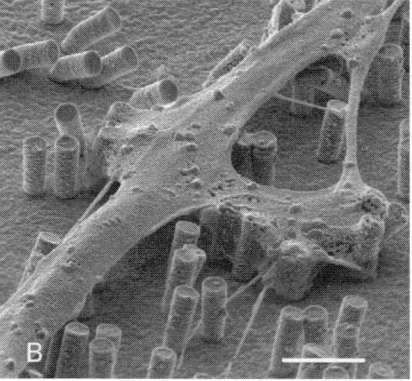

Fig. 20.5 Primary mouse astroglia cells growing on a nanolawn (Pt). (**a**) SEM image of cell-typical spreading toward confluency (bar: 20 μm), (**b**) Close-up of cell-substrate contacts (bar: 2 μm). Cell culture and sample preparation by U. Gimsa and L. Jonas, University of Rostock, Germany

primary mouse astroglial cells very close cell–cell and cell–substrate contacts have been observed [80] (Fig. 20.5). Considering the growing importance of interfaces of microelectronic systems with living tissue [210], those contacts can be interesting for neuroprosthetic device development – one more area of rising importance for microelectronic packaging.

20.14 Conclusion

NWs can be regarded as promising materials to supplement existing packaging applications such as flip-chip bonding, anisotropically adhesive foils, and electromagnetic shielding. Moreover, they open up new vistas for future applications.

The properties of a single NW will be defined by the grains or even by the single grains it consists of. Therefore, their/its orientation and existing grain boundaries have to be taken into account for practical application and design.

The reliability of future nanostructure-based components will depend on the knowledge of the thermomechanical behavior of single metal grains. To obtain the necessary data, NWs are well-suited objects to study.

The characterization and measurement of typical electrical properties such as electrical conductivity or resistance and thermomechanical properties require reliable techniques for electrical connection and mechanical clamping.

The sorting, manipulation, and placement, i.e., assembly of individual NWs demands touchless and contact-free handling to avoid their damage, contamination, or destruction. External electromagnetic fields are well suited to generate necessary forces for doing so.

Because of the high aspect ratio, i.e., the anisotropy of NW properties interacting with electromagnetic fields, NWs and NW arrays offer applications in signal processing and storage.

The collective behavior of NWs in regular arrays or statistically arranged brushes and nanolawn opens up new applications in photonics, EMF shielding, sensorics, and biomedicine.

To summarize, focused research with NWs can pave the way for low-energy bonding techniques, high-density interconnects, high-density (spin) data storage, molecular electronics, biomedical electronics, and life sciences – what promising prospects for microelectronic packaging!

References

1. Collins PG, Hersam M, Arnold M, Martel R, Avouris P (2001) Current saturation and electrical breakdown in multiwalled carbon nanotubes. Phys Rev Lett 86:3128–3131
2. Muller CJ, van Ruitenbeek JM, de Jongh LJ (1992) Conductance and supercurrent discontinuities in atomic-scale metallic constrictions of variable width. Phys Rev Lett 69:140–143
3. Li CZ, Bogozi A, Huang W, Tao NJ (1999) Fabrication of stable metallic nanowires with quantized conductance. Nanotechnology 10:221–223

4. Landman U, Barnett RN, Scherbakov AG, Avouris P (2000) Metal-semiconductor nanocontacts: Silicon nanowires. Phys Rev Lett 85:1958–1961
5. Grüter L (2005) Mechanical controllable break junction in liquid environment: A tool to measure single molecules. Inauguraldissertation Universität Basel, Basel
6. Okawa Y, Aono M (2001) Linear chain polymerization initiated by a scanning tunneling microscope tip at designated positions. J Chem Phys 115:2317–2322
7. Gu J, Li D, Lederman D, Timperman A (2005) Self-assembly of fibrous proteins for molecular electronics. American Physics Society Meeting, March abstract W35.8. http://meetings.aps.org/link/BAPS.2005.MAR.W35.8
8. Gu Q, Cheng C, Gonela R, Suryanarayanan S, Anabathula S, Dai K, Haynie DT (2006) DNA nanotechnology. Nanotechnology 17:R14–R25
9. Wang Z, Medforth CJ, Shelnutt JA (2004) Self-metallization of photocatalytic porphyrin nanotubes. J Am Chem Soc 126:16720–16721
10. Tans SJ, Devoret MH, Dai H, Thess A, Smalley RE, Geerligs LJ, Dekker C (1997) Individual single-wall nanotubes as quantum wires. Nature 386:474–477
11. Bachtold A, Hadley P, Nakanishi T, Dekker C (2001) Circuits with carbon nanotube transistors. Science 294:1317–1320
12. Javey A, Qi P, Wang Q, Dai H (2004) Ten- to 50-nm-long quasi-ballistic carbon nanotube devices obtained without complex lithography. Proc Natl Acad Sci 101:13408–13410
13. Li J, Ye Q, Cassell A, Ng HT, Stevens R, Han J, Meyyappan M (2003) Bottom–up approach for carbon nanotube interconnects. Appl Phys Lett 82:2491–2493
14. Hoenlein W, Kreupl F, Duesberg GS, Graham AP, Liebau M, Seidel RV, Unger E (2004) Carbon nanotube applications in microelectronics. IEEE Trans Compon Packag Technol 27:629–634
15. Kreupl F, Graham AP, Duesberg GS, Steinhögl W, Liebau M, Unger E, Hönlein W (2002) Carbon nanotubes in interconnect applications. Microelectron Eng 64:399–408
16. Cui Y, Lieber CM (2001) Functional nanoscale electronic devices assembled using silicon nanowire building blocks. Science 291:851–853
17. Huang Y, Duan X, Cui Y, Lauhon LJ, Kim K-H, Lieber CM (2001) Logic gates and computation from assembled nanowire building blocks. Science 294:1313–1317
18. Melosh NA, Boukai A, Diana F, Gerardot B, Badolato A, Petroff PM, Heath JR (2003) Ultrahigh-density nanowire lattices and circuits. Science 300:112–115
19. Thelander C, Agarwal P, Brongersma S, Eymery J, Feiner LF, Forchel A, Scheffler M, Riess W, Ohlsson BJ, Gösele U, Samuelson L (2006) Nanowire-based one-dimensional electronics. Mater Today 9:28–35
20. Li Y, Qian F, Xiang J, Lieber CM (2006) Nanowire electronic and optoelectronic devices. Mater Today 9:18–27
21. Law M, Goldberger J, Yang P (2004) Semiconductor nanowires and nanotubes. Ann Rev Mater Res 34:83–122
22. Lu W, Lieber CM (2006) Semiconductor nanowires. J Phys D Appl Phys 39:R387–R406
23. Shankar KS, Raychaudhuri AK (2005) Fabrication of nanowires of multicomponent oxides: Review of recent advances. Mater Sci Eng C25:738–751
24. Lieber CM (1998) One-dimensional nanostructures: Chemistry, physics and applications. Solid State Commun 107:607–616
25. International Technology Roadmap for Semiconductors 2001. Semiconductor Industry Association, San Jose, CA. www.itrs.net
26. Xia YN, Yang PD, Sun YG, Wu YY, Mayers B, Gates B, Yin YD, Kim F, Yan YQ (2003) One-dimensional nanostructures: Chemistry, physics and applications. Adv Mater 15:353–389
27. Kovtyukhova NI, Mallouk TE (2002) Nanowires as building blocks for self-assembling logic and memory circuits. Chem Eur J 8:4354–4363
28. Wang ZL (Ed.) (2006) Nanowires and nanobelts: Materials, properties and devices. I. Metal and semiconductor nanowires. 1(st ed., 2003)Springer, Berlin
29. Rao CNR, Govindaraj A (2005) Nanotubes and nanowires. Series RSC Nanoscience and Nanotechnology Series. Royal Society of Chemistry, London

30. Banerjee S, Dan A, Chakravorty D (2002) Synthesis of conducting nanowires. J Mater Sci 37:4261–4271
31. Wanekaya AK, Chen W, Myung NV, Mulchandani A (2006) Nanowire-based electrochemical biosensors. Electroanalysis 18:533–550
32. He H, Tao NJ (2003) Electrochemical fabrication of metal nanowires. In: Encyclopedia of nanoscience and nanotechnology, Nalwa HS (Ed.). American Scientific Publishers, Stevenson Ranch, CA, 2:755–772
33. Natelson D (2002) Fabrication of metal nanowires. In: Recent research developments in vacuum science and technology, Dabrowski J (Ed.). Research Signpost, Kerala, Chapter 9
34. Kline TR, Tian M, Wang J, Sen A, Chan MWH, Mallouk TE (2006) Template-grown metal nanowires. Inorg Chem 45:7555–7565
35. Dresselhaus MS, Lin Y-M, Rabin O, Black MR, Dresselhaus G (2004) Nanowires. In: Springer handbook of nanotechnology, Bhushan B (Ed.). Springer, Berlin, 99–145
36. Taczak MD, Rolfe B (2005) Nanowire research and development: A survey of selected research from 2001 to 2005. MITRE McLean, VA (MP 05W0000171). www.mitre.org/work/tech_papers/tech_papers_05/05_1223/05_1223.pdf
37. Possin GE (1970) A method for forming very small diameter wires. Rev Sci Instrum 41:772–774
38. Fleischer RL, Price PB, Walker RM (1965) Tracks of charged particles in solids. Science 149:383–393
39. Petrov AV, Fink D, Richter G, Szimkowiak P, Chemseddine A, Alegaonkar PS, Berdinsky AS, Chadderton LT, Fahrner WR (2003) Creation of nanoscale electronic devices by the swift heavy ion technology. In: Proceedings of Fourth Siberian Russian Workshop and Tutorials EDM'2003, Section I, 1–4 July, Erlagol, 40–45
40. Fraflex®, distributed by FRACTAL AG Staßfurt (Germany)
41. Hulteen JC, Martin CR (1997) A general template-based method for the preparation of nanomaterials. J Mater Chem 7:1075–1087
42. Martin JI, Nogues J, Liu K, Vicent JL, Schuller IK (2003) Ordered magnetic nanostructures: Fabrication and properties. J Magn Magn Mater 256:449–501
43. Granström M, Carlberg JC, Inganäs O (1995) Electrically conductive polymer fibres with mesoscopic diameters. II. Studies of polymerization behaviour. Polymer 36:3191–3196
44. Kumar S, Zagorski DL, Kumar S, Chakarvarti SK (2004) Chemical synthesis of polypyrrole nano/microstructures using track etch membranes. J Mater Sci 39:6137–6139
45. Despic A, Parkhutik VP (1989) Electrochemistry of aluminum in aqueous solutions and physics of its anodic oxide. In: Modern aspects of electrochemistry, Bockris JO'M, Conway BE, White RE (Eds.). Plenum, New York, 20:401–503
46. Sauer G, Brehm G, Schneider S, Nielsch K, Wehrspohn RB, Choi J, Hofmeister J, Gösele U (2002) Highly ordered monocrystalline silver nanowire arrays. J Appl Phys 91:3243–3247
47. Moon J-M, Wei A (2005) Uniform gold nanorod arrays from polyethylenimine-coated alumina templates. J Phys Chem B 109:23336–23341
48. Rabin O, Herz PR, Lin Y-M, Akinwande AI, Cronin SB, Dresselhaus MS (2003) Formation of thick porous anodic alumina films and nanowire arrays on silicon wafers and glass. Adv Funct Mater 13:631–638
49. Zhang Z, Lai C, Xu N, Ren S, Ma B, Zhang Z, Jin Q (2007) Novel nanostructured metallic nanorod arrays with multibranched root tails. Nanotechnology 18:1–6
50. Gelves GA, Murakami ZTM, Krantz MJ, Haber JA (2006) Multigram synthesis of copper nanowires using ac electrodeposition into porous aluminium oxide templates. J Mater Chem 16:3075–3083
51. Wu B, Boland JJ (2006) Synthesis and dispersion of isolated high aspect ratio gold nanowires. J Colloid Interface Sci 303:611–616
52. Yin AJ, Li J, Jian W, Bennett AJ, Xu JM (2001) Fabrication of highly ordered metallic nanowire arrays by electrodeposition. Appl Phys Lett 79:1039–1041
53. Karthaus J (2000) Galvanische Abscheidung von Metallen aus nichtwäßrigen Elektrolyten für die Mikrosystemtechnik. Dissertation, Universität Karlsruhe, Karlsruhe, Germany

54. Martin CR (1994) Nanomaterials: A membrane-based synthetic approach. Science 266:1961–1966
55. Apel PYu, Schulz A, Spohr R, Trautmann C, Vutsadakis V (1997) Tracks of very heavy ions in polymers. Nucl Instrum Methods Phys Res B 130:55–63; Compare: Apel P, Spohr R. Introduction to ion track etching in polymers. http://www.iontracktechnology.de
56. Ferain E, Legras R (2003) Track-etched templates designed for micro- and nanofabrication. Nucl Instrum Methods Phys Res B 208:115–122
57. Ferain E, Legras R (2001) Pore shape control in nanoporous particle track etched membrane. Nucl Instrum Methods Phys Res 174:116–122
58. Spohr R (1990) Ion tracks and microtechnology. Principles and applications. Vieweg, Braunschweig
59. Apel PYu, Korchev YuE, Siwy Z, Spohr R, Yoshida M (2001) Diode-like single-ion track membrane prepared by electro-stopping. Nucl Instrum Methods B 184:337–346
60. Toimil-Molares ME, Chtanko N, Cornelius TW, Dobrev D, Enculescu I, Blick RH, Neumann R (2004) Fabrication and contacting of single Bi nanowires. Nanotechnology 15:S201–S207
61. Chittrakarn T, Bhongsuwan T, Wanichapichart P, Nuanuin P, Chongkum S, Khonduangkaew A, Bordeepong S (2002) Nuclear track-etched pore membrane production using neutrons from the Thai research reactor TRR-1/M1. Songklanakarin J Sci Technol 24 (Suppl): 863–870
62. Lindeberg M (2003) High aspect ratio microsystem fabrication by ion track lithography. PhD Dissertation, Uppshala Universitet; Lindeberg M, Jaccard Y, Ansermet JP, Hjort K (2001) In: TRANSDUCERS'01, EUROSENSORS XV 2001, Munich, Germany (Springer), 172–175
63. Shorin VS (2002) Analytical description of the hole overlapping process on the nuclear-track membrane surface. High Energy Chem 36:294–299
64. Masuda H, Yamada H, Satoh M, Asoh H, Nakao M, Tamamura T (1997) Highly ordered nanochannel-array architecture in anodic alumina. Appl Phys Lett 71:2770–2772
65. Lee W, Ji R, Ross CA, Gösele U, Nielsch K (2006) Wafer-scale Ni imprint stamps for porous alumina membranes based on interference lithography. Small 2:978–982
66 Fiedler S, Zwanzig M, Schmidt R, Auerswald E, Klein M, Scheel W, Reichl H (2006) Evaluation of metallic nano-lawn structures for application in microelectronic packaging. In: First Electronics System Integration Technology Conference, Dresden, September, 2:886–891
67. Hong BH, Bae SC, Lee C-W, Jeong S, Kim KS (2001) Ultrathin single-crystalline silver nanowire arrays formed in an ambient solution phase. Science 294:348–351
68. Schüth F (2003) Endo- and exotemplating to create high-surface-area inorganic materials. Angew Chem Intl Ed 42:3604–3622
69. Braun E, Eichen Y, Sivan U, Ben-Yoseph G (1998) DNA-templated assembly and electrode attachment of a conducting silver wire. Nature 391:775–778
70. Richter J, Mertig M, Pompe W, Mönch I, Schackert HK (2001) The construction of highly conductive nanowires on a DNA template. Appl Phys Lett 78:536–538
71. Richter J, Seidel R, Kirsch R, Mertig M, Pompe W, Plaschke J, Schackert HK (2000) Nanoscale palladium metallization of DNA. Adv Mater 12:507–510
72. Gu Q, Cheng C, Gonela R, Suryanarayanan S, Anabathula S, Dai K, Haynie DT (2006) DNA nanowire fabrication. Nanotechnology 17:R14–R25
73. Ford WE, Harnack O, Yasuda A, Wessels JM (2001) DNA nanowire fabrication. Adv Mater 13:1793–1797
74. Scheibel T, Parthasarathy R, Sawicki G, Lin X-M, Jaeger H, Lindquist SL (2003) Conducting nanowires built by controlled self-assembly of amyloid fibers and selective metal deposition. Proc Natl Acad Sci 100:4527–4532
75. Maubach G, Csáki A, Seidel R, Mertig M, Pompe W, Born D, Fritzsche W (2003) Controlled positioning of a DNA molecule in an electrode setup based on self-assembly and microstructuring. Nanotechnology 14:1055–1056
76. Liu Y, Meyer-Zaika W, Franzka S, Schmid G, Tsoli M, Kuhn H (2003) Gold-cluster degradation by the transition of B-DNA into A-DNA and the formation of nanowires. Angew Chem Int Ed 42:2853–2857

77. Nam KT, Kim DW, Yoo PJ, Chiang CY, Meethong N, Hammond PT, Chiang YM, Belcher AM (2006) Virus-enabled synthesis and assembly of nanowires for lithium ion battery electrodes. Science 312:885–888
78. Patolsky F, Weizmann Y, Willner I (2004) Actin-based metallic nanowires as bio-nanotransporters. Nat Mater 3:692–695
79. Hornung J, Müller T, Fuhr G (1996) Cryopreservation of anchorage-dependent mammalian cells fixed to structured glass and silicon substrates. Cryobiology 33:260–270
80. Gimsa U, Igli A, Fiedler S, Zwanzig M, Kralj-Igli V, Jonas L, Gimsa J (2007) Actin is not required for nanotubular protrusions of primary astrocytes grown on metal nano-lawn. Mol Membr Biol 24:243–255
81. Evans E, Bowman H, Leung A, Needham D, Tirrel D (1996) Biomembrane templates for nanoscale conduits and networks. Science 273:933–935
82. Karlsson A, Karlsson R, Karlsson M, Cans A-S, Strömberg A, Ryttsén F, Orwar O (2001) Networks of nanotubes and containers. Nature 409:150–152
83. Lobovkina T, Dommersnes P, Joanny J-F, Bassereau P, Karlsson M, Orwar O (2004) Mechanical tweezer action by self-tightening knots in surfactant nanotubes. Proc Natl Acad Sci 101:7949–7953
84. Schnur JM (1993) Lipid tubules: A paradigm for molecularly engineered structures. Science 262:1669–1676
85. Yang Y, Constance BH, Deymier PA, Hoying J, Raghavan S, Zelinski BJJ (2004) Electroless metal plating of microtubules: Effect of microtubule-associated proteins. J Mater Sci 39:1927–1933
86. Ng HT, Li J, Smith MK, Nguyen P, Cassell A, Han J, Meyyappan M (2003) Growth of epitaxial nanowires at the junctions of nanowalls. Science 300:1249
87. Vasilev K, Zhu T, Wilms M, Gillies G, Lieberwirth I, Mittler S, Knoll W, Kreiter M (2005) Simple, one-step synthesis of gold nanowires in aqueous solution. Langmuir 21:12399–12403
88. Wei Z, Mieszawska AJ, Zamborini FP (2004) Synthesis and manipulation of high aspect ratio gold nanorods grown directly on surfaces. Langmuir 20:4322–4326
89. Jana NR, Gearheart L, Murphy CJ (2001) Wet chemical synthesis of high aspect ratio cylindrical gold nanorods. J Phys Chem B 105:4065–4067
90. Chen C, Wang L, Jiang G, Zhou J, Chen X, Yu H, Yang Q (2006) Study on the synthesis of silver nanowires with adjustable diameters through the polyol process. Nanotechnology 17:3933–3938
91. Huang L, Wang H, Wang Z, Mitra A, Yan Y (2001) Silver nanowires eelectrodeposited from reverse hexagonal liquid crystals. Mater Res Soc Symp Proc 676:Y3321–Y3326
92. Hamley IW (2003) Nanostructure fabrication using block copolymers. Nanotechnology 14:R39–R54
93. Goolaup S, Singh N, Adeyeye NO (2005) Coercivity variation in Ni80Fe20 ferromagnetic nanowires. IEEE Trans Nanotechnol 4:523–526
94. Gubbiotti G, Tacchi S, Carlotti G, Vavassori P, Singh N, Goolaup S, Adeyeye AO, Stashkevich A, Kostylev M (2005) Magnetostatic interaction in arrays of nanometric permalloy wires: A magneto-optic Kerr effect and a Brillouin light scattering study. Phys Rev B 72:224413–224420
95. Cheung CK, Nikolic RJ, Reinhardt CE, Wang TF (2006) Fabrication of nanopillars by nanosphere lithography. Nanotechnology 17:1339–1343
96. Martensson T, Carlberg P, Borgström M, Montelius L, Seifert W, Samuelson L (2004) Nanowire arrays defined by nanoimprint lithography. Nano Lett 4:699–702
97. Tikhodeev SG, Gippius NA, Christ A, Zentgraf T, Kuhl J, Giessen H (2005) Waveguide-plasmon polaritons in photonic crystal slabs with metal nanowires. Phys Status Solidi C 2:795–800
98. Ozbay E (2006) Plasmonics: Merging photonics and electronics at nanoscale dimensions. Science 311:189–193
99. Menke EJ, Thompson MA, Xiang C, Yang LC, Penner RM (2006) Lithographically patterned nanowire electrodeposition. Nat Mater 5:914–919
100. Prober DE, Feuer MD, Giordano N (1980) Fabrication of 300-Angstrom metal lines with substrate-step techniques. Appl Phys Lett 37:94–96

101. Zach MP, Ng KH, Penner RM (2000) Molybdenum nanowires by electrodeposition. Science 290:2120–2123
102. Favier F, Walter EC, Zach MP, Benter T, Penner RM (2001) Hydrogen sensors and switches from electrodeposited palladium mesowire arrays. Science 293:2227–2231
103. Adelung R, Aktas OC, Franc J, Biswas A, Kunz R, Elbahri M, Kanzow J, Schürmann U, Faupel F (2004) Strain-controlled growth of nanowires within thin-film cracks. Nat Mater 3:375–379
104. Mützel M, Müller M, Haubrich D, Rasbach U, Meschede D, O'Dwyer C, Gay G, Viaris de Lesegno B, Weiner J, Ludolph K, Georgiev G, Oesterschulze E (2005) The atom pencil: Serial writing in the sub-micrometre domain. Appl Phys B 80:941–944
105. Hochleitner G, Fischer M, Wanzenboeck H, Heerb R, Brueckl H, Bertagnolli E (2006) Electron beam-induced direct-deposition of magnetic nanostructures. Proceedings of Micro- and Nano-Engineering MNE06, 17–20 September, Barcelona, Spain, 165–166
106. Piner RD, Zhu J, Xu F, Hong S, Mirkin CA (1999) "Dip-pen" nanolithography. Science 283:661–663
107. Yang D-Q, Sacher E (2007) Accurate assembly and size control of Cu nanoparticles into nanowires by contact atomic force microscope-based nanopositioning. J Phys Chem C 111:10105–10109
108. Winograd GI, Han L, McCord MA, Pease RFW, Krishnamurthi V (1998) Multiplexed blanker array for parallel electron beam lithography. J Vac Sci Technol B 16:3174–3176
109. Bullen D, Chung S-W, Wang X, Zou J, Mirkin CA, Liu C (2004) Parallel dip-pen nanolithography with arrays of individually addressable cantilevers. Appl Phys Lett 84:789–791
110. Guo W, Wang ZB, Li L, Whitehead DJ, Luk'yanchuk BS, Liu Z (2007) Near-field laser parallel nanofabrication of arbitrary-shaped patterns. Appl Phys Lett 90:243101
111. Moormann C, Bolten J, Kurz H (2004) Spatial phase-locked combination lithography for photonic crystal devices. Microelectron Eng 73–74:417–422
112. Kretschmer R, Fritzsche W (2004) Pearl chain formation of nanoparticles in microelectrode gaps by dielectrophoresis. Langmuir 20:11797–11801
113. Bhatt KH, Velev OD (2004) Control and modeling of the dielectrophoretic assembly of on-chip nanoparticle wires. Langmuir 20:467–476
114. Guan L, Shi Z, Li H, Youb L, Gu Z (2004) Super-long continuous Ni nanowires encapsulated in carbon nanotubes. Chem Commun 1988–1989
115. Schuchert IU, Toimil-Molares ME, Dobrev D, Vetter J, Neumann R, Martin M (2003) Electrochemical copper deposition in etched ion track membranes. Experimental results and a qualitative kinetic model. J Electrochem Soc 150:C189–C194
116. Kazeminezhad I, Barnes AC, Holbrey JD, Seddon KR, Schwarzacher W (2007) Templated electrodeposition of silver nanowires in a nanoporous polycarbonate membrane from a non-aqueous ionic liquid electrolyte. Appl Phys A 86:373–375
117. Jana NR, Gearheart L, Murphy CJ (2001) Wet chemical synthesis of high aspect ratio cylindrical gold nanorods. J Phys Chem B 105:4065–4067
118. Chtanko N, Toimil-Molares ME, Cornelius TW, Dobrev D, Neumann R (2005) Ion-track based single-channel templates for single-nanowire contacting. Nucl Instrum Methods Phys Res B 236:103–108
119. Gelves GA, Murakami ZTM, Krantz MJ, Haber JA (2006) Multigram synthesis of copper nanowires using ac electrodeposition into porous aluminium oxide templates. J Mater Chem 16:3075–3083
120. Fert A, Piraux L (1999) Magnetic nanowires. J Magn Magn Mater 200:338–358
121. Whitney TM, Jiang JS, Searson PC, Chien CL (1993) Fabrication and magnetic properties of arrays of metallic nanowires. Science 261:1316–1319
122. Zabala N, Puska MJ, Nieminen RM (1998) Spontaneous magnetization of simple metal nanowires. Phys Rev Lett 80:3336–3339
123. Krauss PR, Fischer PB, Chou SY (1994) Fabrication of single-domain magnetic pillar array of 35 nm diameter and 65 Gbits/in2 density. J Vac Sci Technol B 12:3639–3642
124. Tehrani S, Chen E, Durlam M, DeHerrera M, Slaughter JM, Shi J, Kerszykowski G (1999) High density submicron magnetoresistive random access memory. J Appl Phys 85:5822–5827

125. Denver H, Hong J, Borca-Tasciuc DA (2007) Fabrication and characterization of nickel nanowire polymer composites. In: Nanowires and carbon nanotubes – Science and applications, Bandaru P, Endo M, Kinloch IA, Rao AM (Eds.). Mater Res Soc Symp Proc 963E, Warrendale, PA
126. RenJen L, Yung YuH, YuChih C, SyhYuh C, Uang R-H (2005) Fabrication of nanowire anisotropic conductive film for ultrafine pitch flip chip interconnection. IEEE Electron Compon Technol Conf 2005 Proc 55:66–70
127. Nicewarner-Peña SR, Freeman RG, Reiss BD, He L, Peña DJ, Walton ID, Cromer R, Keating CD, Natan MJ (2001) Submicrometer metallic barcodes. Science 294:137–141
128. Mock JJ, Oldenburg SJ, Smith DR, Schultz DA, Schultz S (2002) Composite plasmon resonant nanowires. Nano Lett 2:465–469
129. Lehmann V (2002) Barcoded molecules. Nat Mater 1:12–13
130. Keating CD, Natan MJ (2003) Striped metal nanowires as building blocks and optical tags. Adv Mater 15:451–454
131. Sioss JA, Keating CD (2005) Batch preparation of linear Au and Ag nanoparticle chains via wet chemistry. Nano Lett 5:1779–1783
132. Qin L, Park S, Huang L, Mirkin CA (2005) On-wire lithography. Science 309:113–115
133. Yan X-M, Kwon S, Contreras AM, Koebel MM, Bokor J, Somorjai G (2005) Fabrication of dense arrays of platinum nanowires on silica, alumina, zirconia and ceria surfaces as 2-D model catalysts. Catal Lett 105:127–132
134. Contreras AM, Yan X-M, Kwon S, Bokor J, Somorjai GA (2006) Catalytic CO oxidation reaction studies on lithographically fabricated platinum nanowire arrays with different oxide supports. Catal Lett 111:5–13
135. Dobrev D, Vetter J, Angert N, Neumann R (2000) Periodic reverse current electrodeposition of gold in an ultrasonic field using ion-track membranes as templates: Growth of gold single crystals. Electrochim Acta 45:3117–3125
136. Moon J-C, Wei A (2005) Uniform gold nanorod arrays from polyethylenimine-coated alumina templates. J Phys Chem B 109:23336–23341
137. Toimil Molares ME, Buschmann V, Dobrev D, Neumann R, Scholz R, Schuchert IU, Vetter J (2001) Single-crystalline copper nanowires produced by electrochemical deposition in polymeric ion track membranes. Adv Mater 13:62–65
138. Wang J, Tian M, Mallouk TE, Chan MHW (2004) Microtwinning in template synthesized single crystal metal nanowires. J Phys Chem B 108:841–845
139. Bietsch A, Michel B (2002) Size and grain-boundary effects of a gold nanowire measured by conducting atomic force microscopy. Appl Phys Lett 80:3346–3348
140. Zhu Y, Ke C, Espinosa HD (2007) Experimental techniques for the mechanical characterization of one-dimensional nanostructures. Exp Mech 47:7–24
141. Landman U (1998) On nanotribological interactions: Hard and soft interfacial junctions. Solid State Commun 107:693–708
142. Johansson J, Karlsson LS, Svensson CPT, Martensson T, Wacaser BA, Deppert K, Samuelson L, Seifert W (2006) Structural properties of <111> B-oriented III–V nanowires. Nat Mater 5:575–580
143. Sabate N, Vogel D, Gollhardt A, Keller J, Michel B, Cane C, Gracia I, Morante JR (2006) Measurement of residual stresses in micromachined structures in a microregion. Appl Phys Lett 88:071910.1–071910.3
144. Wunderle B, Mrossko R, Kaulfersch E, Wittler O, Ramm P, Michel B, Reichl H (2006) Thermo-mechanical reliability of 3D-integrated structures in stacked silicon. Proceedings of MRS Fall Meeting, November 2006, Boston, MA, 0970-Y02-04
145. Kraft O, Freund LB, Phillips R, Arzt E (2002) Dislocation plasticity in thin metal films. MRS Bull 27:30–37
146. Wu B, Heidelberg A, Boland JJ (2005) Mechanical properties of ultrahigh-strength gold nanowires. Nat Mater 4:525–529
147. Uchic MD, Dimiduk DM, Florando JN, Nix WD (2004) Sample dimensions influence strength and crystal plasticity. Science 305:986–989

148. Gilman JJ, Uchic MD, Dimiduk DM, Florando JN, Nix WD (2004) Oxide surface films on metal crystals. Science 306:1134–1135
149. Uchic MD, Dimiduk DM, Wheeler R, Shade PA, Fraser HL (2006) Application of microsample testing to study fundamental aspects of plastic flow. Scripta Mater 54:759–764
150. Huber CA, Huber TA, Sadoqi M, Lubin JA, Manalis S, Prater CB (1994) Nanowire array composites. Science 263:800–802
151. Buffat P, Borel J-P (1976) Size effect on the melting temperature of gold particles. Phys Rev A 13:2287–2298
152. Toimil-Molares ME, Balogh AG, Cornelius TW, Neumann R, Trautmann C (2004) Fragmentation of nanowires driven by Rayleigh instability. Appl Phys Lett 85:5337–5339
153. Jakobsen B, Poulsen HF, Lienert U, Almer J, Shastri SD, Sørensen HO, Gundlach C, Pantleon W (2006) Formation and subdivision of deformation structures during plastic deformation. Science 312:889–892
154. US Patent 020060057354A1, March 16, 2006
155. US Patent 000006359288B1, March 19, 2002
156. Kovtyukhova NA, Martin BR, Mbindyo JKN, Mallouk TE, Cabassi M, Mayer TS (2002) Layer-by-layer self-assembly strategy for template synthesis of nanoscale devices. Mater Sci Eng C 19:255–262
157. Vila L, Vincent P, Dauginet-De Pra L, Pirio G, Minoux E, Gangloff L, Demoustier-Champagne S, Sarazin N, Ferain E, Legras R, Piraux L, Legagneux P (2004) Growth and field-emission properties of vertically aligned cobalt nanowire arrays. Nano Lett 4:521–524
158. Dobrev D, Vetter J, Neumann R, Angert N (2001) Conical etching and electrochemical metal replication of heavy-ion tracks in polymer foils. J Vac Sci Technol B 19:1385–1387
159. Sides CR, Martin CR (2005) Nanostructured electrodes and the low-temperature performance of Li-ion batteries. Adv Mater 17:125–128
160. Lindeberg M, Öjefors E, Rydberg A, Hjort K (2003) 30 GHz litz wires defined by ion track lithography. In: TRANSDUCERS'03, 12th International Conference on Solid State Sensors, Actuators and Microsystems, Boston, MA, June 8–12, 887–890
161. Toimil ME, Höhberger EM, Schäflein C, Blick RH, Neumann R, Trautmann C (2003) Electrical characterization of electrochemically grown single copper nanowires. Appl Phys Lett 82:2139–2141
162. Barati M, Sadeghi E (2001) Study of the ordinary size effect in the electrical conductivity of Bi nanowires. Nanotechnology 12:277–280
163. Fusil S, Piraux L, Mátéfi-Tempfli S, Mátéfi-Tempfli M, Michotte S, Saul CK, Pereira L, Bouzehouane K, Cros V, Deranlot C, George J-M (2005) Nanotechnology nanolithography based contacting method for electrical measurements on single template synthesized nanowires. 16:2936–2940
164. Cronin SB, Lin YM, Koga T, Sun X, Ying JY, Dresselhaus MS (1999) Thermoelectric investigation of bismuth nanowires. In: 18th International Conference on Thermoelectrics, Baltimore, MD, 554–557
165. Valizadeh S, Abid M, Rodríguez AR, Hjort K, Schweitz JÅ (2006) Template synthesis and electrical contacting of single Au nanowires by focused ion beam techniques. Nanotechnology 17:1134–1139
166. Penate-Quesada L, Mitra J, Dawson P (2007) Non-linear electronic transport in Pt nanowires deposited by focused ion beam. Nanotechnology 18:215203
167. Li Q, Koo S, Richter CA, Edelstein MD, Bonevich JE, Kopanski JJ, Suehle JS, Vogel EM (2007) Precise alignment of single nanowires and fabrication of nanoelectromechanical switch and other test structures. IEEE Trans Nanotechnol 6:256–262
168. Walton AS, Allen CS, Critchley K, Gorzny MŁ, McKendry JE, Brydson RMD, Hickey BJ, Evans SD (2007) Four-probe electrical transport measurements on individual metallic nanowires. Nanotechnology 18; doi:101088/0957–4484/18/6/065204
169. Keebaugh S, Kalkan AK, Nam WJ, Fonash SJ (2006) Gold nanowires for the detection of elemental and ionic mercury. Electrochem Solid State Lett 9:H88–H91

170. Fuhr G, Müller T, Schnelle Th, Hagedorn R, Voigt A, Fiedler S, Arnold WM, Zimmermann U, Wagner B, Heuberger A (1994) Radio-frequency microtools for particle and live cell manipulation. Naturwissenschaften 81:528–535
171. Schnelle Th, Müller T, Fiedler S, Fuhr G (1999) The influence of higher moments on particle behaviour in dielectrophoretic field cages. J Electrostat 46:13–28
172. Fiedler S .(2004) Nano-bio-packaging – Ansätze, Chancen und Trends, Part I and Part II. PLUS (Produktion von Leiterplatten und Systemen) 8:1169–1178; 1362–1368 (ISSN 1436-7505, B 49475)
173. Fiedler S, Shirley SG, Schnelle Th, Fuhr G (1998) Dielectrophoretic sorting of particles and cells in a microsystem. Anal Chem 70:1909–1915
174. Müller T, Pfennig A, Klein P, Gradl G, Jäger M, Schnelle T (2003) The potential of dielectrophoresis for single-cell experiments. Eng Med Biol Mag IEEE 22:51–61
175. Smith PA, Nordquist CD, Jackson TN, Mayer TS, Martin BR, Mbindyo J, Mallouk TE (2000) Electric-field assisted assembly and alignment of metallic nanowires. Appl Phys Lett 77:1399–1401
176. Boote JJ, Evans SD (2005) Dielectrophoretic manipulation and electrical characterization of gold nanowires. Nanotechnology 16:1500–1505
177. Papadakis SJ, Gu Z, Gracias DH (2006) Dielectrophoretic assembly of reversible and irreversible metal nanowire networks and vertically aligned arrays. Appl Phys Lett 88:2331181–2331183
178. Hangarter CM, Rheem Y, Yoo B, Yang E-H, Myung NV (2007) Hierarchical magnetic assembly of nanowires. Nanotechnology 18:205305; doi:101088/0957–4484/18/20/205305
179. Lapointe C, Hultgren A, Silevitch DM, Felton EJ, Reich DH, Leheny RL (2004) Elastic torque and the levitation of metal wires by a nematic liquid crystal. Science 303:652–655
180. Boote JJ, Critchley K, Evans SD (2006) Surfactant mediated assembly of gold nanowires on surfaces. J Exp Nanosci 1:125–142
181. Martin BR, Dermody DJ, Reiss BD, Fang M, Lyon LA, Natan MJ, Mallouk TE (1999) Orthogonal self-assembly on colloidal gold-platinum nanorods. Adv Mater 11:1021–1025
182. Chen C, Yan L, Kong ES-W, Zhang Y (2006) Ultrasonic nanowelding of carbon nanotubes to metal electrodes. Nanotechnology 17:2192–2197
183. Yoo B, Rheem Y, Beyermann WP, Myung NV (2006) Magnetically assembled 30 nm diameter nickel nanowire with ferromagnetic electrode. Nanotechnology 17:2512–2517
184. Scheel W, Fiedler S, Krause F, Schütt J (2000) Vorrichtung zur elektrischen und mechanischen Fügung von flächigen Anschlussstrukturen. DE 10002182.4:19.01.2000/09.08.2001, Fraunhofer Ges z Förderung d angew Forschung, München
185. Lin R-J, Hsu Y-Y, Chen Y-C, Cheng S-Y, Uang R-H (2005) Fabrication of nanowire anisotropic conductive film for ultra-fine pitch flip chip interconnection. Electronic Components and Technology Concerence ECTC Proceedings IEEE, Orlando, FL, 66–70
186. Langford RM, Wang T-X, Thornton M, Heidelberg A, Sheridan JG, Blau W, Leahy R (2006) Comparison of different methods to contact to nanowires. J Vac Sci Technol B 24:2306–2311
187. Gu Z, Ye H, Gracias DH (2005) The bonding of nanowire assemblies using adhesive and solder. JOM 57:60–64
188. Sharma G, Chong CS, Ebin L, Kripesh V, Gan CL, Sow CH (2007) Patterned micropads made of copper nanowires on silicon substrate for application as chip to substrate interconnects. Nanotechnology 18:305306
189. Sharma G, Chong SC, Ebin L, Hui C, Gan CL, Kripesh V (2007) Fabrication of patterned and non-patterned metallic nanowire arrays on silicon substrate. Thin Solid Films 515:3315–3322
190. Pang YT, Meng GW, Fang Q, ZhangLD (2003) Silver nanowire array infrared polarizers. Nanotechnology 14:20–24
191. Wehrspohn RB, Schilling J (2001) Electrochemically prepared pore arrays for photonic-crystal applications. MRS Bull 26:623–626
192. Saib A, Vanhoechenacker-Janvier D, Raskin J-P, Crahay A, Huynen I (2001) Microwave tunable filters and nonreciprocal devices using magnetic nanowires. Proceedings of the First IEEE Conference on Nanotechnology, IEEE-NANO 2001, 28–30 October, Maui, HI, 260–265

193. Ren B, Yao JL, She CX, Huang QJ, Tian ZQ, Surface Raman spectroscopy on transition metal surfaces. Internet J Vib Spectrosc 4, 2nd Ed. www.ijvs.com/volume4/edition2/section3.html
194. Papadakis SJ, Miragliotta JA, Gu Z, Gracias DH (2005) Scanning surface-enhanced Raman spectroscopy of silver nanowires. In: Plasmonics: Metallic nanostructures and their optical properties. III. Stockman MI (Ed.). Proc SPIE, Bellingham, WA, 59271:H1–H8
195. Choi J, Sauer G, Nielsch K, Wehrspohn RB, Gösele U (2003) Hexagonally arranged monodisperse silver nanowires with adjustable diameter and high aspect ratio. Chem Mater 15:776–779
196. Mühlschlegel P, Eisler H-J, Martin OJF, Hecht B, Pohl DW (2005) Resonant optical antennas. Science 308:1607–1609
197. Podolskiy VA, Sarychev AK, Shalaev VM (2003) Plasmon modes and negative refraction in metal nanowire composites. Optics Express 11:735–745
198. Neubrech F, Kolb T, Lovrincic R, Fahsold G, Pucci A, Aizpurua J, Cornelius TW, Toimil-Molares ME, Neumann R, Karim S (2006) Resonances of individual metal nanowires in the infrared. Appl Phys Lett 89:253104–1 –253104–3
199. Girard C, Dujardin E (2006) Near-field optical properties of top-down and bottom-up nanostructures. J Opt A Pure Appl Opt 8:S73–S86
200. Schider G, Krenn JR, Hohenau A, DitlbacherH, Leitner A, Aussenegg FR, Schaich WL, Puscasu I, Monacelli B, Boremann G (2003) Plasmon dispersion relation of Au and Ag nanowires. Phys Rev B 68:1555427/1 – 155427/4
201. van der Zande BMI, Koper GJM, Lekkerkerker HNW (1999) Alignment of rod-shaped gold particles by electric fields. J Phys Chem B 103:5754–5760
202. Chiu J-C, Chang C-M, Jou W-S, Cheng W-H (2007) Electromagnetic shielding performance for a 2.5 Gb/s plastic transceiver module using dispersive multiwall carbon nanotubes. In: Proceedings of 57th ECTC Electronic Components and Technology Conference, 29 May – 1 June, Reno, NV, 183–187
203. Burke PJ, Li S, Yu Z (2006) Quantitative theory of nanowire and nanotube antenna performance. IEEE Trans Nanotechnol 5:14–334
204. Rheem Y, Yoo B-Y, Beyermann WP, Myung NV (2007) Electro- and magneto-transport properties of a single CoNi nanowire. Nanotechnology 18:015202
205. Tang X-T, Wang G-C, Shima M (2006) Perpendicular giant magnetoresistance of electrodeposited Co/Cu-multilayered nanowires in porous alumina templates. J Appl Phys 99:033906–1 – 033906–7
206. Menon VP, Martin CR (1995) Fabrication and evaluation of nanoelectrode ensembles. Anal Chem 67:1920–1928
207. Dalby MJ, Berry CC, Riehle MO, Sutherland DS, Agheli H, Curtis ASG (2004) Attempted endocytosis of nano-environment produced by colloidal lithography by human fibroblasts. Exp Cell Res 295:387–394
208. Tan JL, Tien J, Pirone DM, Gray DS, Bhadriraju K, Chen CS (2001) Cells lying on a bed of microneedles: An approach to isolate mechanical force. Proc Natl Acad Sci 100:1484–1489
209. Katsen-Globa A, Peter L, Pflueger S, Doerge T, Daffertshofer M, Preckel H, Zwanzig M, Fiedler S, Schmitt D, Zimmermann H (2006) Cell behaviour on nano-and microstructured surfaces: From fabrication, treatment and evaluation of substrates towards cryopreservation. Cryobiology 53:445–446
210. Töpper M, Klein M, Buschick K, Glaw V, Orth K, Ehrmann O, Hutter M, Oppermann H, Becker K-F, Braun T, Ebling F, Reichl H, Kim S, Tathireddy P, Chakravarty S, Solzbacher F (2006). Biocompatible hybrid flip chip microsystem integration for next generation wireless neural interfaces. In: 56th Electronic Components and Technology Conference ECTC, 30 May – 2 June, San Diego, CA, 705–708

Chapter 21
Design and Development of Stress-Engineered Compliant Interconnect for Microelectronic Packaging

Lunyu Ma, Suresh K. Sitaraman(✉), Qi Zhu, Kevin Klein, and David Fork

21.1 Introduction

Power and latency are fast becoming major bottlenecks in the design of high performance microprocessors and computers. Power relates to both consumption and dissipation, and therefore, effective power distribution design and thermal management solutions are required. Latency is caused by the global interconnects on the integrated circuit (IC) that span at least half a chip edge due to the resistance–capacitance (RC) and transmission line delay [1]. Limits to chip power dissipation and power density and limits on hyper-pipelining in microprocessors threaten to impede the exponential growth in microprocessor performance. In contrast, multicore processors can continue to provide a historical performance growth on most consumer and business applications provided that the power efficiency of the cores stays within reasonable power budgets. To sustain the dramatic performance growth, a rapid increase in the number of cores per die and a corresponding growth in off-chip bandwidth are required [2]. Thus, it is projected by the Semiconductor Industry Association in their International Technology Roadmap for Semiconductors (ITRS) (Table 21.1) that by the year 2018, with the IC node size shrinking to 22 nm by 2016 and 14 nm by 2020, the chip-to-substrate area-array input–output interconnects will require a pitch of 70 μm [3]. Furthermore, to reduce the RC and transmission line delay, low-K dielectric/Cu and ultra-low-K dielectric/Cu interconnects on silicon will become increasingly common. In such ICs, the thermo-mechanical stresses induced by the chip-to-substrate interconnects could crack or delaminate the dielectric material causing reliability problems.

Flip chips with solder bumps are being increasingly used today to address these needs because of their several advantages: higher I/O density, shorter leads, lower inductance, higher frequency, better noise control, smaller device footprint, and lower profile [4]. Flip-chips on board (FCOB) are gaining increased acceptance both for cost-performance as well as high-performance applications. Epoxy-

---

S.K. Sitaraman
Computer-Aided Simulation of Packaging Reliability (CASPaR) Lab, The George W. Woodruff School of Mechanical Engineering, Georgia Institute of Technology, 813 Ferst Drive, Atlanta, GA 30332-0405, USA
suresh.sitaraman@me.gatech.edu

Table 21.1 ITRS 2005 roadmap for assembly and packaging [3]

Year of production	2014	2015	2016	2017	2018	2019	2020
DRAM ½-pitch (nm) (contacted)	28	25	22	20	18	16	14
MPU/ASIC metal 1 (M1) ½-pitch (nm) (contacted)	28	25	22	20	18	16	14
MPU physical gate length (nm(μm))	11	10	9	8	7	6	6
Wire bond pitch – single in-line (μm)	20	20	20	20	20	20	20
Flihip area array pitch (μm)	80	80	80	80	70	70	70

based underfills are often used in such FCOB assemblies to accommodate the coefficient thermal expansion (CTE) mismatch among different materials (e.g., silicon IC on an organic substrate) and to enhance the solder joint reliability against thermo-mechanical fatigue failure [5, 6]. However, additional underfill process steps, material and processing costs, reworkability, delamination, and cracking are some of the concerns with the use of underfills. Also, as the pitch size decreases, the cost and the difficulties associated with underfill dispensing will increase dramatically [7, 8]. Conductive adhesives are an alternative given the move of industry to lead-free solders, but processing difficulties restrict them to low I/O density applications.

Furthermore, when low-K dielectric (ultra low-K dielectric in the future) is used in the IC and when such ICs are assembled on organic substrates, the chip-to-substrate interconnects are subjected to extensive differential displacement due to the CTE mismatch between the die and the substrate under thermal excursions. On the one hand, these interconnects, especially stiff solder bumps, could crack or delaminate the low-K dielectric material in the die. On the other hand, if the solder bumps are not underfilled, they will fatigue crack and fail prematurely. Therefore, it is necessary to explore alternate interconnects that are compliant so that they will not crack or delaminate the low-K dielectric, that will not fatigue fail prematurely without an underfill, that are easy to fabricate and assemble using existing infrastructure, that they are scalable, that are wafer-level, and that will meet the electrical, thermal, and mechanical requirements for next-generation microsystems.

21.2 Literature Review on Compliant Interconnects

With the advent of the area array packages and the increasing concern of thermo-mechanical reliability, a promising solution is to increase the mechanical compliance of interconnect to accommodate more differential displacement due to the CTE mismatch. There are several different types of the first-level compliant interconnects already available in commercial market or currently under development. In this section, each of them will be briefly reviewed (Table 21.2).

21 Design and Development of Stress-Engineered 467

Table 21.2 Summary of current compliant off-chip interconnect technologies

Technology and company name	Attributes
μBGA® and WAVE® (Tessera) [9–11]	* S-shaped Cu/Au ribbon bonds + No underfill + Batch fabrication + Reliable − Elastomer encapsulating − Limited planar compliance
MOST ® (FormFactor) [12, 13]	* Au wire bonds + No underfill + Good compliance − Serial fabrication − Limited pitch and interconnect size
Sea of Leads (SOL) (IFC, Georgia Tech) [14–16]	* Electroplated Au lead * Low-modulus polymer with air-gap + No underfill + Batch fabrication − High temperature processing − Limited in-plane compliance
Helix Interconnects (CASPaR, Georgia Tech) [17–19]	* Electroplated Cu interconnect * Layer-by-layer fabrication + No underfill + Batch fabrication + Good planar and out-of-plane compliance + Varying compliance designs

Tessera's compliant μBGA™ technologies have been developed since 1995 [9, 10]. The metal compliant interconnects (or ribbons) are formed by patterning the metal layer on a flexible organic tape. Then the compliant interconnects are bonded to the die and vertically expanded into the free-standing positions by injecting the low-modulus elastomer layer between the die and the flexible tape. The solder joints are bumped on the other side of the flexible tape. One of the advantages of this technology is that all the fabrication processes can be made on the wafer level. Therefore, the fabrication cost and time per unit can be significantly reduced. WAVE™ (wide area vertical expansion) technology is the second generation of compliant package from Tessera. The low modulus compliant layer absorbs the majority of package deformation caused by CTE mismatch. Thus, the stresses on the solder balls are reduced. This reduced stress level eliminates the need for an underfill [11]. However, since the similarity between this technology and TAB, only peripheral array can be achieved. The pitch size and the I/O density are limited. Additionally, the dispensing of elastomer layer will induce more cost and time.

MicroSpring™ interconnect has been developed by FormFactor Inc. The 3D microspring interconnects are formed in out-of-the-plane direction using wire bonding method. Each microspring is formed first, by placing a specially designed and shaped wire bond at the desired location, and then plating up the wire bond, transforming it into a spring. The plating alloy provides the spring strength, while a finish layer of gold ensures a stable electrical contact [12]. The MicroSpring™ technology offers low normal force, fine pitch, high pin count arrays and no-underfill, all of which are needed for current and future microelectronic packaging [13]. However, because of the limits of the wire bonding process, the pitch size is limited to be in the range of 100 μm. Another drawback is that as the fabrication is a sequential process, and therefore, the high cost and the long fabrication cycle need to be improved.

The Interconnect Focus Center (IFC), Georgia Institute of Technology, has proposed a compliant interconnect structure called Sea of leads (SOLs) [14–16]. It enables a compliant wafer level packaging (CWLP) at low cost. In the fabrication, a layer of overcoat polymer is deposited over a patterned sacrificial polymer. The sacrificial polymer is later thermally decomposed to form the air gaps in the overcoat polymer. The air gaps can enhance the vertical compliance of the leads. The curved Au leads are then patterned by photolithography and deposited by electroplating. The fabrication of SOLs is compatible with standard IC fabrication. It can achieve small pitch size and does not need an underfill material for thermo-mechanical reliability.

Helix interconnect is a spiral compliant interconnect that is under development at the Computer Aided Simulation of Packaging Reliability (CASPaR) Lab, Georgia Institute of Technology [17–19]. The fabrication is based on the MEMS-type high aspect ratio via formation and electroplating. The structure is built up layer by layer. The geometry of each layer is designed on each photolithography mask. First, the cavity of the first layer is formed in photoresist by photolithography. The metal is then grown in the cavity by electroplating. After the first layer is finished, a same process is repeated for the second layer, and so on. The solder material and its barrier layer can be deposited in the electroplating of the last layer. Finally, after all layers are formed, the photoresist is removed by dry-etching process and the structure becomes free-standing. The helix-type structures can be achieved through the wafer level fabrication. The fabrication processes are completely compatible with the standard IC processes. Photolithography enables the control of fine pitch and small dimension during the fabrication. Because of the good mechanical compliance, the underfill material is not required to relieve the thermally induced stress/strain concentration in the structure. Both the vertical and in-plane compliance of the helix-type interconnect can be greater than 10 mm N^{-1} [17]. A modified version of the helix interconnects called *FlexConnects* that requires less number of masking steps is also under development [20].

21.3 Stress-Engineered Compliant Interconnects

This chapter presents the fabrication and design of stress-engineered compliant interconnects. The fabrication of the stress-engineered compliant interconnect is based on the standard IC fabrication and stress-engineering theory. The primary goal of stress-engineered compliant interconnects is to achieve higher compliance than the conventional C4 solder joints and obtain an improved thermo-mechanical reliability. The stress-engineered, compliant interconnect for high density application, developed by a consortium of Georgia Institute of Technology, XEROX Palo Alto Research Center

(PARC) and Nanonexus Inc [21, 22], is innovative and uses the intrinsic stress gradient induced during the DC-sputtering deposition to create the compliant structure.

21.3.1 Fabrication of Stress-Engineered Compliant Interconnects

Stress-engineered compliant interconnects are fabricated using processes such as DC sputtering, photolithography, and wet etching, which are highly compatible with the front-end (IC) processes in semiconductor industry. The fabrication can be done in an area array format at the wafer level, and has important advantages. First, the interconnect fabrication can be easily integrated into the standard semiconductor front-end processes. Additionally, the area array wafer-level fabrication of interconnect can greatly reduce the fabrication cost per unit and fabrication time, and is highly consistent with the wafer-level packaging (WLP). Finally, the standard IC fabrication enables the good control of interconnect geometry and pitch size.

Hoffman and Thornton [23] reported the stress-engineering process in the thin film metals that the intrinsic stresses can be gradually transitioned during DC magnetron sputtering by changing the sputtering condition, such as the Argon (Ar) pressure. The impurities/atomic peening model and the grain boundary (GB) relaxation model were used to explain the stress-engineering effect [24]. Smith and Alimonda [21] first reported the fabrication of the stress-engineered compliant interconnects with 80 µm pitch for microelectronic packaging applications.

A detailed description of fabrication process is given here. A schematic picture of fabrication processes is shown in Fig. 21.1. A bare wafer is used as the substrate for the compliant interconnect fabrication. A thin layer of Titanium (Ti), about 0.5 µm, is deposited on the substrate using DC sputtering. It is known that Ti has good interfacial adhesion with most of the materials used in IC fabrication, especially between the metals and ceramics. Therefore, the Ti layer can prevent the highly stressed thin film metal layer from undesired peeling-off during the fabrication processes. The Ti layer is also called the release layer or the adhesion layer.

A layer of $Mo_{80}Cr_{20}$ alloy (by weight) thin film, 1.5 µm, is sputtered onto the Ti layer. During the DC sputtering process, the argon pressure is carefully manipulated to obtain the desired intrinsic stress conditions. At low argon pressure (less than 0.5 Pa), the possibility of collision between target metal atoms and argon atoms is low. The target metal atoms can be deposited in a condensed formation on the substrate because of less scattering effect. Thus, the interatomic distance between two neighboring metal atoms is smaller than the equilibrium distance. In this case, a compressive intrinsic stress is present in the sputtered metal layer. On the contrary, if the argon pressure is higher (greater than 0.5 Pa), the chance of collision between target metal atoms and argon atoms is higher. The target metal atoms are deposited in a coarse formation, and thus a tensile intrinsic stress (attractive interatomic force) is present in the deposited metal layer. The intrinsic stress magnitude can be varied from −1 GPa at the bottom of the Mo_{80}-Cr_{20} layer to +1 GPa at the top of the

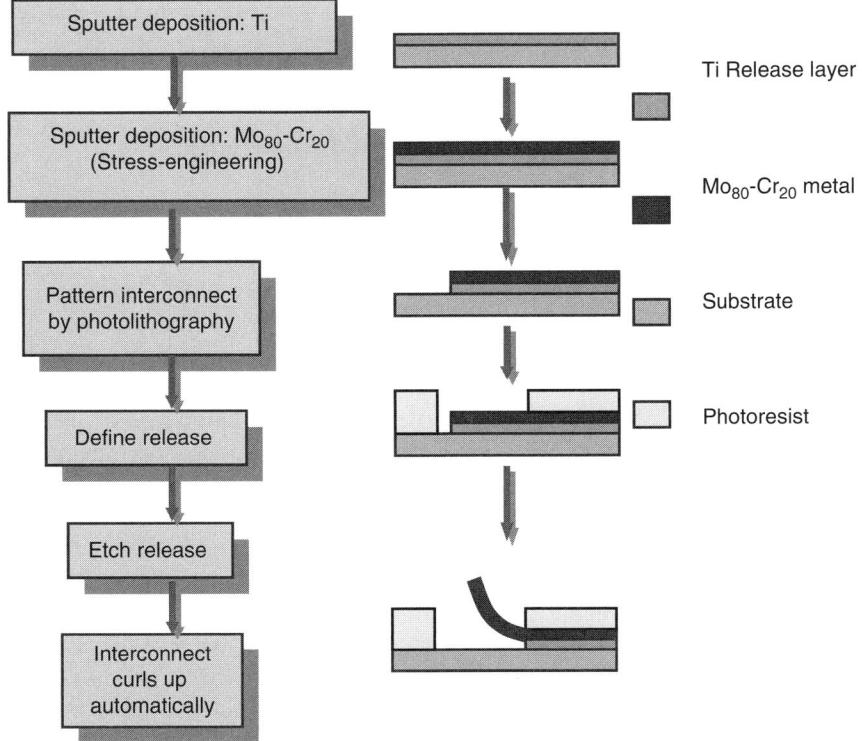

Fig. 21.1 Fabrication process of stress-engineered compliant interconnect

$Mo_{80}Cr_{20}$ layer, by gradually changing the argon pressure. The intrinsic stress gradient throughout the thickness can form an intrinsic up-bending moment.

To achieve a very uniform intrinsic stress gradient throughout the whole substrate, a rotational planetary system is used in the sputtering tool. The thin film metals used in stress-engineered compliant interconnects are deposited in a sputter deposition system in which substrates travel on a planetary system. Analogous to the motion of celestial bodies, the substrates revolve around their own centers, as the substrate also orbits the center of the deposition system. The sputter gun, which holds the metal target for deposition, is typically located on the orbit of the substrate's center.

Although Mo_{80}-Cr_{20} alloy has very good mechanical property for the stress-engineering application, its electrical property, like the conductivity, is not as good as the electrically conductive materials, such as gold (Au) and copper (Cu). To improve the conductivity of interconnect, a layer of Au, usually about 1.0 μm, is sputtered over the Mo_{80}-Cr_{20} thin film. The lower elastic modulus and the small thickness of Au will ensure that the interconnects remain compliant.

After the DC sputtering, the geometry shape of the stress-engineered compliant interconnects is patterned using photolithography. Thus far, the stress-engineered Mo_{80}-Cr_{20} layer is held down onto the substrate by the Ti adhesion layer. Next, the stress-engineered compliant interconnect should be released by etching the Ti layer. To prevent the stress-engineered compliant interconnects from peeling off from the substrate, one end of the interconnect structure will be anchored on the substrate by the Ti release layer, while the other end is released from the substrate. To have this released structure, release windows are to be defined using a second photolithography process.

A layer of photoresist is first dispensed on the substrate wafer. The photoresist is patterned to create release windows for the stress-engineered interconnects. The anchoring end is still encapsulated in the photoresist to protect the underlying Ti layer from etching away. A selective etching solution is used to remove only the Ti layer within the release window. After the wet etching, the stress-engineered compliant interconnect is released from the substrate wafer and curls up automatically due to the relaxation of its intrinsic stress gradient, as seen in Fig. 21.1. Following the release process, the rest of photoresist on the substrate is stripped away.

Figure 21.2 presents stress-engineered compliant interconnects fabricated at 6 and 80-μm pitch. These interconnects are called linear or straight interconnects, as their unreleased geometry is along a straight line.

21.3.2 J-Spring Compliant Interconnects

The linear spring discussed thus far has excellent out-of-plane compliance. However, the in-plane compliance, especially along the axis of the linear spring, can be enhanced by suitable design modification. Accordingly, a new-structure called a "J-spring" is designed, with a J-like shape in the unreleased stage. The J-spring has both good in-plane compliance and excellent out-of-plane compliance. The J-spring compliance can be altered by changing various geometric parameters such as the length of the linear segment (L), the width (W), the inner radius of arc segment (R), and the subtended angle of arc segment (α), as illustrated in Fig. 21.3. An array of fabricated J-springs with a subtended arc angle of 90° is also shown in Fig. 21.4.

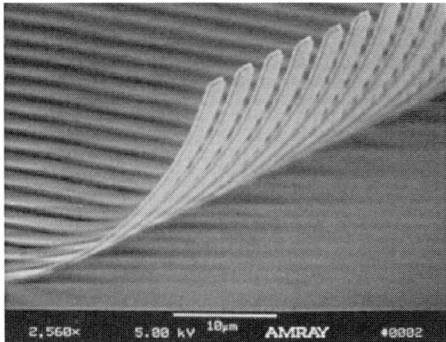

Fig. 21.2 Scanning electron microscopy (SEM) pictures of stress-engineered compliant interconnect arrays at 6 μm and 80 μm [courtesy: Don Smith] pitches

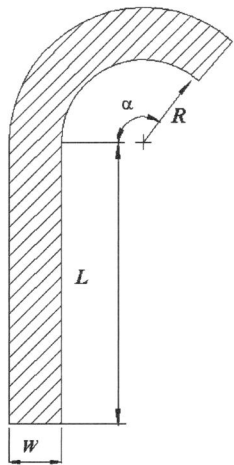

Fig. 21.3 Schematic of J-spring

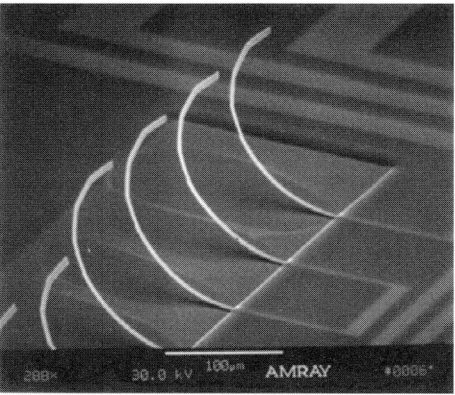

Fig. 21.4 SEM image of released J-springs

21.4 Compliance Analysis

The compliance of linear spring and J-spring was computed using a finite-element model. In the model, the substrate was modeled as a rigid substrate, and the released spring geometry was obtained starting with the residual stress gradient in the sputtered metal. A force load was then applied to the free end of the released spring geometry, and the compliance in the force direction was calculated as the ratio between the displacement and the force in the same direction. Compliance analysis was carried out by changing the various geometry parameters of the spring structures. For the linear spring, the geometric parameters that can be changed are the length and the width of the spring. For the J-spring, the geometric parameters that can be changed are the length, width, arc radius, and the subtended angle. For the sake of brevity, selected results are presented here. Additional details can be found in [25]. Figure 21.5 shows the variation of compliance with the linear spring length, and Fig. 21.6 presents J-spring compliance variation with the inner radius of the arc segment.

It is seen that the curves of the X and Y compliance are almost parallel to each other, or the magnitudes of the increase are about the same. Nevertheless, because L can also contribute to the Y compliance, thus the Y compliance is always larger than the X compliance in this case. The maximum Y compliance is about four times larger than its original value, while the maximum of the X compliance is about eight times larger than its initial value. This is because the Y

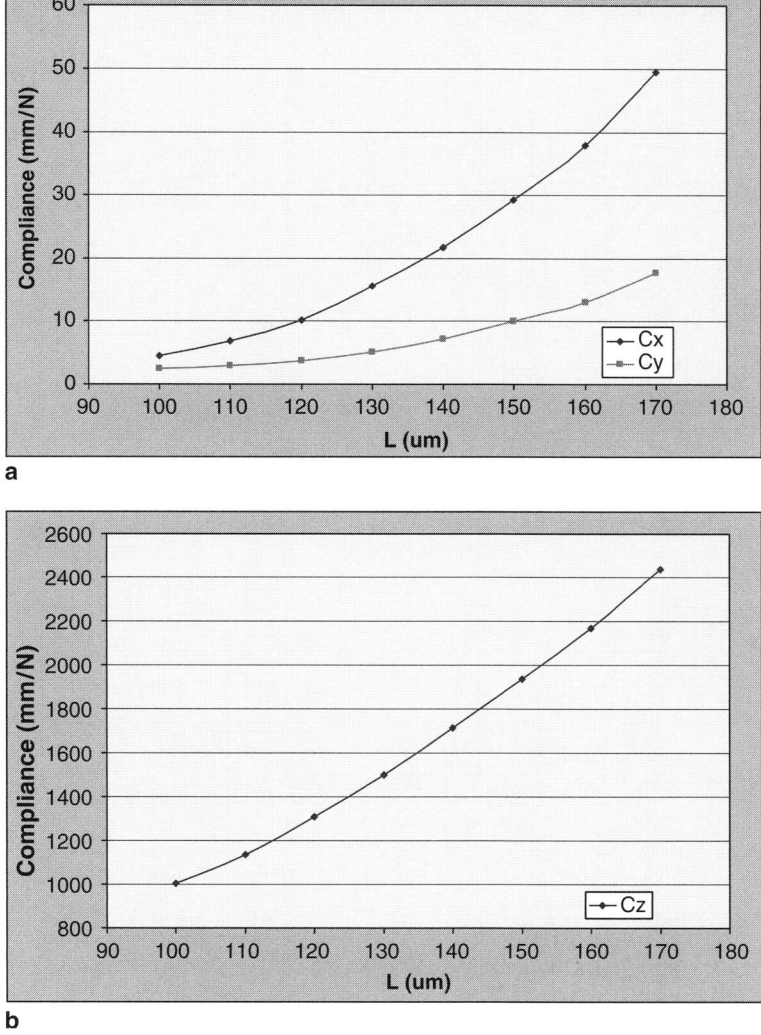

Fig. 21.5 (a) In-plane and (b) out-of-plane compliance variation with length for the linear spring

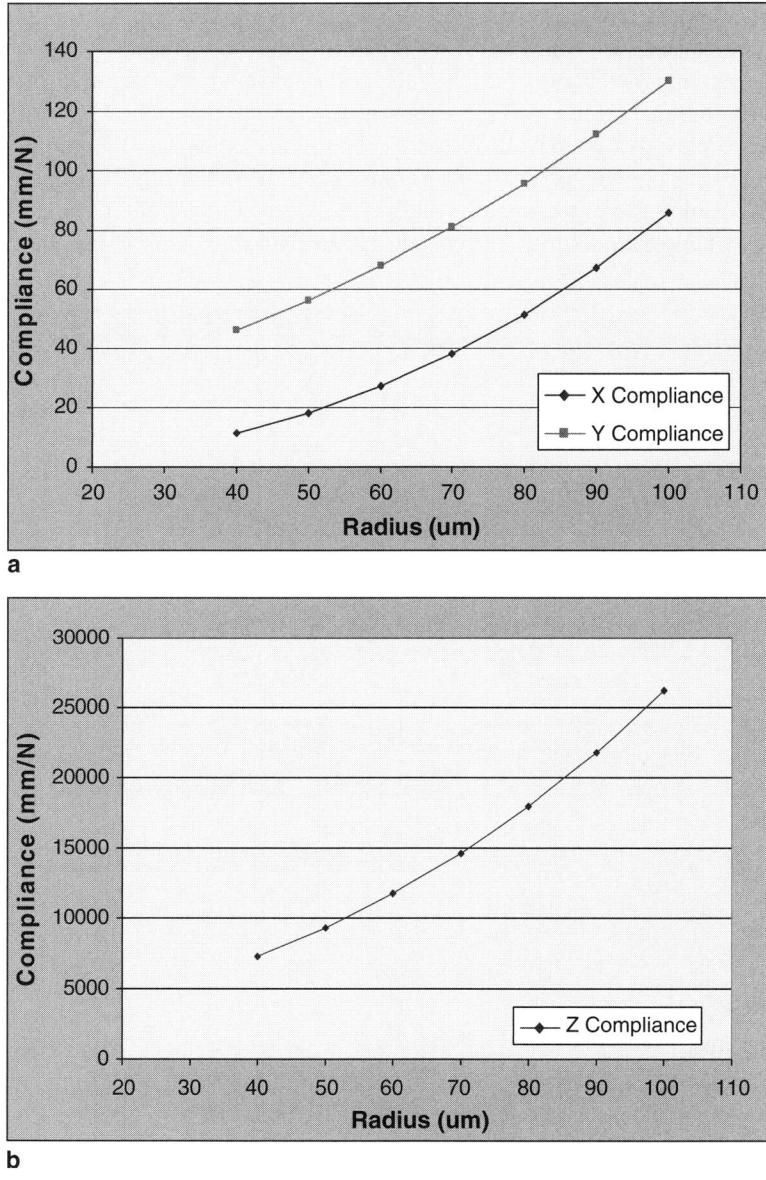

Fig. 21.6 (a) In-plane and (b) out-of-plane compliance variation with radius for the J-spring

compliance value is influenced by L and R, while the X compliance is primarily influenced by R. Although the Z compliance magnitude and rate of increase are much higher than the planar compliance, its maximum value is still approximately four times larger than its original value, similar to the Y compliance. When only R increases, it increases not only the moment arm, but also the release height. Because of these effects, R also has significant influence on the J-Spring compliance.

21.5 Compliant Interconnect Assembly Process

The thermo-mechanical reliability of compliant interconnects is assessed through accelerated thermal cycling tests. Test vehicles of stress-engineered compliant interconnect arrays were fabricated and assembled. Optoelectronic devices, such as the high density vertical cavity surface-emitting laser (VCSEL) array, will be one of the important applications for this interconnection technology [26], so the pitch size of the stress-engineered compliant interconnects was designed to be 21 μm in the test vehicle. The width of the linear compliant interconnects is 17 μm. The compliant interconnects were fabricated on transparent substrates, such as quartz and glass. The test vehicle layout is given in Fig. 21.7.

In the test vehicle, there is one long horizontal array of compliant interconnects oriented along X direction in the center, and six short vertical arrays oriented along Y direction with some offset from X axis. The metal traces are fanned out from the compliant interconnects to the probing pads on the edges of the test vehicle. Between every two adjacent probing pads, the compliant interconnects form a daisy chain with the bonding pads on the die. A complete daisy chain loop between every two adjacent probing pads is called a channel. In the horizontal array, there are 11 channels with 100 compliant interconnects in each channel. Each vertical array is a channel containing about 110 compliant interconnects. Because of their small size, the compliant interconnects cannot be seen in the layout, so zoomed-schematics of the compliant interconnects and bonding pads are also shown in Fig. 21.7.

The compliant interconnects were assembled on substrates using two different techniques: (1) The compliant interconnects were in sliding contact with the bonding pads, and the substrate was held in place using adhesive at the corners; (2) An underfill was used to assemble the substrate on the die. Both of these two assembly processes can be carried out at room temperature, and thus can eliminate thermal expansion induced misalignment between the interconnect and the bonding pad. The assembly processes use UV curing resulting in short assembly time.

21.5.1 Sliding Contact Package

The substrates with the released stress-engineered compliant interconnects and die are cleaned before assembly. After rinsing in acetone, methanol, and isopropyl alcohol, the substrate and die are cleaned in oxygen plasma to remove organic contamination. As shown in Fig. 21.8, the quartz substrate with the compliant intercon-

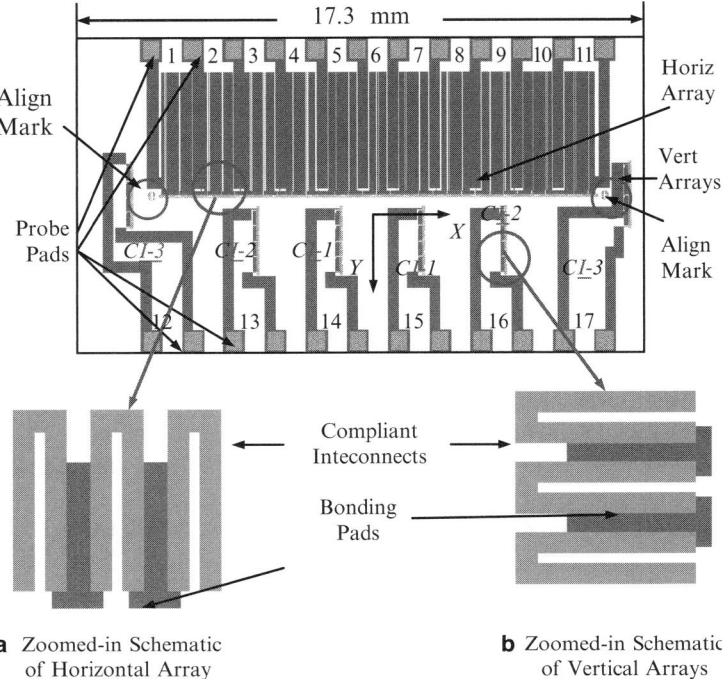

Fig. 21.7 Test vehicle layout

nects is first mounted onto the stationary stage and held by the vacuum. Before the silicon die is mounted onto the movable stage, a UV-curable adhesive is applied to four corners of the rectangular silicon die. After the die is mounted, a coarse alignment is performed so the compliant interconnect tips do not contact the bonding pads. The compliant interconnects and bonding pads are registered with the aid of two alignment marks. After the coarse alignment, the substrate is moved down toward the die to make electrical contact between the compliant interconnects and the bonding pads. The movement is controlled to avoid over-flattening by observing the optical image through the microscope. At the same time, a fine alignment is performed to ensure the electrical contact. During the downward movement, the adhesive applied onto the die is also touching the substrate.

Once the fine alignment is completed, a UV light is shined over the assembly for 1–2 min to register the final position. The intensity of UV exposure is 70 mW cm^{-2}. In this assembly, the compliant interconnect tips contact the bonding pads without any encapsulation. Therefore, they can slide freely on the pads.

21.5.2 Nonsoldered Underfilled Package

The second method of assembly deals with underfilling of the interconnects to make contact between the interconnects and the bonding pads. There are two methods to assemble using an underfill, capillary underfilling assembly and no-flow assembly.

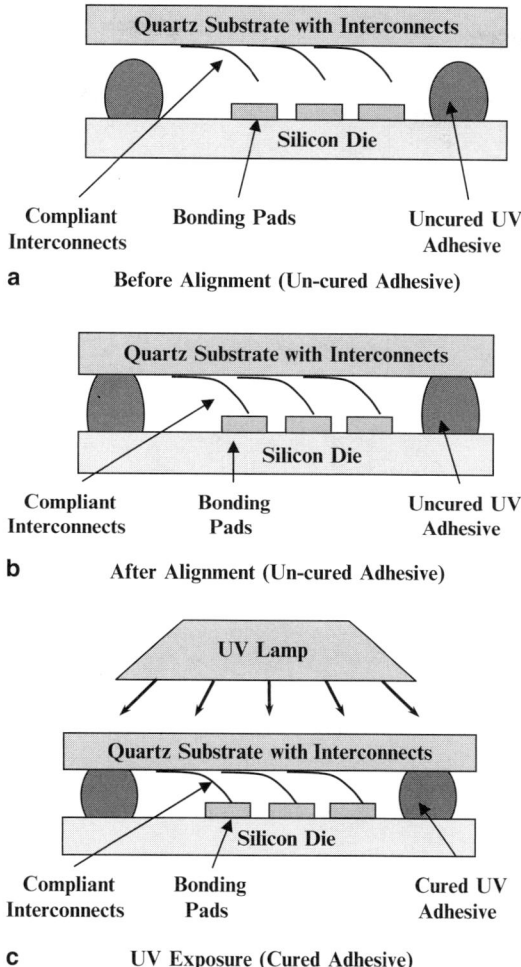

Fig. 21.8 Sliding contact assembly process

The capillary underfilling assembly is similar to the free-air sliding contact assembly. After the final registration with UV exposure, an underfill material is dispensed along one or two edges of package. Under capillary force, the underfill flows into the gap between the die and the substrate, and encapsulates all compliant interconnects. A second UV exposure is done to cure the underfill.

In the no-flow assembly, a transparent UV curable underfill is used to comply with the optical alignment. The underfill is first applied onto the die before alignment. After the coarse alignment, the substrate is moved toward the die, and the compliant interconnects are pressed into the underfill. The fine alignment is conducted as the compliant interconnects are completely immersed into the underfill. After the fine alignment, the package is exposed to UV radiation. The UV intensity is about the same as that used in the free-air sliding contact package. The exposure time is about 3 min. The assembly process is schematically shown in Fig. 21.9.

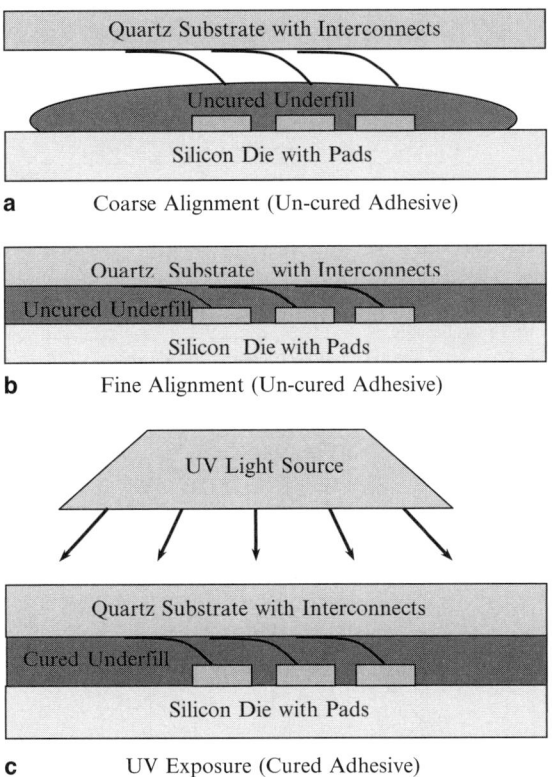

Fig. 21.9 Nonsoldered and underfilled assembly process

21.6 Accelerated Thermal Cycling Test of Underfilled Package

After the assembly, the packages were thermal cycled to assess their thermo-mechanical reliability. Prior to thermal cycling, the electrical resistance of different daisy chains was measured at room temperature. This electrical resistance was taken as the baseline resistance. The air-to-air thermal cycling test was based on JESD22-A104-B [27]. On the basis of the test standard guidelines, the maximum dwell temperature (T_{max}) was selected to be 125°C, which was lower than the glass-transition temperature (T_g) of the underfill. Two temperature ranges were initially selected for the thermal cycling tests; one was from 30 to 125°C, and the other was from −55 to 125°C. The temperature ramping rate was 15°C min^{-1}. The dwell time at the maximum and minimum temperature was 10 min each. The electrical resistance was measured as the thermal cycling test was conducted to assess the thermomechanical reliability. The resistance change in the accelerated thermal cycling test is shown in Fig. 21.10. The results from the accelerated thermal cycling tests are summarized below:

The electrical resistivity of a metal changes with temperature (T), with the thermal coefficient of resistivity (TCR) of pure metals typically about 4×10^{-3} K^{-1}, and generally lower for alloys [28]. From the data in Fig. 21.10, the TCR (α) can be estimated as:

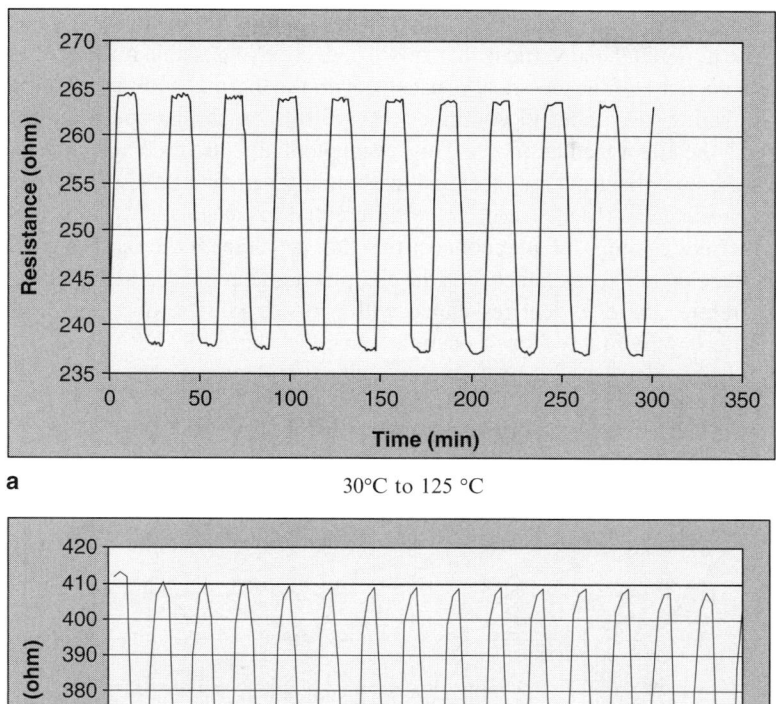

Fig. 21.10 Resistance change with temperature during thermal cycling

$$\alpha = \frac{\Delta R}{R_0 \Delta T} \approx 1.2 \times 10^{-3} \sim 2.0 \times 10^{-3} / K,$$

where α is the thermal coefficient of resistivity (TCR), R_0 is the electrical resistance at room temperature in the beginning of experiment, ΔR is the increase in electrical resistance from the lowest to the highest temperature, and ΔT is the temperature range of the thermal cycle.

As seen, the measured TCR is within the range reported by others.

The average electrical resistance decreased gradually in the first several cycles of test. As seen in Fig. 21.11, this decrease was prominent in the first 1,500 min (about 50 cycles). But as the thermal cycling test continued, the rate of decrease became less and less, and finally the electrical resistance value stabilized. This trend was observed in both the horizontal and vertical channels. There are two possible reasons to explain the initial decrease of the electrical resistance with the thermal cycling tests: (a) with thermal cycling, the underfill continues to cure further, shrink, and bring the substrate and the die together more. This can potentially increase the contact area between the compliant interconnect and the bonding pad, and thus will decrease the electrical resistance. (b) Because of the differential expansion of various materials during thermal cycling, the interconnect tips flex and scratch the die pad surface to make better electrical contact between the compliant interconnect and the die bonding pad. Therefore, the electrical resistance will be lowered. The observed resistance decrease during the first few cycles can be termed as "burn-in".

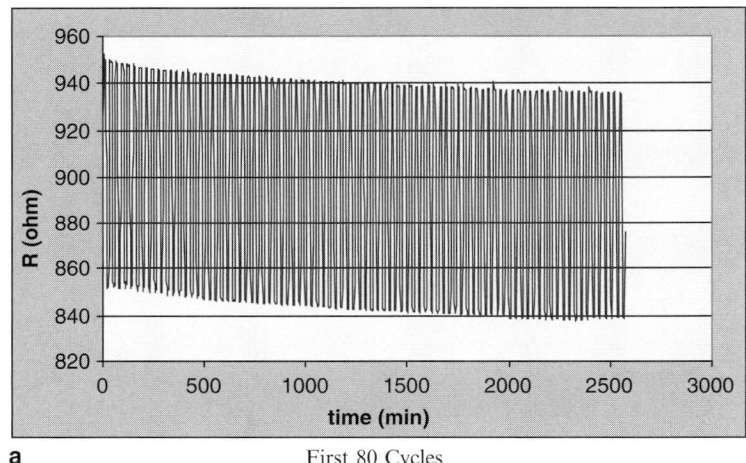

a   First 80 Cycles

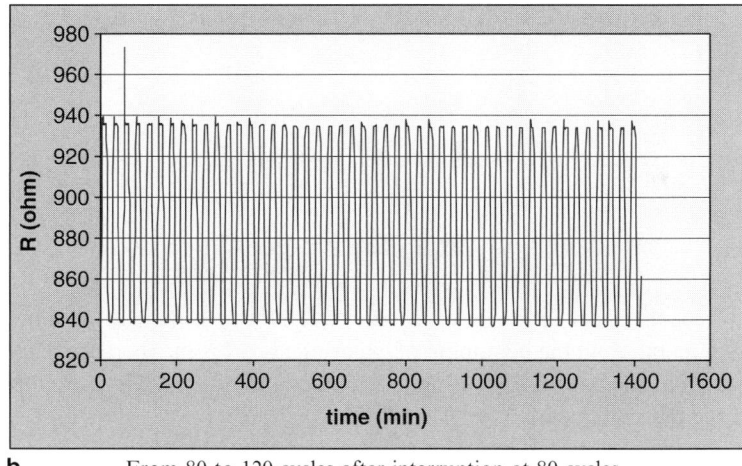

b   From 80 to 120 cycles after interruption at 80 cycles

Fig. 21.11 "Burn-in" behavior of resistance

Another characteristic of the "burn-in" is that it occurs irreversibly in the thermal cycling test, even if there is an interruption. Figure 21.11 shows the electrical resistance change in a thermal cycling test with an intentional interruption of a few hours between the first 2,500 min (80 cycles) and the next 1,500 min (40 cycles). As seen, the electrical resistance remains unchanged during interruption of thermal cycling, and continues to decrease once the thermal cycling is resumed during the burn-in phase

21.7 Thermal Cycling of Free-Air Sliding Contact Packages

To prevent the deterioration of electrical connection in free-air sliding contact packages from contamination, the packages were tested in a sealed container. The test results are similar to those obtained from the nonsoldered/underfilled packages. When the pads were examined after thermal cycling, scratch marks were seen on the gold bonding pads, as shown in Fig. 21.12. On each pad, there are two scratch marks. These marks are caused by the tips of the compliant interconnects scrubbed on the pads, and have the same width as the stress-engineered compliant interconnects, located at the contact regions on the pads.

The cause of these scratch marks can be explained as the following. There were two types of mechanical loads applied along the contact interfaces between the compliant interconnects and the bonding pads. The first load is compressive in the vertical direction (normal to the bonding pad surface) caused by the compression of the substrate toward the die in the assembly operation. The second load is in tangential direction, parallel to the pad surface. This load is caused by the differential displacement due to the coefficients of thermal expansion (CTE) mismatch in the packages at a temperature different from the stress-free condition. In the thermal cycling test, the tangential load is cyclic and leads to gradual degradation of contact integrity. Once the conductive

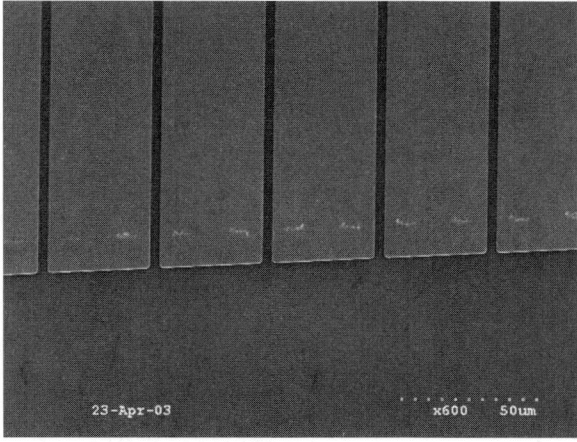

Fig. 21.12 Scratch marks on Au-bonding pads (by SEM)

material between the compliant interconnect tips and bonding pads is completely or partially worn out, the channels in the free-air sliding contact package fail.

21.8 Nanocantilever Fabrication for Sensing Applications

Thus far, we have discussed the application of stress-engineering principle to microscale interconnects. Now, we will discuss the potential extension of this to nanoscale sensing devices. In the ongoing work, we use electron-beam lithography and liftoff processes to fabricate the nanocantilever arrays. Liftoff processes have the capability to provide high-fidelity patterns with very fine features. These processes work well with unidirectional deposition methods that do not provide significant sidewall coverage, such as filament or e-beam evaporation. Sputter deposition, which is capable of providing a high internal stress gradient is multidirectional, provides sidewall coverage, and therefore, complicates the liftoff processes. The sidewall coverage in the sputter deposition process is due to the fact that the sputtered atoms deposit randomly at various angles on the rotating substrate. Two liftoff methods are being pursued for the fabrication of stress-engineered cantilevers: a single-layer resist approach and a bilayer resist approach, as illustrated in Fig. 21.13. Prior to spinning the resist, a sacrificial metal layer is deposited, which will later be selectively etched to release the patterned cantilever. The bilayer process uses a thin layer of polymethylmethacrylate (PMMA) resist spun over a thicker base layer of methylmethacrylate (MMA) resist. The single layer process uses only a positive resist, e.g., PMMA. In both processes, the total resist height to deposited metal height is desired to be 7:1 or greater; a ratio less than 7:1 makes the liftoff process difficult [29].

As illustrated in Fig. 21.13, on a bare wafer (a), a thin layer of release metal is first deposited (b). In the single layer process, PMMA is spun on the wafer, patterned, and developed. In the bilayer process, a thick layer of MMA is spun on the wafer, followed by a thinner layer of PMMA (c). After patterning of the bilayer with e-beam lithography tool, the MMA base layer is undercut during developing, because of greater exposure sensitivity when compared with the PMMA top layer (d). In both the single-layer and bilayer resist cases, the cantilever metal film, Cr, is deposited with intrinsic stress (e). Liftoff is performed using acetone in an ultrasonic bath. For smaller aspect ratios, a mechanical polish, e.g., a gentle wipe with an acetone soaked wiper, is needed to complete liftoff (f). Once the spring metal deposition is complete, a second layer of resist is spun and patterned to define a release window, which allows the selective etching the release layer (g). After selectively etching the release layer, the stressed metal stack curls up to form a free-standing cantilever (h).

21.8.1 Fabrication Results

The pattern illustrated in Fig. 21.14 was investigated. The pattern consists of a series of cantilevers, 10 μm in length, that vary in width from 10 to 100 nm in steps

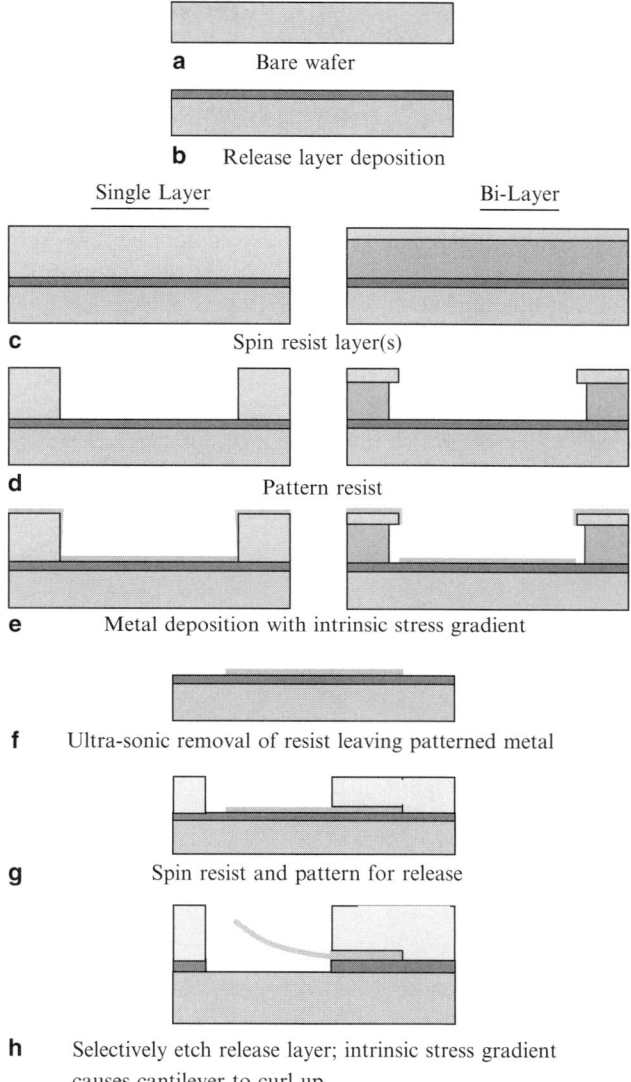

Fig. 21.13 Fabrication steps for nanocantilevers with liftoff processes

of 10 nm and from 100 to 1,000 nm in steps of 100 nm, where each width is repeated five times. Two layers of chromium sputtered at a low and a high argon pressure were used to fabricate the cantilevers in thicknesses varying from 25 to 250 nm. In some cases, a thin gold layer <50 nm was deposited on the top surface of the cantilevers, and this gold layer could be used to increase surface's binding affinity for proteins and antibodies.

Thin-film deposition was performed using PVD-300 Unifilm™ magnetron sputter system, which is capable of routinely and reproducibly depositing very uniform (>99%) and homogeneous films. A computer-controlled planetary system with two degrees of freedom (orbit and spin) is used in conjunction with a calibration of the deposition rate with respect to position. Calibrated intrinsic stress vs. Ar pressure for Cr sputter film is shown in Fig. 21.15. Although stress is primarily dependent on the Ar pressure, stress values showed a slight variation depending on the deposition power. The stress measurements in Fig. 21.15 were performed by measuring the curvature of standard 4″ Si wafers, before and after the deposition of a stressed metal of a particular film thickness [30]. From Fig. 21.3, it can be seen that Unifilm is capable of developing only tensile stresses for Cr, and therefore, the cantilever stress gradient is developed by depositing two layers with an increasing magnitude of tensile stress.

Figure 21.16 illustrates selected images of nanocantilevers after the release layer is selectively etched. The images were taken with a video microscope (Fig. 21.16a) and SEM (Fig. 21.16b). As seen, the nanocantilevers curl up as designed.

Fig. 21.14 The 1 μm cantilever array mask designs with widths ranging from 10 to 1,000 nm

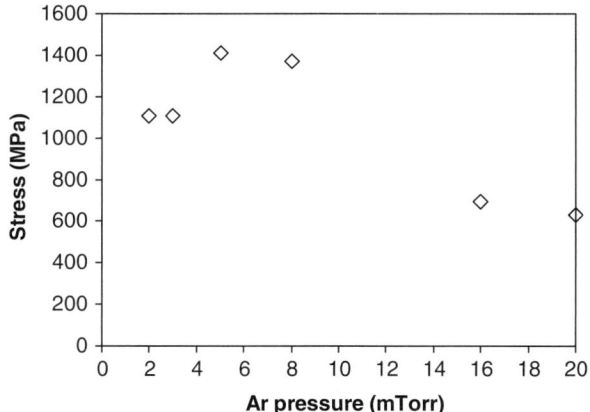

Fig. 21.15 Intrinsic stress vs. Ar pressure for sputter-deposited Cr

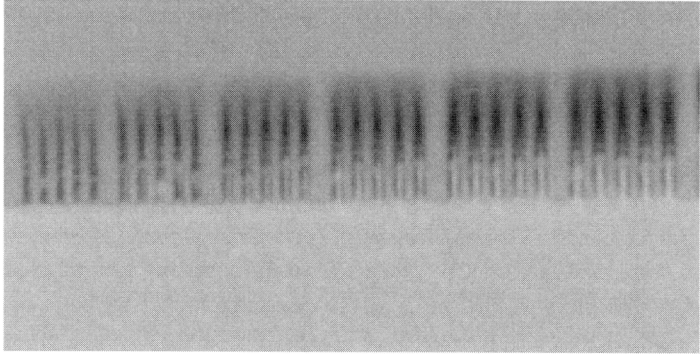

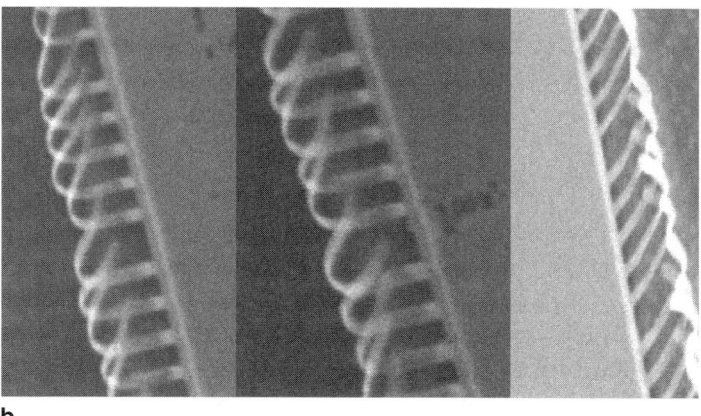

Fig. 21.16 Video microscope (**a**) and SEM (**b**) images of the released free-standing nanocantilevers

21.9 Summary

The stress-engineered compliant interconnect is a novel technology that is based on the thin film metal deposition by DC magnetron sputtering with an engineered stress gradient through the thickness of the thin film. Because of its photolithography-based fabrication, the stress-engineered compliant interconnect can satisfy the fine pitch requirement for high density chip-to-next level interconnection in the next-generation microelectronic packages. Additionally, its fabrication process is compatible with the front-end wafer level IC fabrication processes. The interconnect's good mechanical compliance and high release clearance enable new packaging configurations, such as nonsoldered packages and the free-air sliding package. The nonsoldered and free-air sliding contact packages can greatly reduce the process time and cost. They are also consistent with the trend of environmentally-friendly packaging. Another potential advantage is that the proposed process steps do not involve high temperature, which might damage the IC devices and induce excessive thermal stresses.

The compliance of the interconnect is important for the thermo-mechanical reliability, as it can accommodate the differential displacement caused by the CTE mismatch. The linear stress-engineered compliant interconnect can provide an excellent out-of-plane compliance that can meet the compliance requirement of the first-level interconnect, while the in-plane compliance is limited and has a strong orientation-dependence. To improve the in-plane compliance, an alternate stress-engineered compliant interconnect, called "J-spring", has been designed and fabricated. The "J-Spring" has been proposed to have a spiral free-standing structure after releasing, which can provide a better in-plane compliance than the linear stress-engineered compliant interconnect.

Two assembly processes have been developed for the stress-engineered compliant interconnect package without using solder: One is a nonsoldered/underfilled package, and the other is a free-air sliding contact package. The assembly has been conducted on a customized optical alignment assembly station with five degrees of freedom and with submicron accuracy. Both of these two assembly processes can be conducted at room temperature, without the potential damage to the IC devices and package. The UV curing can greatly reduce the assembly time. The assembled packages have been subjected to the thermal cycling tests to assess the thermo-mechanical reliability.

The geometries have been scaled to nanoscale. When bioconjugated, the nanoscale studies can be used for biosensing applications. Monoclonal antibodies have been immobilized on the nanoscale structures, and are intended to detect extremely low levels of circulating antigens associated with human cancer. Thus, microscale and nanoscale structures can be fabricated using stress-engineering principle, and can be used for microelectronic packaging, microelectronic probing, bio-sensing, and other applications.

References

1. Ho, R., Mai, K., and Horowitz, M., "The Future of Wires," *Proceedings of the IEEE*, vol. 89, no. 4, pp. 490–504, 2001
2. Hofstee, H.P, "Future Microprocessors and Off-Chip SOP Interconnect," *IEEE Transactions on Advanced Packaging*, vol. 27, no. 2, pp. 301–303, 2004
3. International Technology Roadmap for Semiconductor – 2002 Update, http://public.itrs.net/, 2002
4. Lau, J.H. and Pao, Y.S., *Solder Joint Reliability of BGA, CSP, Flip Chip, and Fine Pitch SMT Assemblies*, New York: McGraw-Hill, ISBN 0-070-36648-9, 1996
5. Tummala, R.R, ed., *Fundamentals of Microsystems Packaging*, New York: McGraw-Hill, 2001
6. Viswanadham, P. and Singh, P., *Failure Modes and Mechanisms in Electronic Packages*, New York: Chapman & Hall, 1998
7. Ghaffarian, R., "Chip-Scale Package Assembly Reliability," *Chip Scale Review*, Nov 1998
8. Chang, C.S., Oscilouski, A., and Bracken, R.C., "Future Challenges in Electronics packaging," *Circuits & Devices*, vol. 14, no. 2, pp. 45–54, 1998

9. Fjelstad, J., "The Evolution of Area Array Packaging from BGA to CSP," Proceedings of SEMICON Europa, Geneva, Switzerland, April, 1998
10. Fjelstad, J., "Wafer Level Packaging of Compliant CSPs Using Flexible Film Interposers," HDI: The Magazine of High-Density Interconnect, April 1999
11. Kim, Y.G., Mohammed, I., Seol, B.S., Kang, T.G., "Wide Area Vertical Expansion (WAVE™) Package Design for High Speed Application: Reliability and Performance." Proceedings of the 51th Electronic Components and Technology Conference (ECTC), 2001, pp. 54–62
12. Novitsky, J. and Miller, C., "MicroSpringTM Contacts on Silicon: Delivering Moore's Law-type Scaling to Semiconductor Package, Test and Assembly", Proceedings of 2000 HD International Conference on High-Density Interconnect and Systems Packaging (SPIE Vol. 4217), Colorado, USA, p. 250, 2000
13. Tracy, N.L., Rothenberger, R., Copper, C., Corman, N., Biddle, G., Mattews, A., McCarthy, S., "Array Sockets and Connects Using MicroSpringTM Technology," 26th IEEE/CPMT International Electronics Manufacturing Technology Symposium, Oct 2000, pp. 129–140
14. Patel, C.S., Martin, K.P., Meindl, J.D., "Performance Issues in High-Density Printed Wiring Board Design for High I/O Compliant Wafer Level Packages," 2nd Annual Semiconductor Packaging Technologies Symposium, July 1999
15. Bakir, M.S. Bakir, H.A. Reed, H.D. Thacker, P.A. Kohl, K.P. Martin, and J.D. Meindl. "Sea of leads (SoL) ultrahigh density wafer-level chip input/output interconnections for gigascale integration (GSI)," IEEE Transactions on Electron Devices, vol. 50, no. 10, October, 2003, pp. 2039–2048
16. Bakir, M.S., Reed, H.A., Kohl, P.A., Martin, K.P., and Meindl, J.D., "Sea of Leads Ultra High-Density Compliant Wafer-Level Packaging Technology", Proceedings of 52nd Electronic Components and Technology Conference, San Deigo, CA, p. 1087, 2002
17. Zhu, Q., Ma, L., and Sitaraman, S.K., "Design and Fabrication of β-Helix – an Off-Chip Interconnect", Proceedings of ASME International Mechanical Engineering Congress (ASME- IMECE), New Orlean, LA, 2002
18. Zhu, Q., Ma, L., and Sitaraman, S.K., "Design Optimization of One-Turn Helix – a Novel Compliant Off-Chip Interconnect," *IEEE Transactions on Advanced Packaging*, vol. 26, no. 2, pp. 106–112, 2003
19. Zhu, Q., Ma, L., and Sitaraman, S.K., "Development of G-Helix Structure as Off-Chip Interconnect," *Journal of Electronic Packaging*, ASME Transactions, vol. 126, pp. 237–246, 2004
20. Kacker, K., Sokol, T., and Sitaraman, S.K., "FlexConnects: A Cost-Effective Implementation of Compliant Chip-to-Substrate Interconnects," Proceedings of 57th Electronic Components and Technology Conference, IEEE-CPMT and EIA, Reno, NV, pp. 1678–1684, 2007
21. Smith, D.L. and Alimonda, A.S., "A New Flip Chip Technology for High-Density Packaging", Proceedings of 46th Electronic Components and Technology Conference, pp. 1069, 1996
22. Ma, D.L.; Zhu, A. Q.; Sitaraman, S.; Chua, C. L.; Fork, D. K., "Compliant cantilevered spring interconnects for flip-chip packaging." Proceedings of the 51st Electronic Components and Technology Conference, May 29-June 1, 2001, Orlando, FL., pp. 761–766
23. Hoffman, D.W. and Thornton, J.A., "Internal Stresses in Cr, Mo, Ta and Pt Films Deposited by Sputtering from a Planar Magnetron Source", *Journal of Vacuum Science Technology*, vol. 20, no. 3, p. 355, 1982
24. Windischmann, H., "Intrinsic Stress in Sputter-Deposited Thin Films", *Critical Reviews in Solid State and Material Sciences*, vol. 17, no. 6, p. 547, 1992
25. Ma, L., Zhu, Q and Sitaraman, S.K., "Mechanical and Electrical Study of Linear Spring and J-Spring," Proceedings of the ASME International Mechanical Engineering Congress and Exposition, November, 2002, New Orleans, LA, IMECE2002/EPP-39683
26. Chua, C.L., Fork, D.K., and Hantschel, T., "Densely Packed Optoelectronic Interconnect Using Micromachined Springs", *IEEE Photonics Technology Letters*, vol. 14, no. 6, p. 846, 2002
27. JESD22-A104-B, "JEDEC Standard – Temperature Cycling", JEDEC Solid State Technology Association, Electronic Industries Alliance, July, 2000

28. Pascoe, K.J., *Properties of Materials for Electrical Engineers*, New York: Wiley, ISBN 0–471–66911–3, p. 163, 1973
29. Klein, K.M., Zheng, J., Gewirtz, A., Sarma, D.S.R., Rajalakshmi, S., and Sitaraman, S.K., "Array of Nano-Cantilevers as a Bio-Assay for Cancer Diagnosis," Proceedings of 55th Electronic Components and Technology Conference, IEEE-CPMT and EIA, Orlando, FL, pp. 583–587, 2005
30. Klein, K.M. and S.K. Sitaraman. "Compliant stress-engineered interconnects for next generation packaging," Proceedings of the ASME International Mechanical Engineering Congress and Exposition, Anaheim, CA, November 13–19, 2004, IMECE2004-61990

Unused references

Archard, J.F., "Contact and Rubbing of Flat Surfaces", *Journal of Applied Physics*, vol. 24, no. 8, p. 981, 1953

Akyuz, F.A. and Merwin, J.E., "Solution of Nonlinear Problems of Elastoplasticity by Finite Element Method", *AIAA Journal*, vol. 6, p. 1825, 1968

Antler, M., "Sliding Wear of Metallic Contacts", *IEEE Transactions on Components, Hybrids, and Manufacturing Technology*, vol. CHMT-4, no. 1, p. 15, 1981

Barber, J.R. and Ciavarella, M., "Contact Mechanics", *International Journal of Solids and Structures*, vol. 37, p. 29, 2000

Begley, M.R. and Hutchinson, J.W., "Plasticity in Fretting of Coated Substrates", *Engineering Fracture Mechanics*, vol. 62, p. 145, 1999

Campbell, D.S., "Mechanical Properties of Thin Films", in *Handbook of Thin Film Technology*, Edited by Maissel, L.I. and Glang, R., New York: McGraw-Hill, p. 12, 1970

CINDAS Micrelectronics Packaging Materials Database (version 2.21), Center of Information and Numerical Data Analysis and Synthesis, Purdue Univeristy, 1999

Challen, J.M., Oxley, P.L.B., and Hockenhull, B.S., "Prediction of Archard's Wear Coefficient for Metallic Sliding Friction Assuming A Low Cycle Fatigue Wear Mechanism", *Wear*, vol. 111, p.275, 1986

Chen, J., http://chenjian.virtualave.net/packaging/, 2003

Chidambaram, N.V., "A Numerical and Experimental Study of Temperature Cycle Wire Bond Failure", Proceedings of 41st Electronic Components Technology Conference, Atlanta, GA, p. 877, 1991

Cuomo, J.J., Harper, J.M.E., Garnieri, C.R., Yee, D.S., Attansio, L.J., Wu, C.T., and Hammond, R.H., "Modification of Niobium Film Stress by Low Energy Ion Bombardment During Deposition", *Journal of Vacuum Science Technology*, vol. 20, p.349, 1982

Dais, J.L. and Howland, F.L., "Fatigue Failure of Encapsulated Gold-Beam Lead and TAB Devices", *IEEE Transaction on Components, Hybrids and Manufacturing Technology*, vol. CHMT-1, no. 2, p. 158, 1978

D'Heurle, F.M., "Aluminum Films Deposited by RF Sputtering" *Metallurgy Transaction*, vol. 1, p. 725, 1970

Dumas, G. and Baronet, C.N., "Elasto-plastic Indentation of a Half-space by a Long Rigid Cylinder", *International Journal of Mechanical Sciences*, vol. 13, p. 519, 1971

Emory, R., Emory, R., "Mechanical and Electrical Considerations of Compliant Interconnect", Internal Project Report, Intel Corporation, 2001

Finegan, J.D. and Hoffman, R.W., "Stress and Stress Anisotropy in Evaporated Iron Films", *Journal of Applied Physics*, vol. 30, p. 597, 1959

Gere, J.M. and Timoshenko, S.P., *Mechanics of Materials*, Third Edition, Boston: PWS, 1990

Giannakopoulos, A.E. and Larsson, P.L., "Analysis of Pyramid Indentation of Pressure-sensitive Hard Metals and Ceramics", *Mechanics of Materials*, vol. 25, p. 1, 1997

Giannakopoulos, A.E. and Suresh, S., "A Three-dimensional Analysis of Fretting Fatigue", *Acta Materialia*, vol. 46, no. 1, p. 177, 1998

Grover, F. W., "Inductance Calculations – Working Formulas and Tables", New York: D. Van Nostrand, p. 164, 1980
Hardy, C., Barnet, C.N., and Tordion, G.V., "Elastoplastic Indentation of a Half-space By a Rigid Sphere", *Journal of Numerical Methods in Engineering*, vol. 3, p. 451, 1971
Harper, J.M.E., Cabral Jr., C., Andricacos, P.C., Gignac, L., Noyan, I.C., Rodbell, K.P., and Hu, C.K., "Mechanisms for Microstructure Evolution in Electroplated Copper Thin Films Near Room Temperature," *Journal of Applied Physics*, vol. 86, no. 5, pp. 2516–2525, 1999
Hertz, H., "On the Contact of Elastic Solids", *J. Reine und Angew. Math.*, vol. 92, p. 156, 1882, (in German)
Hills, D.A., Nowell, D., and Sackfield, A., *Mechanics of Elastic Contacts*, Oxford: Butterworth-Heinemann, ISBN 0–750–60540–5, 1993
Hoffmann R.W., "The Mechanical Properties of Nonmetallic Thin Films", in *Physics of Non-Metallic Thin Films*, NATO Advanced Study Institute Series, Edited by Dupuy, C.H. and Cachard, A., vol. B-14, New York: Plenum, p. 273, 1976
Hoffman, D.W. and Gaerttner, M.R., "Modification of Evaporated Chromium by Concurrent Ion Bombardment", *Journal of Vacuum Science Technology*, vol. 17, p.425, 1980
Iscoff, R., "Demands For Higher Speed and Greater Accuracy Are Driving the Die Placement Equipment Market", *Chip Scale Review*, Jan–Feb, 2001
Johnson, K.L., *Contact Mechanics*, U.K.: Cambridge University Press, ISBN 0-521-34796-3, 1985
Kapoor, A., "A Re-evaluation of the Life to Rupture of Ductile Metals By Cyclic Plastic Strain", *Fatigue Fracture Engineering Material Structure*, vol. 17, no. 2, p. 201, 1994
Kasap, S.O., *Principles of Electronic Materials and Devices*, Second Edition, New York: McGraw-Hill, ISBN 0-072-45636-1, p. 111, 2001
Kim, W., Madhavan, R., Mao, J., Choi, J., Choi, S., Ravi, D., Sundaram, V., Sankararaman, S., Gupta, P., Zhang, Z., Lo, G., Swaminathan, M., Tummala, R., Sitaraman, S., Wong, C.P., Iyer, M., Rotaru, M., and Tay, A., "Electrical Design of Wafer Level Package on Board for Gigabit Data Transmission," Proceedings of Electronics Packaging and Technology Conference, Singapore, pp. 150–159, 2003.
Kragelsky, I.V., *Friction and Wear*, London: Butterworths, 1965
Lau, J.H., *Flip Chip Technologies*, New York: McGraw-Hill, ISBN 0-07-036609-8, 1995
Lau, J.H., ed., *Flip Chip Technologies*, New York: McGraw-Hill, 1996.
Lau, J.H., Rice D.W., and Avery, P.A., "Elastoplastic Analysis of Surface-Mount Solder Joints", *IEEE Transaction on Components, Hybrids and Manufacturing Technology*, vol. CHMT-10, no. 3, p. 346, 1987
Lau, J.H., Rice, D.W., and Harkins, C.G., "Thermal Stress Analysis of Tape Automated Bonding Packages and Interconnections", *IEEE Transaction on Components, Hybrids and Manufacturing Technology*, vol. CHMT-13, no. 1, p. 182, 1990
Lau, J.H. and Lee, S.W.R., "Reliability of 96.5Sn-3.5Ag Lead-Free Solder-Bumped Wafer Level Chip Scale Package (WLCSP) on Build-Up Microvia Printed Circuit Board", Proceedings of 2001 International Conference on High-Density Interconnect and System Packaging, Santa Clara, CA, p. 314, 2001.
Lee, C.H., Masaki, S., and Kobayashi, S., "Analysis of Ball Indentation", *International Journal of Mechanical Sciences*, vol. 14, p. 417, 1972
Li, D.L., Light, D., Castillo, D., Beroz, M., Nguyen, M., and Wang, T., "A Wide Area Vertical Expansion (WAVETM) Packaging Process Development", Proceedings of 51st Electronic Components and Technology Conference, Lake Buena Vista, FL, p. 367, 2000
Lingk, C., Gross, M.E., and Brown, W.L., "Texture development of blanket electroplated copper films," *Journal of Applied Physics*, vol. 87, no. 5, pp. 2232–2236, 2000
Loctite Corporation, "Product 3335 – Technical Data Sheet", Nov 2000
Material Database of MEMSnet, http://www.memsnet.org/material/, 2003
Meyers, R.H. and Montgomery, D.C., *Response Surface Methodology: Process and Product Optimization Using Designed Experiments*, New York: Wiley, 1995
Modi, M., "Fracture in Stress-Engineered, High Density, Thin Film Interconnects", Ph.D. Dissertation, Woodruff School of Mechanical Engineering, Georgia Institute of Technology, 2003

Muskhelishvili, N.I., *Singular Integral Equations, Boundary Problems of Function Theory and their Application to Mathematical Physics*, Gronigen: Noordhoff, 1953

Nordic Electronics Packaging Guideline, http://extra.ivf.se/ngl/, developed by Danish Electronics, Lights and Acoustics (DELTA), IVF – Swedish Institute of Production Engineering Research, SINTEF – Norway, VTT – Technical Research Center of Finland

Pan, T., "Thermal Cycling Induced Plastic Deformation in Solder Joints I: Accumulated Deformation in Surface Mount Joints", Transactions of the ASME. *Journal of Electronic Packaging* vol. 113, no. 1, p. 8, 1991

Pan, T., "Thermal Cycling Induced Plastic Deformation in Solder Joints I: Accumulated Deformation in Through-hole Joints", *IEEE Transactions on Components, Hybrids and Manufacturing Technology*, vol. 14, no. 4, p. 824, 1991

Pang, J.H.L. and Tan, C.K., "Thermal Analysis of a Wire Bond Chip-On-Board Package", ITherm'98, The Sixth InterSociety Conference on Thermal and Thermomechanical Phenomena in Electronic Systems, Seattle, Washington, p. 481, 1998

Perez-Prado, M.T. and Vlassak, J.J., "Microstructural evolution in electroplated Cu thin films," *Scripta Materialia*, vol. 47, no. 12, pp. 817–823, 2002

Qin, I.W., "Wire Bonding: The Preferred Interconnect Method", Chip Scale Review, Nov–Dec, 2002

Rabinowicz, E., "The Least Wear", *Wear*, vol. 100, p. 533, 1984

Semiconductor Industry Association, International Technology Roadmap for Semiconductors: Assembly and Packaging, 2005.

Skalski, K., "Contact Problem Analysis of an Elastoplastic Body", Prace Naukowe Mechanica, z. 67, Warsaw Polytechnic, 1979

Smith, L.D., Anderson, R.E., Forehand, D.W., Pelc, T.J., and Roy, T., "Power Distribution System Design Methodology and Capacitor Selection for Modern CMOS Technology," *IEEE Transactions on Advanced Packaging*, vol. 22, pp. 284–291, 1999

Suresh, S., *Fatigue of Materials*, Second Edition, U.K.: Cambridge University Press, ISBN 0-521-57847-7, 2001

Swaminathan, M., Kim, J., Novak, I., and Libous, J., "Power Distribution Networks for System on a Package: Status and Challenges," *IEEE Transactions on Advanced Packaging*, vol. 27, no. 2, pp. 286–300, 2004

Syed, A.R. and Doty, M., "Are We Over Designing for Solder Joint Reliability? Field vs. Accelerated Conditions, Realistic vs. Specified Requirements", Proceedings of 49th Electronic Components and Technology Conference, San Diego, CA, p. 111, 1999

Tian, H., Saka, N. and Rabinowicz, E., "Friction and Failure of Electroplated Sliding Contacts", *Wear*, vol. 142, p. 57, 1991

Timoshenko, S. and Goodier, J.N., *Theory of Elasticity*, Third Edition, New York: McGraw-Hill, 1951

Torrance, A.A. and Buckley, T.R., "A Slip-line Field Model of Abrasive Wear", *Wear*, vol. 196, p. 35, 1996

Truman, C.E., Sackfield, A. and Hill, D.A., "Contact Mechanics of Wedge and Cone Indenters", *International Journal of Mechanics Science*, vol. 37, no. 3, p. 261, 1995

Tummala, R.R., *Fundamentals of Microelectronic Packaging*, New York: McGraw-Hill, ISBN 0-071-37169-9, 2001

Tummala, R.R., Orientation Presentation at Industrial Advisory Board Meeting, Georgia Institute of Technology – Packaging Research Center, 2003

Tummala, R.R, "Special Issues on System on a Package (SOP)," IEEE Transactions on Advanced Packaging, vol. 27, no. 2, 2004.

Chapter 22
Flip Chip Packaging for Nanoscale Silicon Logic Devices: Challenges and Opportunities

Debendra Mallik(✉), Ravi Mahajan, and Vijay Wakharkar

22.1 Introduction

After decades of following the roadmap laid out by Moore's law [1], silicon features have reached the nanoscale, which is below 100 nm in dimension, as illustrated in Fig. 22.1. The first logic products with 90-nm transistors, using the traditional silicon dioxide insulator and polysilicon gate, went into volume production in 2003. More recently in 2007, 45-nm devices using a revolutionary high-k metal gate transistor technology have been introduced [2, 3]. These nanoscale devices enable higher performance circuits, which in turn drive advanced features in their packaging. These devices can significantly lower the power consumption of high-performance logic products creating new applications in the fast-growing ultramobile market (Fig. 22.2) and thus requiring packaging to support the demands of these form factors. This chapter will discuss the challenges and opportunities in flip chip packaging for these nanoscale devices.

In the early days of the semiconductor industry, microelectronics packaging primarily provided space transformation, and structural and environmental protection of the small but expensive integrated circuit (IC) devices so that they could be connected to relatively large electronic system boards. The role of microelectronics packaging has continued to expand over the past few decades to include electrical and thermal performance management, as well as enabling system miniaturization. Containing the cost of packaging and meeting environmental regulations have been critical constraints during this evolution.

Electronics packaging technology has made significant advances over the years. To effectively describe these advances and future challenges, we will primarily focus on logic ICs in computing systems such as mobile PCs (personal computers), desktop PCs, and servers, in this chapter. That is because such systems generally need advanced packaging, driven mainly by the high-performance levels of the microprocessors within them. We will also use examples of ultramobile systems typified by the cell phone to describe some of the form-factor-driven packaging technology.

D. Mallik
Intel Corporation, Chandler, AZ, USA

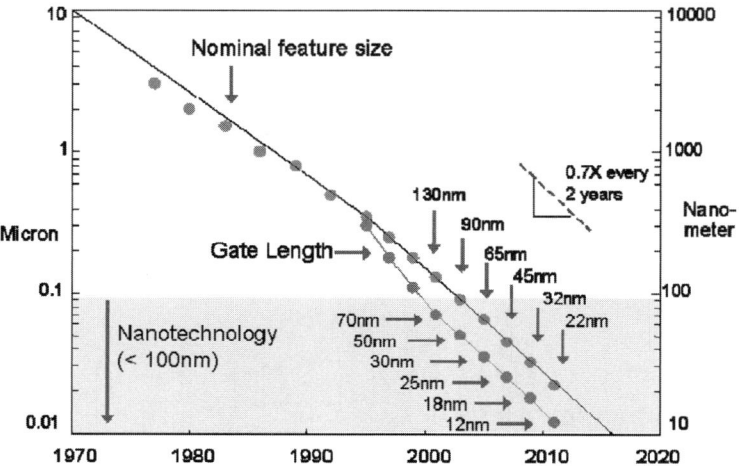

Fig. 22.1 Feature size and transistor gate length scaling [4]

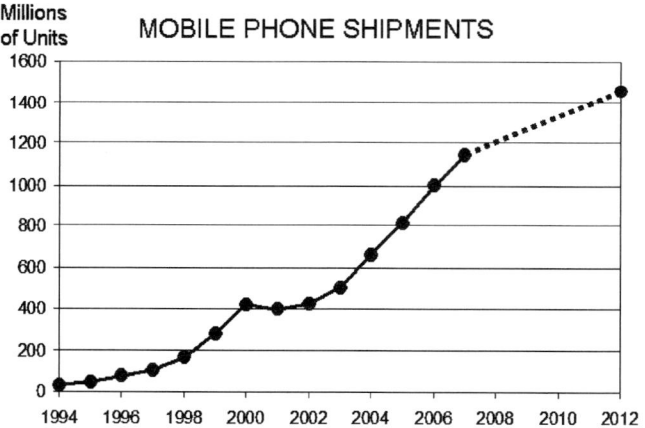

Fig. 22.2 Mobile phone market trend (source: Prismark Partners LLC)

The evolution of packages for the desktop PC is shown in Fig. 22.3. In the early 1980s, the 8088 processor chip was housed in a ceramic dual in-line package (CDIP). It used gold wire-bonds to interconnect the silicon chip to the conducting leads on the ceramic package. This 800-mm² package had a total of 40 leads placed along its two long sides. With an operating frequency of only 5 MHz and low power of about 1–3 watts, fewer than 10% of the leads were needed to supply power to the chip. The remainder of the leads was used for signal transfer in and out of the microprocessor. The primary function of this package was to provide space transformation and environmental protection. By 1994, the Pentium® Pro processor used a 3,000-mm² ceramic pin grid array (CPGA) package with 387 pins, a

---

® Pentium is a registered trademark of Intel Corporation and its subsidiaries in the United States and other countries.

large copper-tungsten heat slug, and two chips – the CPU chip and a separate large SRAM cache chip. Over 40% of the pins were dedicated to deliver power to the chips, illustrating the increasing importance of power delivery. By the mid-1990s, cost and conductor resistance of the ceramic packages drove a significant shift in packaging technology. CPU packages for desktop PCs migrated to the plastic pin grid array (PPGA) technology, which used copper conductors on an organic substrate, unlike the tungsten conductors on a ceramic substrate. PPGA technology continued to use wire bonds to interconnect the silicon to the package. The wire bonding process requires the interconnect pads to be placed only on the periphery of the die. The inductance of the long wires and resistance of the long on-chip interconnects degraded power delivery performance. Additionally, placing both power and signal pads near the die periphery limited the ability to shrink the size of the chip. These limitations drove the other significant technology transitions in the 1990s. By 1999, advanced processors such as the Pentium III migrated to organic flip chip ball grid array (FCBGA) [5, 6] and later to flip chip pin grid array (FCPGA) packages. In 2004, the flip chip land grid array (FCLGA) package was introduced to eliminate the fragile package pins in market segments that could incorporate LGA.

In addition to the changes in interconnect, increase in power consumption and power density nonuniformity has led to significant improvement in power delivery and thermal management technologies. Figure 22.4 shows a specific example of how package attributes have advanced to keep pace with Moore's law-driven product requirements.

Ultramobile systems, by nature, need miniaturized devices. As a consequence, the key components in these systems, such as the logic, memory, and wireless devices, have evolved to provide more computing, communication, and memory content within the package volume. The evolution of key packages for such systems is shown in Fig. 22.5. The traditional scaling of the packages' horizontal dimension has continued throughout this evolution. Additionally, there has been significant progress in leveraging the packages' vertical dimension to achieve high density of IC content.

Fig. 22.3 Evolution of microprocessor packaging for the desktop PC

	1999		2007
Silicon Process	: 180 nm	Silicon Process	: 45 nm
Package Size	: 49.5 mm (sq.)	Package Size	: 37.5 mm (sq.)
Package Line/Space	: 73 µm	Pkg Line/Space	: 14 µm
Thermal Demand	: 23 W	Thermal Demand	: 130 W
Power Delivery Impedance	: 4.5 mΩ	Power Delivery Impedance	: 1.2 mΩ
CPU Frequency	: 700 MHz	CPU Frequency	: 3 GHz
Front Side Bus	: 100 MHz	Front Side Bus	: 1333 MHz
Die bump	: Pb-Sn	Die bump	: Cu
Package bump	: Pb-Sn Eutectic	Package bump	: SnAgCu

Fig. 22.4 Illustrative example showing how packaging has kept pace with silicon and microprocessor advances

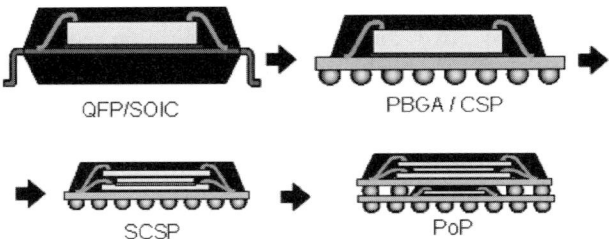

Fig. 22.5 Evolution of packaging for the ultramobile electronic systems

Material technologies play a significant role in packaging. Performance requirements, such as thermal management, power delivery, and signal integrity, as well as structural integrity, environmental and manufacturing considerations, have driven improvement in material technologies. This trend is expected to continue with progress along Moore's law and with the scaling down of system form factors in ultramobile systems. Materials technology evolution is also affected by environmental regulations. In recent years, the electronic packages have been engineered to be environmentally friendly by complying with restrictions on hazardous materials. The environment-friendly initiative primarily consists of providing Pb-free packaging material solutions, as well as enabling of halogen-free substrate technology to eliminate the use of brominated flame retardants. The industry has responded quite well by eliminating Pb from high-performance microprocessor packaging [3, 7] and continues to set the strategy for implementation of halogen-free packaging.

In the following sections we will examine the trends for key characteristics of flip chip logic packaging discussed in this introduction in greater detail.

22.2 Space Transformation

One primary function of the package in an electronic system is to provide cost-effective space transformation. The package transforms the high density of terminals on the small IC chip to a lower density of terminals on the larger package body. The lower density package terminals allow the system board to have larger feature sizes in order to manage its cost. Thus, the package becomes part of the interconnect between one IC component and another. There are three distinct areas of interconnect related to packaging. The portion of the interconnect that enables a transition from the die level to the package level is known as the first level interconnect (FLI). The interconnect within the package that creates the space transformation is referred to as routing. The portion of the interconnect that enables a transition from the package to the next level, which is typically the motherboard, is referred to as the second level interconnect (2LI). Note that the number of 2LI terminals does not have to be equal to the number of FLI terminals. That is because package routing may combine a group of FLI terminals to a group of 2LI terminals based on electrical and manufacturing considerations. Historically, there has been continued increase in demand for the numbers of contact terminals for both FLI and 2LI with every succeeding product generation. This is due to the inclusion of more functionality on the chip, higher data bandwidth between chips, and the need for smaller variation in supply voltage for the chip. The increased interconnect count leads to scaling of feature sizes across all domains of packaging. This trend is expected to continue as the semiconductor industry pursues the Moore's law-driven roadmap.

22.2.1 Die-Package Interconnect

The die-package interconnection for a flip chip package is typically made by a process called controlled collapse chip connection, or C4. In the original version of C4, solder bumps on the die were aligned on top of corresponding metal pads of the ceramic package and were then reflown to form joints with a controlled standoff height [8]. A variation of this chip joining process is predominantly used for organic packaging, and connects nonmelting die bumps to reflowable solder bumps on the package substrate. This process is also loosely referred to as a C4 process. Initially, die bumps for organic packages used high-melt (nonmelting) Pb–Sn solder (3–5% Sn and 97–95% Pb) with a melting temperature near 312°C. The substrate bumps used eutectic Sn–Pb solder with a melting temperature of 183°C (Fig. 22.6a). Environmentally friendly considerations resulted in the use of Pb-free

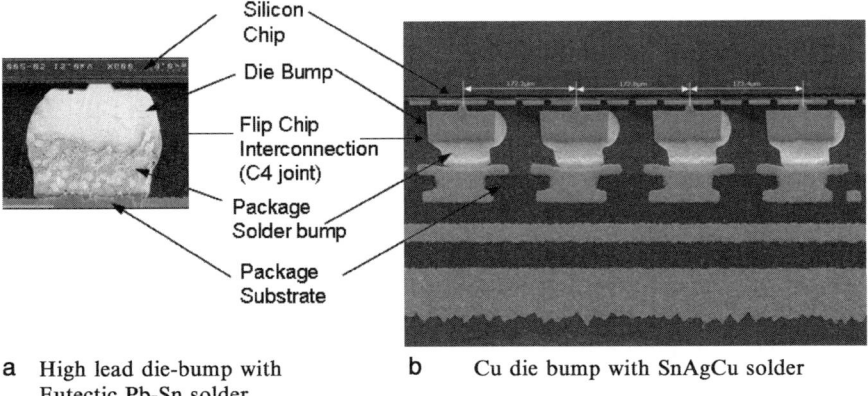

a High lead die-bump with Eutectic Pb-Sn solder

b Cu die bump with SnAgCu solder

Fig. 22.6 Examples of flip chip C4 joints. (**a**) High lead die bump with (**b**) Cu die bump with SnAgCu (SAC) solder eutectic Pb–Sn solder

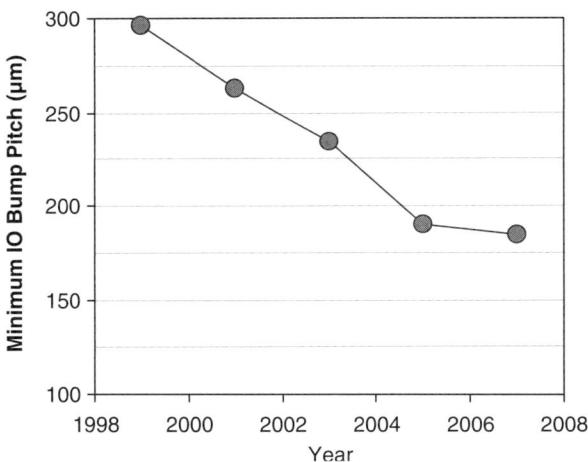

Fig. 22.7 Minimum I/O bump pitch trend for Intel microprocessors

material such as copper die bumps and SnAgCu (SAC) substrate solder as shown in Fig. 22.6b [3].

The increased number of terminals on a given chip size requires reduction in the die bump pitch to accommodate all the bumps. Figure 22.7 shows minimum input/output (I/O) bump pitch trend for mainstream microprocessors. Finer bump pitch reduces C4 joint size and the space between the joints, creating technology challenges related to the FLI. During the package assembly process, the finer bump pitch requires advanced equipment and material to precisely place the die and hold it in place until the C4 joints are formed. The smaller joint size reduces the solder volume and consequently the amount of solder collapse during the chip join reflow process. This demands good coplanarity for die bumps and for the mating package substrate contact area to reduce the risk of electrically open C4 joints.

The finer bump pitch also impacts the choice of underfill material and process due to the reduction in horizontal spacing between the C4 joints and the reduction in vertical standoff height between the die and the package. The reduced space restricts the size of the filler particles, which are used in the underfill material to lower its coefficient of thermal expansion (CTE) and to increase its fracture toughness. Also, the narrow spacing reduces control over the underfill flow dynamics, since the flow speeds may become more sensitive to manufacturing and design variations. Therefore, the finer bump pitch makes optimization of underfill process and material properties more challenging. At the same time, the reduced distance between the flip chip joints makes the FLI less tolerant to any voids and other defects, which may lead to increased material migration between neighboring joints.

The small C4 joints need to deal with high current densities through them. These high current densities are a result of increasing total current (even at the same power consumption due to decreasing device voltage) and/or due to nonuniformity of on-chip current driven by variation in activity levels of devices at different locations on the die. If not addressed properly through design, material, and process choices for the FLI, the high current density, in the presence of high silicon-chip temperature, may lead to electromigration-driven open failures [9].

Another important consideration for the design of the C4 joint is its influence on thermomechanical stresses in the on-chip interconnect. Over the past few generations, silicon process engineers have focused on reducing the dielectric constant of the dielectric material (low k) used in the on-chip interconnect. This has also resulted in lowering the mechanical strength of the on-chip interconnect making it more susceptible to failures induced by thermomechanical stresses due to packaging. The C4 joint needs to be compliant enough to minimize packaging-induced stresses on the silicon.

In summary, the material and process choices for the FLI need to ensure assembly process scalability to smaller dimensions, while still providing compliancy between the silicon and the package, managing high current density through the joint, and being environmentally friendly.

22.2.2 Within Package Interconnect

The bump pitch scaling drives the need for finer line width and spacing to route the signals from the on-chip terminals to the 2LI terminals. The high routing density is typically needed to escape the signal lines out of the congested area near the chip. Before we discuss the trend in within-package interconnect, we briefly describe the organic flip chip package substrate structure. Figure 22.8 shows a schematic of the cross section of an organic flip chip package substrate and a top view of its primary routing layer.

The substrate is generally built around a core layer introduced to provide structural rigidity to the package. The core is made up of glass-fiber-reinforced polymer with mechanically drilled plated through-hole (PTH) vias, and conducting copper layers

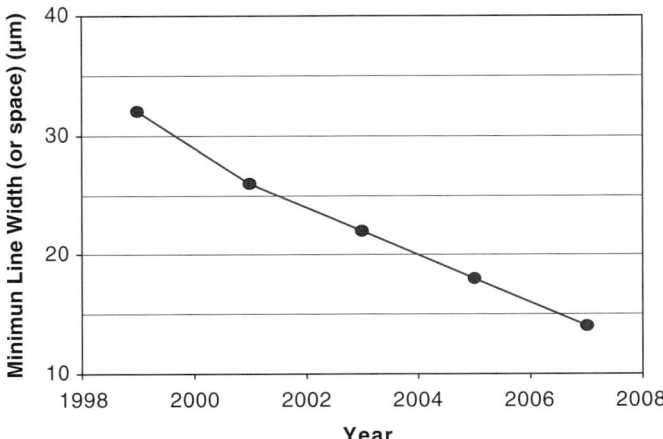

Fig. 22.8 Sketch of a typical organic flip chip package substrate

fabricated by a subtractive photolithography process, which is similar to the widely used process for printed wiring board (PWB) technology. Typical thickness of the core is about 800 μm and the PTH diameter is about 250 μm. Buildup layers are added on both sides of the core layer, one symmetric pair at a time. A buildup layer is typically fabricated by first adding a relatively thin dielectric layer, generally, around 30-μm thick. Then, microvias, i.e., sub-100-μm diameter vias, are created by laser or other means. Copper patterns of about 15-μm thick are then added to form the interconnect layer and via connections to the underlying metal layer. The predominant process for forming the buildup layer conductors is the semiadditive process (SAP). Compared with the traditional subtractive process, the SAP enables formation of finer line widths and spaces on the package layers. The buildup layer pairs are sequentially added until the desired layer count is achieved. A protective solder resist layer is then added to the outer surfaces. This is followed by finish metallization such as electroless NiAu or NiPdAu for the C4 pads on the front side and 2LI pads on the backside. Finally, solder bumps are formed on the C4 pads to create an organic FCLGA package substrate. To create a FCPGA substrate, pins are soldered to the 2LI pads prior to chip assembly. For FCBGA, solder balls are attached to the 2LI pads after chip assembly.

Typically, the C4 pad size is relatively large when compared with routing-line width and spacing. A large C4 pad diameter is used for various reasons. A large C4 pad size increases the joint size, which, in turn, improves resistance to joint cracking failure. A large pad lowers current density through the joint and enables high current carrying capability for the C4 joint by lowering electromigration. Also, a large C4 pad allows large solder volume. During the C4 joint formation process, the increased solder volume leads to higher solder collapse and helps achieve a high yielding C4 assembly process. Figure 22.8c shows a representative example of a routing scheme and dimensions for a flip chip package.

The bump pitch scaling has consistently driven the line width and space to get finer (Fig. 22.9), and the trend is expected to continue. Finer features on the

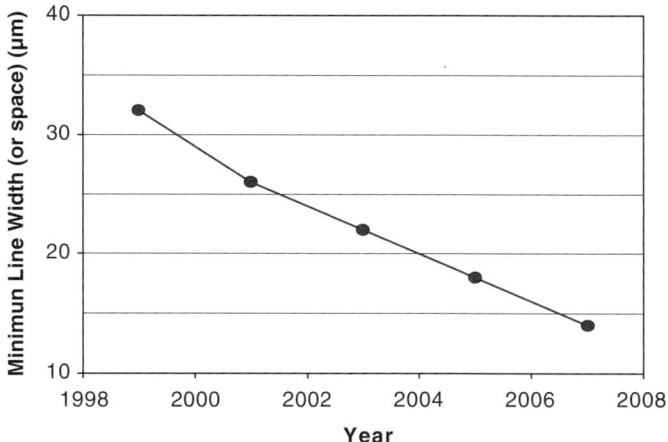

Fig. 22.9 Minimum line width and space trend for microprocessor packages

substrate bring a host of challenges to package substrate technology. These challenges include technology to maintain adhesion of the resist and conductor materials during the photolithography process, and a clean process environment to minimize foreign material-related yield and reliability issues. From a different perspective, the demand for fine line and space can be mitigated by aggressive reduction in the microvia and C4 pad sizes. This would require innovation in substrate technology as well as assembly material and process technologies to produce mechanically robust via and C4 joint with high current-carrying capability. Innovative design techniques can also mitigate the demand on line and space scaling to some extent. One example of this would be to optimize the bump placement as demanded by high-density routing instead of placing the bumps on a fixed grid pattern. Another approach is to reduce the number of lines between the pads. This technique may require additional package layers, and hence higher cost, to do the escape routing, but would lower the demand for line and space scaling.

PTH size is another feature that needs attention. Typical PTH diameter is 250 μm on a 400-μm diameter pad. The relatively large dimensions of the PTH impede high-speed signal transmission due to electrical impedance discontinuities created primarily by significant changes in dimensions between the routing lines and the PTH [10]. Additionally, they can be a limiter to interconnect scaling within the package. PTH size reduction or elimination may be needed in the future to support interconnect density scaling and high-speed signaling.

22.2.3 Package-Board Interconnect

There are three major 2LI contact technologies in use for flip chip logic products. These are the ball grid array (BGA), pin grid array (PGA), and the land grid array (LGA) technologies (Fig. 22.10). The BGA package is directly surface mounted

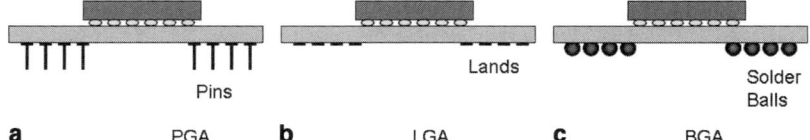

Fig. 22.10 Flip chip package types for package-board interconnect

to the motherboard, whereas the PGA or LGA package is inserted in a socket, which, in turn, is surface mounted to the motherboard. High-end components, like microprocessors, typically use the PGA or LGA format to enable field upgradability and to meet other business requirements. The PGA package is electrically connected to its socket through a clamping mechanism that makes the individual socket contacts grip the package pins. For the LGA package, electrical connection is made by pressing the package down on to the spring-like socket contact probes. Vertical force to maintain this connection is applied by a retention mechanism.

The 2LI density has been increasing over time to accommodate larger interconnect count on a given package surface area. The BGA pitch for flip chip logic products has reduced from 1.27 to 0.8 mm over the last decade. This is achieved by scaling the 2LI contact pads and solder ball diameters, changes to solder metallurgy such as various formulations of SAC solder, and managing BGA joint reliability through advanced board assembly processes such as corner glue and other board-level underfill processes. Denser escape routing on the PCB, driven by finer BGA pitch, is enabled by improved design techniques, such as placing the solder balls anywhere on the package, modest reduction in feature sizes of conventional boards, or use of a high-density board such as those with microvias [11] where it is affordable. Typical challenges with finer pitch BGA are board assembly yield, thermomechanical stress from power cycles and temperature cycles, and mechanical shock. Material and design choices have significant impact on BGA joint assembly yield and reliability [12] and are critical to continued BGA pitch shrink.

The PGA pin pitch has reduced from 1.8 to 1.27 mm over the last decade. On the one hand, further pin pitch reduction is expected to be slow due to challenges in socket to pin contact technology. On the other hand, the LGA is relatively more capable of achieving 2LI pitch reduction due to its simpler contact probe structure. Besides the board routing-driven requirements, the LGA pitch reduction is generally limited by the socket contact resistance, maximum retention load capability of the system, and package–socket alignment capabilities. Innovation in material for package land surface finish as well as design and material for socket contacts are critical in reducing the contact resistance and the load per contact.

22.3 Electrical Performance

22.3.1 Power Delivery

One of the key requirements for efficient operation of the IC device is to maintain a steady voltage level across its power supply rails (Vcc and Vss rails). A change to the power rail voltage level changes the drain-source voltage as well as the gate voltage. Too low a voltage across the power rails lowers the drain and gate voltages and thus can lead to erroneous switching of the transistor, causing failure of the entire circuit operation. Increasing the supply voltage to the power rail is theoretically a way to mitigate the low-voltage-related issues. However, the higher input voltage causes higher power dissipation and has significant impacts on the transistor reliability. The requirement to deliver a stable voltage has driven power delivery management as a major discipline within the packaging field, especially for packaging high-performance logic devices such as microprocessors.

Challenges in microprocessor power delivery fall into three areas as described below:

(a) As the transistor feature-size scales, the device operating voltage continues to scale downward (Fig. 22.11a) to maintain transistor reliability and to minimize power consumption. As microprocessor designs take advantage of silicon size and speed advances to provide improved performance, the electrical current to the microprocessor chip has increased over time, although at a diminishing rate (Fig. 22.11b). The net impact of both of these trends is that the impedance targets for the power delivery path, which is the ratio of the supply voltage to the supply current, continue to trend lower (Fig. 22.12).

(b) The power supply current draw, in typical microprocessor applications, can change suddenly and the power delivery network must be able to quickly respond to provide the charge; otherwise, the supply voltage may drop below

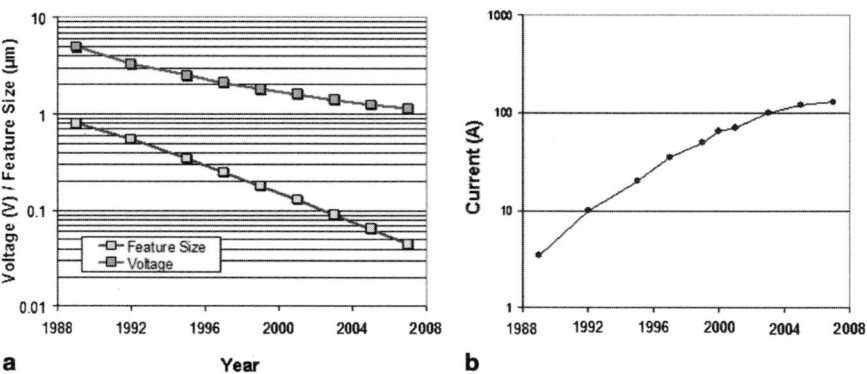

Fig. 22.11 Gate length, voltage, and current trends for Intel microprocessors

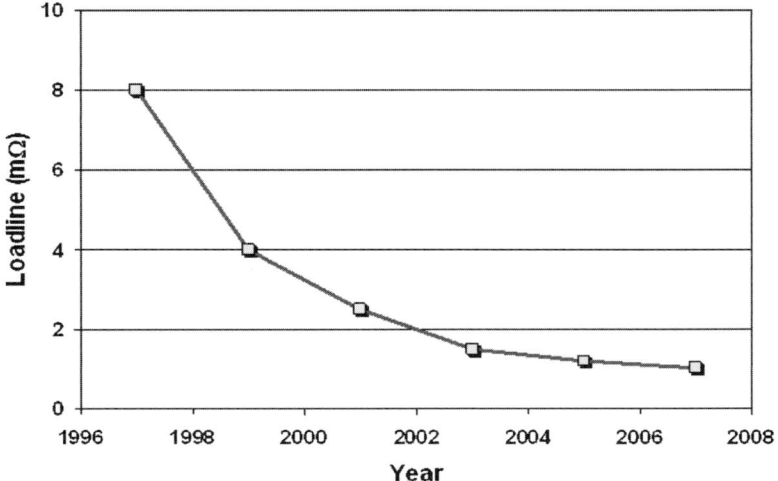

Fig. 22.12 Power-delivery target impedance (loadline) trend for Intel microprocessors

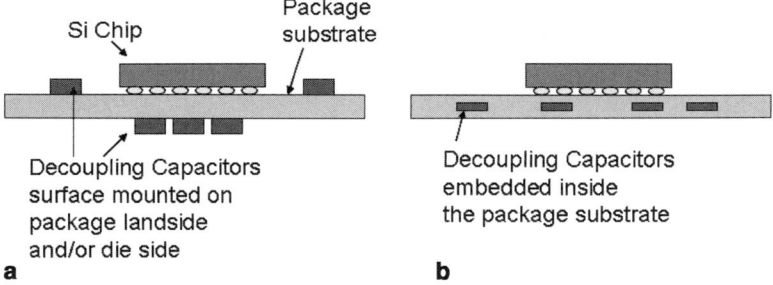

Fig. 22.13 Technologies for high-frequency decoupling

the minimum required value. This means that the power delivery network must be designed to meet the impedance targets across all frequency components of the supply current, which range from DC to high frequencies of several hundred megahertz. This task is accomplished by designing a multistage decoupling solution that has capacitors for high frequency mounted on the package (Fig. 22.13), capacitors for mid-frequency mounted on the motherboard close to the package, and capacitors for low frequency located at the output of the voltage regulator on the motherboard [13].

(c) One of the unwanted consequences of device scaling has been the increase in transistor leakage power [13]. To continue to provide increased performance within a constrained power budget, microprocessors are undergoing an architectural transition from single core to multicore designs [14]. Additionally, the use of low-power sleep states has allowed chip designers to reduce power consumption. Both these architectural transitions require careful design of the power delivery network to ensure that the power delivery paths to each microprocessor core in its various operational states are optimized for minimal impedance.

Capacitor technologies for high-frequency decoupling in microprocessor packages have evolved from the simple two-terminal (2T) capacitors to the more sophisticated interdigitated capacitors (IDC) [13]. IDCs have reduced effective Vcc-Vss loop inductance due to multiple low-inductance current paths within the capacitor. The low-inductance current paths are a result of the innovative placement of Vcc and Vss terminals this technology offers. Therefore, IDCs are more effective in reducing high-frequency noise than the 2T capacitors. Further reduction in capacitor inductance is achievable by array capacitor technology that has area array contact terminals on the surface of the capacitor and connections to the multiple conducting plated-through vias. Note that the high-frequency decoupling performance of any capacitor type can be enhanced by using many of them in parallel. However, available package surface for capacitor attachment and the total cost of decoupling need to be considered in making a selection between any capacitor type and a larger quantity of the less advanced capacitors.

As the inductance of the capacitor is reduced, the inductance of the electrical path between the capacitor and the microprocessor becomes the next major challenge. The manner in which the capacitors are connected to the chip plays a major role in the effectiveness of managing the high-frequency load line of the package power delivery path. For example, capacitors placed on the package landside, directly under the silicon die, generally provide shorter and thus better power supply performance compared with die-side capacitors. Further power delivery performance improvement can be achieved by embedding discrete or buildup capacitors within the package substrate and closest to the dispersed high-load regions of the die (Fig. 22.13b). Embedding the capacitor would require improvements to capacitor, substrate materials, and processing technologies [15].

Low-frequency impedance control is accomplished by reducing the DC resistance of the power delivery path between the motherboard voltage regulator and the microprocessor chip. Typical solutions involve reducing the resistance of sockets, increasing the number of package power and ground pins, and increasing the copper thickness for the package power and ground planes. To contain the potential increases in package size (and hence cost) due to increasing pin count, socket pin pitches have also been reduced over time. However, there are practical limits on reducing socket pitch and increasing pin count, which require exploration of alternate concepts such as separating the voltage-regulator power path and bringing it closer to the die.

22.3.2 Signal Bandwidth

Another key requirement to achieve high performance of the nanoscale electronics is to provide the computing machine with adequate data bandwidth. Increased transistor count, increased frequency, and improved design, such as multicore architectures, drive the computing performance higher. With increase in the processing capability of the CPU, the interconnect links to the CPU need to have exponentially higher data bandwidth to meet the demands of high-performance applications such as graphics. Figure 22.14 shows trends in CPU core frequency and memory bandwidth of traditional CPU and many-core architectures [11]. The graph indicates that the memory bandwidth requirement for the many-core architecture is significantly higher than the requirement for traditional CPU architectures. Therefore, revolutionary solutions are needed to synchronize data bandwidth with CPU performance while staying within the power and cost budgets.

Typically, bandwidth can be increased by increasing the data transfer rate and/or the IO count. Enabling high data rates brings big challenges to package and socket technologies. It is well known that at low frequencies, the package and socket can be treated as R, L, and C elements because their electrical length is much shorter than a wavelength. At high frequencies, the package and socket interconnects are treated as transmission lines. The material properties, manufacturing tolerances, and design scheme for signal lines, whose impact on signal integrity can be ignored at low-frequency data transfer, can have substantial impact at high-frequency data transfer.

To increase data transfer rate, signal frequency is typically increased by managing the interconnect loss and controlling the impedance variation. The primary loss in package transmission lines is due to conductor losses. Since the signal lines today already use copper material, the conductor loss minimization would require improving its surface roughness and maximizing its cross-section area, despite the interconnect density increase described in the previous section. Increasing

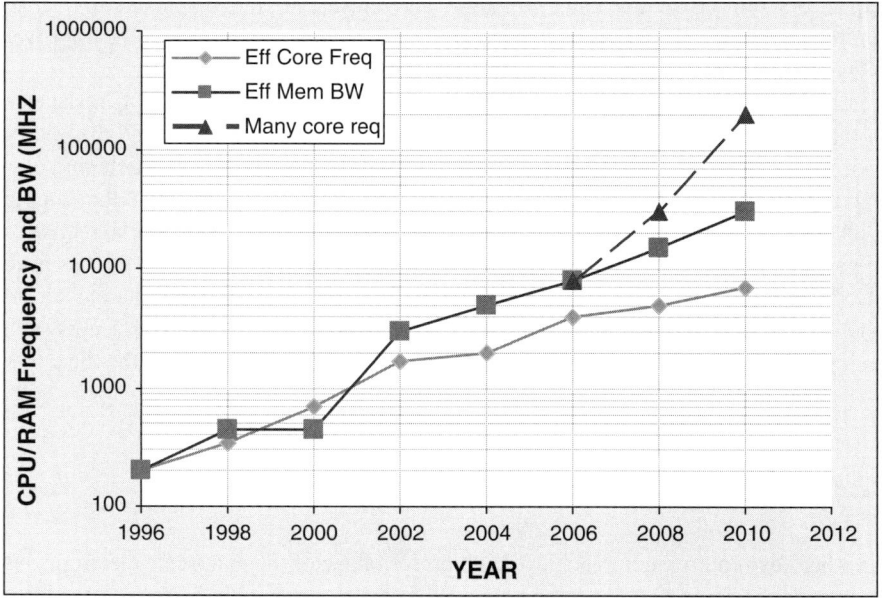

Fig. 22.14 CPU core frequency and bandwidth trend

the conductor cross section leads to high aspect ratio signal lines requiring improvements in substrate material and processing technologies. The other opportunity for higher speed signaling is impedance control. A key optimization approach is to reduce signal path capacitance so that the *characteristic* impedance can be moved up to match system impedance, resulting in a smaller return loss. This is typically achieved through design and package dielectric material improvements. As the high-speed signal travels through the package interconnect, the signal waves encounter various discontinuities, or changes in characteristic impedance. Key discontinuities within a flip chip package occur at the transitions from the signal line to the large PTH and 2LI pad [10, 16]. Each discontinuity causes a reflection and contributes to total signal delay and signal loss. As the signal speed becomes faster, the delay and loss can cause error in data transfer. Solutions for PTH discontinuities can include smaller PTHs or a coreless substrate that eliminates them. Discontinuity at the 2LI pad can be minimized by a smaller 2LI pad and/or creating a void on the metal plane above the pad to lower the pad capacitance.

Bandwidth increases from optimization of traditional interconnect paths, or from positioning of CPU, chipset, and RAM packages closer together on the motherboard may not be enough to meet the demands of high-performance computing and multimedia applications in the future. Increasing the I/O count is another approach to increase bandwidth. This drives the interconnect density increase as described in the previous section, as well as a possible increase in package size and cost. As a result, solutions significantly different from today's paradigm are needed. We show two such approaches: one uses the traditional multichip package (MCP) configuration for the CPU and the memory chips, or places CPU and the RAM side by side (2D configuration) on a single package, and the other uses a 3D stacking approach, or mounts the CPU on the RAM containing through-silicon vias (TSV). Figure 22.15 shows the basic idea of both approaches. The traditional MCP can provide a few hundred gigabytes/second bandwidth based on projected RAM performance around 2010. However, to achieve an order of magnitude of higher bandwidth, 3D stacking of the chips would be needed. The 3D stacking provides very high interconnect density and delivers bandwidth efficiency through higher data rate and improved power consumption. Significant materials development is needed to enable 3D stacking using TSV technology. Optical interconnect is yet another revolutionary approach to achieving high bandwidth [17, 18].

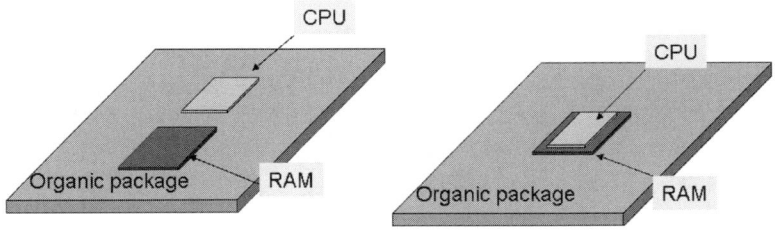

Fig. 22.15 Approaches for high bandwidth memory connection

22.4 Thermal Management

The need for thermal management is not necessarily driven by silicon scaling, however, it is an important consideration that must be thoroughly understood for all nodes on Moore's law. The thermal management problem for electronic components in a computer is one of transporting the steady state thermal design power (TDP) from the die surface, where the temperature of the hottest spot on the die is maintained at or below a certain specified temperature (typically referred to as the junction temperature T_j), to the ambient air at temperature T_a. Using a simple thermal resistance model, the required thermal resistance is represented as the ratio of $(T_j - T_a)$ over TDP.

The temperature difference $(T_j - T_a)$ is expected to trend lower over time since T_j will likely be forced lower by device reliability concerns and product performance expectations, while T_a will be forced higher due to heating of the air inside the chassis caused by increased integration and shrinking system sizes. The thermal problem could become increasingly difficult to contend with either due to increases in TDP, reductions in $(T_j - T_a)$, or a combination of both. Thus the thermal solution designer is faced with the challenge of developing a thermal solution that has a thermal resistance at or below the required thermal resistance.

The thermal management challenge may be described in terms of the following elements:

- Historically, increasing silicon performance for CMOS was accompanied by increasing TDP and was one of the main issues in thermal management [17]. However, with the transition to multicore microprocessor architectures, dramatic increases in TDP do not occur when processor performance increases, and the thermal engineer today does not have to deal with unconstrained TDP levels.
- In addition to an increase in TDP, thermal designers need to account for areas of thermal nonuniformity (typically power densities of 300 + W/cm² are possible). These high thermal flux regions are caused by a nonuniform distribution of power dissipation on the die. The thermal impact of nonuniform power distribution is schematically illustrated in Fig. 22.16.

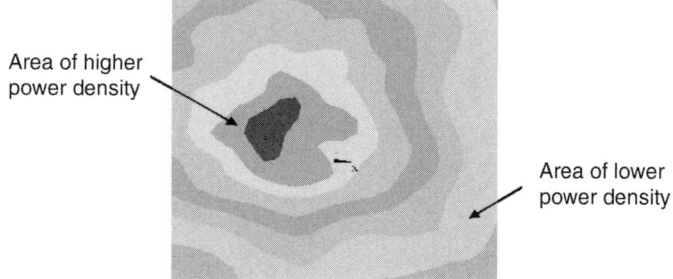

Fig. 22.16 Typical die power map and the hot spots on the corresponding die temperature map. The dark central region represents the highest temperature spot

The general strategy for thermal management focuses on the following:

- Minimizing the impact of local hot spots by improved heat spreading
- Increasing the power dissipation capability of the thermal solutions
- Expanding the thermal envelopes of systems
- Developing thermal solutions that meet business-related cost constraints
- Developing solutions that fit within form factor considerations of the chassis

In this section, we will focus only on package-level thermal solutions. As discussed in [19] and illustrated in Fig. 22.17, there are two thermal design architectures. Architecture I is one where a bare die interfaces to the heatsink solution through a thermal interface material (TIM), and Architecture II is one where an integrated heat spreader (IHS) is attached to the die through the use of a TIM and the heatsink interfaces to the IHS through a second TIM. Architecture I typically has a lower profile compared with Architecture II, and is often used for microprocessors in mobile and handheld computers. Architecture II is typically used for microprocessors in desktop and server applications.

The goal of package-level cooling in Architecture II is to use the IHS to spread the heat while transporting it from the die to the heatsink. The heatsink in turn dissipates heat to the local environment. In Architecture I, the base of the heatsink serves the function of an IHS in terms of spreading the heat. Since Architecture II serves to better illustrate the cooling strategy, we will use it in most of the discussion in this section. The TIM between die and IHS is referred to as TIM1, and the TIM2 is the interface material between IHS and the heatsink.

To increase cooling capability, the strategy is to even out the temperature profiles due to nonuniform power distributions, as close to the source as possible, by spreading out the heat. Heat generated at the device is mostly conducted through the thickness in TIM1 with a minimal amount of spreading. The focus in optimizing TIM1 thermal performance is to minimize the thermal resistance to heat transfer through its thickness. This is accomplished by managing three parameters: (a) the intrinsic thermal conductivity of the TIM, (b) the thermal contact resistance of the die/TIM1 and TIM1/IHS interfaces, and (c) the thermal-interface thickness, also referred to as the bond-line thickness. At the IHS level, the heat spreads and some of the peaks in the power profile are smoothed out. The considerations in designing the IHS are to optimize two factors: the thermal conductivity of the IHS

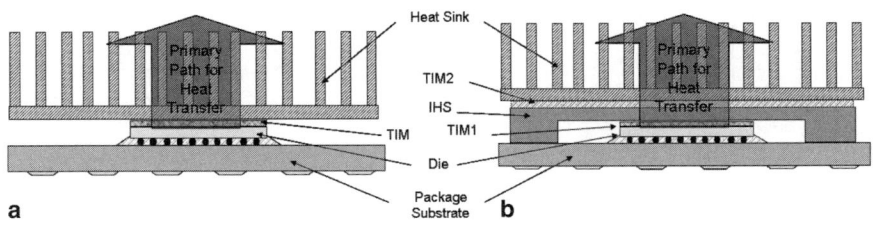

Fig. 22.17 Schematic of the two basic thermal architectures, illustrating their primary heat transfer path. (**a**) Architecture I (**b**) Architecture II

material and the thickness of the IHS, while ensuring that the weight to the package is within acceptable limits. A thicker IHS of high thermal conductivity material will enhance heat spreading. Advances in package thermal resistance are traditionally measured in terms of an area normalized resistance R_{jc}, defined as

$$R_{jc} = ((T_{jmax} - T_{case})/P) \times A_{die}, \quad (22.1)$$

where P is the power dissipated uniformly on the die, T_{jmax} is the maximum die temperature, and T_{case} is a typical point on the IHS or heatsink. In the case of Architecture I, T_{case} is typically a point on the heatsink base over the center of the die, and in the case of Architecture II, it will be a point on the top surface of the IHS over the center of the die. Figure 22.18 shows some of the advances in R_{jc} over the past decade accomplished by materials and process improvements in the control of the TIM1 bond-line thickness.

As seen in Fig. 22.18, incremental improvements in R_{jc} are slowing down and future improvements must be accomplished through breakthrough materials technologies.

Copper is the primary material of choice for IHS because it is abundant, has high thermal conductivity, and is a cost-effective choice. Breakthroughs are needed in developing materials that can replace copper as the material of choice for the IHS. Additionally solders, epoxies, and greases filled with thermally conductive fillers are the most commonly used TIM materials. The discovery that carbon nanotubes (CNTs) have the highest thermal conductivity of any known material has created the anticipation of novel materials for use in microelectronic applications (Fig. 22.19) to further improve the thermal performance of packages in both IHSs and TIMs. In both these applications, enhanced thermal performance can potentially be achieved by embedding the CNTs in a metallic or organic matrix. However, their use in thermal applications has not been realized due to poor interfacial properties (e.g., contact resistance) at the CNT–matrix interface. This issue remains a key technical challenge for enabling the use of CNTs for thermal applications in microelectronic packaging. Overcoming this technical challenge will require a detailed understanding of the CNT interface as well as transport across CNT interfaces.

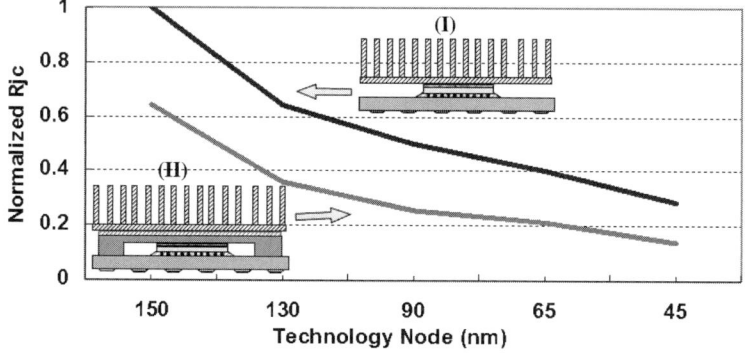

Fig. 22.18 Advances in *Rjc* for Architectures I and II

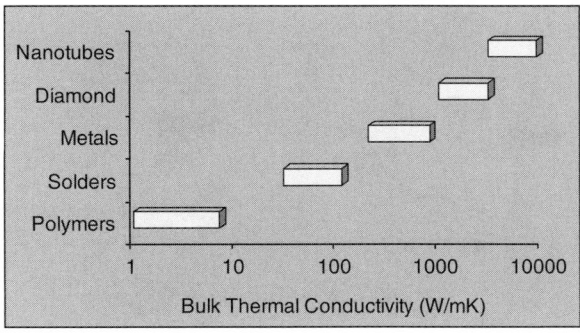

Fig. 22.19 Thermal conductivity ranges show the potential for CNTs as a TIM

Innovation of new CNT surface treatments will be required that provide both low interfacial resistance and good wetting and adhesion to matrices of choice. For a theoretical understanding of the interface, phenomenological relations and computational models will have to be developed to enable the evaluation of interface resistances of nanotubes coated with a metal, ceramic, or polymer, closely simulating the bulk environment. It will be important to evaluate which of the modeling techniques (e.g., MD, Monte Carlo, etc.) can enable development of predictive models of nanotube interface structures and phonon transport along them.

Diamond is another material that has potential for use in spreaders and in TIM materials if it can provide superior performance to copper at equivalent cost.

Another significant area of interest is in the development of nanostructured thermoelectric materials due to the promise of highly efficient thermoelectric energy conversion [20]. This interest is additionally sparked by the potential to integrate thermoelectric devices in the package or silicon to provide local on-demand cooling. However, there are significant challenges that need to be overcome to realize the potential of these materials [21]. Challenges exist in integrating the thermoelectric materials in energy efficient modules that can be integrated on the silicon or in the package. A key area of concern is reducing electrical parasitic resistance in the contact technology for the thermoelectrics. Focus is also needed to ensure that the devices operate at high efficiencies for multiple operating conditions of the electronic components. Finally, the process for integrating the devices in silicon or the package needs to be developed.

22.5 Structural Integrity

Advanced microprocessors depend upon transistor scaling to get improved performance, generation over generation. The accompanying interconnect scaling on the die back end causes interconnect propagation delays to be a significant portion of the clock cycle time, impacting the chip performance. The transition from aluminum die back end interconnects to copper in the 1990s was the first step toward

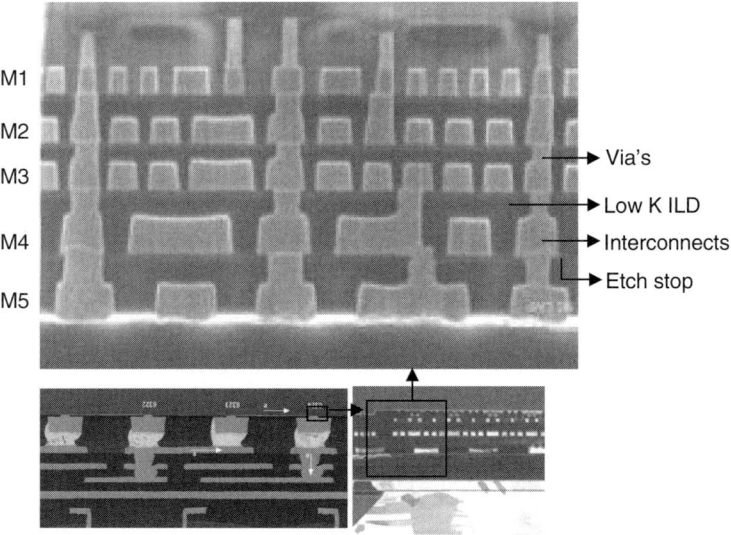

Fig. 22.20 Cross sections showing the interconnect structures in an SRAM

reducing interconnect propagation delay by reducing the interconnect resistance. Further reduction in interconnect delay in today's technologies is achieved by reducing the capacitance in the interlayer dielectrics (ILD) (see Fig. 22.20 for a typical interconnect structure) through the introduction of materials with low dielectric constant, k.

One of the consequences of creating lower k ILD materials, typically accomplished by increasing their porosity, is that the ILD materials become increasingly fragile [22]. Elastic modulus and fracture toughness of these films decrease rapidly as porosity increases. It has been observed that after the die is assembled to the package, the ILD materials with lower adhesive and cohesive strengths experience ILD cracking underneath the flip chip bumps. The root cause of this failure is from two primary effects:

1. Stress on the ILD is created due to the mismatch in the CTE between the silicon (3 ppm/K) and the package substrate. The CTE mismatch in organic packages is significantly higher (~16 ppm/K) compared with ceramic packages (~6 ppm/K).
2. The die to package interconnection, i.e., the flip chip bump transmits the CTE-induced mismatch stresses directly to the ILD.

One approach to reduce packaging stress is to make FLI solders more compliant so as to dissipate more stress through solder deformation, thereby protecting the low-k dielectrics. Another approach to reduce stress transfer to the ILD layers is through the use of geometrically compliant interconnects where mechanical compliance in the interconnect joints is achieved by structural design of the joints. An example of compliant interconnects is shown in Fig. 22.21 [23]. This type of interconnect can be fabricated with standard lithography and electroplating technologies,

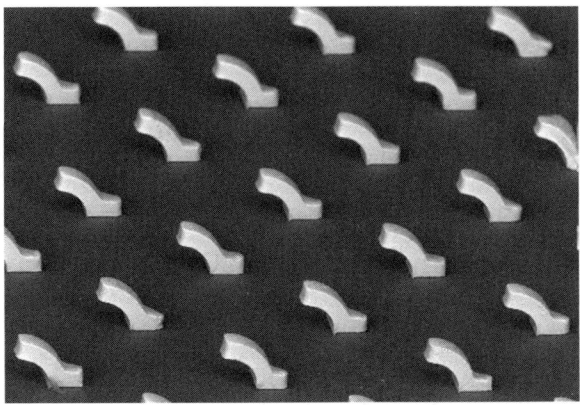

Fig. 22.21 Spring-like compliant interconnects used to minimize stress coupling between the silicon and package

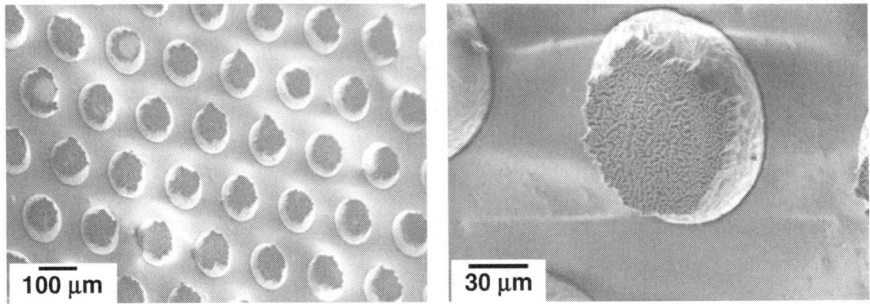

Fig. 22.22 CNT interconnects on solder bumps: potential FLI solution [25]

thus allowing manufacturability, and can potentially be scaled down to ultrahigh-density interconnections. Some of the key challenges that this technology faces are the trade-offs between the electrical performance vs. mechanical compliance and the risk of fatigue due to the geometry of the joints.

A further extension of this approach with novel materials is to consider using CNTs (Fig. 22.22) or metallic nanowires to construct the above interconnect structures. Although potentially both metallic nanostructures and CNTs can be used as building blocks for these nanointerconnects, CNTs can be a better choice because of their ballistic conductance, high electromigration (EM) resistance as well as the excellent mechanical properties (e.g., supercompressibility) [24]. However, significant challenges will need to be overcome before some of these potential opportunities can be realized. These challenges include the low-temperature growth of highly aligned and high-quality CNTs, and understanding and resolving the CNT–metal interface contact resistance issues.

A key challenge for packaging is to develop packaging technologies and assembly processes that minimize the stress induced in the on-die interconnect [22].

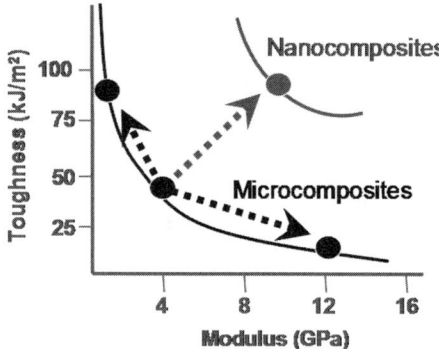

Fig. 22.23 Schematic illustrating the potential benefit of nanocomposites [26]

Solution strategies implemented today to successfully deliver low-k ILD require close collaboration between silicon back-end engineers and packaging engineers, and involve considerable integration challenges in areas of materials, design, and process development. The complex material, design, and process integration have also highlighted the need for predictive modeling of the stress states in the silicon back end to be able to make the right choices.

Typical package polymer materials require a combination of mechanical properties, e.g., CTE of < 16 ppm/K, controlled mechanical modulus, high toughness, and low moisture absorption. Significant improvements in several composite material properties, such as its modulus, CTE, and toughness, may be effected by the addition of certain nanoparticles to the polymeric formulation (Fig. 22.23). These changes can be accomplished with minimal impact to manufacturing aspects of the material such as its rheology. The key challenge is to develop and integrate these materials and to achieve the required properties simultaneously. Significant research will be required for designing and tailoring the nanoparticle–matrix interface, achieving the desired nanoparticle dispersion within the polymer matrix and developing an appropriate *rule of mixtures* for nanocomposites. Since the nanoparticle–matrix interface appears to control a nanocomposite's properties, novel and appropriate chemistries are needed to optimize the design of a nanoparticle–matrix interface to obtain a specified combination of composite properties. Finally, empirical rules of mixtures exist for only a few nanocomposite systems, but global, predictive, rule of mixture guidelines for nanocomposites are needed to enable designed nanomaterials with optimized sets of specified properties.

22.6 Form Factor Management

Smaller form factor systems have been a general trend in the electronics industry for quite some time. This trend is important in mobile computers and is critical in ultramobile systems such as the cell phone (Fig. 22.24).

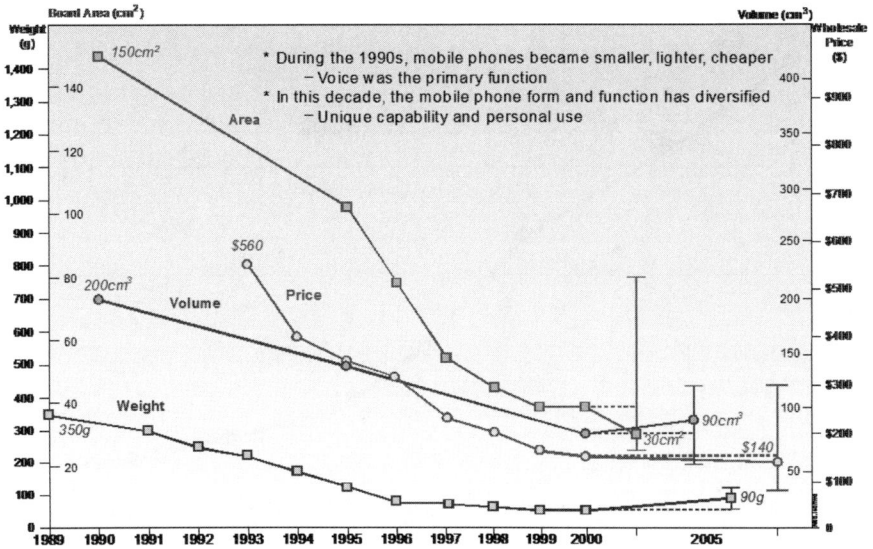

Fig. 22.24 Evolution of mobile phones (source: Prismark Partners LLC)

The packaging industry has been responding to this demand very well. In the early phase of package form factor reduction, smaller body size was achieved by simply reducing the second level interconnect pitch. Later, package thickness was also reduced by thinning the die (Fig. 22.25) and other dimensions such as thickness of mold compound over the die. That eventually led to the chip scale package (CSP) concept, where the package area was reduced to less than 1.2 times the die area [27] as compared with ~2–10 times the die area with the traditional plastic ball grid array (PBGA) technology. CSP was achieved by reducing BGA pitch, wire bond lengths and wire loop heights, thinning the die, and improving the mold material and process. A molded matrix array package (MMAP) assembly process facilitated high-volume CSP production. Packaging concepts such as quad flat pack no-lead (QFN) and wafer level package (WLP) also fall into the CSP category.

The conventional CSP was followed by stacked-die CSP (SCSP) with multiple wire bonded dice stacked inside the same package. This required further thinning of the die, extreme control over wire bonding, die attachment and molding processes, and thinning of the substrate. Further reduction of package thickness or larger numbers of stacked die may require widespread use of flip chip packaging and TSV processes [28].

For stacked die packaging, die test yield is an important consideration. Compounding of die test yield loss limits the maximum number of dice that can be affordably stacked [29]. In addition to this limitation, business considerations such as multiple supply sources and inventory management led to the adoption of the package-on-package (PoP) concept (Fig. 22.5). PoP allows the system manufacturer to source the top and bottom packages from different suppliers and then

Fig. 22.25 Minimum die thickness trend (source: Prismark Partners LLC)

assemble the components with only a small penalty to system volume when compared with the SCSP. Note that PoP may include stacked die within its top and/or bottom packages.

Both SCSP and PoP enable chips from the same and/or different process technologies to be placed inside the same package. Thus, logic, DRAM, and RF chips can be assembled in the same package. Such a package performs like a system or subsystem and is commonly known by the term system-in-a package (SiP). There are many types and variations of SiP [30], the majority of which play key roles in form factor reduction of the ultramobile system.

Challenges for SiP are manyfold. These include creating appropriate industry standards to manage PoP interfaces and defining die bond pad locations and assignments to efficiently interconnect the various chips inside the package. It is also important to find an affordable means of assuring known good die (KGD) before committing it to the SiP. Arguably, the most critical challenge for SiP and other thin packages is managing technical issues associated with extremely thin die. The thin die challenges include the thinning process, stress-induced electrical property changes, and package warpage. Wafer thinning down to 40 μm and in some cases to 10 μm or less [30] would require good control over thickness variation, including surface roughness, relieving residual stresses caused by the thinning process, and protecting the die from chipping and cracking during wafer thinning and die singulation. Handling of the significantly warped thin wafer and die is another key challenge. Improved carriers as well as pick and place equipment would be required to keep the wafer and die flat during the assembly process. It is shown that mechanical stress may induce changes to the electrical properties of the transistors [31]. The choice of die configurations and changes in material properties and dimensions of the various constituents of a thin package can have significant impact on the stress level of the thin die. A thin package with thin die is prone to warpage due to its low stiffness. Future decreases in die thickness are expected to make this problem worse, impacting the board assembly yield. Making the problem even more challenging is the need for a more coplanar BGA ball field to enable board assembly with finer BGA pitch. Solutions for these challenges would need innovation in packaging material and process technologies.

22.7 Summary

In this chapter, we have illustrated some of the key drivers that have influenced the evolution of the high-performance logic packaging in consumer and business computing systems as well as in ultramobile devices. As the silicon features go deeper into the nanoscale, packaging technologies will be challenged to extend the trends in physical density scaling, power delivery, signal bandwidth, thermal management, structural integrity, and form factor reduction.

Delivering innovative packaging solutions to meet these challenges would require successfully accelerating the transition of emerging technologies, like nanotechnology, from research to affordable implementation. However, the trend of increasing complexity has the tendency to increase the development cycle time and also increase the cost of packaging. Opportunities to successfully leverage emerging technologies to address complexity and cost will continue to present themselves as significant challenges, as well as huge competitive advantages.

References

1. G.E. Moore, "Cramming More Components onto Integrated Circuits," *Electronics*, Apr. 1965, pp. 114–117.
2. M.T. Bohr, R.S. Chau, T. Ghani, and K. Mistry, "The High-k Solution," *IEEE Spectrum*, Vol. 44, Oct. 2007, pp. 29–35.
3. K. Mistry, et al., "A 45 nm Logic Technology with High-k + Metal Gate Transistors, Strained Silicon, 9 Cu Interconnect Layers, 193 nm Dry Patterning, and 100% Pb-Free Packaging," *IEDM Technical Digest*, Dec. 2007, pp. 247–250.
4. S. Chou, "Integration and Innovation in the Nanoelectronics Era," *IEEE Solid-State Circuits Conference Digest*, 2005, pp. 36–41.
5. R. Shukla, et al., "Flip Chip CPU Package Technology at Intel: A Technology and Manufacturing Overview," *Electronic Components and Technology Conference*, 1999, pp. 945–949.
6. V.K. Nagesh, et al., "Challenges of Flip Chip on Organic Substrate Assembly Technology," *Electronic Components and Technology Conference*, 1999, pp. 975–978.
7. M. Renavikar, et al., "Materials Technology for Environmentally Green Microelectronic Packages," Intel Technology Journal, Vol. 12, No. 1.
8. R.R. Tummala, ed., *Fundamentals of Microsystems Packaging*, McGraw-Hill, New York, 2001, p. 368.

9. A. Yeoh, et al., "Copper Die Bumps (First Level Interconnect) and Low-K Dielectrics in 65 nm High Volume Manufacturing," *Electronic Components and Technology Conference*, 2006, pp. 1611–1615.
10. N. Na, J. Audet, and D. Zwitter, "Discontinuity Impacts and Design Considerations of High Speed Differential Signals in FC-PBGA Packages with High Wiring Density," *Electrical Performance of Electronic Packaging*, Oct. 2005, pp. 107–110.
11. D. Mallik, K. Radhakrishnan, J. He, et al., "Advanced Package Technologies for High-Performance Systems," *Intel Technology Journal*, Nov. 2005, pp. 259–271.
12. R. Darveaux, et al., "Solder Joint Fatigue Life of Fine Pitch BGAs – Impact of Design and Material Choices," *Microelectronics Reliability*, 2000, pp. 1117–1127.
13. K. Aygun, M. Hill, K. Eilert, K. Radhakrishnan, and A. Levin, "Power Delivery for High-Performance Microprocessors," *Intel Technology Journal*, Nov. 2005, pp. 273–283.
14. Intel Corporation, available online at ftp://download.intel.com/technology/computing/multi-core/multi-core-revolution.pdf.
15. C.M. Garner, et al., "Challenges for Dielectric Materials in Future Integrated Circuit Technologies," *Microelectronics Reliability*, 2005, pp. 919–924.
16. U.A. Shrivastava, et al., "Package Electrical Specifications for Giga Bit Signaling I/Os," *Electronic Components and Technology Conference*, 2003, pp. 1452–1458.
17. A. Alduino and M. Paniccia, "Wiring Electronics with Light," *Nature Photonics*, 2007, pp. 153–155.
18. L. Schares, et al., "Terabus: Terabit/Second-Class Card-Level Optical Interconnect Technologies," *IEEE Journal of Selected Topics in Quantum Electronics*, 2006, pp. 1032–1044.
19. R. Mahajan, et al., "Advances and Challenges in Flip-Chip Packaging," *IEEE Custom Integrated Circuits Conference*, 2006, pp. 703–709.
20. R. Venkatasubramanian, et al., "Thin-Film Thermoelectric Devices with High Room-Temperature Figures of Merit," *Nature*, 2001, pp. 597–602.
21. S. Krishnan, et al., "Towards a Thermal Moore's Law," *IEEE Transactions on Advanced Packaging*, 2007, pp. 462–474.
22. B. Chandran, R. Mahajan, M. Bohr, and Q. Vu, "The Mechanical Side of Ultralow k: Can It Take the Strain?" *FUTURE FAB International*, 2004, pp. 121–124.
23. S. Muthukumar, et al., "High-Density Compliant Die-Package Inter-connects," *Electronic Components and Technology Conference*, 2006, pp. 1233–1238.
24. A. Cao, et al., "Super-Compressible Foamlike Carbon Nanotube Films," *Science*, 2005, pp. 1307–1310.
25. A. Kumar, P.M. Ajayan, R. Baskaran, and A. Camacho, "Novel Low Temperature Contact Transfer Methodology for Multiwalled Carbon Nanotube Bundle Applications," *MRS Spring Meeting*, San Francisco, CA, Apr. 2007.
26. V. Wakharkar and C. Matayabas, "Opportunities and Challenges for Use of Nanotechnology in Microelectronics Packaging," *MRS Spring Meeting*, San Francisco, CA, Apr. 2007.
27. Joint Industry standards on "Implementation of Flip Chip and Chip Scale Technology," IPC/EIA J-STD-012.
28. M. Kawano, et al., "A 3D Packaging Technology for 4 Gbit Stacked DRAM with 3 Gbps Data Transfer," *IEDM Technical Digest*, 2006, pp. 1–4.
29. K. Brown, "System in Package – The Rebirth of SIP," *IEEE Custom Integrated Circuits Conference*, 2004, pp. 681–686.
30. Assembly and Packaging Section, International Technology Roadmap for Semiconductors, 2007.
31. A. Hamada, et al., "A New Aspect of Mechanical Stress Effects in Scaled MOS Devices," *IEEE Transactions on Electron Devices*, Apr. 1991, pp. 895–900.

Chapter 23
Nanoelectronics Landscape: Application, Technology, and Economy

G.Q. Zhang

23.1 Introduction

The semiconductor industry and its suppliers are the cornerstones of today's high-tech economy. Representing a worldwide sales value of $340 billion in 2005, the sector supported a global market of more than $1.3 trillion in terms of electronic systems and an estimated value of $6 trillion in services, with applications ranging from transportation to health care, and from general broadcasting to electronic banking.

The shift from the past era of microelectronics where semiconductor devices were measured in microns (one millionth of a meter) to the new era of nanoelectronics where they shrink to dimensions measured in nanometers (one billionth of a meter) will make the semiconductor sector even more pervasive than it is today. It will allow much more intelligence and far greater interactivity to be built into many more everyday items around us, with the result that silicon chip technology will play a part in virtually every aspect of our lives, from personal health and traffic control to public security.

This chapter will briefly summarize the current and potential application domains, highlight the major technology challenges and research agenda, and discuss the economic implications of micro/nanoelectronics.

23.2 Applications

The society of the future expects environments that are sensitive and responsive to the presence and needs of people, characterized by many invisible devices distributed and connected throughout the environment – devices that know about their situational state, that can recognize individual users, tailor themselves to each

---

G.Q. Zhang
Department of Precision and Microsystems Engineering Delft University of Technology
Delft, The Netherlands, NXP Semiconductors, Eindhoven, The Netherlands

user's needs, and anticipate each user's desires without conscious mediation. In other words, environments will be of more and more Ambient Intelligence.

Although micro/nanoelectronics is relevant for nearly all aspects of human life, the following six application domains, each of them driven by clearly recognizable societal needs, will be highlighted in this chapter [1–4].

23.2.1 Health

The continuous rise of healthcare costs, in most highly developed countries, substantially higher than the rise of the GDP, calls for a reform of the healthcare systems to increase the efficiency and to improve the quality. One way to achieve this is the move from *how to treat patients* to *how to keep people healthy and prevent illness*.

The current health reform going on in several countries, including the Netherlands, leads to an increased responsibility for individuals with respect to choice (which doctor, which hospital, which insurance and insurance level). This results in a situation that individuals need to be informed how to invest in effort as well as money. Also the organization of healthcare outside hospitals (primary care in group practices and week-end service, care after discharge, and care pathways or "ketenzorg") leads to a situation that several care professionals are involved in the care of patients. All these professionals should have access to up-to-date information of the patient involved, not only in their offices but also on the move. In the situation of care after discharge and prevention of recurrent hospitalization of chronic patients, telemonitoring and remote support of patients will be important.

The proper means to do this is however not available yet, or only rudimentary implementations exist. Examples of means in this area are sensors or sensor networks that capture relevant signs from individuals and send them to a health application, which can give personalized advice taking into account the health information stored at several locations (including the electronic health records at hospitals and general practitioners). An approach like this would not only support patients recovering from diseases or having a chronic disease but also individuals who want to maintain healthy condition by maintaining a healthy lifestyle.

Requirements in this domain are sensors and actuators (including the connection to sensor networks for home and mobile monitoring, security and privacy), fine-grained role-based access control, interoperability standards, data exchange protocols on the application level, use of multiple access networks to an internet backbone, dynamic configuration of services, web services technology, and ecosystems that allow multiple electronics health record services (from hospitals, GPs, other care providers) to collaborate.

Also, many improvements are necessary for the care cycle itself, which can be described as a healthcare workflow involving technologies inside and outside the hospital to prevent, diagnose, treat, and monitor diseases. To improve the life expectation of the patient, the time between the patient entering the hospital and the treatment should be as short as possible. This also needs minimal invasive treatment as the standard way of medical intervention. In addition, optimization of the information flow around the patient is a key so that all necessary information is available and can be assessed when needed, including patient history, anamnesis, medication, previous images, etc.

To support the reduction of the time in the hospital it is important to be able to track the movement of patient through the hospital. Relevant information of the patient should preferably be prefetched at each place of contact with the personnel. This implies the availability of a safe and secure ambient identification system. In addition, fast access to all relevant historical and current medical information of the patient is important. Diagnostic imaging, being a key technology for obtaining actual patient information, should be processed and result in short times to well-structured information in all relevant aspects. This involves massive image processing in short times using segmentation to isolate organs and associated metabolistic processes based on molecular imaging technologies such as PET and SPECT. Multimodality interventional rooms (X-MR, X-CT, X-Spect, PET-CT) are important to monitor and assess the staging and treatment progress of the patient. The availability of multiple modalities and multimodality processing will speed up the whole treatment process considerably, improve patient outcome, and reduce costs. Clinical outcome can also be improved in a more cost-efficient manner by predicting treatment planning through the use of advanced numerical simulation tools.

Moving the monitoring of patients after treatment to outside the hospital can further reduce hospital time. Patients may be managed remotely at home, or in less expensive *recovery homes*. This needs reliable, safe, and secure network connections between the medical personnel and the patient. The monitoring system should be mainly autonomic relating the monitored information to the profile of the patient. Alarms should be precise and in time to enable direct treatment of endangering complications. In urgent situations, ambulances should be connected to the network to enable precise treatment already during the transport of patients.

Minimal invasive treatment needs actual image information during surgery. Virtual simulation of endovascular devices prior to treatment will become a standard method for determination of device dimension, position of its deployment, and determination of potential complications. This enables the doctor to see what is happening without operating the patient. As a consequence, in even shorter times, the same well-structured information about the present state of the patient should be available. Therefore, improved medical equipment, mainly diagnostic medical imaging as well as catheters and associated equipment, should be developed to address these needs. In particular, earlier obtained medical data of the patient should be related to the present position of the patient. This should not be disturbed by intrinsically moving parts of the patients (heart, lungs). To reduce the amount of damage caused by the treatment, the medical equipment should be as intelligent as possible to find its way semiautomatically through the patient via the best possible way.

Reduction of failures can be achieved in several ways. Formalizing the medical know-how in accessible databases enables the selection of the right protocols for the right situation. Semiautomatic support guides the course of diagnosis and treatment according to these protocols. In deviant situations, new protocols should be selected or added by the medical specialist. Advanced information technology is necessary to support decisions and to plan and execute the workflow in real-time situations. As the patient is moving within and outside the hospital, wireless devices are needed to ensure that the right treatment is given to the right patient.

Another domain is prevention. This can be achieved by population screening (for high-impact diseases such as breast, prostate, and colon cancer). However, the analysis of information and the low number of exceptions requires technologies like computer-aided detection to improve the productivity of radiologists.

IT services, like long- and medium-term storage of patient data, will be outsourced to reduce the management costs. External medical information needs to be added to improve the diagnosis and the treatment plan. Hospital infrastructures need to support the incorporation of real-time safe and secure access to external services at any time.

Fast, highly sensitive DNA/protein assays made possible by innovative new biosensors will allow many diseases to be diagnosed "in vitro" from simple fluid samples (blood, saliva, urine, etc.) even before sufferers complain of symptoms. Similar tests will identify those predisposed to certain diseases, allowing them to enter screening programs that will identify early onset of the disease. Conventional and molecular imaging, increasingly combined with therapy, will pinpoint and eradicate diseased tissue. Early diagnosis will lead to earlier treatment, and earlier treatment to better prognoses and aftercare. By *nipping disease in the bud* it will make many therapies either noninvasive or minimally invasive. Equipped with body sensors that continuously monitor their state of health and report significant changes through telemonitoring networks, patients will be able to return home sooner and enjoy a faster recovery. Nanomedicine will also revolutionize prosthetics, with bioimplants restoring sight to the blind and hearing to the deaf. Automated drug-delivery implants will prevent conditions such as epileptic fits.

Nanoelectronics will pose many challenges, such as the biocompatibility of the materials, reliability, and the need for very low power dissipation. Developing implants in biocompatible packages will push system-in-package (SiP) miniaturization to the limits, while at the same time having to cope with the integration of devices such as biological sensors, nanoelectro-mechanical system (NEMS)/mechatronics devices, optical devices, energy scavenging systems, and RF interfaces. At the same time, many of these highly complex heterogeneous systems will have to provide life-support system reliability.

23.2.2 Mobility/Transport

The following aspects are essential to achieve sustainable road transportation: safety (increased safety for all road users), mobility (meeting the significant growth and changes in mobility and the transport of goods), and sustainable powertrain (reducing environmental impact and managing the efficient use of energy resources). As the volume of traffic on our roads continues to increase, there will be an increased demand for safety management systems that outperform their drivers in terms of speed control and collision avoidance through drive-by-wire systems. At the same time, there will be a need to transfer an increasing amount of data to and from moving vehicles, not only for driver information, navigation, and entertainment systems, but also for vehicle tracking and road toll applications. The world's limited oil and energy resources will stimulate the development of far more fuel-efficient vehicles as well as new alternative energy (battery or fuel-cell powered) vehicles. Nanoelectronics will be at the heart of many of these advances.

With the current revolution in electronics and electronic applications, the automotive industry is faced with a new technology: HMI (human machine interaction). This is the field of expectations and acceptance of, and attitude and behavior toward, new intelligent vehicle systems (e.g., drive-by-wire systems). Although HMI is not new in itself, other requirements are requiring new research. HMI is becoming increasingly important with respect to driver safety and safety of all road users, efficiency, and productivity in the field of professional transport and comfort. Electronics allow for most comfort improvements; infotainment is probably the strongest growth segment within this area.

Throughout the automotive industry, a digital revolution is taking place that has prompted the emergence of a new innovation area: Dependability and Information and Communication Technology. Electronics and, subsequently, software, are key challenges for R&D and for the automotive industry. In 2002, the value creation per car on electrics/electronics including software was 20% – of which some two-thirds were in software alone. More than 90% of all future automotive innovations will be driven by electronics and embedded software, including other drivers for more complex embedded systems – in addition to active safety systems – such as for more comfort: with in-car information, navigation, and entertainment, and helping the environment: less pollution, lower fuel consumption, new fuels, alternative propulsion technologies, etc. In 2015, the value creation of electronics will be 35–40% of total value creation per car.

Electronic automotive systems have to withstand very harsh environments, including high temperatures, humidity, vibration, fluid contamination, and EMC (electromagnetic compatibility). While these problems have largely been solved for conventional IC-style packaged devices, a whole new set of challenges will have to be faced when these packages also contain integrated sensor, actuator, mechatronics or optoelectronics functions. Some systems, such as collision avoidance radars and engine-assist/traction motor drives, will push the performance limits of current high-volume low-cost semiconductor solutions in terms of frequency capability or power/thermal handling. In addition, the critical role in drive-by-wire systems will require extreme reliability, with failure rates down to a few parts per billion, which means zero defects for the whole production volume of a device. On top of this, the automotive industry imposes special constraints such as parts warranty for up to 20 years and conformance with EU directives.

23.2.3 Security

Security reflects itself in public demand for personal emergency and home security systems, and government-led protection from crime and terrorism. There is a need for massive deployment of personal protection without restriction of liberty, which means that safety and security systems need to be both reliable and easy to use. We need to improve the recognition of individuals and detection of threats such as explosives and chemical warfare agents (CWA). Nanoelectronics will provide the necessary sensors, computing power, and reliability at cost levels that allow safety and security to be built into the fabric of our environment. These system will includes advanced algorithms within the embedded systems that fit in an infrastructure and standards necessary for an integrated solution.

Massive deployment for military, civil security, and personal safety requires the costs of these functionalities to decrease tenfold. For example, the department of defense (DOD–USA) needs for miniaturized detectors amount close to US$500 millions, and the personal markets in the EU and US alone following the military and civil market will increase to more than US$5 billions.

Safety and security systems can be divided into three groups: first, low-cost personal emergency and home protection systems, which are affordable for consumers, second, high-performance, high-efficiency systems for applications such as airport, transportation, seaports, shopping malls, etc, and third, distributed information systems such as electronic banking, electronic ordering and payment, health information systems within hospitals, between multiple care providers and patients outside hospitals. For these applications, it is essential that traditional security and privacy solutions of the IT system of closed infrastructures be extended to include general purpose and dedicated devices (remote monitoring) at multiple locations, authentication, and identification of persons. To enable massive deployment, these systems need to be unobtrusive and of low cost. Therefore, these systems must be small and easy to use, and only high levels of miniaturization are able to serve this unmet market demand. Yet their requirement to be highly reliable also means that they must be complex and multifunctional so that they make decisions based not on a single parameter but on combinations of parameters (fingerprint, voice, iris pattern, etc.) This will involve the integration of a wide range of sensors, MEMS/mechatronics, and photonic devices. Such devices will also need to communicate reliably by wired and wireless networks, and they must be made tamper-resistant and able to withstand environmental conditions that might affect their performance (radiation, chemical corrosion, shock, etc.).

23.2.4 Communication

People are becoming used to having easy access to friends, relations, and information services, and more frustrated when that access is not available to them. Making information available anywhere at any time relies on connectivity and communication, increasingly via the use of wireless-based networks (cellphones, Wi-Fi networks, etc.) to meet the *anywhere* requirement. In future, such communication systems must be easier to use, even to the point where specific connectivity channels become irrelevant to the user. Information will simply tunnel itself to its destination by whatever communication channels are available. At the same time, the bandwidth of systems will increase dramatically to cope with the increasing amount of data that people want to move around (voice, pictures, video, file transfer, etc.). We need to ensure that they will become much more secure against eavesdroppers and hackers.

Nanoelectronics will be required not only to meet the miniaturization requirements of handheld portable communications devices but also to allow much more functionality, in terms of the number of different communication channels, to be packed inside them. The *multiband multimode* devices that this will enable will be the key to decoupling communication from specific communication pipes, heralding a whole new era of seamless communications. At the same time, wireless communication channels will move to higher frequencies to increase data rates and maximize spectrum usage. This will require the increasing integration of RF MEMS and new RF architectures that allow circuitry to be reused across many different RF channels and modulation schemes.

Embedded systems technology is needed to bind the systems together in end-to-end services to deploy this into easy-to-use systems complying with the needs of the users, fitting with their day-to-day tasks and their personal lifestyle.

As portable communications devices pack more functionality, low-power consumption will become an even more critical requirement. The need to keep devices active for long periods of time between battery recharges or even autonomous in terms of energy supply will require the integration of energy scavenging devices that pull and store power from the local environment. At the same time, affordability, reliability, and environmental compatibility (disposability, recycling, and reuse) will be other major drivers.

23.2.5 Education/Entertainment

Content for education and entertainment must not only be accessible anywhere and anytime, but it must also be of the right quality and accessible in the right format. Access to information will be required in many different locations (at home, in the car, in the street, in hotels, etc.) and delivered through a variety of channels (terrestrial, satellite, cable, phone line, wireless, discs, etc.). Yet in each location the rendering device for that content, and people's expectations of it, will be different (e.g., what is expected from a flat-panel TV set compared with a videophone). Digital media, such as DVDs and HDTV, have increased people's expectations of video quality, yet this video quality will have to be delivered through existing networks.

The need to deliver high-quality media through a range of different communication channels while maintaining the required quality will need new developments in multi-format encoder/decoders, data compression and transmission systems, with media broadcasters (e.g., internet servers) automatically tailoring transmission to the capabilities of the rendering device on which the content will be experienced. Storage and distribution (CD, DVD, digital home networks, etc.) will need to be developed that are compatible with the digital rights management requirements of content providers.

Content producers will require new equipment (e.g., HDTV studio equipment and lightweight portable HDTV cameras) to capture content, and content providers will need advanced video compression and transmission schemes to distribute it.

The demand for users to create, manage, and even publish (sometimes to a selected audience) their own content will also require significant advances in areas such as image capture (digital cameras, camcorders, etc.), image analysis, and picture quality improvement at affordable consumer-product prices.

Interactive content presented in, for example, virtual worlds and games needs the development of new standardized content formats, presentation devices (with enhanced experience beyond the current TV and PC, such as light effects, advanced audio, 3D video, and experience movement), and new interaction means (gazing, gestures, etc.). This will need new sensors and actuators as well embedded systems technology to deploy them.

23.2.6 Energy/Environment

Among the key concerns set out in the renewed European Sustainable Development Strategy, those connected to Energy are at the first place. The expansion of energy consumption, related to better life conditions, and to the increased need for mobility, is putting the world energy resources under a serious stress. The rise of new economic powers, like China and India, will probably further accelerate the request for energy, at a moment when the consumption of fossil fuels is already raising serious concerns about global warming and resulting climate changes.

Tapping new energy sources will no longer suffice to meet the future energy demand. Even the most optimistic evaluations for the introduction of renewable energy sources do not see more than 10% coverage of future needs. The main new energy source that will be readily available is energy saving. Already in the last 20 years, the electronic control of industrial processes, of engines, of illumination has produced significant savings in energy consumption. However, there is still ample space for improvement.

In a recent document, the European Commission estimated that around 20% of energy usage is wasted due to lack of efficiency. It is in this field that nanoelectronics can play a major role. Among the areas where nanoelectronics can drive the energy saving, we can mention engine management, for cars and planes, household appliances, illumination and heating, and traffic coordination. The final result will be not only a drastic reduction in energy consumption, but also a significant contribution to the protection of the environment.

Nanoelectronics can support the fight to protect the environment also in other ways: networked autonomous sensor can keep track of critical data, such as air and water pollution, forest fires, earthquakes, and volcanic eruptions. An integrated network of sensors supported by distributed computing capabilities can improve our knowledge of meteorological and geophysical phenomena, thus preventing large-scale disasters, while the introduction of bioelectronic devices, capable of providing real-time DNA analysis will contribute to prevent the spread of diseases.

The deployment of these applications will require significant progress in the field of sensors, but also in the fields of RF communications, energy scavenging, and cheap and low-energy computation power. Care for the environment also needs production technologies, materials, and product properties that allow recycling and reuse.

23.3 Technology

Explosive technology development has been playing a vital role for the success of the semiconductor industry. The future development of semiconductor technologies can be outlined in the following six domains [1, 3, 5–7].

23.3.1 More Moore

More Moore covers the development of logic CMOS and memory technology, essentially controlled by scaling, and following Moore's Law. The key technologies are as follows [1, 6]:

Lithography: The challenging issues are developing technology and infrastructures to extend immersion lithography to 45 and 32 nm, developing a technology and infrastructure for extreme ultraviolet (EUV) lithography for below 32 nm, developing strong infrastructures for resist development and understanding of the limiting mechanisms, developing maskless lithography (ML2) for early development/prototyping/low-volume production, and clarifying the importance/potential of imprint and other nonevolutionary lithography strategies.

Logic Technologies: The challenging issues are developing coengineered substrates/materials/devices, developing technology and devices for high-k/metal-gate stacks for 32-nm generation, developing technology and devices for three-dimensional structures (multiple gates and channels) for 32-nm generation, developing physical understanding of the limits of the transistors, e.g., transport physical mechanism, device matching, impact of atomic level statistical fluctuations, assessing limits of the low-k/Cu-interconnect scheme, and developing innovative solutions (such as air gap, 3D).

Memories: The challenging issues are integration density, which translates into an aggressive scaling path, followed by nonvolatility, speed, and energy consumption.

Dynamic Random Access Memories: The challenging issues are developing new materials for capacitors, developing new memory structures for beyond 30 nm, and introduction of 450-mm wafers.

Nonvolatile Memories: The challenging issues are new materials: high-k as interpoly, discrete trap layer for charge storage and low-k, 3D cell structure exploration.

23.3.2 Beyond CMOS

"Beyond CMOS" covers the most advanced research activities to allow scaling of logic and memory functions to continue beyond the physical limits of silicon-based CMOS technology.

After more then 30 years of scaling according to Moore's law, we are rapidly approaching the CMOS scaling limit because we are reaching a point where an increase in power consumption coincides with an insufficient increase in operating speed. These highly undesirable effects are caused by a decrease in channel mobility and an increase in the interconnection resistance for smaller process geometries. The power consumption is largely due to increased leakage currents, short channel effects, source-drain tunneling, and p/n junction tunneling. Moreover, interconnections are increasingly becoming a limiting factor: the decrease in the pitch of interconnections and in the size of contacts and vias is causing an increase in overall resistance, while the reduced spacing is increasing propagation capacitances. The consequence is an increase in propagation delays, and in the power consumption related to charging and discharging of interconnects to a point that already in today's 90-nm logic devices, a significant portion of transistors is dedicated only to driving interconnection lines, without playing any computational role. Physical limits of existing materials have been reached, and no large progress can be expected in this area.

On top of that, the increasing impact of defects and the high level of complexity in both lithography and design have resulted in manufacturing costs rising dramatically. At the same time, the variability induced by the process variation at this nanoscale impacts also the yield and thus the cost. Even without taking into account physical limits, all these combined effects push us closer to a point of reaching the limit of CMOS scaling.

Development beyond CMOS technology can be realized by a gradual and evolutionary approach, introducing nanotechnology concepts and structures into the classical CMOS technology. Examples are developing disruptive technologies for replacing CMOS. Examples of architectures are quantum cellular automata, cellular networks, reconfigurable computers, quantum computing, and biologically inspired architectures. Examples of devices are D-logic devices, i.e., semiconductor nanowire devices and carbon nanotube FETs, resonant tunneling devices (RTT, RTD-FET), nanofloating gate memory, nanowire memory, molecular memory, molecular switch, single-electron transistors (SET), spin logic and memory devices, and ferromagnetic logic devices.

23.3.3 More than Moore

In recent years, we have witnessed the emergence of an increasingly diverse area of microelectronics that goes beyond the boundaries of Moore's law into the area of "More than Moore" (MtM). From a technology perspective, MtM refers to all technologies based on or derived from silicon technologies, but that do not simply scale with Moore's law. A typical example is the ongoing integration of passive components such as inductors, capacitors, and resistors onto silicon to meet the integration requirements of today's multiband multimode mobile phones. Other examples are high-voltage, power, analogue, RF devices, solid-state lighting, sensors, and actuators.

Current nontraditional CMOS semiconductor process technologies exist, which implement high-voltage, low power, analogue, and radio frequency devices, and solid-state lighting. In addition to this, dedicated new technologies are needed to realize mechanical, thermal, acoustic, chemical, and optical functions. Combined nano and biotechnology is also around the corner, both elements of which will require different new wafer-processing technologies. Heterogeneous integration will then be the key enabler for integrating these diverse semiconductor technologies into smart and multifunctional products. In the world of Ambient Intelligence, it will eventually lead to a direct interface with the human body or the environment, embedding power sources with the electronics, and enhancing electronics with nonelectronic functions. Therefore, from an application perspective, MtM enables functions equivalent to the eyes, ears, noses, arms, and legs of a human being, along with the brain provided by microprocessor and memory subsystems.

The technology challenges for MtM are application and product specific, such as RF, sensors/actuators, biofluidics, HV and power, MEMS/NEMS, etc. Some generic issues are as follows:

- Integrated process development via mastering the requirements and interaction among IC, package, and heterogeneous systems
- Establishing reusable design platforms, processes, and assembly environments for cost-effective mass implementation of a wide range of CMOS-compatible sensor/actuator/MEMS
- Multiscale simulation, modeling, and characterization for multiphysical behavior of MtM processes and products
- Designing for reliability, testability, and manufacturability

23.3.4 Heterogeneous Integration

The future of ICT will see a combination of MM and MtM components, combined in one package (*System-in-Package* or SiP). SiP refers to (multi)functional systems built up using semiconductors and/or in combination with other technologies in an electronic package dimension. SiP focuses on achieving the highest value for a single system package, by extreme miniaturization, heterogeneous function (such as electrical, optical, mechanical, bio, etc.), integration, short time-to-market, and competitive function/cost ratio. Its concept applies to quite diverse technologies, such as semiconductors, sensors, actuators, power, RF modules, solid-state lighting, and to various healthcare devices. To distinguish between various SiPs, one can characterize SiPs into three categories. The first refers to packages with multidies, such as McM, PiP, and PoP. The second refers to subsystems built up using more than just IC process, such as passive integration. The last, the most challenging one, refers to subsystem with more than electric functions, built up using multitechnologies and heterogeneous integration.

With such a SiP solution, the application benefits from a comparable level of miniaturization to that achievable with a *System on Chip* (More Moore) solution, from the enhanced functionality of MtM solutions, and from having each part of the system fabricated in an optimum process technology. The technology behind such SiPs is heterogeneous integration. Heterogeneous integration will not only bring all these components together into one package but also provide an interface to the application environment. It therefore represents the glue between the world of micro/nanoelectronic devices and systems that humans can interact with. Heterogeneous integration has to ensure the integration of components based on different technologies and materials. For example, an ultraminiature single-package biosensor could contain photonic components for detection, RF components (using InP or GaAs) for communication, logic components for data compression and communication protocols, and energy scavenging or energy storage components (thermoelectrics, fuel cells, thin-film batteries) for power supply.

The major technology challenges can be defined from three integration levels:

Wafer-Level Integration: Research issues are integrated passives (by layer deposition) and thin devices, integration of coupling caps, vertical chip integration by through-silicon-vias/power-vias, thin wafer technologies such as dicing and handling, thin interconnects and impact on overall device performance, and secure package technologies.

Module Integration: Research issues are embedding of devices, flexible substrates (for reel-to-reel manufacturing), integration of optical interconnects, and photonic packaging.

Interconnect, Assembly, and Packaging: To cope with the integration needs of MM and MtM, interconnect, packaging, and assembly become the bottlenecks in system cost and performance, and the assembly and packaging role is expanding to include system level integration functions. These factors drive an unprecedented pace of innovation in new materials (Cu/low-k, green materials, high-k dielectrics), new technologies (wafer thinning, 3D, and heterogeneous integration, WLP, MEMS packaging, passive/active integration), heterogeneous systems integration including ultrahigh integration density, integration of different functions in one module/package, integrated on-chip/off-chip design as well as system partitioning/modularization, optical inter and intrachip communication, integrated power conversion and storage, nanopackaging and nanoassembly, short time to market, and low cost. Other research issues are microbumps, high-temperature interconnects, package and die stacking, self-alignment/alignment support, integration of advanced cooling concepts, and materials.

23.3.5 Equipment and Materials

The challenges that semiconductor-related equipment are facing are similar to the challenges of the device manufacturers: shrinking dimensions of the structures, more three-dimensional structures, higher demand on the specifications of the substrates, the requirement to achieve high yields, and requirements for heterogeneous integration. The main technology challenges for equipment are as follows:

Lithography: EUV lithography is on its way for industrial application at the 32-nm node. To fulfill the industrial requirements and to be competitive with the EUV tool, improvements in wafer throughput, lifetime of components, and source power are necessary, all aiming at a reduction of the cost of ownership. Also in the area of EUV mask production, especially with respect to contamination control, further efforts are necessary. As a major advantage of EUV lithography at the 32-nm node, with respect to competing technologies, extendability to the 22-nm node and beyond was always claimed. At the same time, immersion lithography is still an option for 32-nm lithography, especially with new high-index immersion fluids and lens material in combination with double patterning methods. Maskless e-beam lithography is already used today for process development and prototyping, although at such low throughput it makes it nonviable for production. It has to be shown whether parallel maskless e-beam operation can improve the productivity sufficiently. New patterning methods, not based on imaging in a photosensitive resist, are visible. Nanoimprint is the most prominent example. Their benefit in IC production still has to be proven; application in the MtM area is more likely.

Wafer Processing Equipment: To meet the future challenges of semiconductor manufacturing, the wafer processing equipment must meet two parallel major requirements: manufacturability and introduction of new processes, combined with new materials, not only to meet More Moore technology requirements, but also especially requirements regarding More than Moore technologies. Examples of further research effort are use of novel wafer etching, cleaning and deposition techniques, use of novel chemistries and materials, minimized process-induced wafer/device damage, lower thermal budgets, innovative chamber in-situ cleaning processes, improved step coverage, and fill properties for high-aspect ratio structures, improved planarization techniques for dielectric and metal structures, and interface engineering.

MtM Equipment: MtM requires that equipment enable heterogeneous integration and heterogeneous assembly. Packaging can be difficult enough when the package contains the entire system integrated onto a single silicon die – the so-called System-on-Chip (SoC) approach. It becomes even more difficult when multiple dies with nondigital components are integrated into a single package to incorporate functions that are technically difficult or commercially inconvenient to incorporate into a SoC. For nanoelectronics development, materials play an essential role. To some extent, the success of nanoelectronics depends on the profound understanding of the properties and behavior of materials and their interfaces under manufacturing, qualification testing, and use conditions, and the capability to tailor the material design for the requirements of specific applications. This issue is already acute in the design of microelectronics. It is even more so for nanoelectronics and MtM technologies, wherein both multiscale size effect and multimaterial compatibility, stability, and reliability will be the keys to success. Among many challenges, characterization and modeling of materials and their interface behavior need more attention, especially for multiscale, multiphysics, and time-dependent situations.

23.3.6 Design

Moving to nanoelectronics technologies enables a tremendous increase in the functionality of electronics systems and in the applications of SoC and SiP products. The price to pay is that design efficiency hardly follows this increase of complexity and functionalities. The demanding solutions must be capable of capturing formal design specifications provided by System Houses, allowing high-level system and architecture exploration within the underlying constraints of available implementation technologies. All aspects of product development must be embraced, including digital, analogue/mixed signal, power electronics, and embedded software in conjunction with nonelectrical components such as MEMS and NEMS. This will require expertise drawn from the many different disciplines involved in product design (System Level Design, HW and SW codesign, circuit design, packaging design, assembly design, verification, and physical implementation with constraints for test, reliability, and manufacturability). Several challenges should be solved with respect to design.

Design of Heterogeneous Systems: The complex and heterogeneous systems assemblies will increasingly be interconnected three-dimensionally (wafer level packaging), and controlled by an ever-increasing amount of software. The design of these heterogeneous systems will require new methods and approaches to compose heterogeneous subsystems. Interfaces linking digital/analogue, hardware/software, electrical/mechanical, etc. need to be handled at different abstraction levels. Another important issue is the design platform and reuse technology. It is economically not sustainable without developing a design platform and reuse technology to account for the needs of more product characteristics.

Design for Manufacturability is aimed to accelerate process ramp-up, and to enhance process yield, robustness, and reliability. To be able to do so, effort is needed to develop and enable random and systematic yield loss estimations from design through yield models and process-aware design flows, enabling yield optimization early in the design flow and reducing costly iterations.

Design for Reliability is aimed to predict, optimize, and design upfront the reliability of products and processes. Achieving this aim will require a broad scope of research activities including basic understanding of material behavior, degradation and failure mechanisms under multiloading conditions, accelerated reliability qualification tests, and advanced failure analysis, in combination with various accurate multiphysical and multiscale simulation models being able to predict the failure evolution. Other issues are increasing occurrence of soft errors, variability, and soft failure mechanisms beyond 90-nm technology, which demand research into self-repairing circuits and self-modifying architectures.

Embedded systems will have more impact on the success of nanoelectronics. Superior hardware technology alone cannot guarantee business success. It is well known that huge effort is made to lower dielectric constant k in the back-end stacks of ICs from 2.3 to 2.0, which is a 15% improvement, but this effort so far leads to serious yield issues while better software optimization can improve power efficiency and performance by factors between 2 and 10 [5]. It becomes obvious that simultaneous changes are needed in process technology, materials, device architectures, and design technologies coping with the need to create software-dominated platforms for future application domains.

It is worth emphasizing that the functionality and market appeal of integrated MM and MtM products are strongly dependent on the contributions of the software that is embedded in it. There is a strong trend for more and more software with greater functional diversity and architectural complexity. Taking the automotive industry as an example, it is estimated that 70% of future innovations will be software-related and most other sectors are moving in a similar direction. The growing need for embedded software is fuelled by a need for additional and heterogeneous functionality, real-time performance, distribution across subsystems, reuse for multiple systems or in a platform, and long life-time reliability needs. As system ability is becoming one of the key success criteria for future technology development, there is urgent need to integrate nanoelectronics technology with embedded systems and system level design. Another trigger for hardware and software codesign is due to the fact that the dramatically increased number of design tasks and their complexity are already leading to a phenomenon known as the "design gaps" – the difference between what should theoretically be integrated into systems and what can practically be designed into them and what should be manufactured vs. what has been designed.

23.4 Economy

Although the semiconductor industry has an instrumental impact for human life, society, and the world economy, unlike other industrial sectors, it is hard to predict the marketing trend due to characteristic fluctuations. Tables 23.1 and 23.2 show key semiconductor application market drivers [8].

It can be seen that the general picture of the semiconductor industry is very bright. While the need for ICs will still be sustainable, a very positive development trend in recent years is the MEMS business. The worldwide MEMS systems market reached $48 billon in 2005, and is expected to rise to $72 billon by 2008, and $95 billon by 2010, for a compound annual growth rate (CAGR) of about 15% over the 5-year period [9]. With the commonly recognized vision and more effort to develop MtM technologies and products, the semiconductor market span will be further broadened, so that more sustainable and stable business growth can be expected for years to come. Some observations on the recent semiconductor economy are as follows:

- Ninety nanometer wafer volumes experienced a rapid ramp-up in 2006, but there is also now the initial ramp-up of 65-nm technology. It is likely that the lifetime of 90-nm technology for digital designs will be short.

Table 23.1 Worldwide annual semiconductor volumes by application

System volume (million units) Application	2005	2010	CAGR 2005–2010
Automotive	63.0	76.2	3.9%
Biometrics	35.0	1,150.0	101.1%
Bluetooth	345.0	1,500.0	34.2%
DAB radio	2.4	62.5	91.6%
Hard disk drives (HDD)	385.0	891.0	18.3%
Digital camcorders	16.6	44.0	21.5%
Digital still cameras	78.9	84.0	1.3%
DVD players	93.2	65.0	−7.0%
DVD recorders	23.8	109.4	35.7%
Global positioning (GPS)	50.8	284.0	41.1%
Graphics	380.5	472.8	4.4%
Integrated flat-panel TVs	19.6	131.0	46.2%
Memory cards	294.5	945.0	26.3%
Mobile phones	805.0	1,536.0	13.8%
Near field communications	0.0	490.0	
PC and servers	207.9	355.0	11.3%
PDA/handheld computers	15.1	23.5	9.2%
RF-ID tags (shop label)	900.0	65,000.0	135.4%
RF-ID tags (nonshop label)	490.0	1,000.0	15.3%
Robotics	1.4	51.5	107.0%
Smartcards	3,020.0	4,100.0	6.3%
TV set-top boxes	56.8	143.0	20.3%
UltraWideBand (UWB)	0.5	190.0	228.1%
USB flash disks	93.3	410.0	34.5%
Video game consoles	53.1	64.6	4.0%
Wi-Fi	129.0	528.0	32.6%
WiMAX	0.8	150.0	188.5%
Zigbee	3	750.0	201.7%
FH sub total million units	7,564	80,607	60.5%
Without shop label RF-ID	6,664	15,607	18.6%

- Expenditures for process technology development have declined from 35.6% of total R&D in 1996 to 19.6% in 2010. Product development R&D, on the other hand, has increased from 55.4 to 73.7% of total R&D in the same time frame. It is critical for the semiconductor industry to continue reducing feature dimensions and become more efficient in using financial resources, which requires new business models for process R&D.
- The cost of developing and ramping up process technology will increase from $278 million for 0.18-μm technologies to $1,592 million for 32 nm.
- The revenues of IC vendors for supporting 32-nm technology will be $19.9 billion if process R&D expenditures are 20% of total R&D expenditures and total R&D represents 20% of product revenues.

Table 23.2 Worldwide total annual semiconductor market revenues by application

Revenue (US$ billion) Application	2005	2010	CAGR 2005–2010
Automotive	16.7	39.6	18.8%
Biometrics	0.2	3.3	77.0%
Bluetooth	1.2	3.6	24.6%
DAB radio	0.1	1.3	79.4%
Hard disk drives (HDD)	3.4	5.6	10.5%
Digital camcorders	1.5	2.5	10.3%
Digital still cameras	2.6	1.9	−6.4%
DVD players	1.4	0.7	−13.4%
DVD recorders	1.1	2.5	18.7%
Global positioning (GPS)	1.5	3.5	18.1%
Graphics	6.8	8.3	4.1%
Integrated flat-panel TVs	2.1	8.0	30.7%
Memory cards	7.1	20.2	23.3%
Mobile phones	36.3	71.3	14.5%
Near field communications	0.0	1.0	
PC and servers	56.8	91.3	10.0%
PDA/handheld computers	2.2	3.3	8.8%
RF-ID tags (shop label)	0.1	1.0	52.8%
RF-ID tags (nonshop label)	1.0	1.0	0.40%
Robotics	1.3	14.6	62.2%
Smartcards	2.4	3.8	9.3%
TV set-top boxes	3.8	8.1	16.4%
UltraWideBand (UWB)	0.0	1.2	199.3%
USB flash disks	2.6	8.0	25.6%
Video game consoles	4.7	7.4	9.5%
Wi-Fi	1.0	4.5	35.1%
WiMAX	0.1	2.3	101.1%
Zigbee	0.0	1.0	151.2%
FH sub total	158.0	320.8	15.20%
Others	34.8	51.8	8.3%
WW IC TAM US$ billion	192.8	372.6	14.1%
Market coverage	81.9%	86.1%	

23.5 Conclusions

In the past several decades, microelectronic industries have been spending tremendous effort in developing and commercializing the Moore's law, leading to not only many breakthroughs and revolution in ICT, but also noticeable changes in the way of living of human being. While this trend will still be valid, reflected in the Eniac [1] technology domains of More Moore and Beyond Moore, there are ever-increasing awareness, R&D effort, and business drivers to push the development and application of MtM that are based upon or derived from silicon technologies but do not simply scale with Moore's law, enabling various nondigital functionalities. The future business opportunities and technology challenges will be the integration of Moore's law focusing mainly on digital functions with MtM focusing on mainly nondigital functions via heterogeneous system integration (Fig. 23.1). This development trend results in the following challenges [7]:

High Level of Heterogeneity: Future micro/nanoelectronics will be made of very different materials, which have to coexist despite their different behaviors (thermal expansion, biochemical interactions). These building blocks will be prepared using various process nodes exposed to a broad range of environmental constraints. Scenarios of use are also leading to heterogeneous requirements for the communication links that will build on various wireless standards. Ultimately, the radio components of SiPs themselves will have to optimize in real time to the best air interface for the targeted application requirements.

Complexity: Most of these systems are designed and built to embed intelligence and to enable products with the ability to react to their environment and to provide relevant and ergonomic information to their users. The amount of multimodal data to be processed by the system is very large. The user interfaces have to cope with complex and variable environments while taking into account the context of use and rapidly changing user behaviors.

Autonomous Solutions: Most of the basic functions embedded in the systems have to be designed according to strong power requirements exploiting the most relevant

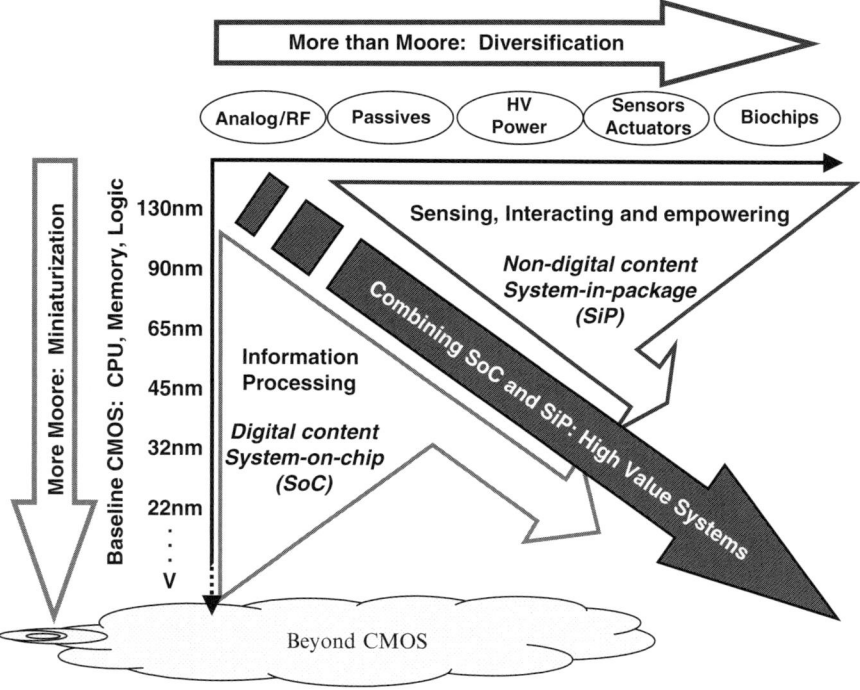

Fig. 23.1 More Moore (MM) and More than Moore (MtM) integration

energy sources. In many applications, energy resources are often the main bottleneck preventing larger market penetration of autonomous devices in both industrial and consumer products.

Multi-scale: Multiscale nature (in both geometric and time domains) will have a very strong impact on the whole value chain of product creation process, from technology development to industrialization. Various constituting elements of SiP cover a very large-scale difference of geometric features, ranging from nanometers to millimeters. Moving from *micro* to *nano* and from nano to multiscale will come with a paradigm shift. Indeed, as can be illustrated by the evolution of inertial sensor generations, each step of miniaturization may require a change of design, technology, and even a change of detection principle.

Multidisciplinary: Micro/nanoelectronics innovation requires a large body of know-how. Taking a biosensor SiP as an example, it needs not only knowledge of electronic engineering, but also chemistry and bioengineering. Chemistry, thermal, metallurgy, physics, electrical, mechanics, optics, electromagnetics, and biology may all be involved in future product creation, which will require both hardware and software engineering.

Stochastic in Nature: For micro/nanoelectronics, it is virtually impossible to design and manufacture products and to process them with deterministic performance. For design parameters, such as material/interface properties, geometric dimensions, process window, and loading intensities, the deviations represented by different statistic characteristics and magnitudes are inevitable. With the IC technology moving fast to the Beyond CMOS domain, control of multivariability at different scales becomes vital, especially if the performance at an atomistic level has to be linked with micro or macrolevel requirements.

Notice that all these characteristics are interacted intrinsically and nonlinearly with each other, wherein there exists a big gap between technological development and the underlying fundamental understanding. Integrated approaches are needed to find the optimal solutions, which will shape the technological environment in the coming two decades, and demand new research fields for the scientific community.

References

1. Strategic Research Agenda of Eniac, European Technology Platform of Nanoelectronics, Eniac, 2006.
2. Zhang GQ, Van Driel W and Fan XJ, Mechanics of Microelectronics Series: Solid Mechanics and Its Applications, Vol. 141, Springer, Dordrecht, 2006.
3. Zhang GQ, Begeer R and Hartman RA, Strategic Research Agenda of Point-One, 2007.
4. Strategic Research Agenda of EPoSS, European Technology Platform of Smart System Integration.
5. Towards and Beyond 2015: Technology, Devices, Circuits and Systems, Medea + Scientific Committee, 2006.
6. International Technology Roadmap of Semiconductors.
7. STW SmartSiP Program Definition, 2006.
8. Various/Future Horizons, 2006.
9. Global MEMS/Microsystems Markets and Opportunities, SEMI, 2006.

Index

A

AAO. *See* Anodic aluminum oxide
ACA. *See* Anisotropic conductive adhesive
ACF. *See* Anisotropic conductive film
Acoustic cavitation, 114, 115
Activated tunneling, 143, 148
Activation energy, 143, 202
Adhesion, 41–46, 69, 71, 79, 80, 83, 125, 126, 197, 203, 214, 224, 225, 249, 252, 288, 384, 423, 453, 469, 470, 499, 509
AFM. *See* Atomic force microscope
Ag-carboxylate, 241, 245
Agglomerate, 128, 199, 254, 380, 430
Agglomeration, 116–118, 198, 216, 239, 250, 254, 290, 423, 430
Aggregation, 216, 240, 258
Air core, 164, 167–174
Alignment, 5, 122, 176, 250, 334, 337, 368, 371, 437, 450, 476–478, 500, 528
$Al(OH)_3$. *See* Bayerite
Al_2O_3. *See* Alumina
AlOOH. *See* Boehmite
Alumina, 4, 258, 340,
Amorphous, 46, 50, 80–84, 147, 151, 168, 180, 183, 325, 327, 372, 397, 447
Anisotropic conductive adhesive (ACA), 189–192, 198–205
Anisotropic conductive film (ACF), 189–192, 198–205
Anisotropy, 167, 168, 174, 176–182, 191
Anodic aluminum oxide (AAO), 204, 443, 444, 446, 449, 451, 452
ANSYS, 17, 24, 301, 305, 387
Antennas, 452
Arc-discharge, 327–331, 334, 341
Architecture, 503, 504, 506–508, 523, 526, 530, 531
Asymmetric film, 144

Atomic force microscope (AFM), 2, 140, 141, 356
Atomistic, 17, 18, 20, 27, 40, 42, 61, 84, 85, 87, 93, 535
Atomization, 112

B

Back-end, 62, 67, 68, 74–76, 509, 512, 530
Ball bond, 71, 75, 449
Ball grid array (BGA), 209, 210, 231, 232, 280, 499, 500, 513, 515
Band bending, 359–361, 367
Barium titanate, 3, 124, 125, 127
$BaTiO_3$. *See* Barium titanate
Bayerite, 84
BCB. *See* Benzocyclobutene
Benzocyclobutene (BCB), 183
Benzotriazole (BTA), 203
BER. *See* Bit-error-rate
BGA. *See* Ball grid array
Biological sensing, 6
Biomolecular detection, 372
Bit-error-rate (BER), 407–410
Black diamond, 82, 83
Blind vias, 211, 228
Block co-polymer, 204
Boehmite, 84
Boltzmann, 53, 79, 143, 144
Breakdown, 110, 111, 133
BTA. *See* Benzotriazole
Buckling, 71–74
Bucky ball, 345
Buildup, 110, 498, 503
Bumps, 4, 210, 259, 260, 317–321, 389–391, 415, 465, 466, 495, 496, 498, 499, 510, 511
Buried vias, 210, 211, 231

C

C4. *See* Controlled collapse chip connection
Cantilever arrays, 445
Capacitor, 3, 121–127, 502, 503, 525, 526
Capillaries, 24, 30, 31, 48, 140, 142, 150, 246, 248, 287, 288, 450, 476, 477
Carbon fibers, 5, 6, 395, 396
Catalyst, 96, 333–341, 385, 386, 390, 397, 399
Catalytic, 326, 332, 334, 346, 447
Cermet, 3, 93, 94, 99, 102, 103, 139, 140, 144, 155
Charging energy, 101, 130, 140, 157, 159, 368
Chemical bonding, 55, 84, 224
Chemical interactions, 80, 85
Chemical vapor deposition (CVD), 5, 96, 289, 325, 334–341, 397, 399, 416, 445, 446
Chirality, 195, 196, 345, 349, 350–352, 353, 371, 372, 416
CIJ. *See* Continuous ink jet
Close-ended CNTs, 332
Coalescence, 94, 98, 99, 140–142
Coefficient of thermal expansion (CTE), 30, 31, 44, 51, 56, 210, 227, 231, 287, 288, 291, 295, 297, 298, 314, 315, 318, 320, 322, 422, 423, 432–434, 466, 467, 481, 486, 497, 510, 512
Coercivity, 164, 168, 176, 177, 179–183
Compliant interconnects, 466–471, 475–477, 481, 510
Compliant wafer level packaging (CWLP), 468
Conductive cooling, 5
Confocal microscope, 2
Contact angle, 95, 418, 425, 426
Continuous ink jet (CIJ), 248–250
Controlled collapse chip connection (C4), 495–499
Convective cooling, 5
Cost, 20, 23, 35, 39, 61, 87, 112, 113, 116, 119, 121, 125, 163, 164, 174, 175, 183, 190, 202, 210, 211, 214, 218, 241, 257, 329, 357, 372, 385, 387, 389, 395, 396, 404, 415, 465–469, 485, 491, 493, 495, 499, 503–509
Coulomb block, 100–102
Coulomb blockade, 2, 101, 129–131, 143, 156, 367, 368
Coulomb staircase, 156, 159
Crack initiation, 65
Crack propagation, 68, 84
Crack tip, 63–65, 68–70
Creep resistance, 4
$Cr_x(SiO)_{1-x}$, 145, 147, 152, 153, 155–157, 160
Cr_xSi_{1-x}, 148–153
$(Cr_xSi_{1-x})_{1-y}N_y$, 148, 149, 152, 153

CTE. *See* Coefficient of thermal expansion
CTE mismatch, 30, 44, 210, 318, 320, 432–434, 466, 467, 481, 486, 510
Cu_2O. *See* Cuprous oxide
Cuprous oxide, 42–45
Cure kinetics, 28
Current crowding, 29, 33, 34
Current densities, 33, 497
CVD. *See* Chemical vapor deposition
CWLP. *See* Compliant wafer level packaging
Cyanate ester, 214

D

De Broglie, 79
Debye–Scherrer, 94
Decoupling, 124, 502, 503, 524
Delamination, 61–76, 83, 85, 226, 232, 317, 318, 320, 465, 466
Deoxyribonucleic acid (DNA), 442
Dendrite, 202
Dicarboxylic acid, 200, 201, 224
Die lift, 69, 71
Die-attach, 62, 69, 71
Dielectric constant, 3, 82, 100, 122–127, 130, 134, 142, 143, 157, 160, 167, 227, 228, 361, 497, 510, 530
Dielectric loss, 122–124, 126–135, 228
Differential scanning calorimetry (DSC), 222, 382, 417, 423, 424
Diffusion barriers, 203
Diffusion bonding, 452
Discontinuous metal film, 139, 140
Discontinuous metal thin film (DMTF), 93, 94, 102
Dislocation density, 433, 434
Dispenser, 241, 246, 248
Dispersion, 49, 109, 119, 125, 128, 131, 135, 147, 178, 193, 197, 205, 213, 216, 221, 240, 244, 290, 352, 353, 396, 398, 399, 411, 416, 420, 422, 429, 443, 452, 512
Dissipation factor, 124, 130, 134
Dithiol, 200, 201
DMA. *See* Dynamic mechanical analysis
DMTF. *See* Discontinuous metal thin film
DNA sensors, 357
DNA. *See* Deoxyribonucleic acid
DOD. *See* Drop-on-demand
Drop test, 265, 266, 276, 280–284
Drop-on-demand (DOD), 248, 250
DSC. *See* Differential scanning calorimetry
Dynamic mechanical analysis (DMA), 307

Index

E

EAM. *See* Embedded atom method
ECA. *See* Electrically conductive adhesive
Eddy current, 164, 167, 169, 171, 175, 177–180
EDX. *See* Energy Dispersive X-Ray
EELS. *See* Electron Energy Loss Spectroscopy
Elastic modulus, 78, 196, 225, 288, 291, 295, 298, 301, 305, 307, 308, 310, 311, 314, 354, 416, 470, 510
Electrically conductive adhesive (ECA), 3, 4, 189, 212, 213, 221–223, 225
Electro-deposition, 25
Electromagnetic immunity, 407
Electromagnetic interference (EMI), 395, 398, 404, 407, 411, 452
Electromagnetic susceptibility (EMS), 396, 407, 409–411
Electron diffraction, 94, 96, 147, 150
Electron microscopy, 96, 146, 348, 349, 372, 417
Electron tunneling, 2, 3, 7, 101, 102, 130, 140, 368, 369
Electron energy loss spectroscopy (EELS), 180
Electrospinning, 378, 382, 385
Electrospun, 5
Ellipsoid, 94, 101, 380
Embedded atom method (EAM), 17
Embedded passives, 121, 122
EMI. *See* Electromagnetic interference
EMC. *See* Epoxy molding compound
EMS. *See* Electromagnetic susceptibility
End-of-life, 20, 113
Energy dispersive X-ray (EDX), 144, 147, 150, 251, 252, 417, 421
Environmental impact, 112, 113, 116
Environmental stress testing, 232
Environmentally friendly, 190, 485, 494, 495, 497
Epoxy, 5, 41–43, 46, 48–51, 69, 84, 125, 127–132, 196, 203, 204, 214, 215, 219, 222, 223, 226, 245, 288, 290, 291, 295, 299, 300, 314, 380, 396
Epoxy molding compound (EMC), 42–49, 402, 521
Exotemplate, 443–447
Eye diagram, 407–409

F

Faraday, M., 103
Far-field, 396, 401, 406
Fatigue failure, 317, 320, 466
Fatigue life, 30, 415
FCOB. *See* Flip-chip on board
Ferrite, 4, 172, 173, 183, 402
Ferromagnetic resonance (FMR), 167
Ferromagnetism, 140
FE-SEM. *See* Field emission SEM
Field emission, 346, 356, 449
Field emission SEM (FE-SEM), 275, 417, 419, 436
Filler load, 379
Filler shape, 379
Fine pitch, 31, 32, 190–192, 197, 205, 218, 241, 255, 258, 291, 390, 415, 438, 452, 468, 485
Finite difference, 79
Finite element, 20, 23, 32, 39, 42, 61, 64–66, 86, 87, 291, 292, 295, 298, 301, 321, 472
First level interconnect, 486, 495
Flexible printed circuits (FPC), 259
Flip chip, 6, 7, 30–34, 209, 210, 213, 227, 231, 245, 287, 288, 317–319, 322, 389–391, 415, 442, 451, 452, 491, 493, 495, 497–500, 505, 510, 513
Flip-chips on board (FCOB), 465
Fluidized bed reactor, 335, 339, 340
FMR. *See* Ferromagnetic resonance
Form factor, 491, 507, 512–515
FormFactor Inc., 467
Four point probe, 243–244, 417
Four point probe bending, 64, 69
FPC. *See* Flexible printed circuits
Fracture, 5, 24, 63–65, 67–69, 85, 87, 197, 229, 245, 266, 279, 280, 284, 431, 434–437, 439, 497, 510
Fullerene, 195, 325–327, 329, 332, 333, 345, 350, 397

G

Gage factor, 102
Galvanic corrosion, 6
Galvanic deposition, 446
Giant magnetic resistance (GMR), 453
Gibbs free energy, 115
Glass transition temperature, 23, 41, 50, 51, 307, 310
Glow discharge, 329, 330
GMR. *See* Giant magnetic resistance
Grain boundaries, 4, 90, 140, 176, 429, 445
Grain boundary scattering, 140
Grain growth, 4, 432, 448

Granular, 99, 140, 142, 143, 148, 156, 157, 159, 179, 218
Graphene, 332, 349–351, 364
Griffith flaws, 433

H
Hall-Pecht, 99
Hall-Petch mechanism, 432
Hardness, 99, 212, 266, 276, 278, 279, 284, 427
Heat removal, 377, 378, 382, 385, 389
Heat transfer coefficient, 377, 380
Helix interconnect, 468
High-k, 122, 124–128, 131, 132, 134, 491, 525, 528
Hopping, 2, 128, 143, 155, 425
Hydrophilic, 117, 256, 290
Hydrophobic, 116, 117, 202, 256, 290, 380
Hygro-swelling, 69
Hygro-thermal-mechanical, 30, 205

I
ICA. *See* Isotropic conductive adhesive
IHS. *See* Integrated heat spreader
IMC. *See* Intermetallic compound
Impedance control, 503, 505
Indium tin oxide (ITO), 191
Ink-jet printing, 4, 241, 245–250, 258, 259
Integrated heat spreader (IHS), 507
Intermetallic compound (IMC), 4, 33, 265
International Technology Roadmap for Semiconductors (ITRS), 465
Intrinsic stresses, 469–471, 482, 484
Ion milling, 168, 175
Islands, 94, 95, 101, 102, 140, 142–144, 147, 148, 150, 151, 153, 156, 157, 159, 368
Isotropic conductive adhesive (ICA), 3, 6, 99, 214, 390
ITO. *See* Indium tin oxide
ITRS. *See* International Technology Roadmap for Semiconductors

J
Jitter, 409

K
Kelvin models, 300
Kirkendall, 4, 265

L
Laminar flow, 381
Laminate, 80, 121, 195, 211–213, 221, 225, 226, 229, 231, 233, 317, 318, 320
Lamination, 211, 225, 228, 229, 233
Land grid array (LGA), 493, 499
Landau-Lifshitz (LL), 177
Latency, 465
LCP. *See* Liquid crystal polymer
Leakage current, 167, 204
Leakage flux, 167, 171
LGA. *See* Land grid array
Life-cycle, 20, 35
Liftoff, 445, 482
Liquid crystal, 5, 190, 395, 450
Liquid crystal polymer (LCP), 395
LL. *See* Landau-Lifshitz
Loss factor, 128
Loss tangent, 124, 126, 128, 134, 227, 228
Low temperature co-fired ceramics (LTTC), 183
LTTC. See Low temperature co-fired ceramics

M
M3D®. *See* Maskless mesoscale material deposition
Magnetic core, 164, 167, 169, 171, 172, 174
Magnetostatic, 167, 169, 175, 181, 182
Magnetostriction, 178, 179, 181
Magnetostrictive, 176
Magnetron, 181, 484
Magnetron sputtering, 168, 174, 182, 469, 485
Maleimide, 214
Maskless mesoscale material deposition (M3D®), 249
Maxwell model, 300, 301, 303, 305, 314
Maxwell-Garnett, J. C., 103
MD. *See* Molecular dynamics
Mean free paths, 140
Mechanical interlocking, 77, 79, 80, 85, 87, 88, 224
Melting point depression, 2, 96–98, 252, 448
Metal-induced gap state, 359
Metallic-insulator transition, 95
Methylmethacrylate (MMA), 482
Micro-channels, 5, 382, 385
MicrospringTM, 467, 468
Microstructure, 126, 139, 146, 147, 181, 217, 252, 256, 257, 419–422, 432
Mie, G., 103
Mitigation, 20
MMA. *See* Methylmethacrylate

Index 541

Moisture
    absorption, 30, 69, 512
    loading, 69
Molecular dynamics (MD), 2, 17, 27, 28, 34, 35, 39, 43, 50, 52, 61, 77–79, 84, 86, 87, 99
Molecular wires, 200, 201, 442
Monopole antenna, 406, 407
Monte Carlo, 19, 41
Mooney-Rivlin, 24
Moore's Law, 7, 163, 491, 493–495, 506, 525, 526, 533, 534
MS. See Metal-induced gap state
Multi-physics, 15, 17, 20, 33, 35
Multi-scale, 15, 20, 27, 535

N

Nanocrystalline, 6, 175, 180
Nanocomposite, 99, 127–132, 134, 135, 178, 183, 205, 222, 358, 416, 423–425, 429, 512
Nano-diamond, 6
Nano-electrode, 6, 7, 204, 453
Nanoelectronics, 6, 7, 85, 450, 517, 518, 520–524, 529–531, 534, 535
Nanograins, 2, 181, 182
Nano-imprinting, 6
Nano-indentation, 448
Nano-interconnect, 6–7, 511
Nanolawn, 442, 444, 447, 451, 453
Nanowires, 96, 192, 204, 205, 212, 217, 441, 443–453, 511
Native oxide, 3, 177, 182
Navier-Stokes, 26, 381, 382
NCA. See Non-conductive adhesive
Near-field, 396, 402–410, 445, 452
Negative differential resistance, 362, 367
Negative magnetic permeability, 452
Negative TCR, 142
Neuroprosthetic, 454
Ni-Zn, 172
No-flow, 476, 477
No-flow underfill, 30, 287, 288, 290
Noise, 163, 178, 210, 402, 409, 465, 503
Non-conductive adhesive, 4
No-Pb, 4, 98
N-type CNT, 363, 366
Nucleation, 93, 112, 115, 118, 251, 326, 334, 431, 436

O

Oblate spheroids, 103
Oblate absorption, 102
Oblate interconnect, 6, 103, 505, 528
Oblate transceiver, 395, 396, 402, 404–407
Organic monolayers, 200
Ostwald ripening, 98, 451
Oxidation, 42, 54, 84, 109, 118, 172, 180, 182, 205, 218, 267, 271, 289

P

Package-on-package (PoP), 513, 514, 527
Particle shape, 3, 94, 103, 289, 379
Passive components, 3, 121, 163, 529
PDF. See Probability distribution function
PECVD. See Plasma enhanced chemical vapor deposition
PEEK. See Poly ether ether ketone
Percolation, 94, 99, 125–128, 135, 140, 142, 143, 151, 180, 181, 189, 193, 194, 196, 213, 217–220, 396, 411, 425, 444
Permalloy, 167, 177, 178, 181, 182
Permeability, 164, 166–168, 172, 173, 175–178, 180–183, 452
Personal health, 517
PGA. See Pin grid array
Photolithography, 164, 168, 174, 385, 386, 390, 445, 468–471, 485, 498, 499
Physical adsorption, 224
Physical bonding, 224
Physical interactions, 80, 85
Piezodroplet, 248
Piezoelectric, 248
Pin grid array (PGA), 492, 493, 499, 500
Plasma enhanced chemical vapor deposition (PECVD), 80, 174, 325, 390
Plated through hole (PTH), 210, 211, 497
PMMA. See Polymethylmethacrylate
Polarization, 99, 102, 122–124, 130, 176, 240
Polyether ether ketone (PEEK), 423
Polyimide, 172, 175, 195, 205, 214, 257, 444
Polymer solder, 189
Polymethylmethacrylate, 482
Polyurethane, 214
Polyvinyl acetate (PVAc), 192
Polyvinyl pyrrolidone (PVP), 115, 216
Polyvinylidene fluoride (PVdF), 126, 194
PoP. See Package-on-package
Potential function, 17, 52, 56, 77, 78, 86, 88
Printable electronics, 256, 258
Probability distribution function (PDF), 18, 19

Prolate spheroids, 103
Prony-Series, 301, 303, 305
Pseudo-inductive, 144
PTH. *See* Plated through hole
P-type CNT, 363, 366
Public security, 517
Pull test, 266, 279, 284
PVAc. *See* Polyvinyl acetate
PVdF. *See* Polyvinylidene fluoride
PVP. *See* Polyvinyl pyrrolidone

Q

Q-factor. *See* Quality factor
Quality factor, 163–170, 172, 173, 175, 178, 183
Quantum resistance, 368
Quantum wires, 346, 441

R

Rayleigh, L., 103
Rayleigh instability, 448
Rectification, 361, 364, 365
Reflow, 25, 28, 219, 226, 229, 232, 233, 266–279, 284, 287, 418, 419, 424, 452, 495, 496
Relaxation, 123, 300, 301, 305–307, 313, 314, 322, 469, 471
Reliability, 15, 18–20, 29, 30, 32, 35, 41, 42, 46, 55, 61, 69, 74, 121, 164, 183, 202, 204, 209, 210, 212, 213, 224–226, 231, 258, 259, 265, 287, 288, 291, 316–317, 322, 357, 372, 382, 415, 442, 443, 445, 448, 450, 454, 465, 466, 468, 475, 478, 486, 499–501, 506, 520–523, 527, 529–531
Restriction of Hazardous Substances (RoHS), 113
Reynolds number, 381, 382
Roadmap, 6, 7, 465, 491, 495
RoHS. *See* Restriction of Hazardous Substances
Rule of mixtures, 512

S

SAC. *See* Sn-Ag-Cu
SAM. *See* Self assembled molecular
Saturation magnetization, 164, 172, 179–183
Scanning tunneling microscopy (STM), 157, 345
Schottky barrier, 358, 360, 361, 364, 366–368

Schottky junction, 361
Sea of leads (SOL), 468
Second level interconnect, 495, 513
Sedimentation, 197, 198, 241, 245, 258, 260
SEL. *See* Step-edge lithography
Self assembled molecular (SAM), 200–202
Self resonant frequency, 171
Shear strength, 192, 196, 217, 433
Shielding, 6, 395–398, 401–411, 445, 452, 454
SiC. *See* Silicon carbide
Signal integrity, 372, 494, 504
Silane, 4, 216, 225, 290
Silicon carbide, 382, 418
Silicone, 214
Siloxane, 203
Single molecule, 356, 357
Sintering, 2, 4, 98–99, 118, 119, 125, 197–200, 202, 212, 213, 215, 217, 220–222, 233, 243, 244, 250–254, 258, 417, 420, 422, 430, 432, 438
SiOC(H), 80, 82
SiP. *See* System-in-package
Sn-Ag, 4, 28, 35, 98
Sn-Ag-Cu, 265, 416, 417, 421–432, 436–439
Sn-Pb, 98, 265, 416, 417, 419, 420, 422–439, 495
SOC. *See* System-on-chip
SOL. *See* Sea of leads
Space charge, 102, 159, 328
Sputter, 21, 145, 470, 484
Sputter deposition, 144, 151, 168, 181, 470, 482
Sputtering, 21–23, 140, 145, 151, 164, 168, 174, 182, 337, 338, 469, 470
Squeegee, 26
Stacked-die, 513
Stencil, 26, 27
Step-edge lithography (SEL), 445
Stiffness, 54, 74, 80, 176, 515
STM. *See* Scanning tunneling microscopy
Stress gradient, 469–472, 482, 484, 485
Superparamagnetism, 180
Surface energy, 3, 41, 79, 94, 96, 202, 216, 256, 426, 429
Surface plasma resonance, 116
Surface plasmonic effect, 109
Surface roughness, 80, 170, 504, 514
Surface tension, 24, 48, 94, 243, 247, 250, 256, 382, 426
Surface/volume ratio, 93
Surfactant, 117–119, 128, 193, 205, 221, 240, 254, 416, 450
Switching, 102, 121, 143, 249, 346, 365, 367, 370, 371, 501

System-in-package (SiP), 121, 520, 527
System-on-chip, 121, 529

T

Tape carrier packages (TCP), 190
TCP. *See* Tape carrier packages
TEM. *See* Transmission electron microscopy
Tensile strength, 5, 196, 224, 225, 233, 354, 385, 399, 416, 419, 428–429
Tessera, 467
Tg. *See* Glass transition temperature
TGA. *See* Thermogravimetric analysis
Thermal composite, 4
Thermal conductivity, 5, 39–41, 52–56, 196, 202, 212, 245, 249, 252, 355, 356, 377–380, 382, 385, 387, 389, 507–509
Thermal cycling, 41–45, 71, 193, 226, 265, 478–482, 486
Thermal resistivity, 385
Thermal interface material (TIM), 5, 39, 377, 378, 382–385, 391, 507
Thermionic emission, 193
Thermoelectric, 509
Thermogravimetric analysis, 382
Thermo-mechanical fatigue, 4, 415, 466
Thermo-mechanical reliability, 288, 322, 466, 468, 475, 478, 486
Thermoplastic, 49, 214, 219
Thermoset, 49, 189, 214
TIM. *See* Thermal interface material
TLPS. *See* Transient liquid phase sintering
Tm. *See* Melting point depression
Toughness, 62–64, 69, 71, 74, 212, 497, 510, 512
Traffic control, 517
Transient liquid phase sintering (TLPS), 220
Transmission electron microscopy (TEM), 115, 116, 131, 146–153, 157, 159, 196, 347–349, 372, 397
Tunneling barrier, 102, 148, 364, 368

U

UBM. *See* Under-bump metallization
Uncertainty, 18–20, 35
Under-bump metallization (UBM), 33, 34
Underfill, 4, 30, 232, 246, 287, 288, 290–292, 295, 297–302, 307–318, 320–322, 466–468, 475–478, 480, 497, 500
UV curing, 475, 486

V

Van der Waals, 40, 41, 44, 46, 49, 52, 80, 85, 349, 358, 370, 388
Vapor phase, 140, 325
Vapor-liquid-solid (VLS), 96, 341, 446
Via filling, 25, 195, 214, 217
Virtual design, 18
Viscosity, 4, 26, 27, 30, 197, 222, 239, 241, 243–247, 249, 258, 260, 290, 381
VLS. *See* Vapor-liquid-solid
Voids, 4, 51, 52, 83, 127, 197, 212, 214, 216, 217, 265, 267, 271, 290, 437, 497
Volume fraction, 31, 151, 155, 181, 189, 190, 192, 213, 288, 291–295, 297–300, 307, 309, 310, 312–315, 322, 379, 380, 432

W

Wafer level chip scale packages (WLCSP), 416
Waste Electrical and Electronic Equipment (WEEE), 113
Wave™. *See* Wide area vertical expansion
WEEE. *See* Waste Electrical and Electronic Equipment
Weibull, 282, 317, 319
Wide area vertical expansion (Wave™), 467
Wire pull, 74, 75
Wiring, 82, 110, 116, 119, 121, 209–211, 228, 231, 233, 258, 259, 498
WLCSP. *See* Wafer level chip scale packages
Work function, 96, 102, 200, 359, 361, 364, 367

Y

Yield strength, 99, 397, 416, 427–429, 434, 439
Young's modulus, 30, 35, 41, 51, 56, 81, 83, 99, 196, 354, 385, 430

Printed in the United States of America